DESIGN OF
ANALOG FILTERS
Passive, Active RC,
and Switched Capacitor

ROLF SCHAUMANN

Department of Electrical Engineering
Portland State University

M. S. GHAUSI

Department of Electrical Engineering
University of California at Davis

KENNETH R. LAKER

Department of Electrical Engineering
University of Pennsylvania

Prentice-Hall International, Inc.

Material in this book adapted in part from MODERN FILTER DESIGN by M. S. Ghausi and K. R. Laker, Prentice Hall 1981.

M. S. Ghausi/K. R. Laker, MODERN FILTER DESIGN: Active RC and Switched Capacitor © 1981 Figures 1-1, 1-6, 1-7, 1-8, 1-9, 1-14, 1-26, 1-27, 1-28, 1-29, 1-30, 1-32, 1-33, 1-36, 1-37, 1-38, 2-3, 5-14, 5-16, 5-25, P1-3, B-1 and tables 1-1, 1-3, B1, B2, B3, B4, and B5 adapted by permission of Prentice Hall, Inc., Englewood Cliffs, NJ.

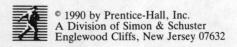

© 1990 by Prentice-Hall, Inc.
A Division of Simon & Schuster
Englewood Cliffs, New Jersey 07632

Printed in the Republic of Singapore.

10 9 8 7 6 5 4 3 2 1

ISBN 0-13-201591-9

Prentice-Hall International (UK) Limited, *London*
Prentice-Hall of Australia Pty. Limited, *Sydney*
Prentice-Hall Canada Inc., *Toronto*
Prentice-Hall Hispanoamericana, S.A., *Mexico*
Prentice-Hall of India Private Limited, *New Delhi*
Prentice-Hall of Japan, Inc., *Tokyo*
Simon & Schuster Asia Pte. Ltd., *Singapore*
Editora Prentice-Hall do Brasil, Ltda., *Rio de Janeiro*
Prentice-Hall, Inc., *Englewood Cliffs, New Jersey*

Contents

The Design
of *LC* Filters

Sensitivity

Operational Amplifiers and
Fundamental Active Building Blocks

Second-Order Active
Filter Sections

FIVE　　　　　　　　　　　　　　　　　　　　　　　　　　　　　　　　　*249*

The Design
of High-Order Filters

SIX　　　　　　　　　　　　　　　　　　　　　　　　　　　　　　　　　*319*

Fully Integrated
Continuous-Time Filters

SEVEN   410

Switched-Capacitor Filters

EIGHT   487

Contents

Laplace Transforms

z-Transforms

Tables of Classical Filter Functions

The contents of the book are as follows:

Chapter 1 covers basic properties and classifications of systems, together with types of filters and the important aspect of mathematically approximating the desired transfer characteristics. Frequency- and time-domain concepts and continuous-time and discrete-time signal transmissions are considered.

Chapter 2 deals mainly with discrete passive LC ladder networks designed for maximum power transfer. Our motivation for including this classical topic, even if only in a very concise form, in a modern filter text is twofold. First, in numerous cases, lossless LC ladders are the preferred filters because of their excellent low passband sensitivities to component tolerances and because they can be used in high-frequency applications where other methods prove inadequate. Second, passive LC ladders are often simulated via active discrete or integrated continuous-time or sampled-data filters in an attempt to inherit their excellent performance in the *active* circuit. Thus, even if the main objectives of a course are active filters, we felt that some understanding of classical LC networks would prove most beneficial.

Sensitivity, one of the main criteria for judging the quality of a filter, is covered in Chapter 3. Various definitions of sensitivity as found in the literature are given. We have included statistical measures, sensitivity optimization, and design centering. Although important for the design of filters with better manufacturability and higher yield, these topics are rarely found in textbooks. A study of this chapter should help the student understand the motivation for pursuing the various design strategies for active realizations that are discussed in later sections of the text.

Chapter 4 deals almost entirely with operational amplifiers (op amps), their properties as far as they are pertinent for this book, and the op amp circuits which form the fundamental building blocks of active filters. Guided by our experience that even graduate students often find op amps a difficult subject and considering that an understanding of op amps is fundamental for active filter work, we have treated the basic op amp behavior in quite some detail. Although the op amp remains to this day the main active element used in active filter design, we have included in our text a discussion of voltage-to-current converters (transconductors) and transconductor circuits which are beginning to be recognized as the more versatile and, in many respects, more appropriate elements for designing active filters. Other elements, such as integrators, gyrators, impedance converters, and frequency-dependent negative resistors (FDNRs), are circuits derived from op amps or transconductors.

Chapter 5 discusses second-order active sections (biquads), which are sometimes used by themselves but are predominantly employed as building blocks in the design of high-order active filters. For the most part, we have derived only those biquads which have proven themselves in practice, but a few others are included to make certain points or to illustrate some design principle. In keeping with the discussion in Chapter 4, the biquadratic sections considered include circuits built with single and multiple op amps and with transconductors.

Chapter 6 provides a treatment of high-order filter design. Cascade and

Preface

This text started out as a revision of the earlier Prentice Hall book, *Modern Filter Design: Active RC and Switched-Capacitor*, by M. S. Ghausi and K. R. Laker. We discovered soon, though, that the changed emphasis and the large number of new topics to be included in the text, together with our wish to make the text more easily accessible for senior-year undergraduate students, necessitated a complete rewriting of the material so that an entirely new book seemed more appropriate. Although we could not possibly hope to treat in one volume all aspects of the design of analog filters, we have nevertheless attempted to produce a reasonably self-contained text which covers all the fundamental concepts of the most important analog filter topics: *discrete* passive *LC* and active *RC* filters as well as *fully integrated* continuous-time and sampled-data switched-capacitor (SC) filters. The book should provide the reader with enough knowledge and understanding of the material to permit the design of analog filters in any of these classifications or at least to facilitate ready access to the relevant literature if more sophisticated designs are called for.

The book is intended as a text for senior undergraduates and/or first-year graduate students. Complete coverage of the material will require a full year (three-quarter or two-semester) course; however, students and instructors will not find it difficult to arrive at suitable selections of topics if a shorter course seems more appropriate or is required. Certain sections or even whole chapters can readily be bypassed depending on the students' needs or backgrounds. Although written in a fairly concise format, the book can also be used for self-teaching and reference by the interested professional or practicing engineer.

multiple-loop feedback methods are considered in some detail, as these configurations, in particular the *cascade* topology, are widely used in practice and provide simple rapid realizations with good or—for an appropriate design—excellent sensitivity performance. A large part of this chapter as well as of the remaining chapters is devoted to various active simulations of passive *LC* ladders because, as mentioned, these topologies result in the best sensitivity performance so that ladder simulation is emerging as the procedure of choice for high-order filters. The merits and disadvantages of various design methods are considered, so that the designer is made aware of the trade-offs between different approaches.

In Chapter 7 we discuss procedures which can be used if a continuous-time filter must be implemented in fully integrated form on an integrated circuit (IC) chip. Apart from some relatively minor differences related to the implementation technology, the procedures will be recognized as fundamentally equal to normal active *RC* techniques. The main problem which makes an integrated realization different pertains to automatic on-chip tuning of the filter against tolerances and errors of various origins. The text presents different techniques for achieving "self-tuning" in considerable detail.

Chapter 8 is concerned with fundamentals and design of a different type of integrated filter: sampled-data switched-capacitor (SC) filters. SC filters have been very popular, as they can be realized with high accuracy and no postdesign tuning in IC form, compatible with digital VLSI technology. As always, we restrict our discussion to proven techniques: biquads and, for realizing higher-order functions, cascade connections of biquads or *LC* ladder simulations.

Each chapter is complemented by numerous examples in the text and by homework problems. The examples are phrased as problems, and solutions are provided. Because it is our belief that more can be learned from nontrivial examples, these are often quite involved and lengthy and are used to make certain points or to illustrate derivations. To gain a better understanding, the reader is strongly encouraged to treat the examples as part of the text to be read and worked through. The homework problems are used to provide the reader with exercise material in order to gain a better understanding of the book's contents and to test his or her design skills. Often, the problems are used to let the reader fill in lengthy but fundamentally straightforward derivations or to extend design methods to cover practical situations.

All references are listed by number at the end of the text. For the most part, only readily available and pertinent references are given which should help the student in further studies. No attempt has been made to cite always the original papers or to be complete.

Three appendices are included for ease of reference or to facilitate the solution of problems. The first two contain abbreviated tables of Laplace and *z*-transform pairs and their properties. Appendix III contains tables of classical filter functions. The appendices are intentionally kept brief in order to avoid duplication of tables found in many books.

Although the art of designing active filters has matured, the need for better performance, lower sensitivity, different implementation technologies, and reduced cost has motivated several recent advances in the state of the art. These include computer-aided design (CAD) methods, better ladder simulations for realizing low-sensitivity high-order filters, designs for low-cost monolithic IC realizations, trans-conductance-C methods for high-frequency filters, and switched-capacitor networks to realize precision filters at low to medium frequencies. Our purpose in writing this book is to provide in a consistent systematic manner for students as well as practicing engineers these and other recent advances which heretofore have appeared only in technical articles scattered in many journals. Some material, especially that dealing with continuous-time fully integrated filters and automatic tuning, appears to the best of our knowledge for the first time in a textbook in a detailed form. Although we have discussed recent developments, we have, of course, included those bread-and-butter active filter realization methods which have stood the test of time. The book provides, therefore, a comprehensive overview of classical as well as highly current practical active filter design methods and technology.

The authors wish to acknowledge the contributions of many individuals whose results were used in this text, but whose names may or may not appear. We would like to thank especially Drs. Geert DeVeirman and Mehmet A. Tan and Professors R. L. Geiger, L. P. Huelsman, H. K. Kim, G. C. Temes, and many other, anonymous individuals who reviewed portions of the manuscript and offered many helpful suggestions and thoughtful advice. Many students made most valuable comments when large sections of the manuscript were class-tested by two of us (R. S. and K. R. L.). The Alexander von Humboldt Foundation Award also supported work by one of the authors (M.S.G.). Some of the authors' original research results appearing in this book were supported by National Science Foundation grants, for which we wish to express our gratitude.

R. Schaumann, *Portland, Oregon*
M. S. Ghausi, *Davis, California*
K. R. Laker, *Philadelphia, Pennsylvania*

Fundamental Concepts and Tools for the Design of Linear Analog Filters

ONE

1.1 INTRODUCTION

An electrical filter consists of an interconnection of components, such as resistors, capacitors, inductors, and active devices (transistors, amplifiers, controlled sources). The filter operates on or processes applied electrical *signals* which are referred to as the *input* or *excitation*. The product of the processing performed by the network on the excitation is referred to as the *output* or *response*. The excitation and response will differ according to the type of processing or filtering performed by the network, and as we shall see shortly, there are several ways by which we can represent and specify this filtering operation.

We know from elementary signal theory that any periodic signal $x(t)$ of period $2\pi/\omega_0$ can be represented by a Fourier series expansion

$$x(t) = a_0 + \sum_{k=1}^{\infty} (a_k \cos k\omega_0 t + b_k \sin k\omega_0 t)$$

$$= A_0 + \sum_{k=1}^{\infty} A_k \cos (k\omega_0 t + \phi_k) = \sum_{k=-\infty}^{\infty} X_k e^{jk\omega_0 t} \qquad (1\text{-}1a)$$

where the real numbers a_k, b_k or A_k, ϕ_k or the complex numbers X_k represent the (discrete) spectrum of $x(t)$. Similarly, if $x(t)$ is not periodic, it can be expressed by a Fourier integral

$$x(t) = \frac{1}{2\pi} \int_{-\infty}^{\infty} X(j\omega) e^{j\omega t} \, d\omega \qquad (1\text{-}1b)$$

where $X(j\omega)$ represents the (continuous) spectrum of $x(t)$. Filtering can then, in general, be understood as a spectrum-shaping process whereby the numbers X_k or the function $X(j\omega)$ are altered in certain ways in order to produce a desired form of the output signal $y(t)$. Expressed differently, the object of filtering is to perform frequency-selective transmission whereby some spectral components (the passband) are passed through to the output whereas others (stopbands) are rejected by the operation of the filter. If the filter is *linear*, the harmonic content of the output signal cannot be richer than that of the input signal (the output contains no frequencies that are not present in the input). The task of the filter designer is then to find a circuit or network with the appropriate interconnection of elements of carefully chosen values that will perform the desired filtering operation.

Throughout the communications and measurement industries, electrical filters in all technologies are being used in huge numbers. Indeed, it becomes difficult to name any electrical system that does not contain some kind of signal filter. The choice of network implementation is almost always determined by economic considerations and technology. Because of advances in technology, what was considered the best economical filter implementation 10 years ago, or perhaps only one year ago, may not be the most economical implementation today. Figure 1-1 shows the evolution of filter technology in the Bell System, for voice-frequency ($f < 4$ kHz) applications over a nearly 60-year span. What is shown in each stage of Fig. 1-1 is the hardware to implement a second-order filtering function.

Although the evolution depicted is specific to the Bell System, it is to a good approximation representative of the evolution industrywide. From 1920 to the latter 1960s the majority of voice-frequency filters were realized as discrete *RLC* networks. However, it was recognized in the 1950s that size and eventual cost reductions could potentially be achieved by replacing the large, costly inductors with active networks. That is, a network composed of resistors, capacitors, and transistors could be made to resonate like a tuned *RLC* network. These active networks, referred to as *active RC networks*, remained essentially research curiosities until the mid-1960s, when good-quality active components such as operational amplifiers became inexpensive and readily available. Although size has not been significantly reduced, filtering and amplification were achieved simultaneously. In the early 1970s the economic potential envisioned for active *RC* filters began to be realized with batch-processed thin-film hybrid integrated circuits (HICs). The HIC shown in Fig. 1-1, composed of two thin-film capacitors, nine thin-film resistors, and one silicon integrated-circuit operational amplifier, is 1.05×1.00 in. This circuit represented about a factor-of-2 cost reduction over the equivalent passive *RLC* realization. By 1975, thin-film technology had advanced such that the 1.05×1.00 in. HIC could be reduced in size to fit into a small 16-pin dual-in-line package (DIP).

Today, using switched-capacitor and digital filter techniques, very high-order filters can be realized as microminiature silicon integrated-circuit chips, and promising methods are being developed for fully integrated continuous-time active *RC* circuits (Chapter 7). As examples we show in Fig. 1-1 two packaged integrated-

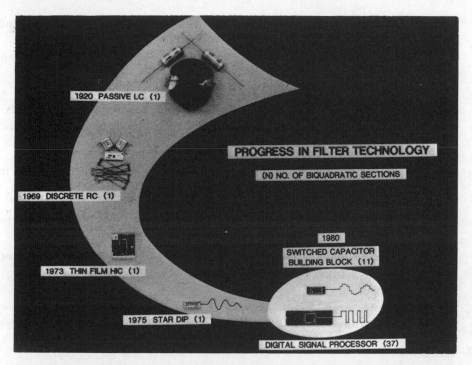

Figure 1-1 Bell System progress in filter technology, depicting the evolution of voice-frequency ($f < 4$ kHz) filters from 1920 to 1980. Shown on the crescent is the hardware required to realize a second-order section with (a) passive LC, (b) discrete-component active RC, and (c) and (d) hybrid thin-film active RC technologies. Shown in the oval are examples of monolithic filters: (e) an analog switched-capacitor building block capable of realizing up to 11 second-order sections and (f) a digital signal capacitor processor capable of realizing up to 37 second-order sections with an 8-kHz sampling rate.

circuit chips; namely, the digital signal processor (DSP) and the switched-capacitor building block (SCBB). The DSP implements up to 37 pole pairs of digital filtering, whereas the SCBB implements up to 11 pole pairs of analog filtering. A detailed description of the SCBB is reserved for Chapter 8. The reader interested in pursuing the area of digital filters is referred to references 1, 2, and 3.

As the evolution depicted in Fig. 1-1 has demonstrated, the use of active networks has enabled engineers to utilize the advances in integrated-circuit technology to implement low-cost, microminiature voice-frequency filters ($f < 4$ kHz); voice-frequency applications alone account for tens of millions of the active filters that are produced yearly throughout the world. Aside from their obvious size and weight advantages over equivalent passive RLC implementations shown in Fig. 1-1, active filters provide the following additional advantages:

1. Circuit reliability is increased because all manufacturing steps can be automated.

2. In large quantities the cost of integrated-circuit active filters is much lower than that of equivalent passive filters.
3. Improved performance is possible because high-quality components can be readily manufactured.
4. Parasitics are reduced because of smaller size.
5. Analog active filters and digital circuitry can be integrated onto the same silicon chip.   This advantage has already been fully realized with active switched-capacitor filters.

In addition to these advantages, which stem from their physical implementation, there are other advantages which are circuit-theoretic in nature, namely:

1. The design and tuning processes are often simpler than those for passive filters.
2. Active filters can realize a wider class of filtering functions than passive filters.
3. Active filters can provide gain; in contrast, passive filters often exhibit a significant loss.

With these advantages there are some drawbacks to active network implementations.   They are:

1. Active components have a finite bandwidth, which tends to limit most active filters to audio-frequency applications.   In contrast, passive filters do not have such an upper-frequency limitation and they can be used up to approximately 500 MHz.
2. Passive filters are less affected by component drift due to manufacturing or environmental changes.   This phenomenon, referred to as *sensitivity*, will be seen to be an important criterion for comparing similar filter realizations. Much of the disadvantage experienced by active-network implementations stems from the relatively large number of components required to realize an active filter as compared to equivalent passive realizations.
3. Active filters require power supplies, whereas passive filters do not.   It is often particularly important that power consumption be reduced; therefore, the reader will be encouraged to design active filters with the minimum number of active devices consistent with precision performance.
4. The signal magnitude that active devices can process without unacceptable distortion is limited, often to levels below 1 V.   In addition, resistors and active elements produce electrical noise.   Thus, the dynamic range of active filters, i.e., the range of signal amplitudes between the noise floor and the upper level where nonlinearities arise, is generally smaller than that of passive circuits.

In voice and data communication systems, the economic and performance advantages of active filters far outweigh these disadvantages. Testimony to this fact is the general acceptance and usage of active filters throughout the telecommunications industry. Nevertheless, passive *RLC* filters are still in very wide use.

It is assumed that the reader has a working knowledge of network theory and has had some exposure to linear system theory. In this chapter we review some of the fundamental concepts and tools needed to characterize a continuous linear filter.

1.2 CLASSIFICATION OF SYSTEMS

A schematic "black box" representation of a single-input, single-output system (the only case considered in this book) is shown in Fig. 1-2. The input to the system, or the excitation, $x(t)$, is operated upon by the system to create the output, or the response, $y(t)$, symbolically represented as

$$x(t) \rightarrow y(t) \quad \text{or} \quad y(t) = f\{x(t)\} \tag{1-2}$$

As indicated, $x(t)$ and $y(t)$ are, in general, functions of time and, for our purposes, are electrical signals, i.e., voltages or currents. In this book, the system itself will always represent an electrical network, or more specifically a *filter*, made up of a finite number of *lumped* active and passive elements and, in the case of sampled-data switched-capacitor filters, of switches. A lumped element is defined as one having physical dimensions small compared to the wavelength of the applied signals.

In order to help in describing the system's operation, let us review briefly the properties and classes of systems germane to the material in this text.

1.2.1 Linear and Nonlinear Systems

A system is *linear* if it satisfies the principle of superposition. Mathematically, if $y_1 = f\{x_1\}$ and $y_2 = f\{x_2\}$ are the responses to two input signals x_1 and x_2, then the response to the input $x = \alpha x_1 + \beta x_2$, where α and β are arbitrary constants, equals

$$y = f\{x\} = f\{\alpha x_1 + \beta x_2\} = \alpha f\{x_1\} + \beta f\{x_2\} = \alpha y_1 + \beta y_2 \tag{1-3}$$

A linear system is governed by a linear differential or difference equation. Only linear systems are considered in this text, and nonlinearities, if any, occurring during the operation of any of our networks—e.g., due to an overdriven amplifier—are imperfections that should be eliminated or minimized.

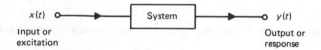

$x(t)$ Input or excitation   System $y(t)$ Output or response

Figure 1-2 Schematic representation of a system.

1.2.2 Continuous-Time and Discrete-Time or Sampled-Data Systems

A system is said to be a *continuous-time* system or a *continuous analog* system if input x and output y are continuous functions of the continuous variable time. We make the fact of continuous change evident by writing x and y as functions of the continuous time variable t; i.e.,

$$x = x(t) \quad \text{and} \quad y = y(t) \tag{1-4}$$

In *discrete-time*, or *sampled-data*, systems, the input and output signals take on continuous values that change at only discrete instants of time (the *sampling instants*) between which values of the signals are of no interest. In sampled-data systems, which are actually analog systems[1], the signals are usually held constant between sampling instants by use of sample-and-hold circuitry. We can represent sampled-data input and output signals as functions of the discrete variable kT:

$$x = x(kT) \quad \text{and} \quad y = y(kT) \tag{1-5}$$

where k is an integer and T is the time interval between samples. The signal formats for each of these system types are shown in Fig. 1-3. Note that the amplitude of the output may be larger, smaller, or equal to the input, depending on the type and design of the filter.

An important mathematical distinction between continuous-time and sampled-data (also discrete-time) systems is the fact that continuous-time systems are characterized by differential equations, whereas discrete-time systems are characterized by difference equations. We shall defer a study of the details of the sampling process and the operation of sampled-data systems until Chapter 8, where these concepts will be used for the first time.

1.2.3 Time-Invariant and Time-Varying Systems

A system is said to be *time-invariant* if the shape of the response to an input applied at any instant of time depends only on the shape of the input and not on the time of application. In mathematical terms, a system is time-invariant if for $x(t) \rightarrow y(t)$ we have

$$x(t - \tau) \rightarrow y(t - \tau) \tag{1-6}$$

for all $x(t)$ and all τ. Similarly, for a discrete-time system, if we have

$$x(kT) \rightarrow y(kT) \tag{1-7}$$

then the system is time-invariant if and only if

$$x[(k - n)T] \rightarrow y[(k - n)T] \tag{1-8}$$

for any $x(kT)$ and n.

---

[1]In digital systems, the signal values at the sampling instants are further encoded or digitized and are no longer continuous but are discrete.

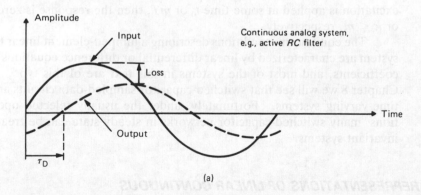

(a)

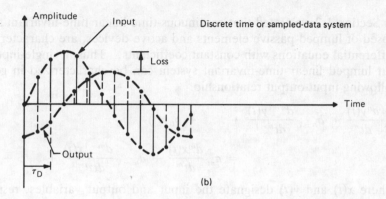

(b)

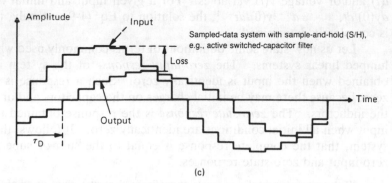

(c)

Figure 1-3 Signal formats for continuous analog (a), sampled-data analog (b), and discrete-time (c) systems with S/H.

For a *causal* system, the response cannot precede the excitation; i.e., if the excitation is applied at some time t_0 or mT, then the response is zero for all $t < t_0$ or $k < m$, respectively.

The equilibrium conditions describing a lumped-element linear time-invariant system are characterized by linear differential or difference equations with constant coefficients, and most of the systems in this text are of this type. However, in Chapter 8 we will see that switched-capacitor sampled-data circuits are, in general, time-varying systems. Fortunately, under the usually selected operating conditions, many switched-capacitor networks in steady state can be treated like time-invariant systems.

1.3 REPRESENTATIONS OF LINEAR CONTINUOUS TIME-INVARIANT SYSTEMS

In Section 1.2 we stated that continuous-time linear time-invariant systems, composed of lumped passive elements and active devices, are characterized by linear differential equations with constant coefficients. Thus, a single-input, single-output lumped linear time-invariant system can be characterized in general by the following input-output relationship:

$$b_n \frac{d^n y(t)}{dt^n} + b_{n-1} \frac{d^{n-1} y(t)}{dt^{n-1}} + \ldots + b_0 y(t)$$

$$= a_m \frac{d^m x(t)}{dt^m} + a_{m-1} \frac{d^{m-1} x(t)}{dt^{m-1}} + \ldots + a_0 x(t) \quad (1\text{-}9)$$

where $x(t)$ and $y(t)$ designate the input and output variables, respectively, and where the coefficients a_i and b_j are real and depend on the network elements and on their interconnections.[2] In electronic networks, $x(t)$ and $y(t)$ will be current $i(t)$ and/or voltage $v(t)$ variables. For a given input and initial conditions $y(0)$, $dy(0)/dt, \ldots, d^{n-1} y(0)/dt^{n-1}$, the solution to Eq. (1-9), namely, the output $y(t)$, is completely determined.

Let us now state some definitions of terms commonly used when referring to lumped linear systems. The *zero-input response* of the system is the response obtained when the input is identically zero. Such a response is not necessarily zero, because there may be initial charges on the capacitors and/or initial fluxes in the inductors. The *zero-state response* is the response obtained for an arbitrary input when all initial conditions are identically zero. It follows, then, for a linear system, that the complete response is equal to the sum or superposition of the zero-input and zero-state responses.

[2]If the network is nonlinear (contains nonlinear elements), some or all of the coefficients are functions of x and/or y; if the network is time-dependent (contains time-dependent elements), some or all of the coefficients are functions of time. If the network is distributed (not lumped), then instead of Eq. (1-9) a partial differential equation would characterize the system.

1.3.1 Frequency-Domain Concepts

For continuous linear time-invariant systems, Laplace transform techniques can be used to transform differential equations with constant coefficients from the time domain into linear algebraic equations in the frequency domain. Noting that the Laplace transform of the nth derivative of some time function $y(t)$ is given by

$$\mathscr{L}\left[\frac{d^n y(t)}{dt^n}\right] = s^n Y(s) - s^{n-1} y(0) - s^{n-2}\frac{dy(0)}{dt} - \ldots - \frac{d^{n-1} y(0)}{dt^{n-1}} \quad (1\text{-}10)$$

we can transform Eq. (1-9) term by term to yield

$$(b_n s^n + b_{n-1}s^{n-1} + \ldots + b_0)Y(s) + IC_y(s)$$

$$= (a_m s^m + a_{m-1}s^{m-1} + \ldots + a_0)X(s) + IC_x(s) \quad (1\text{-}11)$$

where in $IC_y(s)$ and $IC_x(s)$ we have lumped together all terms involving the initial conditions for y and x. In Eq. (1-11), $X(s)$ and $Y(s)$, respectively, are the *Laplace transforms* of the excitation and the zero-state response. A brief table of Laplace transforms is included in Appendix I.

A useful concept in the analysis and synthesis of linear networks is the *network function*, which is the ratio of two Laplace-transformed terminal or port variables. When the two variables are the input $X(s)$ and the output $Y(s)$, the network function is referred to as the *transfer function $H(s)$*. Thus, the transfer function can be defined in terms of the Laplace-transformed excitation $X(s)$ and the zero-state response $Y(s)$:

$$H(s) = \frac{\mathscr{L}[\text{zero-state response } y(t)]}{\mathscr{L}[\text{excitation } x(t)]} = \frac{Y(s)}{X(s)}$$

$$= \frac{a_m s^m + a_{m-1}s^{m-1} + \ldots + a_0}{b_n s^n + b_{n-1}s^{n-1} + \ldots + b_0} \quad (1\text{-}12)$$

$$= \frac{N(s)}{D(s)}$$

where $m \leq n$ for any realizable practical network and N and D are the numerator and denominator polynomials, respectively.

In the only case of importance for filters, where $Y(s)$ and $X(s)$ are voltage transforms [i.e., $V_{out}(s)$ and $V_{in}(s)$, respectively], $H(s) = V_{out}/V_{in}$ is referred to as the *voltage transfer function*. However, in other situations, *current transfer functions* [$Y(s) = I_{out}(s)$, $X(s) = I_{in}(s)$], *transfer impedances* [$Y = V_{out}$, $X = I_{in}$], and *transfer admittances* [$Y = I_{out}$, $X = V_{in}$] are also encountered. Further important network functions of particular use in the design of passive *LCR* filters are the *driving-point* functions obtained when $Y(s)$ and $X(s)$ are voltage and current transforms at the *same* network port. Thus if $Y(s) = V_{in}(s)$ and $X(s) = I_{in}(s)$, the *driving-point impedance* and driving point *admittance* functions, respectively, are

$$Z_{in}(s) = \frac{V_{in}(s)}{I_{in}(s)} \quad \text{and} \quad Y_{in}(s) = \frac{I_{in}(s)}{V_{in}(s)} \tag{1-13}$$

Evidently $Y_{in} = 1/Z_{in}$.

In the development presented above we have shown how the network function $H(s)$ can be obtained by Laplace-transforming the differential equation [Eq. (1-9)] and assuming zero initial conditions. Undoubtedly, the reader will recall from elementary circuits courses that there is no need to first perform a *time-domain* analysis to arrive at differential equations which are then Laplace-transformed to obtain the network's *frequency-domain* description $H(s)$. Rather, it is much more convenient to "Laplace-transform the circuit" (e.g., $i = C\, dv_c/dt \rightarrow I(s) = sCV_c(s)$, that is, $C \rightarrow sC$) and then to analyze the circuit in the *frequency domain* using any convenient well-known manual or computer-based tools. As a matter of fact, even if the differential equation is needed for calculating a time-domain response, it is easier to first derive $H(s)$ in the frequency domain and then to obtain the differential equation by tracing the steps from Eq. (1-12) back to Eq. (1-9). The transformation into the frequency domain greatly facilitates the analysis of a system in addition to providing immediate insight into the system's behavior.

Regardless of how $H(s)$ is obtained, Eq. (1-12) shows that for continuous linear time-invariant lumped networks, a network function is a ratio of two polynomials in s with real coefficients, a so-called *real rational function*. If these polynomials are factored, $H(s)$ can be written in the following alternative representation:

$$H(s) = \frac{N(s)}{D(s)} = \frac{a_m(s - z_1)(s - z_2) \cdots (s - z_m)}{b_n(s - p_1)(s - p_2) \cdots (s - p_n)} \tag{1-14}$$

In Eq. (1-14) the zeros $z_1, z_2, \ldots, z_m$ of the numerator polynomial $N(s)$ are referred to as the *zeros* of $H(s)$ because $H(s) = 0$ when $s = z_i$. The zeros $p_1, p_2, \ldots, p_n$ of the denominator polynomial $D(s)$ are referred to as the poles of $H(s)$ because $H(s) = \infty$ when $s = p_i$. The poles and zeros can be plotted in the complex s-plane where $s = \sigma + j\omega$; a typical set is shown in Fig. 1-4. Note that, since the coefficients of $H(s)$ are all real, imaginary and complex poles and zeros occur in conjugate pairs. For stability all poles must lie in the left half plane. Recall from elementary transform theory that networks with only left half-plane poles have a zero-input response which decays with time. When the poles lie on the $j\omega$-axis and are simple, the network oscillates; and when the poles lie in the right half plane, the responses grow exponentially with time. Thus, for stability reasons, the *characteristic polynomial $D(s)$* of a realizable system must have only left half-plane zeros; i.e., $D(s)$ is a *Hurwitz* polynomial. When zeros of $N(s)$ lie on or to the left of the $j\omega$-axis (i.e., there are no right half-plane zeros), $H(s)$ is referred to as a *minimum-phase function*.

The term *minimum phase* requires a further explanation: for a *prescribed magnitude*, it does not matter whether a transfer function has zero locations in the

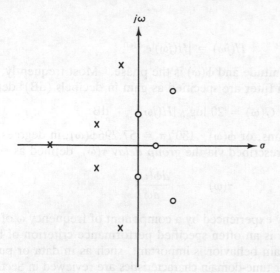

Figure 1-4

left half or at their mirror images in the right half of the s-plane. But for these two situations, the function with only left half-plane zeros has the smaller phase. Let us illustrate this fact with a simple example. With σ and p assumed positive, the two stable transfer functions

$$H_1(s) = \frac{s + \sigma}{s + p} \quad \text{and} \quad H_2(s) = \frac{s - \sigma}{s + p}$$

have the same magnitude

$$|H_1(j\omega)| = |H_2(j\omega)| = \sqrt{\frac{\sigma^2 + \omega^2}{p^2 + \omega^2}}$$

but their phases are different:

$$\phi_1 = \tan^{-1}\frac{\omega}{\sigma} - \tan^{-1}\frac{\omega}{p} \quad \text{and} \quad \phi_2 = \pi - \tan^{-1}\frac{\omega}{\sigma} - \tan^{-1}\frac{\omega}{p}$$

so that $\phi_2 \geq \phi_1$ because $\tan^{-1}(\omega/\sigma) \leq \pi/2$:

$$\phi_2 - \phi_1 = \pi - 2\tan^{-1}\frac{\omega}{\sigma} \geq 0$$

As mentioned, filter performance is typically specified by the voltage transfer function $H(s) = V_{out}(s)/V_{in}(s)$, sometimes more specifically called the *voltage gain function*. Under steady-state conditions (i.e., $s = j\omega$), the voltage gain function[3]

---

[3]It is not uncommon to see filters specified by a loss function $X(s)/Y(s)$ [in terms of voltages: $V_{in}(s)/V_{out}(s)$] or in steady state $X(j\omega)/Y(j\omega)$. The use of loss functions is a carry-over from passive filter design. In this book we will work exclusively with gain functions.

can be written as

$$H(j\omega) = |H(j\omega)|e^{j\phi(\omega)} \qquad (1\text{-}15)$$

where $|H(j\omega)|$ is the magnitude and $\phi(\omega)$ is the phase. Most frequently, the magnitude requirements of a filter are specified as gain in decibels (dB)[4] defined as

$$G(\omega) = 20 \log_{10} |H(j\omega)| \qquad \text{dB} \qquad (1\text{-}16)$$

The phase $\phi(\omega)$, in radians, or $\phi(\omega) \cdot 180°/\pi = 57.296\phi(\omega)$, in degrees, is either specified directly or is prescribed via the *group delay* $\tau(\omega)$, defined as

$$\tau(\omega) = -\frac{d\phi(\omega)}{d\omega} \qquad (1\text{-}17)$$

$\tau(\omega)$ represents the delay experienced by a component of frequency ω of the input spectrum. Group delay is an often specified performance criterion of filters, especially when time-domain behavior is important, such as in data or pulse transmission systems. Further time-domain characteristics are reviewed in Section 1.3.2.

Example 1-1

Analyze the active filter in Fig. 1-5 to determine the transfer function $H(s) = V_2/V_1$. The circuit uses a noninverting amplifier of gain $K = 1.56$, $R_1 = R_2 = 1$ kΩ, and $C_1 = C_2 = 5$ nF. Calculate and sketch the magnitude $|H(j\omega)|$, the gain and phase functions $G(\omega)$ and $\phi(\omega)$, and the delay $\tau(\omega)$. Also give $H(s)$ in factored form and identify the poles and zeros. In which frequency band is the gain reduced from its maximum by less than 3 dB? What type of filter is realized by this circuit?

Solution: Writing the node equations for nodes a and b leads to

$$H(s) = \frac{V_2}{V_1} = \frac{K}{1-K} \frac{sG_1/C_2}{s^2 + s\frac{G_2}{C_2}\left(1 + \frac{C_2}{C_1} + \frac{G_1/G_2}{1-K}\right) + \frac{G_1G_2}{C_1C_2}} \qquad (1\text{-}18)$$

where $G_i = 1/R_i$, $i = 1, 2$, are conductances. Inserting the element values yields

$$H(s) = -\frac{31}{11} \frac{(0.2 \cdot 10^6 \text{ rad/s})s}{s^2 + \left(\frac{0.2}{5.5} \cdot 10^6 \text{ rad/s}\right)s + (0.2 \cdot 10^6 \text{ rad/s})^2}$$

In order to avoid having to manipulate physical dimensions and large powers of 10, let us use a new frequency parameter $s_n = s/(0.2 \cdot 10^6 \text{ rad/s})$ which amounts to scaling (*normalizing*) the frequency axis by $\Omega_0 = 0.2$ Mrad/s. (Section 1.4). As a result of this normalization we can deal with the considerably more manageable expression

$$H(s_n) = -\frac{31}{11} \frac{s_n}{s_n^2 + s_n/Q + 1} \qquad (1\text{-}19)$$

[4]The alternative representation, $H(j\omega) = -\exp[\alpha(\omega) + j\beta(\omega)]$, where $\alpha(\omega) = -\ln |H(j\omega)|$ is the loss in nepers [1 neper = (20 log e) dB = 8.686 dB] and $\beta(\omega) = -\phi(\omega)$ is the phase, is used less frequently.

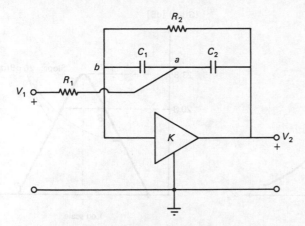

Figure 1-5 Single-amplifier active filter.

where we introduce the parameter $Q = 5.5$. From Eq. (1-19) with Eqs. (1-15) to (1-17), the magnitude, gain, phase, and delay functions can be derived as

$$|H(j\omega_n)| = \frac{31}{11} \frac{\omega_n}{\sqrt{(1 - \omega_n^2)^2 + (\omega_n/Q)^2}} \tag{1-20a}$$

$$G(\omega_n) = 20 \log \left[\frac{31}{11} \frac{\omega_n}{\sqrt{(1 - \omega_n^2)^2 + (\omega_n/Q)^2}} \right] \tag{1-20b}$$

$$\phi(\omega_n) = -\frac{\pi}{2} - \tan^{-1} \frac{\omega_n/Q}{1 - \omega_n^2} \tag{1-20c}$$

and

$$\tau(\omega) = -\frac{d\phi(\omega)}{d\omega} = -\frac{d\phi(\omega_n)}{d\omega_n} \frac{d\omega_n}{d\omega} = -\frac{1}{\Omega_0} \frac{d\phi(\omega_n)}{d\omega_n} \tag{1-20d}$$

Thus, calling the normalized delay $\tau_n(\omega_n)$, we get:

$$\tau_n(\omega) \equiv \Omega_0 \tau(\omega) = \frac{(1 + \omega_n^2)/Q}{(1 - \omega_n^2)^2 + (\omega_n/Q)^2} \tag{1-20e}$$

Figure 1-6 shows sketches of $|H(j\omega_n)|$, $G(\omega_n)$, $\phi(\omega_n)$, and $\tau_n(\omega_n)$. Note that the magnitude has its maximum value $31Q/11 = 15.5$ at $\omega_n = 1$ and that the -3 dB frequencies are therefore obtained from $|H(j\omega_n)| = 15.5/\sqrt{2}$; the resulting quadratic equation in ω_n, $1 - \omega_n^2 = \pm\omega_n/Q$, yields

$$\omega_{n1,2} = \sqrt{1 + \frac{1}{4Q^2}} \mp \frac{1}{2Q}$$

and

$$B = \omega_{n2} - \omega_{n1} = \frac{\Delta\omega}{\Omega_0} = \frac{1}{Q} \tag{1-21}$$

This result is general: the 3-dB bandwidth B is the bandcenter frequency, here Ω_0, divided

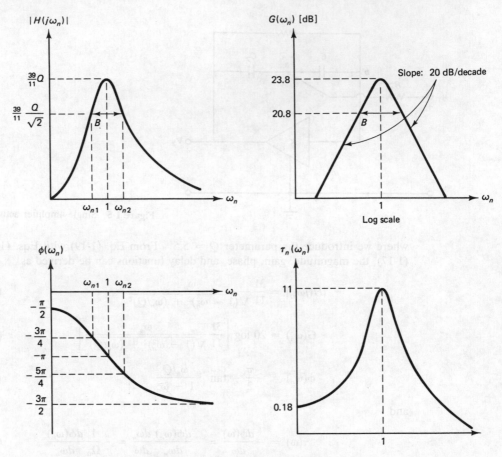

Figure 1-6 Performance of the circuit in Fig. 1-5.

by the *quality factor Q*. Note also that at the two frequencies ω_{n1} and ω_{n2}, the phase ϕ is 45° off the phase $\phi(1) = -\pi/2$ at bandcenter; that is, $\phi(\omega_{n1}) = -3\pi/4$ and $\phi(\omega_{n2}) = -5\pi/4$. In factored form $H(s_n)$ becomes

$$H(s_n) = -\frac{31}{11} \frac{s_n}{\left(s_n + \dfrac{1}{2Q} + j\dfrac{\sqrt{4Q^2 - 1}}{2Q}\right)\left(s_n + \dfrac{1}{2Q} - j\dfrac{\sqrt{4Q^2 - 1}}{2Q}\right)}$$

That is, $H(s_n)$ has zeros at $z_{n1} = 0$ and $z_{n2} = \infty$ and poles at $p_{n1,2} = (-1 \pm j\sqrt{120})/11$ as shown in Fig. 1-7. Note from $H(s_n)$ or from the plot of $|H(j\omega_n)|$ or $G(\omega_n)$ that the gain goes to zero at $s_n = 0$ and at $s_n = \infty$ ($s_n = 0$ and ∞ are *transmission zeros*); i.e., signals at very low and at very large frequencies are *rejected*, whereas signals with frequencies in a band of width B around $\omega_n = 1$ are *passed*; consequently, the circuit is a *bandpass filter* with a normalized 3-dB passband of width $B = Q^{-1}$.

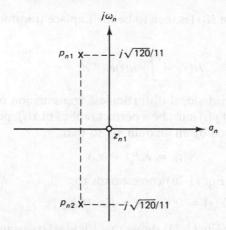

Figure 1-7 Pole-zero diagram for the circuit of Fig. 1-5. z_{n2} is at ∞.

1.3.2 Time-Domain Concepts [4, 5, 6, 19]

Recall that, via a method equivalent to the differential Eq. (1-9), in a linear time-invariant system the response $y(t)$ can be obtained by convolving the excitation $x(t)$ with the *impulse response* $h(t)$. Mathematically, this leads to the convolution or superposition integral

$$y(t) = \int_0^t h(\lambda)x(t - \lambda)\, d\lambda \qquad (1\text{-}22a)$$

where λ is a dummy integration variable and where we assume that the system is causal and $x(t) = 0$ for $t < 0$. Alternatively, we can obtain $y(t)$ by convolving the *step response* $a(t)$ with the derivative of the input signal, $dx(t)/dt$, as

$$y(t) = \int_0^t a(\lambda) \frac{dx(t - \lambda)}{d\lambda}\, d\lambda \qquad (1\text{-}22b)$$

That $h(t)$ is indeed the impulse response is easily verified by assuming that the excitation is an impulse, $x(t) = \delta(t)$, and by making use of the sifting property of impulse functions and of the definition $\int_0^t \delta(\lambda)\, d\lambda = 1$ with $\delta(t - \lambda) = 0$ for $\lambda \neq t$:

$$y(t) = \int_0^t h(\lambda)\delta(t - \lambda)\, d\lambda = h(t) \int_0^t \delta(t - \lambda)\, d\lambda = h(t) \qquad (1\text{-}23)$$

A similar method can be used to verify that $a(t)$ is the step response (Problem 1.7). The connection between the convolution integral and the frequency-domain representation [Eq. (1-12)] is established by Laplace-transforming Eq. (1-22a):

$$Y(s) = \mathscr{L}\{y(t)\} = \int_0^\infty e^{-st}\left[\int_0^\infty x(t - \lambda)h(\lambda)\, d\lambda\right] dt = H(s)X(s) \qquad (1\text{-}24)$$

where the network function $H(s)$ is seen to be the Laplace transform of the impulse response:

$$H(s) = \int_0^\infty h(t)e^{-st}\, dt \qquad (1\text{-}25)$$

If a system is to provide ideal distortionless transmission of an input signal $x(t)$, it stands to reason that $y(t)$ must be a perfect replica of $x(t)$, possibly multiplied by a constant K and delayed by an amount τ_0, so that

$$y(t) = Kx(t - \tau_0) \qquad (1\text{-}26)$$

In the frequency domain, Eq. (1-26) corresponds to

$$Y(s) = \mathscr{L}\{y(t)\} = Ke^{-s\tau_0}X(s) \qquad (1\text{-}27)$$

which when compared with Eq. (1-24) shows that for ideal transmission the network function equals

$$H(s) = Ke^{-s\tau_0} \qquad (1\text{-}28)$$

It has a constant magnitude K, a *linear* phase $\phi = -\omega\tau_0$, and, with Eq. (1-17), a constant delay $\tau(\omega) = \tau_0$. Clearly, $H(s)$ in Eq. (1-28) is not a real rational function and consequently is not realizable as a lumped network with a finite number of elements. In Section 1.6 we will discuss a popular method for *approximating $H(s)$* of Eq. (1-28) by a rational function, but any physical network has in practice frequency-dependent magnitude and delay and therefore causes transmission errors when compared with the ideal situation in Eq. (1-26).

In the following subsections we consider some terminologies in connection with two important system responses that are used to define deviations from an ideal transmission: the step and the impulse response.

1.3.2.1 Step response. In Eq. (1-22b) we defined the *step response $y(t)$* $= a(t)$ for a unit step input $x(t) = u(t)$. By Eq. (1-26), in a distortionless transmission system, $a(t)$ is given by

$$a(t) = Ku(t - \tau_0) \qquad (1\text{-}29)$$

which requires an unrealizable transfer function of the form of Eq. (1-28). In contrast, a typical response of a *real* system, a lowpass filter, is shown in Fig. 1-8. The quantity γ, which is the difference between the peak value and the final value, is called the *overshoot* and is given in percent. τ_d is the *delay time*, defined as the time required for the step response to reach 50% of its final value. τ_r, the *rise time*, is defined as the time required for the step response to rise from 10% to 90% of its final value. For example, consider the simple one-pole case

$$H(s) = \frac{1}{s + \sigma_1}$$

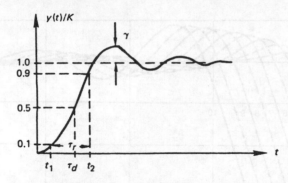

Figure 1-8 Typical step response for a stable linear system.

The step response is given by

$$y(t) = (1 - e^{-\sigma_1 t})u(t) \tag{1-30}$$

In this case there is no overshoot ($\gamma = 0$), and we find $\tau_d \simeq 0.69 /\sigma$ and $\tau_r \simeq 2.2/\sigma_1$. In general, for a filter with negligible overshoot ($\gamma \leq 5\%$), the following empirical result holds:

$$\tau_r \omega_{3dB} \simeq 2.2 \quad \text{or} \quad \tau_r f_{3dB} \simeq 0.35 \tag{1-31}$$

Popular lowpass filters are the Butterworth, Chebyshev, and Bessel-Thomson filters discussed in Section 1.6. Their step responses for different degrees are shown in Fig. 1-9a–c, respectively.

1.3.2.2 Impulse response.

When the input is an impulse, i.e., $x(t) = \delta(t)$, the ideal impulse response is given by Eq. (1-23). For an ideal transmission system,

$$h(t) = \mathcal{L}^{-1}\{Ke^{-s\tau_0}\} = K\delta(t - \tau_0) \tag{1-32}$$

Evidently, since in Eq. (1-32) $h(t)$ is the derivative of the response in Eq. (1-29), one can obtain the impulse response from the step response, and vice versa. Impulse responses, corresponding to those in Fig. 1-9a–c, are given in Fig. 1-10a–c.

Precise calculations of τ_r and τ_d for a given filter response are usually quite time-consuming. A more convenient method for such calculations that results in considerable simplification in characterizing time-domain responses was proposed by Elmore [6, 7]. His definitions of delay time and rise time are valid if $a(t)$ has negligible overshoot or none:

$$\tau_D = \int_0^\infty th(t)dt \tag{1-33}$$

$$\tau_R = \left[2\pi \int_0^\infty (t - \tau_D)^2 h(t) \, dt \right]^{1/2} \tag{1-34a}$$

$$= \sqrt{2\pi} \left[\int_0^\infty t^2 h(t) \, dt - \tau_D^2 \right]^{1/2} \tag{1-34b}$$

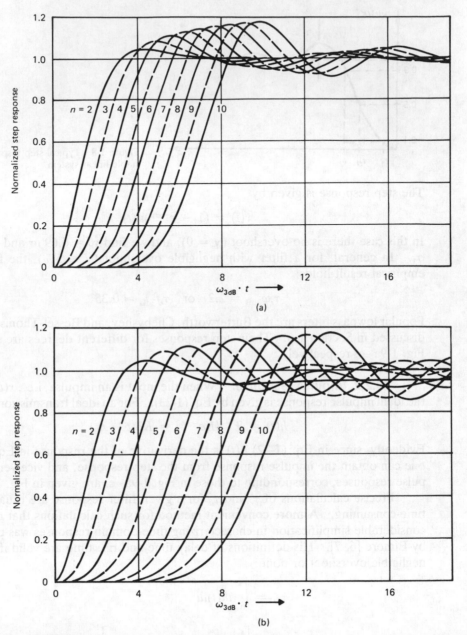

Figure 1-9 Step responses for (a) Butterworth filters, (b) Chebyshev filters (1/2 dB ripple), and (c) Thomson (Bessel) filters for $n \leq 10$. (From Henderson and Kautz [27]). ©IEEE, Dec. 1958 used with permission.

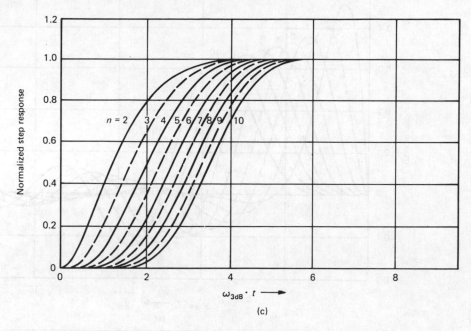

(c)

Figure 1-9 (*Continued*)

where $h(t)$ is the impulse response normalized such that $\int_0^\infty h(t)\,dt = 1$. Capital-letter subscripts are used to differentiate these definitions from those given in Section 1.3.2.1.

The interpretation of Eqs. (1-33) and (1-34) is shown in Fig. 1-11a and b. The ease of application of these definitions is illustrated below. Consider the transfer function, normalized such that $H(0) = 1$:

$$H(s) = \frac{1 + a_1 s + a_2 s^2 + \ldots + a_m s^m}{1 + b_1 s + b_2 s^2 + \ldots + b_n s^n} \quad n \geq m \tag{1-35}$$

The transfer function is related to the normalized impulse response by Eq. (1-25), which can be expanded into a power series as:

$$H(s) = \int_0^\infty h(t) \left(1 - st + \frac{s^2 t^2}{2!} - \ldots \right) dt$$

$$= 1 - s\tau_D + \frac{s^2}{2!} \int_0^\infty t^2 h(t)\,dt - \ldots \tag{1-36}$$

$$= 1 - s\tau_D + \frac{s^2}{2!} \left(\frac{\tau_R^2}{2\pi} + \tau_D^2 \right) - \ldots$$

From Eq. (1-35) one gets, by direct division,

$$H(s) = 1 - (b_1 - a_1)s + (b_1^2 - a_1 b_1 + a_2 - b_2)s^2 + \ldots \tag{1-37}$$

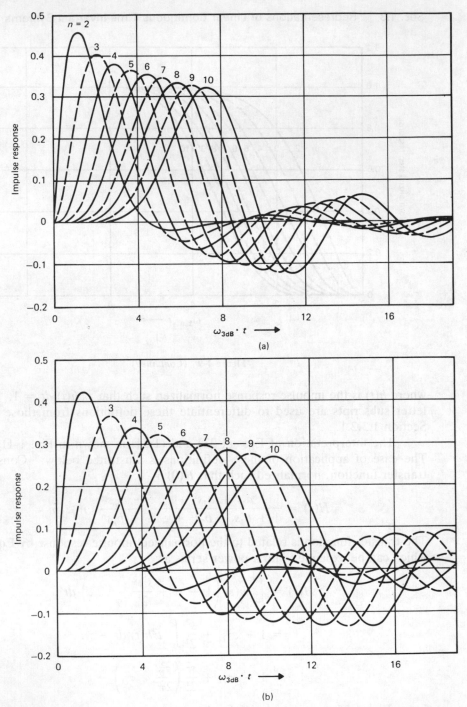

Figure 1-10 Impulse responses for (a) Butterworth filters, (b) Chebyshev filters (1/2 dB ripple), and (c) Thomson (Bessel) filters for $n \leq 10$. (From Henderson and Kautz [27]). ©IEEE, Dec. 1958 used with permission.

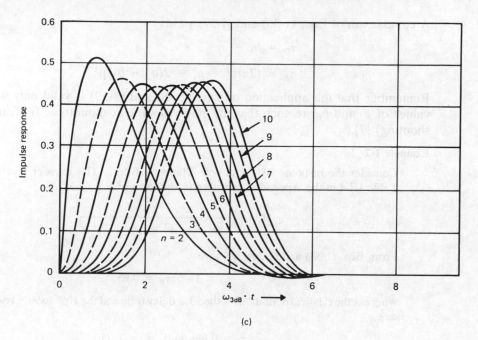

(c)

Figure 1-10   (*Continued*)

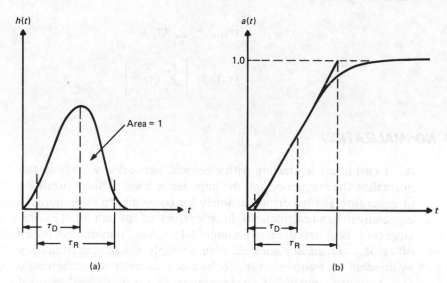

Figure 1-11   Impulse and step responses illustrating Elmore's definitions of rise and delay times.

A comparison of Eqs. (1-36) and (1-37) yields

$$\tau_D = b_1 - a_1 \tag{1-38}$$

$$\tau_R = \{2\pi[b_1^2 - a_1^2 + 2(a_2 - b_2)]\}^{1/2} \tag{1-39}$$

Remember that the application of Eqs. (1-38) and (1-39) is valid only when the values of a_i and b_i are such that the step response is monotonic (i.e., nonovershooting) [7].

Example 1-2

Consider the response of a four-pole Thomson filter. The transfer function from Table III-4 in the Appendix, rewritten in the form of Eq. (1-35), is

$$H(s) = \cfrac{1}{1 + s + \frac{3}{7}s^2 + \frac{2}{21}s^3 + \frac{1}{105}s^4}$$

From Eqs. (1-38) and (1-39), we have

$$\tau_D = 1 \qquad \tau_R = 0.95$$

whereas the classical definitions—the 50% delay time and the 10% to 90% rise times—are

$$\tau_d = 0.98 \quad \text{and} \quad \tau_r = 1.05$$

The ease of computation of these quantities using Elmore's definitions should be apparent. For higher-order systems this advantage becomes even more attractive. It can be shown that for k monotonic cascaded stages, the overall delay and rise times are given by

$$(\tau_D)_o = \sum_{i=1}^{k} (\tau_D)_i \tag{1-40}$$

$$(\tau_R)_o = \left[\sum_{i=1}^{k} (\tau_R)_i^2 \right]^{1/2} \tag{1-41}$$

1.4 NORMALIZATION

As is customary in dealing with electrical networks, we will in this book always normalize the frequency and the impedance level. Normalization causes no loss of generality and is performed solely for convenience in numerical computations, especially in hand calculations, in order to avoid the tedium of having to manipulate large powers of 10 (compare Example 1-1). Also, it minimizes the effects of round-off errors. *Frequency normalization* is simply a change in frequency scale effected by *dividing* the frequency variable by a conveniently chosen normalizing frequency Ω_0. Similarly, *impedance-level normalization* is performed by *dividing* all impedances in the circuit by a normalizing resistance R_0. Thus, the *normalized* frequency parameter, identified by a subscript n, is

$$s_n = \frac{s}{\Omega_0} \tag{1-42a}$$

and the *normalized* impedances which are represented by a resistor R, an inductor L, and a capacitor C, respectively, become

$$R_n = \frac{R}{R_0} \qquad s_n L_n = \frac{(s/\Omega_0)\Omega_0 L}{R_0} \quad \text{and} \quad \frac{1}{s_n C_n} = \frac{1}{s/\Omega_0} \frac{1}{\Omega_0 C R_0}$$

The normalized element values are, therefore,

$$R_n = \frac{R}{R_0} \qquad L_n = L\frac{\Omega_0}{R_0} \qquad C_n = C\Omega_0 R_0 \tag{1-42b}$$

Obviously, the actual unnormalized physical parameters s, R, L, and C are obtained by inverting the equations of (1-42). Two comments are in order:

1. One effect of normalization is to remove the dimensions from the circuit variables (s_n, R_n, L_n, and C_n are dimensionless numbers). This fact makes it easy to remember the relations in Eq. (1-42): for example, to denormalize a given value C_n, it must be multiplied by $1/(R_0\Omega_0)$ with the farad as the unit.

2. *Dimensionless* network functions, such as current ratios or the voltage transfer function $H(s) = V_{out}/V_{in}$ used in the active-filter work in this text, are independent of impedance level. Thus, the normalizing resistor R_0 is always a free parameter that the designer should choose to obtain convenient and practical element values.

Since in this book we will always consider circuit components and frequency to be normalized, the subscript n will be deleted to simplify notation as long as no confusion can arise.

Example 1-3

To illustrate normalization, consider a second-order Butterworth filter $H_L(s) = H_L(0)/(s^2 + \sqrt{2}s + 1)$ (see Section 1.6) to be realized by the circuit of Fig. 1-12. The cutoff frequency is $f_{3dB} = 10$ kHz, and a noninverting amplifier of gain K is used. s is normalized with respect to ω_{3dB}, and the voltage gain at dc, $H_L(0)$, is not specified. The voltage transfer function of the circuit can be readily determined as

$$H(s) = \frac{V_o}{V_i} = \frac{K/(R_1 R_2 C_1 C_2)}{s^2 + s\left(\dfrac{1}{R_1 C_1} + \dfrac{1}{R_2 C_1} + \dfrac{1-K}{R_2 C_2}\right) + \dfrac{1}{R_1 R_2 C_1 C_2}}$$

Since there are more circuit parameters than needed to realize $H(s)$, we choose $R_1 = R_2 = R$ and $C_1 = C_2 = C$ to get

$$H(s) = \frac{K/(R^2 C^2)}{s^2 + s\dfrac{3-K}{RC} + \dfrac{1}{R^2 C^2}} \tag{1-43}$$

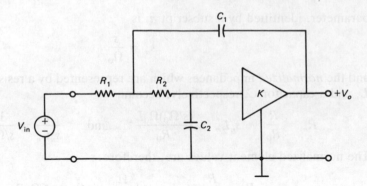

Figure 1-12 Example active *RC* circuit.

Equating the coefficients of Eq. (1-43) and the prescribed function $H_L(s)$ gives

$$\frac{1}{R^2 C^2} = 1 \quad \text{and} \quad \frac{3 - K}{RC} = \sqrt{2}$$

Because, as was stated earlier, the impedance level is a free parameter, we arbitrarily choose $R = 1 \ k\Omega$ to get, with the equations of (1-42),

$$C = \frac{1}{R\omega_{3dB}} = 15.9 \ nF \quad \text{and} \quad K = 1.586$$

The realized gain at dc is $H_L(0) = K = 1.586$.

1.5 TYPES OF FILTERS

As discussed in Section 1.1, filters are electrical networks that provide frequency-weighted transmission and are typically categorized according to the filtering function performed. If magnitude (gain, attenuation) requirements are of primary importance, the filters are classified as lowpass, highpass, bandpass, or bandreject networks. If, on the other hand, we are mainly concerned with phase or delay specifications with no change in magnitude, the filters typically will be allpass networks or delay equalizers. In any event, the coefficients of the transfer function [Eq. (1-12)]

$$H(s) = \frac{N(s)}{D(s)} = \frac{a_m s^m + \ldots + a_1 s + a_0}{s^n + \ldots + b_1 s + b_0} \tag{1-44}$$

have to be determined such that the desired filter specifications are met. Recall that $D(s)$ is a Hurwitz polynomial that determines the location of the poles in the left half-plane and that $N(s)$ determines the location of the transmission zeros of the filter. Also, the *order* of the filters n must satisfy $n \geq m$. Note that without loss of generality we have normalized the coefficient b_n to unity.

In this section we define each of the filter types and the manner in which they are specified, and in Section 1.6 we discuss how to obtain a function $H(s)$ that meets the prescribed specifications.

1.5.1 Filter Magnitude Specifications

The function of the *lowpass* (LP) *filter* is to pass low frequencies from dc to some desired cutoff frequency and to attenuate high frequencies. This filter is specified by its cutoff frequency ω_c, stopband (SB) frequency ω_s, dc gain, passband (PB) ripple, and stopband attenuation. The filter passband is defined as the frequency range $0 \leq \omega \leq \omega_c$, the stopband as the frequency range $\omega \geq \omega_s$, and the transition band (TB) as the frequency range $\omega_c < \omega < \omega_s$. These specifications are shown graphically in Figure 1-13: note that the specified filter must lie within the unshaded region.

The function of the *highpass* (HP) *filter* is to pass high frequencies above some specified cutoff frequency ω_c and to attenuate low frequencies from dc to some specified stopband frequency ω_s. The highpass filter is specified in much the same manner as the lowpass filter, as is shown graphically in Fig. 1-14. Again the desired filter response is to lie within the unshaded area. In principle, the passband of the highpass filter extends to $\omega = \infty$. In practice, however, it is limited in active filters by the finite bandwidth of the active devices and by parasitic capacitances. As a result, the gain of the highpass filter will eventually roll off at high frequencies, as illustrated in Fig. 1-14.

The function of the *bandpass* (BP) *filter* is to pass a finite band of frequencies while attenuating both lower and higher frequencies. This filter has both a lower stopband, SB_L, and an upper stopband, SB_H, as illustrated in Fig. 1-15. In general, the bandpass filter will not be symmetrical, and attenuation in the lower and upper SBs will be different. Similarly, the upper and lower transition bands TB_L and

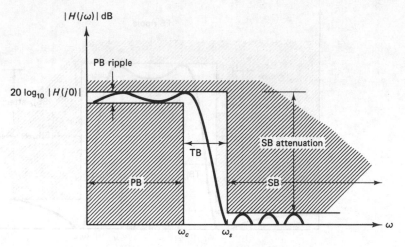

Figure 1-13 Lowpass filter characteristic.

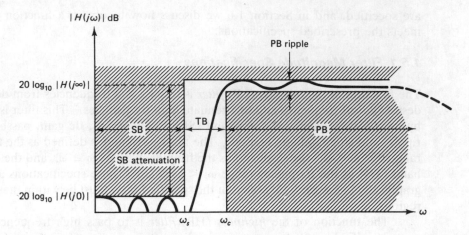

Figure 1-14   Highpass filter characteristic.

TB_H need not be the same (i.e., in general, $\omega_{sH}/\omega_{cH} \neq \omega_{cL}/\omega_{sL}$). Of course, such a specification can be met with a symmetric bandpass filter; however, a symmetrical filter will provide an overdesign in one stopband, which typically implies that the nonsymmetrical specification can be met with a lower-order nonsymmetrical filter. But as we will see in Chapter 6, it is often much easier to design a geometrically symmetrical bandpass filter.

The function of the *bandreject* (BR) *filter* is to attenuate a finite band of frequencies while passing both lower and higher frequencies. As a result, this filter has both lower- and upper-frequency passbands, PB_L and PB_H. The spec-

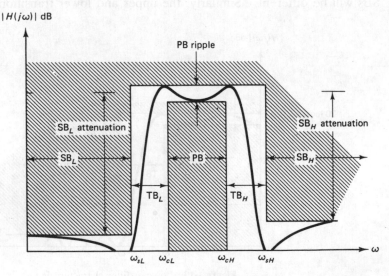

Figure 1-15   Bandpass filter characteristic.

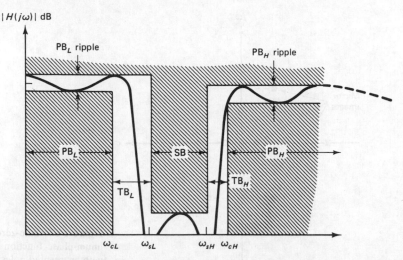

Figure 1-16   Band-rejection filter characteristic.

ification, shown graphically in Fig. 1-16, is similar to the bandpass case.   As in the highpass filter, in reality the upper passband is limited due to the band-limited active devices and parasitics.   The resulting high-frequency roll-off is shown by a dashed curve in Fig. 1-16.

1.5.2 Filter Phase or Delay Specifications

In order to realize the magnitude specifications most efficiently, i.e., with a transfer function $H(s)$ of lowest order n, one normally places all *transmission zeros* [the roots of $N(s) = 0$ in Eq. (1-44)] on the $j\omega$-axis in the s-plane.   It can be shown [9] that *in such a minimum-phase function magnitude and phase are not independent*; i.e., once the magnitude has been specified, one has to live with the resulting phase. This is usually of little concern in voice or audio-frequency applications because the human ear is very insensitive to changes in phase or delay with frequency. However, in video or digital transmission applications, the phase changes introduced by a filter can cause intolerable distortion in the shape of the time-domain video or digital signal.   In situations such as these, when magnitude *and* phase or delay performance are prescribed, the network function must, therefore, be *non-minimum*-phase; i.e., $H(s)$ must have zeros in the right half s-plane as shown in Fig. 1-4.   For ease of realization, such a non-minimum-phase function $H_N(s)$ is usually split into the product of a minimum-phase function $H_M(s)$ and a so-called *allpass* function $H_{AP}(s)$:

$$H_N(s) = H_M(s)H_{AP}(s) \qquad (1\text{-}45)$$

This separation is accomplished by augmenting the prescribed non-minimum-phase function $H_N(s)$ by a pole *and* a zero in the left half-plane at any location that is

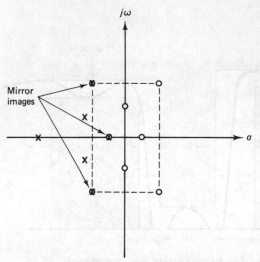

Figure 1-17 Pole-zero plot of non-minimum-phase function as in Fig. 1-4 but with augmented poles and zeros.

the mirror image of a right half-plane zero. The process is illustrated in Fig. 1-17, where the pole-zero pattern of $H_N(s)$ of Fig. 1-4 is redrawn but augmented by pole and zero mirror images as discussed above. The allpass function $H_{AP}(s)$ $= N_{AP}(s)/D_{AP}(s)$ is then formed by collecting into $N_{AP}(s)$ all right half-plane zeros and into $D_{AP}(s)$ all poles which are mirror images of the right half-plane zeros. The minimum-phase function $H_M(s)$ contains all remaining poles and zeros. Example 1-4 should illustrate the process.

Example 1-4

Express the non-minimum-phase function

$$H_N(s) = \frac{N_N(s)}{D(s)} = \frac{(s^2 + 2s + 6)(s^2 - 4s + 8)(s - 3)}{(s^2 + s + 4)(s^2 + 3s + 7)(s + 1)}$$

as a product of a minimum-phase function $H_M(s)$ and an allpass function $H_{AP}(s)$.

Solution. $H_N(s)$ has three zeros in the right half-plane, one at $s_1 = 3$ and a pair at $s_{2,3} = 2 \pm j2$. Thus, we augment $H_N(s)$ by three poles and three zeros which are mirror images of s_1, s_2, and s_3:

$$H_N(s) = \frac{N_N(s)}{D(s)} = \frac{(s - 3)(s^2 + 2s + 6)(s^2 - 4s + 8)(s + 3)(s^2 + 4s + 8)}{(s + 1)(s^2 + s + 4)(s^2 + 3s + 7)(s + 3)(s^2 + 4s + 8)}$$

This augmentation does not change $H_N(s)$, because the new pole and zero factors just cancel! Now, according to the rule stated above, we have

$$H_{AP}(s) = \frac{N_{AP}(s)}{D_{AP}(s)} = \frac{(s - 3)(s^2 - 4s + 8)}{(s + 3)(s^2 + 4s + 8)}$$

$$H_M(s) = \frac{N_M(s)}{D_M(s)} = \frac{(s + 3)(s^2 + 4s + 8)(s^2 + 2s + 6)}{(s + 1)(s^2 + s + 4)(s^2 + 3s + 7)}$$

and $H_N = H_M H_{AP}$. Note that the price paid for this process is an increase in the order of the prescribed function by the number of right half-plane zeros, three, in our case.

We still have to explain the nomenclature *allpass* function for $H_{AP}(s)$: Observe that H_{AP} is formed by assigning to $N_{AP}(s)$ roots in the right half-plane which are the mirror images of the roots of $D_{AP}(s)$ in the left half-plane. Consequently, it should be apparent that

$$N_{AP}(s) = \pm D_{AP}(-s)$$

so that an allpass function has the form

$$H_{AP}(s) = \pm \frac{D_{AP}(-s)}{D_{AP}(s)} \qquad (1\text{-}46)$$

and $|H_{AP}(j\omega)| \equiv 1$ for all ω, that is, signals at *all* frequencies *pass* equally well. Thus, $H_{AP}(j\omega) = \exp j[\phi_{AP}(\omega)]$, where

$$\phi_{AP}(\omega) = -2 \tan^{-1} \frac{D_I(\omega)}{D_R(\omega)} \qquad (1\text{-}47)$$

and where $D_R(\omega) = \text{Re}\,[D_{AP}(j\omega)]$ is the real part and $D_I(\omega) = \text{Im}\,[D_{AP}(j\omega)]$ is the imaginary part of $D_{AP}(j\omega)$.[5] We observe then that an allpass network can realize an arbitrary nontrivial phase function given by Eq. (1-47) or, equivalently, a designable delay $\tau_{AP}(\omega) = -d\phi_{AP}(\omega)/d\omega$ without any effect on the magnitude transmission.

This characteristic of allpass networks allows us in general to find an efficient transfer function $H(s)$, appropriate for the magnitude specifications, and to implement the corresponding filter. If the resulting phase or delay is unsatisfactory, we simply find a suitable allpass function $H_{AP}(s)$ to multiply $H(s)$ such that the total phase ϕ_T or total delay τ_T is as required. The two networks are connected in *cascade*, and the individual phases and delays simply add:

$$\phi_T(\omega) = \phi(\omega) + \phi_{AP}(\omega) \qquad (1\text{-}48a)$$

$$\tau_T(\omega) = \tau(\omega) + \tau_{AP}(\omega) \qquad (1\text{-}48b)$$

The cascaded allpass can, of course, only increase the phase and delay of $H(s)$; this is normally no problem, because for distortionless transmission, by Eq. (1-28), only the linearity of ϕ_T, i.e., the constancy of τ_T, in the frequency range of interest is important, not its actual size. Depending on whether one likes to work with Eq. (1-48a) or Eq. (1-48b), the allpass is labeled a *phase* or *delay equalizer*; it is determined such that when the compensation $\tau_{AP}(\omega)$ is added to the unacceptable delay $\tau(\omega)$ of the gain-shaping filter or of other parts of the transmission system,

[5]Note that the factor ± 1 in Eq. (1-46) has no effect on $|H_{AP}|$ and only adds possibly a constant angle of $180°$ to ϕ_{AP}, usually of no consequence.

the sum is $\tau_T(\omega) \simeq$ const in the frequency range of interest. See Example 1-5 in Section 1.5.3 for an illustration.

1.5.3 Second-Order Filters

A particularly important class of active networks realizes the second-order transfer function

$$H(s) = \frac{a_2 s^2 + a_1 s + a_0}{s^2 + b_1 s + b_0} = \frac{a_2(s + z_1)(s + z_2)}{(s + p_1)(s + p_2)} \tag{1-49}$$

Commonly referred to as the *biquadratic function*, it serves as the building block for a wide variety of active filters. In fact, Chapter 5 is devoted entirely to active filter realizations of this function.

For complex poles and zeros,[6] where $z_2 = z_1^*$ and $p_2 = p_1^*$ (superscript * denotes a complex conjugate), we can express Eq. (1-49) as

$$H(s) = K \frac{s^2 + [2 \text{ Re } (z_1)]s + \text{Re } (z_1)^2 + \text{Im } (z_1)^2}{s^2 + [2 \text{ Re } (p_1)]s + \text{Re } (p_1)^2 + \text{Im } (p_1)^2} \tag{1-50a}$$

$$= K \frac{s^2 + (\omega_z/Q_z)s + \omega_z^2}{s^2 + (\omega_p/Q_p)s + \omega_p^2} \tag{1-50b}$$

where Re ( ) and Im ( ) denote the real part and the imaginary part, respectively.

Equation (1-50b) is the standard notation for biquadratic functions because it identifies clearly the important filter performance parameters. The dc gain and the asymptotic gain as $\omega \to \infty$ are given by

$$20 \log_{10} |H(j0)| = 20 \log_{10} \left(K \frac{\omega_z^2}{\omega_p^2} \right) \tag{1-51a}$$

and

$$20 \log_{10}|H(j\infty)| = 20 \log_{10} (K) \tag{1-51b}$$

The gain function reaches its maximum value approximately at the pole frequency

$$\omega_p = \sqrt{\text{Re } (p_1)^2 + \text{Im } (p_1)^2} \tag{1-52a}$$

which is the radial distance from the origin of the s-plane to the pole location. The zero frequency ω_z determines approximately the point at which the gain function is minimum. ω_z is related to the zero location by the expression

$$\omega_z = \sqrt{\text{Re } (z_1)^2 + \text{Im } (z_1)^2} \tag{1-52b}$$

which is the radial distance from the origin to the zero location. The sharpness

---

[6]Although real zeros do occasionally occur in second-order active *RC* filters, real poles are of no interest, because they can be implemented with *passive RC* circuits.

of the maximum or bump at ω_p is determined by the *pole quality factor* Q_p, defined as

$$Q_p = \frac{\omega_p}{2 \text{ Re } (p_1)} = \frac{\sqrt{\text{Re } (p_1)^2 + \text{Im } (p_1)^2}}{2 \text{ Re } (p_1)} \tag{1-53a}$$

and the depth of the minimum at $s \simeq j\omega_z$ is determined by the *zero quality factor*

$$Q_z = \frac{\omega_z}{2 \text{ Re } (z_1)} = \frac{\sqrt{\text{Re } (z_1)^2 + \text{Im } (z_1)^2}}{2 \text{ Re } (z_1)} \tag{1-53b}$$

In many cases one has $Q_z = \infty$, that is, Re $(z_1) = 0$, and $\omega_z = \text{Im } (z_1)$ defines a null in gain (i.e., attenuation is infinite). Also, one can show that for large values of Q_p and Q_z, and when $\omega_p \ll \omega_z$ or $\omega_p \gg \omega_z$, the position of the gain maximum is virtually unaffected by the zeros.

Several special cases of the general biquadratic function [Eq. (1-49)] are of importance:

If $a_2 = a_1 = 0$, $H(s)$ is a second-order *lowpass* (LP) function commonly denoted by

$$H_{\text{LP}}(s) = \frac{K\omega_p^2}{s^2 + (\omega_p/Q_p)s + \omega_p^2} \tag{1-54}$$

A sketch of the corresponding gain function is given in Fig. 1-18a. Note that $H_{\text{LP}}(s)$ has a double zero at $s = \infty$, that the dc gain $|H_{\text{LP}}(j0)|$ is K, and that for $\omega \gg \omega_p$, $|H_{\text{LP}}(j\omega)|$ decreases as $1/\omega^2$ or -40 dB/decade. We can extrapolate this observation for an nth-order all-pole lowpass function, where $|H(j\omega)|$ rolls off for high frequencies at a rate of $-n \times 20$ dB/decade.

If $a_1 = a_0 = 0$, Eq. (1-49) results in a *highpass* (HP) function written as

$$H_{\text{HP}}(s) = \frac{Ks^2}{s^2 + (\omega_p/Q_p)s + \omega_p^2} \tag{1-55}$$

where K is the high-frequency gain $|H_{\text{HP}}(j\infty)|$. A sketch of the corresponding gain function is given in Fig. 1-18b. Here $|H_{\text{HP}}(j\omega)|$ increases as ω^2 for low frequencies, which corresponds to a low-frequency slope of 40 dB/decade.

If, in Eq. (1-49), $a_0 = a_2 = 0$, we obtain a *bandpass* (BP) function, denoted by

$$H_{\text{BP}}(s) = \frac{K(\omega_p/Q_p)s}{s^2 + (\omega_p/Q_p)s + \omega_p^2} \tag{1-56}$$

where $K = |H_{\text{BP}}(j\omega_p)|$ is the midband gain. Note that $H_{\text{BP}}(s)$ has a simple zero at $s = 0$ and at $s = \infty$, so that for $\omega \ll \omega_p$ the gain rises and for $\omega \gg \omega_p$ the gain falls by 20 dB/decade and $H_{\text{BP}}(s)$ has infinite attenuation at dc and at $s = \infty$.

A sketch of the corresponding gain function is given in Fig. 1-18c. For high Q, that is, $Q_p \gg 1$, $|H_{\text{BP}}(j\omega)|$ in Eq. (1-56) is approximately symmetrical about ω_p.

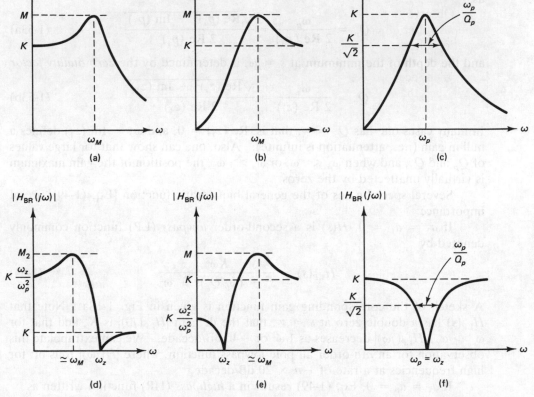

Figure 1-18 Gain curves for second-order filters: (a) lowpass $[M = KQ/\sqrt{1 - 1/(4Q^2)}$, $\omega_M = \omega_p\sqrt{1 - 1/(2Q^2)}]$; (b) highpass $[M = KQ/\sqrt{1 - 1/(4Q^2)}$, $\omega_M = \omega_p/\sqrt{1 - 1/(2Q^2)}]$; (c) bandpass; (d) lowpass notch and (e) highpass notch $[M \simeq KQ_p|1 - (\omega_z/\omega_p)^2|$, $\omega_M \simeq \omega_p\sqrt{1 + \dfrac{1}{[1 - (\omega_z\omega_p)^2]2Q_p^2}}]$; (f) symmetrical notch.

A second-order function that realizes a *bandreject* (BR) gain characteristic is obtained from Eq. (1-49) by setting $a_1 = 0$; it is given by

$$H_{\text{BR}}(s) = \frac{a_2 s^2 + a_0}{s^2 + (\omega_p/Q_p)s + \omega_p^2} = \frac{K(s^2 + \omega_z^2)}{s^2 + (\omega_p/Q_p)s + \omega_p^2} \qquad (1\text{-}57)$$

where $K = |H_{\text{BR}}(j\infty)|$ is the high-frequency gain. Observe that this bandreject, or *notch*, function provides infinite attenuation (a transmission zero) at $\omega = \omega_z$, and that the sharpness of the notch as well as the height of the adjacent bump is controlled by Q_p. One distinguishes between *lowpass* notch (LPN), *highpass* notch (HPN), and *symmetrical* notch gain responses depending on whether $\omega_z > \omega_p$,

$\omega_z < \omega_p$, or $\omega_z = \omega_p$, respectively.   The corresponding functions are plotted in Fig. 1-18d, e, and f.

Observe that a combination of an LPN and an HPN yields a fourth-order bandpass filter like that sketched in Fig. 1-15.   Also, in practice, high-order band-rejection filters usually use LPNs and HPNs.

If we want a second-order phase correction network (*allpass* or *delay equalizer*), then, according to Eq. (1-46), we need to determine the coefficients in Eq. (1-57) such that

$$H_{AP}(s) = K \frac{s^2 - s(\omega_p/Q_p) + \omega_p^2}{s^2 + s(\omega_p/Q_p) + \omega_p^2} \tag{1-58a}$$

$$= K \frac{s_n^2 - s_n/Q_p + 1}{s_n^2 + s_n/Q_p + 1} \tag{1-58b}$$

where K is the flat (frequency-independent) gain of the allpass function and s_n in Eq. (1-58b) is the normalized frequency s/ω_p.   Phase and delay of $H_{AP}(s)$ in Eq. (1-58b) are, with $\omega_n = \omega/\omega_p$,

$$\phi_{AP}(\omega_n) = -2\tan^1 \frac{\omega_n/Q_p}{1 - \omega_n^2} \tag{1-59a}$$

$$\tau_{n,AP}(\omega_n) = \omega_p\tau_{AP}(\omega_n) = \frac{2}{Q_p} \frac{1 + \omega_n^2}{(1 - \omega_n^2)^2 + (\omega_n/Q_p)^2} \tag{1-59b}$$

Figure 1-19a shows a sketch of magnitude and phase of the allpass function for a certain value of Q_p, and in Fig. 1-19b a few normalized delay curves of $\tau_{AP}(\omega_n)/2$ are plotted for $Q_p = 0.02, 0.1, 0.3, 1/\sqrt{3}, 2, 5, 20,$ and 50.   For $Q_p = 1/\sqrt{3}$, the delay curve is *maximally flat* (see Section 1.6.1.3); i.e., it is a good approximation of a *constant* delay in the frequency range $0 \leq \omega_n < 1$, whereas the delay curves are peaking for $Q_p > 1/\sqrt{3}$, with a peak value of $\tau_{AP,max} \simeq 4Q_p/\omega_p$ at $\omega_n \simeq \sqrt{1 - 1/(4Q_p^2)}$.

The student should verify (Problem 1-13) from Eqs. (1-54) to (1-57) with Eq. (1-17) that for second-order LP, HP, BP, and BR functions

$$\tau_{n,LP}(\omega_n) = \tau_{n,HP}(\omega_n) = \tau_{n,BP}(\omega_n) = \tau_{n,BR}(\omega_n) = \frac{1}{2}\tau_{n,AP}(\omega_n) \tag{1-60}$$

Equation (1-59b) and the curves in Fig. (1-19b) may, therefore, be regarded as universally valid for these second-order filters.

From Fig. 1-19b it should be apparent that *high-Q_p* gain-shaping filters will result in severe delay distortion, i.e., in a significant departure from the ideal flat delay asked for in Eq. (1-28).   The purpose of a delay equalizer is then to introduce delay shaping appropriate for making the total delay as flat as possible in the frequency band of interest, as indicated in Eq. (1-48b).   To find delay equalizers having a total filter or system delay with arbitrary prescribed tolerances usually requires computer aids.   Nevertheless, uncritical designs of low order (say,

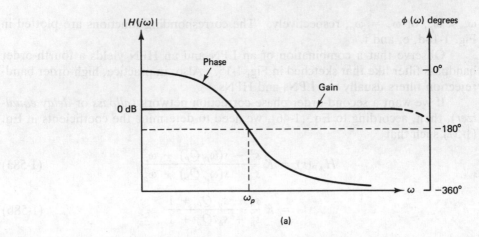

(a)

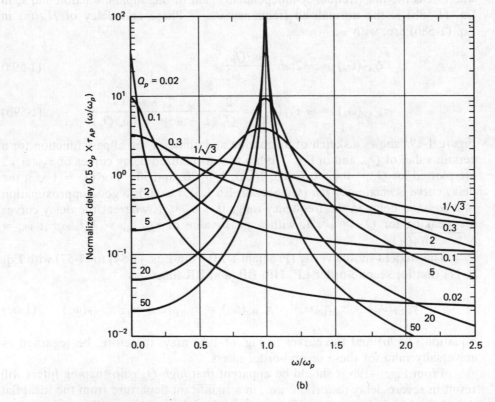

(b)

Figure 1-19a Gain and phase responses for a second-order allpass network. b One half delay of second-order allpass filter as a function of Q_p.

$\Delta\tau/\tau \simeq 10–20\%$) can be performed manually with the aid of the curves in Fig. 1-19b.   Example 1-5 illustrates the process.

Example 1-5

Assume that the curve marked τ_0 in Fig. 1-20, with $\omega_{p0} = 10$ krad/s, is the delay of a filter realized to satisfy some prescribed gain-shaping requirement.   Evidently, the delay is far from constant in the frequency band of interest, $0 \leq \omega_p \leq 1$.   Find two second-order delay equalizers to be connected to the given filter such that the total delay is "reasonably" constant.

Solution.   Observe that the given delay data are unnormalized and plotted on a linear scale so that the curves in Fig. 1-19b must be converted correspondingly.   Note first from Eq. (1-59b) and Fig. 1-19b that

$$\tau_{AP}(0) = \frac{2}{\omega_p Q_p} \tag{1-61a}$$

and, for $Q_p \geq 1.5$,

$$\tau_{AP,max} \simeq \frac{4Q_p}{\omega_p} \quad \text{at} \quad \omega \simeq \omega_p \tag{1-61b}$$

At low frequencies, the delay τ_0 should be increased by $\tau_1(0) \simeq 1$ ms, which with Eq. (1-61a) leads to allpass number 1 with $\omega_{p1} = \omega_{p0} = 10$ krad/s and

$$Q_{p1} \simeq \frac{2}{\omega_{p1}\tau_1} = \frac{2}{10^4 \cdot 10^{-3}} = 0.2$$

The dashed curve in Fig. 20, drawn with the help of Fig. 1-19b, shows the delay τ_1 of this second-order allpass number 1.   Mentally adding τ_1 to τ_0 shows that a peaking

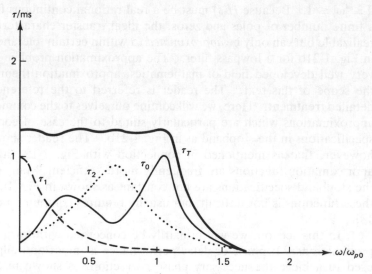

Figure 1-20   Delay curves for Example 1-5.

delay curve with a peak of approximately 0.9 ms at $\omega/\omega_{p0} \simeq 0.7$ must be added to flatten out the total delay. Since the peak occurs at $\omega/\omega_{p2} \simeq 1$, we find that $\omega_{p2} \simeq 0.7\omega_{p0}$ is the normalizing frequency for allpass number 2. Then, from Eq. (1-61b), we obtain

$$Q_{p2} \simeq \frac{1}{4}\omega_{p2}\tau_2(1) = \frac{0.7 \cdot 10^4 \cdot 0.9 \cdot 10^{-3}}{4} = 1.575$$

The dotted line in Fig. 1-20 shows $\tau_2(\omega/\omega_{p0})$. Thus, with $s_n = s/\omega_{p0}$, the two allpass functions

$$H_1(s_n) = \frac{s_n^2 - 5s_n + 1}{s_n^2 + 5s_n + 1} \quad \text{and} \quad H_2(s_n) = \frac{s_n^2 - 0.444s_n + 0.49}{s_n^2 + 0.444s_n + 0.49}$$

in cascade with the initial filter result in a total delay $\tau_T(\omega_n)$ that clearly is approximately constant in $0 \leq \omega_n \leq 1$, as desired.

1.6 APPROXIMATION METHODS

In filter design, usually passband, stopband, and transition-band requirements for magnitude or phase are prescribed, such as the ones indicated in Figs. 1-13 to 1-16. The first task of the filter designer is then to obtain a realizable network $H(s)$ of the form given in Eq. (1-44) which satisfies these specifications. Ideally, one would like to realize perfect transmission (no loss) in the passband(s), infinite attenuation (zero gain) in the stopband(s), and transition band(s) of zero width, as indicated for a lowpass filter in Fig. 1-21a.

In Fig. 1-21 and throughout this section, we assume that the frequency is normalized such that $\omega = 1$ at the edge of the passband and max $|H(j\omega)| = 1$ in $0 \leq |\omega| \leq 1$. Because $H(s)$ must be a real rational continuous function of ω with a finite number of poles and zeros, the ideal transfer characteristic is clearly not realizable but can only be *approximated* to within certain tolerance bands as shown in Fig. 1-21b for a lowpass filter. The approximation problem is addressed by a very well developed field of mathematics, approximation theory, that is beyond the scope of this text. The reader is referred to the references [10–15] for a detailed treatment. Here, we will confine ourselves to the commonly used classical approximations which are particularly suited to the case of constant attenuation specifications in the stopband as in Fig. 1-21b. The reader should keep in mind, however, that as mentioned in connection with Fig. 1-15, other (nonclassical) approximation functions are frequently more "efficient," i.e., of lower order, if the stopband specifications are not constant as shown in Fig. 1-21c. To arrive at these functions is not difficult, but usually requires computer aids [11, 14, 16, 17, 18].

In this section, we will primarily be concerned with lowpass magnitude approximations. If phase or delay performance is important, allpass filters can be used to achieve the necessary phase correction, as shown in Section 1.5.2 and

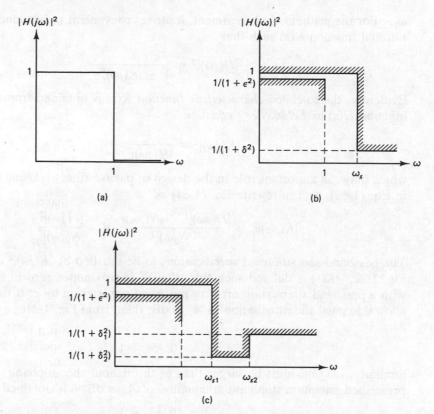

Figure 1-21 Lowpass specifications: (a) nonrealizable ideal lowpass; (b) realizable lowpass specifications with passband tolerances given by $\varepsilon < 1$ and stopband tolerances given by $\delta > 1$; (c) lowpass with nonconstant stopband specifications.

Example 1-5. However, we will discuss one type of lowpass approximation, leading to the Bessel, or Thomson, filter, which is derived specifically to achieve a constant delay. In the second part of this section, we introduce the method of *frequency transformation* for obtaining highpass, bandpass, or band-rejection functions from a given *prototype lowpass* function.

1.6.1 The Lowpass Magnitude Approximation

The squared magnitude of $H(j\omega)$ of Eq. (1-44) is obtained from

$$|H'j\omega)|^2 = \frac{N(j\omega)N(-j\omega)}{D(j\omega)D(-j\omega)} = \frac{|N(j\omega)|^2}{|D(j\omega)|^2} \triangleq \frac{P(\omega^2)}{E(\omega^2)} \qquad (1\text{-}62)$$

Clearly, $|H(j\omega)|^2$ is an *even* rational function that, according to Fig. 1-21b, must approximate unity in the passband, $0 \le |\omega| \le 1$, and zero in the stopband, $|\omega| \ge$

ω_s. For the mathematical treatment, it proves convenient to introduce a new real rational function $K(s)$ such that

$$|H(j\omega)|^2 = \frac{1}{1 + |K(j\omega)|^2} \tag{1-63}$$

Evidently, the so-called *characteristic function* $K(s)$ is obtained from the transfer function $H(s)$ by *Feldtkeller's equation*,

$$|K(j\omega)|^2 = \frac{1}{|H(j\omega)|^2} - 1 \tag{1-64}$$

which plays an important role in the design of passive filters. Using the notation in Eq. (1-62), we can rewrite Eq. (1-64) as

$$|K(j\omega)|^2 = \frac{|D(j\omega)|^2 - |N(j\omega)|^2}{|N(j\omega)|^2} \triangleq \varepsilon^2 \frac{|F(j\omega)|^2}{|N(j\omega)|^2} \tag{1-65}$$

The passband and stopband specifications to be satisfied by $|K(j\omega)|^2$ are shown in Fig. 1-22. $K(s)$ is defined such that $|K(j\omega)|^2$ approximates zero in the passband with a passband attenuation error or *passband ripple* given by ε: If the maximally allowable passband attenuation[7] is A_p in dB, then, from Fig. 1-21b, ε is calculated as

$$\varepsilon^2 = 10^{0.1A_p} - 1 \tag{1-66a}$$

Similarly, $|K(j\omega)|^2$ must be larger than δ^2 throughout the stopband, where for a prescribed minimum stopband attenuation of A_s in dB, δ is obtained from

$$\delta^2 = 10^{0.1A_s} - 1 \tag{1-66b}$$

In Eq. (1-65) we have defined a new polynomial, the *reflection zero* polynomial $F(s)$, whose roots, the *reflection zeros* ω_{ri}, are the zeros of $|K(j\omega)|$. Thus, at $s = j\omega_{ri}$ we have perfect transmission: $|H(j\omega_{ri})| = 1$. Also, recall that the roots ω_{zi} of the polynomial $N(s)$ are the *transmission zeros*, that is, $H(j\omega_{zi}) = 0$; the values ω_{zi} are also called *losspoles*, where $|K(j\omega_{zi})| = \infty$. Further, in Eq. (1-65), we have for convenience in later calculations explicitly shown the parameter ε so that, evidently, we have $|F(j\omega)/N(j\omega)| \le 1$ in $0 \le |\omega| \le 1$. A typical example function $|K(j\omega)|^2$ with losspoles at ω_{z1}, ω_{z2}, ω_{z3}, and ∞, and reflection zeros at 0, ω_{r1}, ω_{r2}, and ω_{r3}, is sketched in Fig. 1-22.

The general lowpass approximation problem should now be clear: we have to find a real rational function $K(s) = \varepsilon F(s)/N(s)$ such that

1. $F(s)$ has all roots on the $j\omega$-axis in the passband.

---

[7]A_p and A_s are defined as

$$\min (20 \log |H(j\omega)|) = -A_p, \text{ dB} \quad \text{for } \omega \text{ in passband}$$

$$\max (20 \log |H(j\omega)|) = -A_s, \text{ dB} \quad \text{for } \omega \text{ in stopband}$$

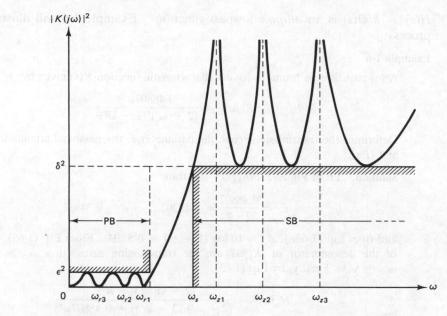

Figure 1-22 Lowpass specifications of Fig. 1-21 to be satisfied by $|K(j\omega)|^2$.

2. $N(s)$ has all roots on the $j\omega$-axis in the stopband.[8]
3. $|F(j\omega)/N(j\omega)| \leq 1$ in $0 \leq |\omega| \leq 1$.
4. $|K(j\omega)| \geq \delta$ in $|\omega| \geq \omega_s$.

After finding $|K(j\omega)|^2$, we employ Eq. (1-63) to obtain $|H(j\omega)|^2$ as a ratio of two *even* polynomials $P(\omega^2)$ and $E(\omega^2)$ as shown in Eq. (1-62). As a final step we use *analytic continuation*, i.e., we replace ω by s/j (or ω^2 by $-s^2$), and factor $P(-s^2)$ and $E(-s^2)$ into $N(s)N(-s)$ and $D(s)D(-s)$, respectively, to obtain the desired $H(s) = N(s)/D(s)$.

Note that $E(-s^2)$, being even, has roots symmetrical with respect to the origin in the s-plane. Because $D(s)$ must be a Hurwitz polynomial, it will simply contain *all the left half-plane roots* of $E(-s^2)$. The polynomial $N(s)$, of course, is already known. It was found explicitly during the determination of $|K(j\omega)|$:

$$N(s) = k \prod_i (s^2 + \omega_{zi}^2)^{m_i} \qquad (1\text{-}67)$$

where k is a constant and the ω_{zi} are the transmission zeros of multiplicity $m_i \geq 0$. If all m_i are zero, there are no finite transmission zeros, $N(s) = k$, and

---

[8]The more general case which has reflection zeros s_{ri} and/or losspoles s_{zi} which are not on the $j\omega$-axis in the s-plane will not be considered in this text. Usually, purely imaginary values for s_{zi} and s_{ri} lead to the most efficient approximations.

$H(s) = k/D(s)$ is an *all-pole* lowpass function. Example 1-6 will illustrate the process:

Example 1-6

A lowpass filter is found to have a characteristic function $K(s)$ given by

$$|K(j\omega)|^2 = \frac{0.48807\omega^{10}}{(2 - \omega^2)^2(3 - \omega^2)^2}$$

Determine the transmission zeros, the parameter ε, the passband attenuation in dB, and $H(s)$.

Solution. From Fig. 1-22, $|K(j1)| = \varepsilon$; thus

$$\varepsilon^2 = \frac{0.48807}{1 \cdot 4} = 0.12202 \qquad \varepsilon = 0.34931$$

and from Eq. (1-66a), $A_p = 10 \log (1 + \varepsilon^2) = 0.5$ dB. From Eq. (1-65), the roots of the denominator of $|K(j\omega)|$ are the transmission zeros; thus $\omega_{z1} = \sqrt{2}$ and $\omega_{z2} = \sqrt{3}$. Finally, by Eq. (1-63),

$$|H(j\omega)|^2 = \frac{(2 - \omega^2)^2(3 - \omega^2)^2}{(2 - \omega^2)^2(3 - \omega^2)^2 + 0.48807\omega^{10}}$$

and, replacing ω^2 by $-s^2$,

$$H(s)H(-s) = \frac{(2 + s^2)^2(3 + s^2)^2}{36 + 60s^2 + 37s^4 + 10s^6 + s^8 - 0.48807s^{10}}$$

This denominator polynomial, $E(-s^2) = D(s)D(-s)$, must now be factored; making use of a computer root-finding routine yields $s_{1,2} = 0.184484 \pm j1.14676$, $s_{3,4} = -0.184484 \pm j1.14676$, $s_{5,6} = 0.787586 \pm j1.33914$, $s_{7,8} = -0.787586 \pm j1.33914$, and $s_{9,10} = \pm 2.63759$.

Note that the roots are distributed symmetrically around the origin as expected. Collecting the left half-plane roots into $D(s)$ results in

$$H(s) = \frac{1.4314(s^2 + 2)(s^2 + 3)}{(s + 2.6376)(s^2 + 0.36897s + 1.3491)(s^2 + 1.5752s + 2.4136)}$$

$$= \frac{1.4314(s^2 + 2)(s^2 + 3)}{s^5 + 4.5817s^4 + 9.4717s^3 + 14.473s^2 + 11.210s + 8.5883}$$

Observe that $H(j0) = 1$ as expected because $K(s)$ has a reflection zero (of multiplicity 5) at the origin. As will become evident shortly, $H(s)$ is a fifth-order *maximally flat* lowpass transfer function with two finite transmission zeros, as sketched in Fig. 1-23.

In the following subsections we will discuss how to obtain the characteristic function $K(s)$ for several different approximation criteria.

1.6.1.1 Maximally flat magnitude (MFM) lowpass filters.

For lowpass filters with maximally flat magnitude, we require, as shown in Fig. 1-23, that the squared magnitude of the transfer function be equal to unity at $\omega = 0$—that is,

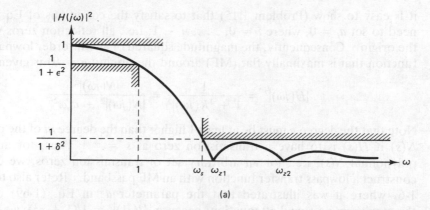

(a)

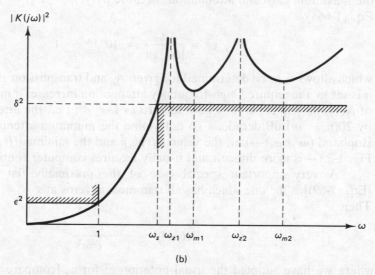

(b)

Figure 1-23 Magnitude squared of a maximally flat lowpass transfer function (a) and characteristic function (b) with finite transmission zeros.

we require ideal transmission at dc—and that all possible derivatives of the transmission error, defined as $\Delta(\omega^2) = 1 - |H(j\omega)|^2$, be equal to zero at $\omega = 0$. Thus, with Eq. (1-63):

$$|K(j0)| = 0$$

and

$$\frac{d^i}{(d\omega^2)^i}\left[\frac{|K(j\omega)|^2}{1 + |K(j\omega)|^2}\right]_{\omega=0} = 0 \qquad i = 1, 2, \ldots \quad (1\text{-}68)$$

Because $|K(j\omega)|^2$ is a ratio of polynomials in ω^2, i.e.,

$$|K(j\omega)|^2 = \frac{a_0 + a_1\omega^2 + a_2\omega^4 + \ldots + a_n\omega^{2n}}{N(j\omega)N(-j\omega)}$$

it is easy to show (Problem 1.15) that to satisfy the conditions of Eq. (1-68) we need to set $a_i = 0$, where $i = 0, \ldots, n - 1$; i.e., all reflection zeros must be at the origin. Consequently, the magnitude squared of an nth-order lowpass transfer function that is maximally flat (MF) around the origin ($\omega = 0$) is given by

$$|H(j\omega)|^2 = \frac{1}{1 + |K(j\omega)|^2} = \frac{|N(j\omega)|^2}{|N(j\omega)|^2 + a_n\omega^{2n}} \tag{1-69}$$

Note that the degree n must be at least 1 higher than the degree m of the polynomial $N(s)$ if $H(s)$ is to have a transmission zero at $s = \infty$. Thus, for an arbitrary polynomial $N(s)$, i.e., for an arbitrary set of transmission zeros, we can always construct a lowpass transfer function with an MF passband. Refer also to Example 1-6, where it was illustrated that the parameter a_n in Eq. (1-69) determines the maximum passband attenuation: because $|H(j1)|^2 = 1/(1 + \varepsilon^2)$, we have, with Eq. (1-66a),

$$\varepsilon^2 = \frac{a_n}{|N(j1)|^2} = 10^{0.1A_p} - 1 \tag{1-70}$$

which allows a_n to be determined for given A_p and transmission zeros. The degree n is set by the required high-frequency attenuation increase: if $m < n$ is the degree of $N(s)$, $|H(j\omega)|$ for $\omega \to \infty$ goes to zero as $1/\omega^{n-m}$; i.e., the attenuation increases by $20(n - m)$ dB/decade. To determine the minimum attenuation level in the stopband ($\omega > \omega_s$)—i.e., the value $|H(j\omega_s)|$ and the minima $|H(j\omega_{mi})|$ identified in Fig. 1-23—is more difficult and usually requires computer routines [11, 18].

A very important special case of the maximally flat lowpass function [Eq. (1-69)] is the one which has all transmission zeros at $s = \infty$, that is, $N(s) = 1$. Then

$$|H(j\omega)|^2 = \frac{1}{1 + \varepsilon^2\omega^{2n}} \tag{1-71}$$

where we have adopted the usual notation ε^2 for a_n [compare Eq. (1-70)]. For this all-pole lowpass, whose attenuation increases monotonically with ω, the degree n is determined from the stopband requirement, $20 \log |H(j\omega_s)| \leq -A_s \, dB$, as

$$n \geq \frac{\log \left[\varepsilon^{-2}(10^{0.1A_s} - 1)\right]}{2 \log \omega_s} \tag{1-72}$$

with ε from Eq. (1-66a).

Example 1-7

Find the degree of an all-pole maximally flat lowpass function to satisfy: passband in $0 \leq f \leq 1.2$ kHz with $A_p = 0.5$ dB maximum attenuation; stopband in $f \geq 1.92$ kHz with $A_s = 23$ dB minimum attenuation.

Solution. Normalize the frequency by $\Omega_1 = 2\pi \cdot 1.2$ kHz so that $\omega = 1$ is the passband edge and $\omega_s = 1.92$ kHz/1.2 kHz $= 1.6$. Then, from Eq. (1-66a),

$$\varepsilon = \sqrt{10^{0.05} - 1} = \sqrt{0.12202} = 0.3493$$

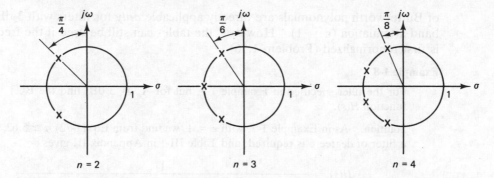

Figure 1-24   Pole locations for Butterworth filters of degree 2, 3, and 4.

and from Eq. (1-72)

$$n \geq \frac{\log [8.196(10^{2.3} - 1)]}{2 \log 1.6} = 7.87$$

Thus, a filter of order $n = 8$ is required.

To determine the location of the poles of an all-pole maximally flat lowpass function, we need to replace ω^2 by $-s^2$ in Eq. (1-71) and factor the polynomial $1 + (-1)^n \varepsilon^2 s^{2n} = 0$.   The roots are found from

$$s^{2n} = \frac{1}{\varepsilon^2} (-1)^{n+1} = \frac{1}{\varepsilon^2} e^{j(n+1+2k)\pi}$$

where k is an integer.   Retaining only the roots in the left half-plane gives the desired poles at

$$s_k = \varepsilon^{-1/n} \exp \left(j \frac{2k + n - 1}{2n} \pi \right) \qquad k = 1, 2, \ldots, n \qquad (1\text{-}73)$$

It is seen that all the poles are located uniformly spaced, $180°/n$ apart, on a circle of radius $\varepsilon^{-1/n}$, as shown for $\varepsilon = 1$ and $n = 2, 3,$ and 4 in Fig. 1-24.

Note from Eq. (1-66a) that $\varepsilon = 1$ implies $A_p = 3$ dB passband attenuation at $\omega = 1$.   The corresponding MF all-pole lowpass filters of the form

$$H(s) = \frac{1}{s^n + b_{n-1}s^{n-1} + \ldots + b_1 s + 1} \qquad (1\text{-}74)$$

are called *Butterworth* (BUT) filters[9] whose denominator polynomials are tabulated (see Appendix III, Table III-1 for Butterworth polynomials up to $n = 10$).   Thus, the transfer function is completely determined as soon as the only parameter, n, is found from Eq. (1-72) for $\varepsilon = 1$ and a prescribed value A_s.   Note that tables

[9]In the literature, frequently all maximally flat all-pole filters, even with $\varepsilon \neq 1$, are referred to as Butterworth filters.

of Butterworth polynomials are directly applicable *only* for filters with 3-dB pass-band attenuation ($\varepsilon = 1$). However, the tables can still be used if the frequency is first renormalized (Problem 1.16).

Example 1-8

For the filter specified in Example 1-7, but for $A_p = 3$ dB, find the BUT transfer function $H(s)$.

Solution. As in Example 1-7, with $\varepsilon = 1$, we find from Eq. (1-72) $n \geq 5.62$. Thus, a filter of degree 6 is required, and Table III-1 in Appendix III gives

$$H(s) = \frac{1}{s^6 + 1 + 3.8637(s^5 + s) + 7.4641(s^4 + s^2) + 9.1416s^3}$$

Note the symmetry in the coefficients.

Example 1-9

Find a third-order allpass transfer function whose group delay is maximally flat around $\omega = 0$. At the origin, the delay must be $\tau(0) = 65$ μs.

Solution. According to Eq. (1-46), a third-order allpass function has the form

$$H(s) = \frac{-s^3 + as^2 - bs + c}{s^3 + as^2 + bs + c}$$

with a phase

$$\phi(\omega) = -2 \tan^{-1} \frac{b\omega - \omega^3}{c - a\omega^2}$$

and, by Eq. (1-20d),

$$\tau(\omega) = -\frac{d\phi}{d\omega} = \frac{2}{\Omega_c} \frac{bc + (ab - 3c)\omega^2 + a\omega^4}{c^2 + (b^2 - 2ac)\omega^2 + (a^2 - 2b)\omega^4 + \omega^6}$$

where we assume that the frequency is normalized with respect to a frequency Ω_c in rad/s. To make $\tau(\omega)$ maximally flat, it has to be of the form of Eq. (1-69), i.e., $\tau(\omega) = N(\omega^2)/[N(\omega^2) + a_n\omega^{2n}]$. Note that maximal flatness is a mathematical property of a function, regardless of whether the function describes magnitude or delay. We need to satisfy, therefore, the restrictions $bc = c^2$, $ab - 3c = b^2 - 2ac$, and $a = a^2 - 2b$.

The solutions, $a = 6$, $b = c = 15$, give

$$H(s) = \frac{-s^3 + 6s^2 - 15s + 15}{s^3 + 6s^2 + 15s + 15}$$

Because $\tau(0) = 2b/(c\Omega_c) = 2/\Omega_c = 65$ μs, the normalizing frequency is $\Omega_c = 30.77$ krad/s.

1.6.1.2 Equiripple all-pole lowpass filters. In the MF all-pole approximation we concentrate all the approximating power at $\omega = 0$ and accept a monotonically increasing error for $\omega \to 1$. It would seem more efficient to distribute the error evenly throughout the passband by finding a characteristic function $K(s)$ such that

$$|K(j\omega)|^2 = \varepsilon^2 C_n^2(\omega) \tag{1-75}$$

is a polynomial which oscillates uniformly between zero and ε^2, i.e., in which $-1 \leq C_n(\omega) \leq 1$, in $0 \leq |\omega| \leq 1$. In this case, in the passband $|K(j\omega)|^2$ would look like the function sketched in Fig. 1-22, but of course with no finite transmission zeros, because $C_n(\omega)$ is a polynomial. There are several rigorous methods for deriving the polynomial $C_n(\omega)$ (e.g., Problem 1-23); a very simple intuitive method is to recognize that a sinusoid

$$C_n(\omega) = \cos[n\phi(\omega)] \tag{1-76a}$$

oscillates uniformly between $+1$ and -1 and with increasing frequency for increasing n. Using well-known trigonometric identities, $\cos(n\phi)$ can be written as a polynomial in $\cos\phi$:

$$\cos n\phi = 2^{n-1}\cos{}^n\phi - \frac{n}{1!}2^{n-3}\cos^{n-2}\phi + \frac{n(n-3)}{2!}2^{n-5}\cos^{n-4}\phi$$

$$- \frac{n(n-3)(n-5)}{3!}2^{n-7}\cos^{n-6}\phi + \dots \tag{1-76b}$$

which is seen to be a polynomial in ω if

$$\phi(\omega) = \cos^{-1}\omega \qquad |\omega| \leq 1 \tag{1-76c}$$

Thus, the *Chebyshev polynomials* of order n are

$$C_n(\omega) = \cos(n\cos^{-1}\omega) \qquad |\omega| \leq 1 \tag{1-77a}$$

In the stopband ($|\omega| > 1$), $\cos^{-1}\omega$ is undefined. It is, however, easy to show (Problem 1.24) that, for $|\omega| \geq 1$,

$$C_n(\omega) = \cosh(n\cosh^{-1}\omega)$$

$$= \frac{1}{2}[(\omega + \sqrt{\omega^2 - 1})^n + (\omega + \sqrt{\omega^2 - 1})^{-n}] \tag{1-77b}$$

which increases monotonically for $\omega > 1$. Thus, the desired gain- and frequency-normalized equiripple all-pole lowpass Chebyshev (CHEB) transfer function $H(s)$ is obtained from

$$|H(j\omega)|^2 = \frac{1}{1 + \varepsilon^2 C_n^2(\omega)} \tag{1-78}$$

The first few Chebyshev polynomials and a recursion formula are given in Table 1-1. It should be noted from this table that $C_n(1) = 1$ for all n and that $C_n(\omega)$ is even for even n and odd for odd n. It follows from Eqs. (1-77) and (1-78), therefore, that $|H(j\omega)|$ oscillates with equal ripple throughout the passband

TABLE 1-1   CHEBYSHEV
POLYNOMIALS

$$
\begin{aligned}
C_0(\omega) &= 1 \\
C_1(\omega) &= \omega \\
C_2(\omega) &= 2\omega^2 - 1 \\
C_3(\omega) &= 4\omega^3 - 3\omega \\
C_n(\omega) &= 2\omega C_{n-1}(\omega) - C_{n-2}(\omega)
\end{aligned}
$$

$0 \leq |\omega| \leq 1$, that $|H(j0)| = 1$ only for odd n but $|H(j0)| = 1/\sqrt{1 + \varepsilon^2}$ for even n, and that $|H(j1)| = \sqrt{1 + \varepsilon^2}$ for any n. Further, $|H(j\omega)|$ decreases monotonically to zero for $|\omega| > 1$ as shown in Fig. 1-25 for $n = 2, 3$, and 4.

The pole locations of a CHEB filter are found from Eqs. (1-78) and (1-77) with $\omega = s/j$ by solving

$$
C_n^2\left(\frac{s}{j}\right) = \cos^2\left(n \cos^{-1}\frac{s}{j}\right) = -\frac{1}{\varepsilon^2} \tag{1-79}
$$

It is easy to show (Problem 1-25) that the roots of Eq. (1-79) lie on an ellipse in the s-plane and that the left half-plane roots, the poles of a Chebyshev lowpass function, are given by $s_k = \sigma_k + j\omega_k$, $k = 1, 2, \ldots, n$, where

$$
\sigma_k = -\frac{1}{2}\left[\left(\frac{1}{\varepsilon} + \sqrt{\frac{1}{\varepsilon^2} + 1}\right)^{1/n} - \left(\frac{1}{\varepsilon} + \sqrt{\frac{1}{\varepsilon^2} + 1}\right)^{-1/n}\right] \sin\left(\frac{2k - 1}{2n}\pi\right) \tag{1-80a}
$$

$$
\omega_k = \frac{1}{2}\left[\left(\frac{1}{\varepsilon} + \sqrt{\frac{1}{\varepsilon^2} + 1}\right)^{1/n} + \left(\frac{1}{\varepsilon} + \sqrt{\frac{1}{\varepsilon^2} + 1}\right)^{-1/n}\right] \cos\left(\frac{2k - 1}{2n}\pi\right) \tag{1-80b}
$$

With the poles known, the transfer function of a CHEB filter becomes[10]

$$
H(s) = \frac{1}{2^{n-1}\varepsilon \displaystyle\prod_{k=1}^{n} (s - s_k)} = \frac{1/(2^{n-1}\varepsilon)}{s^n + b_{n-1}s^{n-1} + \ldots + b_1 s + b_0} \tag{1-81}
$$

The denominator polynomials up to $n = 10$ are listed in Tables III-2a, b, and c in Appendix III for 0.5 dB, 1 dB, and 2 dB passband ripple. Extensive tables for various ripple values are listed in the literature [e.g., 10, 18, 20, 21, 22].

There are two parameters in the design of Chebyshev filters: the ripple (ε) and the amount of attenuation (A_s) at the frequency ω_s at the edge of the stopband. The parameter ε is found from Eq. (1-66a): $\varepsilon = \sqrt{10^{0.1A_p} - 1}$. To determine n, we evaluate

$$
10 \log \frac{1}{1 + \varepsilon^2 C_n^2(\omega_s)} = -A_s \quad \text{dB}
$$

[10]Note from Eqs. (1-76b) and (1-78) that the coefficient $b_n = 2^{n-1}\varepsilon$.

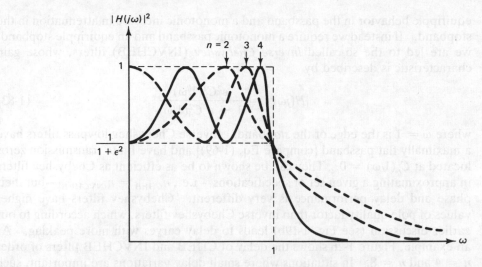

Figure 1-25 Chebyshev lowpass functions for $n = 2, 3, 4$.

or, with Eq. (1-77b),

$$n \geq \frac{\cosh^{-1} \sqrt{(10^{0.1A_s} - 1)/\epsilon^2}}{\cosh^{-1} \omega_s} \tag{1-82a}$$

A convenient approximate formula (Problem 1.26) for n is

$$n \geq \frac{\ln \sqrt{4(10^{0.1A_s} - 1)/\epsilon^2}}{\ln (\omega_s + \sqrt{\omega_s^2 - 1})} \tag{1-82b}$$

Example 1-10

Find the transfer function of a Chebyshev lowpass filter to satisfy the requirement of Example 1-7.

Solution. As before, from Eq. (1-66a), $\epsilon = 0.3493$; then, from Eq. (1-82b), $n \geq 3.82$. Thus, a fourth-order filter is needed. Finally, from Table III-2a in Appendix III with Eq. (1-81):

$$H(s) = \frac{0.358}{s^4 + 1.197s^3 + 1.717s^2 + 11.025s + 0.379}$$

A comparison with Example 1-7 shows that the CHEB filter is more efficient than the MF filter for realizing the same passband and stopband requirements. Problem 1.27 shows that this is always true. Note also that $H(0) = 0.944$, corresponding to -0.5 dB.

We have seen that the attenuation of an all-pole MF approximation increases monotonically with increasing frequency, and that a CHEB approximation has an equiripple behavior in the passband and a monotonic increase in attenuation in the stopband. If instead we require a monotonic passband and an equiripple stopband, we are led to the so-called *inverse Chebyshev* (INVCHEB) filters, whose gain characteristic is described by

$$|H(j\omega)|^2 = \frac{\varepsilon^2 C_n^2(1/\omega)}{1 + \varepsilon^2 C_n^2(1/\omega)} \tag{1-83}$$

where $\omega = 1$ is the edge of the *stop*band. Inverse Chebyshev lowpass filters have a maximally flat passband [compare Eq. (1-69)] and have finite transmission zeros located at $C_n(1/\omega) = 0$. They can be shown to be as efficient as Chebyshev filters in approximating a given set of specifications—i.e., $n_{\text{CHEB}} = n_{\text{INVCHEB}}$—but their phase and delay performance is very different. Chebyshev filters have higher values of pole quality factor than inverse Chebyshev filters, which according to our earlier discussion (see Fig. 1-19b) leads to delay curves with more peaking. As an example, Figure 1-26 shows the delay of CHEB and INVCHEB filters of order $n = 4$ and $n = 8$. In situations where small delay variations are important, such as in video or data transmission systems, inverse Chebyshev filters are therefore preferable.

The derivation of Eq. (1-83) and of some of the properties of inverse Chebyshev filters is left to the problems (Problem 1-34).

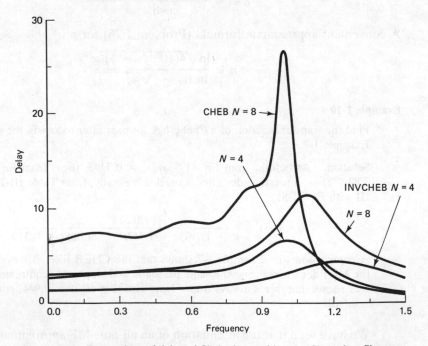

Figure 1-26 Comparison of delay of Chebyshev and inverse Chebyshev filters or order 4 and 8. Note the delay ripples in the passband of the Chebyshev filter.

1.6.1.3 Elliptic lowpass filters.   We saw in Section 1.6.1.2 that the approximation becomes more efficient, i.e., of lower order, if the acceptable error is evenly distributed in an equiripple manner throughout the passband.   It stands to reason, therefore, that even higher efficiency can be obtained if, as shown in Fig. 1-22, we also distribute the stopband error equally throughout the stopband instead of letting the attenuation increase monotonically as $\omega \to \infty$.   The resulting characteristic function $K(s)$ must be a rational function with finite reflection zeros ω_{ri} and transmission zeros ω_{zi}.   The mathematical derivation of

$$|K(j\omega)|^2 = \varepsilon^2 R_n^2(\omega) \tag{1-84}$$

which has equal maxima ε^2 in the passband and equal minima δ^2 in the stopband, as shown in Fig. 1-22, leads to elliptic functions.   The resulting filters have, therefore, been called *elliptic (ELL) filters* (or *Cauer filters*, after the engineer who first introduced them).   The *Chebyshev rational function $R_n(\omega)$* can be shown to be [10, 12, 15]

$$R_n(\omega) = k \prod_{i=1}^{n/2} \frac{\omega^2 - (\omega_s/\omega_{zi})^2}{\omega^2 - \omega_{zi}^2} \qquad \text{for } n \text{ even} \tag{1-85}$$

$$R_n(\omega) = k\omega \prod_{i=1}^{(n-1)/2} \frac{\omega^2 - (\omega_s/\omega_{zi})^2}{\omega^2 - \omega_{zi}^2} \qquad \text{for } n \text{ odd} \tag{1-85a}$$

Recall that, as always, all frequencies are normalized such that the prescribed passband edge $\omega_p = 1$.   The constant k in Eq. (1-85) is determined such that the ripple maxima of $R_n(\omega)$ in the passband $0 \le \omega \le 1$ equal unity so that ε in Eq. (1-84) again determines the filter's passband ripple and is calculated from Eq. (1-66a).   Note that the reflection zeros $\omega_{ri} = \omega_s/\omega_{zi}$ and the transmission zeros ω_{zi} are geometrically symmetrical about the frequency $\sqrt{\omega_s}$; i.e., $\omega_{ri}\omega_{zi} = \omega_s$.   Of the four filter requirements—ε and δ (or, equivalently, A_p and A_s), ω_s, and the degree n—three can be specified independently.   For example, for given values of A_s and ω_s, we need to find, therefore, only the losspoles ω_{zi} and their number such that $R_n(\omega)$ has equal minima of sufficient height in the stopband.   By Eq. (1-85), $R_n(\omega)$ is then completely known and, from Eqs. (1-84) and (1-63), we obtain

$$|H(j\omega)|^2 = \frac{1}{1 + \varepsilon^2 R_n^2(\omega)} \tag{1-86}$$

Just as Eqs. (1-72) and (1-82) were derived for maximally flat and Chebyshev transfer functions, respectively, one can derive an equation for the required degree n of elliptic filters.   However, the expression involves complete elliptic integrals whose evaluation is difficult without available tables [23].   It is much more convenient to note that in the stopband $\varepsilon^2 R_n^2(\omega) \gg 1$.   Consequently, if A_s attenuation (in dB) is required at $\omega = \omega_s$, it follows from Eq. (1-86) that [11]

$$A_s + 20 \log \frac{1}{\varepsilon} \simeq 20 \log |R_n(\omega_s)| \tag{1-87}$$

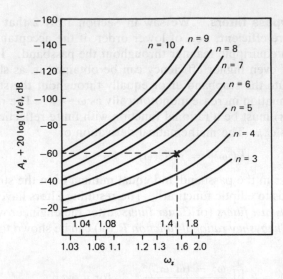

Figure 1-27 Curves for determining required eliptic filter order (n) from out-of-band attenuation (A_s) and in-band ripple (ε) specifications. (From Sedra and Brackett [11], with permission.)

Plotting $20 \log |R_n(\omega_s)|$ versus ω_s leads to a set of design curves [11] as shown in Fig. 1-27, from which the degree n may be estimated for given values of A_s, ε, and ω_s. For example, if $A_p \leq 0.5$ dB, $A_s \geq 50$ dB, and $\omega_s = 1.5$, $\varepsilon = 0.349$ and the required filter order is $n = 5$. With all parameters known, the transmission zeros and the poles of $H(s)$ can be found, an exercise that generally requires computer aids and is beyond the scope of this book. Tables of poles and zeros for a wide variety of cases can be found in references 18 and 21. A sample of normalized elliptic lowpass filter functions is given in Appendix III, Tables III-3a, b, and c. From Eqs. (1-85) and (1-86), the form of the transfer function for an elliptic filter is observed to be

$$H(s) = \frac{H \displaystyle\prod_{i=1}^{m} (s^2 + a_i)}{s^n + b_{n-1}s^{n-1} + b_{n-2}s^{n-2} + \ldots + b_1s + b_0} \tag{1-88}$$

where, in Appendix III, H is chosen such that the peak gain is unity.

Example 1-11

Find an elliptic filter transfer function satisfying the specifications given in Example 1-7.

Solution. As before, we have $\omega_s = 1.6$ and $\varepsilon^2 = 0.122$. To use the tables in Appendix III, we need the numbers $A_1 = \min H(j\omega)$ in $\omega \leq 1$ and $A_2 = \max H(j\omega)$ in $\omega \geq \omega_s$. From Eq. (1-86), we have $A_1 = 1/\sqrt{1 + \varepsilon^2} = 0.944$; and from Eq. (1-66b),

$$A_2 = 1/\sqrt{1 + \delta^2} = 10^{-0.1A_s/2} = 0.0708$$

It is seen readily that the second-order elliptic filter in Table III-3a does not meet specifications, because for $\omega_s = 1.6$ and $A_1 = 0.95$ ($> 0.944!$) we find $A_2 = 0.654$.

However, a third-order filter (Table III-3b) has $A_2 = 0.066 < 0.0708$ for $\omega_s = 1.6$ and $A_1 = 0.95$. Therefore, to four decimal places,

$$H(s) = \frac{0.2816(s^2 + 3.2236)}{(s + 0.7732)(s^2 + 0.4916s + 1.1742)}$$

with $H(0) = 1$ because the degree n is odd.

Note that a *third-order* elliptic filter realizes the specifications for which a maximally flat (MF) function of order $n = 8$ (Example 1-7) and a Chebyshev filter of order $n = 4$ (Example 1-10) was required. The difference in filter order becomes much more pronounced if the specifications are more stringent; for example, let $A_p = 0.05$ dB ($\varepsilon = 0.1076$), $A_s = 80$ dB, and $\omega_s = 1.2$; then, from Fig. 1-27, $n_{\text{ELL}} = 10$, whereas Eq. (1-72) gives $n_{\text{MF}} = 63$ and Eq. (1-82b), $n_{\text{CHEB}} = 20$.

The examples have shown that, for given magnitude specifications, the elliptic approximation is most efficient, followed by the Chebyshev and the maximally flat approximations. The main reasons for this difference are the equiripple distribution of the approximation error in the passband together with the availability of finite transmission zeros, i.e., $\omega_{zi} < \infty$, which permit the elliptic filter function to have a very steep transition region. The most demanding specification is, therefore, a narrow transition band. Note, for instance, that for the same specifications as in Examples 1-7, 1-10, and 1-11, but $\omega_s = 1.1$ instead of 1.6, we obtain $n_{\text{ELL}} = 4$, $n_{\text{CHEB}} = 10$, and $n_{\text{MF}} = 39$.

A further item of interest in the comparison of these types of approximations is their *phase* or *delay* behavior. Because CHEB and ELL functions have higher pole quality factors than all-pole MF (Butterworth) functions, their delay curves *for equal degrees* are more peaking. This observation can, however, be very misleading, because it compares MF (BUT), CHEB, INVCHEB, and ELL functions of *equal degree*, whereas the comparison should be based on these functions with degrees chosen to satisfy *equal attenuation* requirements. In that case, the comparison may turn out very different, as is demonstrated in Fig. 1-28, where the delays of MF, CHEB, INVCHEB, and ELL filters are plotted which meet the attenuation specifications $A_p = 0.5$ dB, $A_s = 23$ dB, and $\omega_s = 1.25$: based on delay variation over the passband $0 \leq \omega \leq 1$, it is seen that the ELL filter has the best delay, closely followed by the INVCHEB filter; whereas the MF and CHEB filters are much worse. Evidently, the outcome of the comparison depends on the degree required to satisfy the prescribed attenuation specifications.

1.6.1.4 Phase or delay approximation: The Bessel-Thomson lowpass filter. From Eqs. (1-15) and (1-17), the phase and delay functions can be expressed as

$$\phi(\omega) = \frac{1}{2j} \ln \frac{H(s)}{H(-s)} \bigg|_{s=j\omega} \tag{1-89}$$

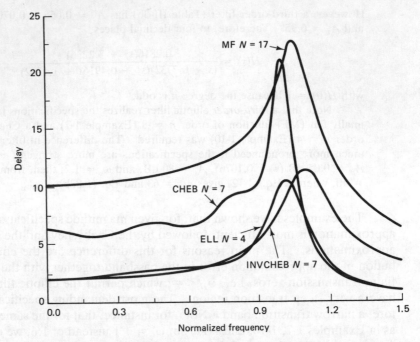

Figure 1-28 Comparison of delay performance of Butterworth, Chebyshev, inverse Chebyshev, and elliptic lowpass filters that meet the same attenuation requirements: $A_p = 0.5$ dB, $A_s = 23$ dB, $\omega_s = 1.25$.

and

$$\tau(\omega) = -\frac{1}{2}\left(\frac{H'(s)}{H(s)} + \frac{H'(-s)}{H(-s)}\right)_{s=j\omega} = -\mathrm{Ev}\left(\frac{H'(s)}{H(s)}\right)_{s=j\omega} \tag{1-90}$$

where Ev{ } means "the even part of { }" and the prime (') stands for the derivative, d/ds. Multiplying $H(s) = N(s)/D(s)$ in Eq. (1-44) in numerator and denominator by $N(-s)$, we obtain

$$H(s) = \frac{N(s)N(-s)}{D(s)N(-s)} \triangleq \frac{Q(s)}{P(s)} \tag{1-91}$$

and can rewrite Eq. (1-90) as follows:

$$\tau(\omega) = \frac{1}{2}\left(\frac{P'(s)}{P(s)} + \frac{P'(-s)}{P(-s)}\right)_{s=j\omega} = \mathrm{Ev}\left\{\frac{P'(s)}{P(s)}\right\}_{s=j\omega} \tag{1-92}$$

The problem of finding a transfer function $H(s)$ with a prescribed delay $\tau(\omega)$ is then reduced to finding a polynomial $P(s) = N(-s)D(s)$ such that Ev $\{P'(s)/P(s)\}$ along the $j\omega$-axis approximates the desired delay. Usually, computer-aided numerical routines will be required for this task. Once $P(s)$ is found, it has to be factored into $N(-s)$ and $D(s)$, where it must be remembered that $D(s)$ is a *Hurwitz polynomial* whose degree n must be larger than or at least equal to the degree m of $N(s)$. $P(s)$ must have, therefore, at least as many roots in the left half-plane as in the right half-plane.

Equation (1-92) will now be used to find an all-pole lowpass approximation for the function

$$H(s) = Ke^{-s\tau_0}$$

that was shown in Section 1.3.2 to provide ideal distortionless transmission, with a linear phase $-\omega\tau_0$, constant delay $\tau(\omega) = \tau_0$, and constant gain K. For convenience in the following development, let us normalize, as always, max $|H(j\omega)|$ to unity (i.e., let $K = 1$) and normalize s by $\Omega_0 = 1/\tau_0$. Our task is then to find a function

$$H(s) = \frac{b_0}{s^n + b_{n-1}s^{n-1} + \ldots + b_1 s + b_0} = \frac{b_0}{D(s)} \tag{1-93}$$

that approximates in some sense e^{-s}. Using Eq. (1-92) with $N(s) = b_0$ we obtain

$$\tau(\omega) = \frac{1}{2}\left(\frac{D'(s)}{D(s)} + \frac{D'(-s)}{D(-s)}\right)_{s=j\omega} = \frac{1}{2}\left.\frac{D'(s)D(-s) + D'(-s)D(s)}{D(s)D(-s)}\right|_{s=j\omega} \tag{1-94}$$

Clearly, for any polynomial $D(s)$ of finite degree n, $\tau(\omega)$ is not constant but can certainly approximate a constant. Selecting maximal flatness as our approximation criterion, we need to bring Eq. (1-94) into the form of Eq. (1-69). Because the numerator of Eq. (1-94) is of order $2n - 1$ and the denominator is of order $2n$ with the coefficient of ω^{2n} equal to $(-1)^n$, this means that, on the $j\omega$-axis, we must satisfy

$$2D(s)D(-s) = D'(s)D(-s) + D'(-s)D(s) + 2(-1)^n s^{2n}$$

or

$$s^{-2n}\left\{\frac{1}{2}[D'(s)D(-s) + D'(-s)D(s)] - D(s)D(-s)\right\} = (-1)^{n+1} \tag{1-95a}$$

Equation (1-95a) can be brought into the form

$$s^{-2n}\{\text{Ev }[D'(s)D(-s)] - D(s)D(-s)\} = (-1)^{n+1} \tag{1-95b}$$

Differentiating Eq. (1-95b) with respect to s leads after a few lines of calculations [24] to the expression

$$\text{Ev }\{[sD''(s) - 2(n + s)D'(s) + 2nD(s)]D(-s)\} = 0$$

which must be true for all values of s if $H(s)$ in Eq. (1-93) is to have an MF delay. Therefore, the term in braces must be identically zero, which, since $D(s) \equiv 0$ is a useless result, leads to

$$sD''(s) - 2(n + s)D'(s) + 2nD(s) \equiv 0 \tag{1-96}$$

Equation (1-96) is a differential equation for $D(s)$. It is shown in reference 24 that the solution is the Hurwitz polynomial

$$D(s) = \sum_{i=0}^{n} b_i s^i \qquad b_i = \frac{(2n - i)!}{2^{n-i} i! (n - i)!}, \, i = 0, \ldots, n - 1 \qquad (1\text{-}97)$$

with $b_n = 1$. The polynomials $D(s)$ are related to Bessel polynomials and were introduced by W. E. Thomson. The resulting maximally flat delay lowpass filters are therefore referred to as *Bessel* or *Thomson* filters. Table III-4 in Appendix III shows the first 10 maximally flat delay polynomials. Observe that for $n = 3$ we have the polynomial found in Example 1-9, $D(s) = s^3 + 6s^2 + 15s + 15$, for an allpass with MF delay. Bessel polynomials of higher order can be obtained from the recurrence relation

$$D_n(s) = (2n - 1)D_{n-1} + s^2 D_{n-2} \qquad (1\text{-}98)$$

It is easy to show that for increasing n, $H(s)$ in Eq. (1-93) gives an increasingly accurate approximation of a constant delay; specifically, note from Eq. (1-97) that [24]

$$\frac{b_i}{b_0} = \frac{(2n - i)!}{2^{n-i} i! (n - i)!} \frac{2^n n!}{2n!} = \frac{n!/(n - i)!}{2n!/(2n - i)! 2^{-i}} \frac{1}{i!}$$

and for $n \to \infty$

$$\frac{b_i}{b_0} \to \frac{n^i}{(2n)^i 2^{-i} i!} = \frac{1}{i!}$$

Thus,

$$\lim_{n \to \infty} H(s) = \lim_{n \to \infty} \frac{b_0}{b_0 \sum_{i=0}^{n} \frac{b_i}{b_0} s^i} = \frac{1}{\sum_{i=0}^{\infty} \frac{1}{i!} s^i} = e^{-s}$$

as desired. Of course, for finite n, $H(s)$ will have finite magnitude and delay errors for which mathematical expressions can be found by relating the MF delay polynomials to Bessel functions of half-integral order (see reference 25 for a good derivation). The single available parameter of Bessel filters, n, must then be chosen such that *both* errors are acceptably small. To help in this choice, design curves have been provided in Fig. 1-29 [26], where the attenuation in dB and the normalized delay error $1 - \tau(\omega)$ in percent are plotted as functions of normalized frequency $\omega = \Omega \tau_0$. Example 1-12 will illustrate their use.

Example 1-12

A lowpass filter has to be designed that has no more than 1.5 dB loss up to $f_1 = 6.8$ kHz and realizes a delay $\tau_0 = 45$ μs at dc with a delay error less than 1.0% up to $f_2 = 9.0$ kHz. Find the appropriate Bessel filter function.

Solution. With $\tau_0 = 45$ μs, normalize the frequency by $\Omega_0 = 1/\tau_0 = 22.22$ krad/s. The two normalized edge frequencies are therefore $\omega_1 = 2\pi f_1 \tau_0 = 1.923$ and $\omega_2 =$

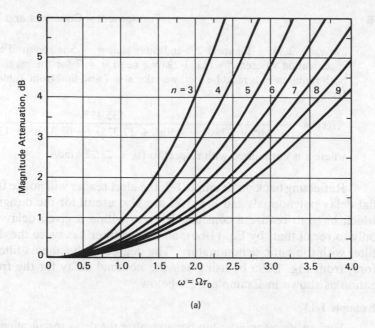

(a)

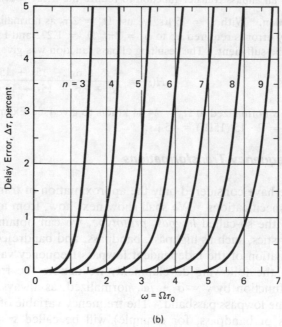

(b)

Figure 1-29 Magnitude and delay error of Bessel filters as functions of normalized frequency $\omega = \Omega\tau_0$ for different degrees n.

$$H(j\omega) = |H(j\omega)|e^{-j\phi(\omega)} \qquad \tau(\omega) = -\frac{d\phi}{d\omega} = \tau_0[1 - \Delta\tau(\omega)]$$

(From W. K. Chen *Passive and Active Filters—Theory and Implementation*. New York, Wiley. [26], ©1986 Wiley, used with permission.)

$2\pi f_2 \tau_0 = 2.545$. Figure 1-29b indicates that $n = 5$ is required to satisfy the delay specification whereas by Fig. 1-29a we need $n = 7$ for the magnitude error. Since both requirements must be met, we take $n = 7$ and find from Table III-4 in Appendix III

$$H(s) = \frac{135,135}{s^7 + 28s^6 + 378s^5 + 3150s^4 + 17,325s^3 + 62,370s^2 + 135,135s + 135,135}$$

where s is normalized with respect to $\Omega_0 = 22.22$ krad/s.

Reflecting back on Example 1-9, the alert reader will notice that the maximally flat delay polynomials and Fig. 1-29 are also useful for the design of allpass filters which have to realize a constant delay to within a given delay error. We have only to recall that, by Eq. (1-60), an allpass filter has twice the delay of a lowpass filter with the same denominator. The degree of the allpass filter is found, therefore, from Fig. 1-29a by using *half* the nominal delay for the frequency normalization as shown in Example 1-13, below.

Example 1-13

Find an allpass transfer function to realize the delay specifications of Example 1-12.

Solution. With $\tau_0 = 45$ μs, we use $\Omega_0 = 2/\tau_0$ as normalization factor. Then 1.0% delay error is required up to $\omega_2 = 2\pi f_2/\Omega_0 = 1.272$ and Fig. 1-29b shows that $n = 3$ will be sufficient. The resulting allpass function was given in Example 1-9:

$$H(s) = \frac{-s^3 + 6s^2 - 15s + 15}{s^3 + 6s^2 + 15s + 15}$$

with s normalized to $\Omega_0 = 44.44$ krad/s to give (see Example 1-9 and Problem 1-41) $\tau(0) = 2 \cdot 15/(15\Omega_0) = 45$ μs.

1.6.2 Frequency Transformations

So far we have considered only the approximation of transfer functions that meet lowpass specifications. We shall show next how, from a known lowpass transfer function, the so-called *lowpass prototype*, we can obtain other types of transfer characteristics, such as highpass, bandpass, and bandreject functions, by a simple transformation of the independent lowpass frequency variable $\bar{s}$. To avoid confusion in the following development, we denote the frequency variable of the *lowpass* function by $\bar{s} = \bar{\sigma} + j\bar{\omega}$, normalized, as always, such that $\bar{\omega} = 1$ at the edge of the lowpass passband. The frequency variable of the target filter function (highpass or bandpass, for example) will be called $s = \sigma + j\omega$ with suitable normalization to be defined as we go along. Also, the unnormalized frequency in radians per second will be denoted by Ω.

The idea behind frequency transformation is then to find a function

$$\bar{s} = F(s) \tag{1-99}$$

such that the passband of the lowpass, $0 \le |\bar{\omega}| \le 1$, is transformed into the passband or passbands of the target filter and, at the same time, the stopband of the lowpass, $|\overline{\omega}| > 1$, is transformed into the stopband(s) of the target filter. Because Eq. (1-99) transforms the $j\omega$-axis into the $j\overline{\omega}$-axis and, further, because the transfer function of the target filter is, of course, also a real rational function of frequency, it follows that $F(s)$ must be an *odd* real rational function, so that

$$\overline{s} = j\overline{\omega} = F(j\omega) \triangleq jf(\omega) \tag{1-100}$$

The function $\overline{\omega} = f(\omega)$ is chosen such that it transforms all passband(s) of the target filter into the range $-1 \leq \overline{\omega} \leq +1$ and all stopbands into $|\overline{\omega}| > 1$. Figure 1-30 shows this requirement for a bandpass with two passbands in $\omega_1 \leq \omega \leq \omega_2$ and $\omega_3 \leq \omega \leq \omega_4$, drawn shaded, and stopbands outside of these ranges. After a little thought, it follows from Fig. 1-30 that $f(\omega)$ must have a simple zero (ω_{ai}) in each passband and a simple pole (ω_{bi}) in each stopband of the target filter; i.e., for the example in Fig. 1-30.

$$\overline{\omega} = f(\omega) = K \frac{(\omega^2 - \omega_{a1}^2)(\omega^2 - \omega_{a2}^2)}{\omega(\omega^2 - \omega_{b1}^2)} \tag{1-101a}$$

or

$$\overline{s} = F(s) = K \frac{(s^2 + \omega_{a1}^2)(s^2 + \omega_{a2}^2)}{s(s^2 + \omega_{b1}^2)} \tag{1-101b}$$

where a possible plus or minus sign has been absorbed in the constant K.[11] Note that $f(\omega)$ has poles at $\omega = 0$ and at $\omega = \infty$ because there are stopbands on the ω-axis around the origin and around infinity. It should be apparent how $F(s)$ is constructed if there are more or fewer passbands in the target filters: we simply have more or fewer pole or zero factors. The available parameters in $F(s)$, i.e., K and the pole and zero frequencies, are chosen such that the passband corner frequencies (ω_1, ω_2, ω_3, and ω_4 in Fig. 1-30) are mapped into the corner frequencies of the prototype lowpass ($\overline{\omega} = \pm 1$). As a final comment we point out that the vertical tolerances of the dependent variable, i.e., gain $|H(j\overline{\omega})|^2$, are, of course, not changed by this transformation of the independent variable $\overline{\omega}$. $|H(j\overline{\omega})|$ is distorted only horizontally, along the ω-axis.

In the following, these concepts will be used to derive the transformation functions for the lowpass-to-highpass (LP-to-HP), lowpass-to-bandpass (LP-to-BP), and lowpass-to-bandreject (LP-to-BR) cases.

1.6.2.1 Lowpass-to-highpass transformation
The passband of a highpass lies in $|\Omega| \geq \Omega_c$ (in rad/s) and the stopband in $|\Omega| \leq \Omega_c$ (in rad/s). Normalizing the passband edge to unity, i.e., taking $\omega = \Omega/\Omega_c$, the highpass has its passband in $|\omega| \geq 1$ as shown in Fig. 1-31a. According to our previous discussion, $f(\omega)$ must

---

[11]Note that this sign is of no consequence, because the transformed function $|H(j\overline{\omega})|^2$ is *even* in $\overline{\omega}$. Thus, we shall always use $+$.

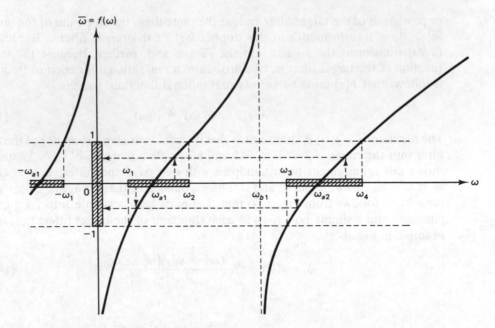

Figure 1-30 Transformation function for a two-passband bandpass into a prototype lowpass. Note that $f(\omega)$ is an *odd* function, so that the graph is *symmetrical* with respect to the origin $\omega = 0$.

have a pole at $\omega = 0$ and a zero at $\omega = \infty$ with $f(1) = 1$. Thus,

$$\bar{\omega} = \frac{1}{\omega} \quad \text{and} \quad \bar{s} = \frac{1}{s} \tag{1-102}$$

is the desired LP-to-HP transformation. Note that the transformation simply interchanges the positions of passbands and stopbands. For example, the third-order 0.5 dB–ripple Chebyshev lowpass filter function

$$H_{\text{LP}}(\bar{s}) = \frac{0.716}{\bar{s}^3 + 1.253\bar{s}^2 + 1.535\bar{s} + 0.716}$$

is transformed into the third-order 0.5 dB–ripple highpass filter

$$H_{\text{HP}}(s) = \frac{0.716s^3}{0.716s^3 + 1.535s^2 + 1.253s + 1}$$

by Eq. (1-102) as shown in Fig. 1-31b.

A simple example of such a transformation, useful in the design of active filters, is the *RC:CR* transformation as shown in Fig. 1-32. In other words, a lowpass circuit that consists of R_i, C_i, and voltage amplifiers can be transformed from a lowpass circuit to a highpass circuit by replacing each resistor R_i in the

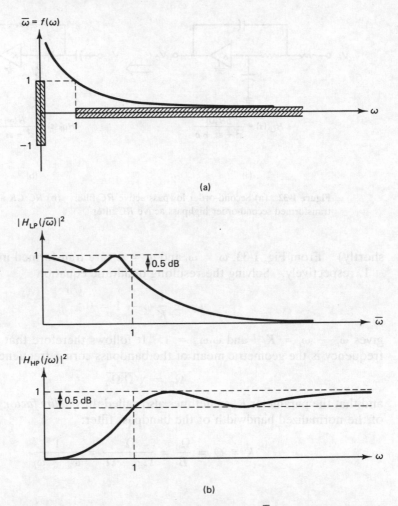

Figure 1-31 Lowpass-to-highpass transformation: (a) $\overline{\omega} = 1/\omega$; (b) transformation of third-order Chebyshev lowpass to highpass.

original network by a capacitor of value $1/R_i$ farads and each capacitor C_j by a resistor of value $1/C_j$ ohms. The verification is left to the reader (Problem 1-48).

1.6.2.2 Lowpass-to-bandpass transformation. The form of the LP-to-BP transformation can be obtained from Fig. 1-33 as

$$\overline{\omega} = K\,\frac{\omega^2 - 1}{\omega}$$

where $\omega = \Omega/\Omega_0$ is normalized to a frequency Ω_0 in the passband (to be identified

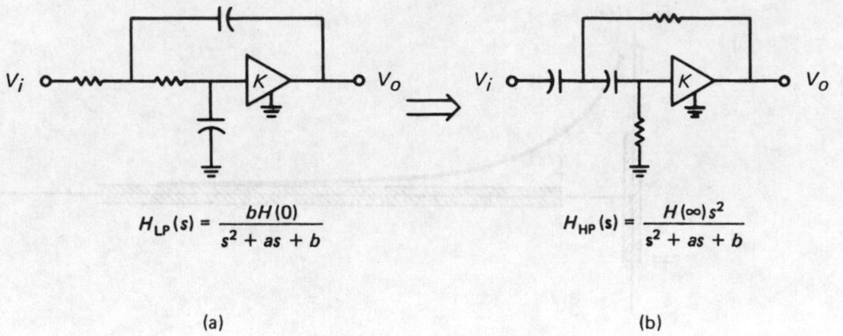

V_i
V_o
K
$H_{LP}(s) = \frac{bH(0)}{s^2 + as + b}$
(a)
V_i
V_o
K
$H_{HP}(s) = \frac{H(\infty)s^2}{s^2 + as + b}$
(b)

Figure 1-32 (a) Second-order lowpass active *RC* filter. (b) *RC*:*CR* admittance-transformed second-order highpass active *RC* filter.

shortly). From Fig. 1-33, $\omega = \omega_l$ and $\omega = \omega_u$ are transformed into $\overline{\omega} = -1$ and $+1$, respectively. Solving the resulting quadratic equation

$$\omega^2 \pm \frac{1}{K}\omega - 1 = 0$$

gives $\omega_u - \omega_l = K^{-1}$ and $\omega_l\omega_u = 1$. It follows therefore that the normalizing frequency is the geometric mean of the bandpass corner frequencies:

$$\Omega_0 = \sqrt{\Omega_l\Omega_u} \tag{1-103a}$$

and that the parameter K, sometimes also called the *quality factor* Q, is the inverse of the normalized bandwidth of the bandpass filter:

$$K \triangleq Q = \frac{\Omega_0}{B} = \frac{\Omega_0}{\Omega_u - \Omega_l} = \frac{1}{\omega_u - \omega_l} \tag{1-103b}$$

$\overline{\omega} = f(\omega)$
+1
−1
ω_l
1
ω_u
ω

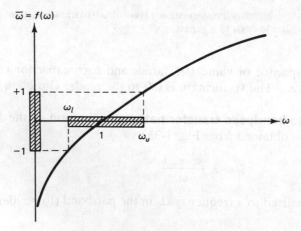

Figure 1-33 Lowpass-to-bandpass transformation function.

where

$$\Omega_u - \Omega_l = B \tag{1-103c}$$

is the bandwidth.

The LP-to-BP transformation is therefore

$$\bar{s} = \frac{\Omega_0}{B} \frac{s^2 + 1}{s} = Q \frac{s^2 + 1}{s} \tag{1-104}$$

It should be noted that this transformation always results in symmetrical bandpass filters; i.e., the upper and lower stopbands have the same attenuation requirement and the product of the stopband corner frequencies equals unity.

Example 1-14

Using the transformation of Eq. (1-104), find a bandpass transfer function with a 1 dB–equiripple passband in 6.6 kHz $< f <$ 8.4 kHz and at least 26 dB attenuation in $f_{s2} \geq 11.5$ kHz and $f_{s1} \leq 5.2$ kHz.

Solution. From Eqs. (1-103), the normalizing frequency (center frequency) Ω_0 and the bandwidth B equal

$$\Omega_0 = 2\pi \cdot 7.446 \text{ krad/s} \quad \text{and} \quad B = 2\pi \cdot 1.8 \text{ krad/s}.$$

Thus, $Q = 4.137$, and from Eq. (1-104),

$$\bar{\omega} = 4.137 \frac{\omega^2 - 1}{\omega}$$

which gives

$$f_l = 6.6 \text{ kHz} \rightarrow \bar{\omega} = -1 \quad \text{and} \quad f_u = 8.4 \text{ kHz} \rightarrow \bar{\omega} = +1$$

Further,

$$f_{s1} = 5.2 \text{ kHz} \rightarrow |\bar{\omega}_1| = 3.03 \quad \text{and} \quad f_{s2} = 11.5 \text{ kHz} \rightarrow |\bar{\omega}_2| = 3.71$$

The specifications of the prototype lowpass are shown in Fig. 1-34; clearly, the lower stopband corner is the more stringent requirement that has to be met by the design.

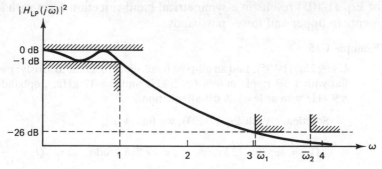

Figure 1-34   Prototype lowpass specifications for Example 1-14.

Thus, from Eqs. (1-66a) and (1-82b),

$$\varepsilon^2 = 10^{0.1} - 1 = 0.2589$$

and

$$n \geq \frac{\ln\sqrt{4(10^{2.6} - 1)/0.2589}}{\ln(3.03 + \sqrt{3.03^2 - 1})} = 2.46$$

The *third-order* Chebyshev lowpass is, from Table III-2 in Appendix III,

$$H_{\text{LP}}(\bar{s}) = \frac{0.491}{\bar{s}^3 + 0.988\bar{s}^2 + 1.238\bar{s} + 0.491}$$

[*Note*: $0.491 = 1/(2^2\varepsilon)$.] Finally, substituting $\bar{s} = 4.137(s^2 + 1)/s$ gives the desired bandpass function

$$H_{\text{BP}}(s) = \frac{0.491 s^3/Q^3}{(s^2 + 1)^3 + \dfrac{0.988}{Q} s(s^2 + 1) + \dfrac{1.238}{Q^2} s^2(s^2 + 1) + \dfrac{0.491}{Q^3} s^3}$$

$$= \frac{0.00693 s^3}{s^6 + 0.239 s^5 + 3.0724 s^4 + 0.485 s^3 + 3.0724 s^2 + 0.239 s + 1}$$

with $|H_{\text{BP}}(j1)| = 1$. Note that the frequency transformation has converted the third-order lowpass into a sixth-order bandpass function.

1.6.2.3 Lowpass-to-bandreject transformation.   We saw in Section 1.6.2.1 that the LP-to-HP transformation in Eq. (1-102) interchanges the position of pass-bands and stopbands. Consequently, if we use this transformation on the LP-to-BP equation [Eq. (1-135)], i.e.,

$$\bar{s} = \frac{1}{Q} \frac{s}{s^2 + 1} = \frac{B}{\Omega_0} \frac{s}{s^2 + 1} \tag{1-105}$$

we interchange passband and stopbands so that the lowpass passband is transformed into the bandreject passbands as shown in Fig. 1-35. The parameters Ω_0, B, and Q are defined as in Eq. (1-103). Again, as in the LP-to-BP transformation, use of Eq. (1-105) results in a symmetrical band-rejection filter with identical require-ments in upper and lower passbands.

Example 1-15

Using Eq. (1-105), find an all-pole bandreject function to satisfy: passbands maximally flat with 1 dB ripple in $0 \leq f \leq 5$ kHz and $f \geq 12$ kHz; stopband in 6.5 kHz $\leq f \leq$ 8.9 kHz with at least 26 dB attenuation.

Solution.   With Eq. (1-103), we find

$$\Omega_0 = 2\pi\sqrt{5 \cdot 12} \text{ krad/s} = 2\pi \cdot 7.746 \text{ krad/r} \quad \text{and} \quad Q = \frac{7.746}{12 - 5} = 1.107$$

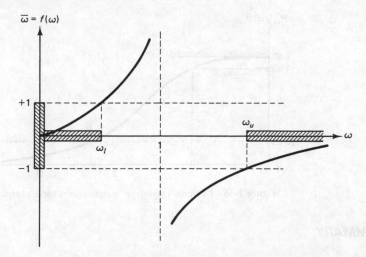

Figure 1-35 Lowpass-to-bandpass transformation function.

Thus, from Eq. (1-105), where on the imaginary axis $\overline{\omega} = \omega/[Q(\omega^2 - 1)]$,

$$f_l = 5 \text{ kHz} \to \overline{\omega} = -1 \quad \text{and} \quad f_u = 12 \text{ kHz} \to \overline{\omega} = +1$$

Further,

$$f_1 = 6.5 \text{ kHz} \to |\overline{\omega}_1| = 2.56 \quad \text{and} \quad f_2 = 8.5 \text{ kHz} \to |\overline{\omega}_2| = 3.24$$

Figure 1-36 shows the lowpass specifications, and Eqs. (1-66a) and (1-72) yield

$$\varepsilon^2 = 10^{0.1} - 1 = 0.2589$$

$$n \geq \frac{\log [(10^{2.6} - 1)/0.2589]}{2 \log 2.56} = 3.90$$

Thus, we obtain, with $n = 4$ and using Eq. (1-71) and the results of Problem 1.21,

$$|H(j\overline{\omega})|^2 = \frac{1}{1 + 0.2589\overline{\omega}^8} = \frac{1}{1 + (0.2589^{1/8}\overline{\omega})^8} = \frac{1}{1 + \hat{\omega}^8}$$

where $\hat{\omega} = \varepsilon^{1/4}\overline{\omega} = 0.84458\overline{\omega}$. As explained in Problem 1.21, if $\varepsilon \neq 1$, it must be absorbed in a frequency normalization factor in order to permit tables for Butterworth polynomials to be used. Then, from Table III-1 in Appendix III,

$$H_{LP}(\hat{s}) = \frac{1}{\hat{s}^4 + 2.613\hat{s}^3 + 3.414\hat{s}^2 + 2.613\hat{s} + 1}$$

or, with $\hat{s} = 0.84458\overline{s}$,

$$H_{LP}(\overline{s}) = \frac{1}{0.5088\overline{s}^4 + 1.5742\overline{s}^3 + 2.4353\overline{s}^2 + 2.2069\overline{s} + 1}$$

Finally, we substitute $\overline{s} = s/[1.107(s^2 + 1)]$ to obtain the final result:

$$H_{BR}(s) = \frac{1.499(s^2 + 1)^4}{1.499(s^8 + 1) + 2.990(s^7 + s) + 8.988(s^6 + s^2) + 10.713(s^5 + s^3) + 15.469s^4}$$

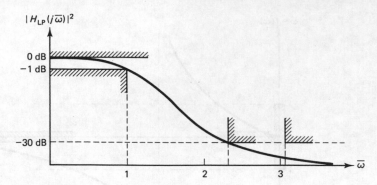

Figure 1-36   Lowpass prototype specifications for Example 1-15.

1.7 SUMMARY

This chapter has served to introduce the notation and to establish a common background useful and necessary for the study of passive LC filters and of continuous-time and sampled-data active filters. We have reviewed the fundamental concepts relevant to the design of such circuits and the methods used for describing system behavior in both the time and the frequency domain. Finally, we have discussed some approximation procedures which lead to realizable system transfer functions that satisfy prescribed system behavior. Although this chapter cannot possibly provide a complete treatment of these subjects, the material should provide sufficient background for understanding and applying the tools given in the forthcoming chapters to the design of practical active and passive filters.

In Chapter 2 we discuss the design of passive lossless LC ladder filters, both because of their continuing importance in practice and because some understanding of LC ladders proves to be very useful in the design of high-order active RC and of switched-capacitor filters.

In Chapter 3 we consider the principal problem faced by the designer of any manufacturable filter, but especially of active filters, namely, the variation of the filter's response due to unintended but unavoidable variations in the component values. Such variations occur because of manufacturing imperfections, changes in the ambient environment, and aging. *Sensitivity analysis* provides techniques for predicting the severity of this problem and points out methods for minimizing it.

PROBLEMS

1.1   Use node analysis to find the transfer function of the circuit in Fig. P1.1.

1.2   An LLF network has an excitation $x(t) = \mathscr{L}^{-1}\{X(s)\}$ and a response $y(t) = \mathscr{L}^{-1}\{Y(s)s)\}$. $X(s)$ and $Y(s)$ are related by $Y(s) = H(s)X(s)$ where $H(s) = N(s)/D(s)$ and N and D are polynomials in s. $D(s)$ is of degree n. Verify and explain the statement, "The

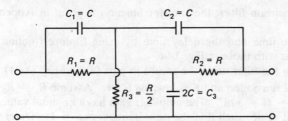

Figure P1.1

mathematical form of the natural response is determined by the poles of $H(s)$, and the mathematical form of the forced response is determined by the poles of $X(s)$."

1.3 Using the results of Problem 1.2, explain how one can construct a network with a zero forced response to the excitation $x(t) = X_0 \cos \omega_0 t$. What would be the practical use of such a network?

1.4 Show that the natural frequencies (the poles) of a network are obtained by analyzing the network with all inputs set to zero.

1.5 Sketch the frequency response (magnitude and phase) of a circuit described by the transfer function

$$H(s) = \frac{s^2 + 1}{s^2 + s/Q + 1}$$

For $Q = 3$. What changes if $Q = 10$?

1.6 Show that the magnitude of a transfer function does not change if a zero is shifted from the left half-plane to the right half-plane or vice versa. Also show that the phase of a transfer function does not change if a denominator factor is placed into the numerator or a numerator factor into the denominator after replacing s by $-s$. In view of these results, find a third-order stable transfer function that has (a) the same magnitude and (b) the same phase as

$$H(s) = \frac{s^2 - 2s + 2}{(s + 1)(s^2 + 4s + 5)}$$

1.7 Use Eq. (1-22b) to show that $a(t)$ is the step response.

1.8 Consider the circuit in Fig. P1.8, where $v_{in}(t) = 2u(t)$.
(a) Use the time-domain approach to calculate $v_o(t)$ for $t \geq 0$.
(b) Use the frequency-domain approach to calculate $|v_o(t)|$ for $t \geq 0$.
(c) Sketch $v_o(t)$.

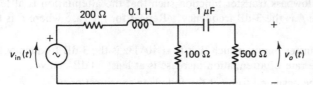

Figure P1.8

1.9 For the transfer function given by

$$H(s) = \frac{s + 2}{(s + 1)(s + 3)}$$

find the step response, the rise time, and the 3-dB bandwidth.

1.10 For the fifth-order Thomson filter, the transfer function is given in Appendix III, Table III-4.

    (a) Determine the rise time and the delay time by using Elmore's definition and compare the results with those in Fig. 1-9c.

    (b) Determine ω_{3dB} for the filter using the results in part (a) and Eq. (1-31).

1.11 The circuit in Fig. 1-12 is a Sallen and Key lowpass filter. Assume $R_1 = R_2 = R = 5.13$ kΩ and $C_1 = C_2 = C = 5$ nF. The amplifier gain has a nominal value $K = 1$. Normalize the element values such that the pole frequency $\omega_0 = 1$. Determine for which values of K the circuit is stable. Plot a root locus to show the position of the poles in the s-plane as a function of K.

1.12 In the functions

$$H_a(s) = 10\,\frac{s^2 + 1}{s^2 + 0.3s + 4}$$

$$H_b(s) = \frac{7s}{s^2 + 0.1s + 2}$$

$$H_c(s) = \frac{8}{s^2 + 0.5s + 1}$$

the frequency s is normalized with respect to $\Omega_0 = 3.5$ krad/s. For each of the functions determine the pole and zero frequencies, the pole Q factor, and the maximum gain. Plot the poles and zeros in the s-plane. Sketch magnitude, phase, and delay versus ω.

1.13 Verify Eq. (1-60).

1.14 Find the transfer functions whose magnitude is given by

    (a) $|H(j\omega)|^2 = \left[1 + \left(\omega\,\frac{3 - \omega^2}{1 - 3\omega^2} \right)^2 \right]^{-1}$

    (b) $|H(j\omega)|^2 = \dfrac{\omega^2 + \omega^4}{1 + \omega^6}$

1.15 Show that a lowpass with maximally flat passband at $\omega = 0$ has a characteristic function $K(s)$ given by $|K(j\omega)|^2 = a_n\omega^{2n}/|N(j\omega)|^2$ where $N(s)$ is the transmission zero polynomial.

1.16 Show that tables for Butterworth filters ($\varepsilon = 1$, 3-dB passband) can be used for maximally flat passband all-pole lowpass filters with $\varepsilon \neq 1$ by suitable frequency scaling. *Hint:* In Eq. (1-71) absorb ε^2 in a frequency scaling factor.

1.17 Find a maximally flat lowpass transfer function such that the attenuation is at least 30 dB at $f = 5f_c$ where f_c is the 3-dB frequency. Repeat for the case where f_c is the 1-dB frequency.

1.18 Find a Butterworth transfer function such that $f_c = 10^4$ Hz is the 3-dB frequency and at high frequencies the rate of attenuation increase is at least 30 dB/octave.

1.19 A lowpass filter is to be designed to meet the following specifications:

    (a) Maximum flat passband in $0 \le f \le 10$ kHz

    (b) Maximum passband attenuation $A_p = 1$ dB

    (c) Transmission zero at $f_1 = 15$ kHz

    (d) Minimum stopband attenuation $A_s = 15$ dB

    Find $|H(j\omega)|^2$ and $H(s)$.

1.20 The maximally flat all-pole functions (Eq. 1-71) derived in the text have $\omega = 0$ as their point of maximum flatness. One can, of course, choose a frequency ω_0 in the passband, with $0 < \omega_0 < 1$, as the point of maximum flatness. The resulting characteristic function is given by

$$K(s) = \varepsilon \left(\frac{s^2 + \omega_0^2}{1 - \omega_0^2} \right)^n$$

For this function $K(s)$, sketch $|K(j\omega)|^2$ and $|H(j\omega)|^2$ and compare with the behavior of Eq. (1-71). What is the effect of n and of ω_0? What restrictions must be satisfied by ω_0 if $|H(j0)| = |H(j1)|$ and if $\min|H(j\omega)|^2 = 1/(1 + \varepsilon^2)$ in $0 \le \omega \le 1$?

1.21 Find a lowpass transfer function $H(s)$ that
   (a) Is maximally flat in the passband with $A_p \le 2$ dB in $0 \le f \le 6$ kHz.
   (b) Has transmission zeros at $f_1 = 12$ kHz and $f_2 = 24$ kHz, and
   (c) For high frequencies, has a rate of attenuation at least 40 dB per decade.

1.22 A transfer function is given by

$$H(s) = \frac{s^2 + s a_1 + a_1 b_1}{s^3 + s^2(1 + a_1) + s b_1 + a_1 b_1}$$

Determine the values of a_1 and b_1 if $|H(j\omega)|$ is specified to have a maximally flat magnitude response.

1.23 Derive a differential equation for the function $C_n(\omega)$. The solution is Eq. (1-77a). *Hint:* The differential equation is easy to derive by noting from Fig. 1-22 with Eq. (1-75) that $1 - C_n = 0$ and $1 + C_n = 0$ where $dC_n/d\omega = 0$, and that, for n even, $1 - C_n = 0$ at $\omega = \pm 1$ and, for n odd, $C_n(1) = 1$, $C_n(-1) = -1$.

1.24 Show that for $\omega \ge 1$, Eq. (1-77a) leads to Eq. (1-77b).

1.25 Show that the roots of Eq. (1-79) lie on an ellipse in the s-plane and that the left half-plane roots are given by Eq. (1-80).

1.26 Use Eq. (1-77b) to derive Eq. (1-82b).

1.27 Show that for the same magnitude specifications, $n_{\text{BUT}} > n_{\text{CHEB}}$.

1.28 Find a lowpass transfer function $H(s)$ to satisfy:
   (a) Equiripple passband, $A_p = 0.5$ dB, in $0 \le f \le 7.8$ kHz
   (b) Monotonic stopband, $A_s \ge 35$ dB, in $f \ge 15$ kHz

1.29 A Chebyshev lowpass is defined by $n = 5$ and $\varepsilon^2 = 0.2$. Determine the maximum and minimum values of the response in the passband, the ripple width in dB, and the 3-dB frequency.

1.30 Find a Chebyshev lowpass to have 0.5-dB passband ripple and 60-dB/decade attenuation increase for high frequencies.

1.31 Find a Chebyshev lowpass function $H(s)$ such that $|H(j\omega)|^2$ has 20% ripple in the passband and a one-octave transition band between the cutoff point and the -10-dB frequency.

1.32 For $\omega \gg 1$, compare the degrees of maximally flat and equiripple lowpass approximations.

1.33 Show that for the same attenuation specifications, Chebyshev and inverse Chebyshev filters have the same degree.

1.34 Derive Eq. (1-83), the magnitude squared of the inverse Chebyshev filter (form and sketch the functions $1 - |H_{CHEB}(j\omega)|^2$ and replace ω by $1/\omega$. Show that $|H_{INVCHEB}(j\omega)|^2$ has a maximally flat passband and that $\omega = 1$ is the stopband edge. Find the frequencies where the attenuation is a minimum. Find an expression for the transmission zeros. Where are the poles of $H_{INVCHEB}(s)$? What are $H_{INVCHEB}(0)$ and $H_{INVCHEB}(j\infty)$? What is determined by ε?

1.35 The magnitude response of a filter must lie within the shaded area shown in Fig. P1.35. Determine the transfer function of the filter that meets the given specifications, using:

(a) A Chebyshev approximation

(b) A Cauer approximation

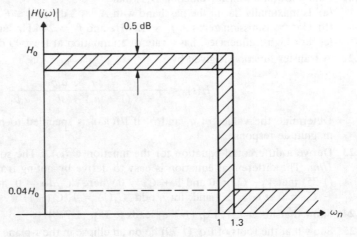

Figure P1.35

1.36 Use Table III-3 in Appendix III to determine the pole-zero locations of an elliptic filter for the following specifications: $\omega_s/\omega_p = 1.4$, minimum stopband attenuation ≥ 19 dB and passband ripple ≤ 0.5 dB.

1.37 Use Table III-3 in Appendix III to find an elliptic lowpass transfer function to satisfy: passband in $0 \leq f \leq 3.2$ kHz with 0.45 dB ripple; stopband in $f \geq 4.2$ kHz with 29 dB minimum attenuation.

1.38 Find an elliptic lowpass transfer function to meet the following specifications: passband magnitude at least 82% of the maximum transmission gain in $0 \leq f \leq 1$; stopband at least 20 dB attenuation for $f \geq 1.22$.

1.39 An elliptic lowpass filter has to be designed that has no more than 1.5 dB passband ripple in $f \leq 11.2$ kHz and a minimum of 31.5 dB attenuation in $f \geq 14$ kHz. Find the transfer function.

1.40 Find the degrees n_{MF}, n_{CH}, and n_{ELL} of an all-pole maximally flat, a Chebyshev, and an elliptic filter, respectively, that satisfy:

(a) $A_p = 1$ dB $A_s = 80$ dB $\omega_s = 1.2$

(b) $A_p = 1$ dB $A_s = 120$ dB $\omega_s = 1.2$

(c) $A_p = 0.1$ dB $A_s = 80$ dB $\omega_s = 1.2$

(d) $A_p = 1$ dB $A_s = 80$ dB $\omega_s = 1.5$

Compare your results.

1.41 Express the delay $\tau(0)$ in terms of the coefficients of the transfer functions in Eqs. (1-44) and (1-93).

1.42 Find a lowpass transfer function such that the attenuation is not larger than 3 dB up to $\Omega = 7000$ rad/s and the delay is maximally flat at $\tau(0) = 0.25$ ms with a delay error of less than 3% up to $\Omega = 6000$ rad/s.

1.43 Find an allpass function with constant delay of 400 μs. The delay error must be no larger than 1.2% until $f = 4.5$ kHz.

1.44 Consider the sixth-order all-pole transfer functions for Butterworth, Chebyshev ($\frac{1}{2}$ dB ripple), and Thomson filters. Determine the attenuation at $\omega = 2$ in each case.

1.45 Repeat Problem 1.22 if $H(j\omega)$ is to have a linear phase.

1.46 The phase of the transfer function for a general linear network can be written as

$$\phi = \arg H(j\omega) = -\tan^{-1} \frac{\alpha_1 \omega + \alpha_3 \omega^3 + \alpha_5 \omega^5 + \dots}{1 + \beta_2 \omega^2 + \beta_4 \omega^4 + \beta_6 \omega^6 + \dots}$$

Show that for a linear phase response, the first two conditions are

$$\alpha_3 - \beta_2 \alpha_1 = \tfrac{1}{3} \alpha_1^3$$

$$\alpha_5 - \beta_4 \alpha_1 = \tfrac{1}{3} \alpha_1^3 (\tfrac{2}{3} \alpha_1^2 + \beta_2)$$

1.47 Find a Thomson filter function to satisfy
(a) $A_p \leq 2$ dB in $f \leq 10$ kHz
(b) Delay of 30 μs with less than 1% error in $f \leq 13$ kHz

1.48 If the voltage ratio transfer of an RC network N_a is given by $H_a(s)$, show that the voltage ratio of another network N_b obtained by $RC{:}CR$ transformation is given by

$$H_b(s) = H_a\!\left(\frac{1}{s}\right)$$

Does this result change if the RC network N_a is active with a voltage amplifier of gain K? If $H_a(s)$ represents the transfer function corresponding to a third-order Butterworth filter, sketch the magnitude response of $H_b(s)$.

1.49 A lowpass prototype is given by the transfer function

$$H(\bar{s}) = \frac{1}{\bar{s}^2 + \sqrt{2}\bar{s} + 1}$$

Obtain the HP, BP, and BR transfer functions and sketch the responses. Assume that $Q = 10$. Show the pole-zero locations of the target filters. What is the 3-dB bandwidth B in each case?

1.50 Find a highpass filter function $H(s)$ such that
(a) Passband: maximally flat, $A_p = 2$ dB, $f \geq 15$ kHz
(b) Stopband: monotonic, $A_s \geq 40$ dB, $f \leq 7$ kHz

1.51 Find $H(s)$ of an all-pole bandpass filter with maximally flat passband to meet the following specifications:
(a) $A_p \leq 1$ dB in 18 kHz $\leq f \leq 23$ kHz
(b) $A_s = 45$ dB in $f \geq 35$ kHz; $A_s > 80$ dB in $f \leq 9$ kHz
Use frequency transformation.

1.52. A filter has to satisfy the specifications
   (a) Equiripple passband with 0.3 dB ripple in 20 kHz $\leq f \leq$ 30 kHz
   (b) At least 40 dB attenuation in $f \leq$ 15 kHz and $f \geq$ 36 kHz
   Use frequency transformation to find the transfer function $H_{\mathrm{BP}}(s)$.

1.53. A filter is to have passbands and stopbands as shown in Fig. P1.53. Derive a frequency transformation equation that transforms this requirement into a prototype lowpass function.

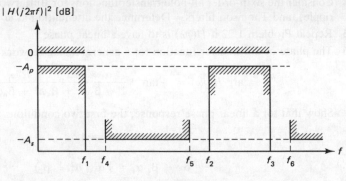

Figure P1.53

1.54. Find a bandpass transfer function to satisfy:
   (a) Maximally flat passband, $A_p = 0.9$ dB, in 35 kHz $\leq f \leq$ 60 kHz
   (b) Equiripple stopbands, $A_s \geq 19$ dB, in $f \leq$ 28 kHz and $f \geq$ 75 kHz

1.55. Find a band-rejection transfer function $H(s)$ to satisfy:
   (a) Equiripple passband, $A_p = 1$ dB, in $f \leq$ 80 kHz and $f > 180$ kHz
   (b) Monotonic stopband, $A_s \geq 20$ dB, in 100 kHz $\leq f \leq$ 150 kHz

1.56. Show that frequency transformation can *not* be used to transform a maximally flat delay (Bessel) lowpass filter into a bandpass filter with maximally flat delay in its passband; i.e., show that frequency transformation destroys the delay properties of the prototype lowpass function. As an example, calculate and sketch the delay in the passband of a bandpass filter with a normalized bandwidth of 10% that was obtained through frequency transformation from a second-order Bessel filter.

   Hint: Find the phase of the Bessel filter and expand it into a power series in ω up to order 5. Use frequency transformation and calculate the delay.

The Design of *LC* Filters

TWO

2.1 INTRODUCTION

Historically, *lossless* (*LC*), four-terminal single-input, single-output networks, also referred to as *twoports*, have played an important role in the design of transmission networks (*filters*) because the availability of *LC resonance* allows impedances to have extremely rapid changes of magnitude and phase with changes in frequency. By appropriate interconnections of these impedances, one can construct twoports with very steep slopes between passbands and stopbands, and series or parallel resonance can be used to block transmission of certain frequencies completely, all in a network which is lossless, i.e., which dissipates no power itself. Although other filtering techniques are now available, such as the active *RC* and switched-capacitor circuits discussed in later chapters of this text, *LC* filters still play an important role in many areas of communication circuits and will continue to do so. This is true especially in applications at higher frequencies ($f > 100$ kHz to 500 kHz) where the operation of active devices becomes less than perfect.

In addition to their continuing importance in filtering practice, understanding the origin and the design of lossless filters provides us with several benefits. First, some insight into the operation of *LC* filters proves helpful in developing the reader's intuitive understanding of the *concept of filtering*, the importance of resonance, the creation and function of transmission zeros, and so forth. Second, many of the best and most practical active *RC* and switched-capacitor designs are based on *active simulations* of passive *LC* filters. This approach is taken because of the very low passband sensitivities of lossless *LC* ladder filters to element tolerances (see Chapter 3). To appreciate and better understand the origin of these approaches, some fundamental knowledge of passive *LC* filters is most helpful.

This brief chapter can, of course, not possibly provide a complete treatment of lossless *LC* filters; numerous books are available to fill in the gaps; e.g., see references 10, 11, 13, 15, 24–26, and 28. We shall provide only a concise discussion of the practically most important aspects of lossless filters, namely, the synthesis of *LC ladders* that are resistively terminated at both ends and are designed for maximum power transfer. Nevertheless, the material should be sufficient for setting a sound basis for our treatment of *LC* ladder simulations in Chapters 6, 7, and 8. In addition, a careful and thorough reading of this chapter should indeed enable the student to design *LC* filters as long as their complexity does not become too great. We should point out here that, building on the results of much research over the last 50 years, the design of an *LC* filter is conceptually not difficult. It is, however, based on some very involved theory and is extremely computation-intensive, a situation made worse by the fact that the problems become numerically quite ill-conditioned. Consequently, for the design of *LC* filters with even moderate complexity it is almost unavoidable to utilize one of the available *LC* filter design programs, such as FILSYN [29] or FILTOR [11].

As the name implies, an *LC* filter is a *lossless transmission network N* consisting of only inductors and capacitors. In normal operation, the network is embedded between a resistive source (V_S, R_S) and a resistive load (R_L) as shown in Fig. 2-1a or, specifically for a *ladder* topology, in Fig. 2-1b. As the reader

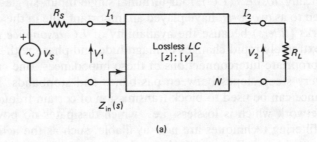

(a)

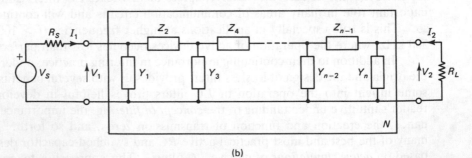

(b)

Figure 2-1 (a) Resistively terminated lossless twoport; (b) resistively terminated ladder structure.

undoubtedly will expect, in the design of *LC* filters one is led into the problem of realizing *LC* impedances or admittances which are the "building blocks" of the network. Section 2.2 is devoted to this topic. Although brief and presented in a way to pertain directly to *LC twoports*, the discussion should enable the student to realize arbitrary *LC oneports*.

2.2 REALIZATION OF LC IMMITTANCE FUNCTIONS[1]

If the network N in Fig. 2-2 is *lossless*, the dissipated power is, of course, zero:

$$P = |I(j\omega)|^2 \, \text{Re} \, \{Z_{LC}(j\omega)\} \equiv 0 \tag{2-1}$$

In Eq. (2-1), Re { } means "the real part of" and $Z_{LC}(s) = V/I$ is the impedance "seen into" the network. Because $I \neq 0$, Eq. (2-1) implies that the real part of an *LC* impedance is identically zero, so that $Z_{LC}(j\omega)$ is purely imaginary and, after replacing ω by s/j, $Z(s) = N(s)/D(s)$ is an *odd* rational function of s (Problem 2.1). Specifically, it can be shown [9, 25] that the most general form of an *LC* impedance of degree $2n + 2$ is

$$Z_{LC}(s) = \frac{N_{2n+2}(s)}{D_{2n+1}(s)} = K \frac{(s^2 + \omega_1^2)(s^2 + \omega_3^2) \cdots (s^2 + \omega_{2n+1}^2)}{s(s^2 + \omega_2^2)(s^2 + \omega_4^2) \cdots (s^2 + \omega_{2n}^2)} \tag{2-2a}$$

where K is a positive constant and

$$0 \leq \omega_1 < \omega_2 < \omega_3 < \ldots < \omega_{2n} < \omega_{2n+1} < \infty \tag{2-2b}$$

It follows, therefore, that all poles and zeros of an *LC* impedance are *simple* and *alternate* on the $j\omega$-axis. The zeros ω_{2i+1}, $i = 0, \ldots, n$, are, of course, those frequencies where the total network N in Fig. 2-2 reduces to a series resonance circuit so that $Z_{LC}(j\omega_{2i+1}) = 0$. Similarly, at the poles ω_{2k}, $k = 1, \ldots, n$, $Z_{LC}(j\omega_{2k}) = \infty$.

As written in Eq. (2-2a), Z_{LC} has poles at the origin and at infinity: $Z_{LC}(0) \to \infty$ and $Z_{LC}(\infty) \to \infty$. Alternative situations are $\omega_1 = 0$, in which case $Z_{LC}(s)$ has a zero at $s = 0$; and/or the absence of the factor $(s^2 + \omega_{2n+1}^2)$, so that the denominator of $Z_{LC}(s)$ is of degree 1 higher than the numerator and Z_{LC} has a zero at $s = \infty$. We note here that, because $Y_{LC}(s) = 1/Z_{LC}(s)$, an *LC* admittance satisfies the same properties as an *LC* impedance. Figure 2-3 shows a typical

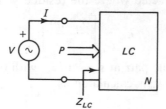

Figure 2-2   *LC* oneport.

[1]*Immittance* is a collective name for *im*pedance and ad*mittance*.

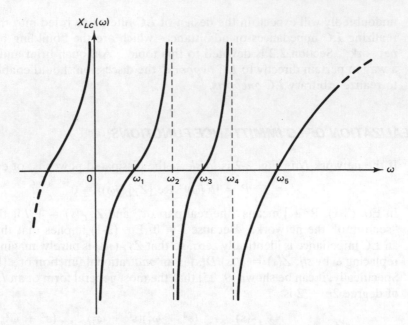

Figure 2-3 Typical reactance function.

sketch of the *reactance* $X_{LC}(\omega) = Z_{LC}(j\omega)/j$ for $n = 2$. The reader is encouraged to compare Eq. (2-2a) and Fig. 2-3 with Eq. (1-101) and Fig. 1-30: it should be apparent that the frequency transformation function introduced in Section 1.6.2 is a reactance function, a fact that will prove useful when we discuss in Section 2.4 the transformation of an *LC* lowpass prototype ladder into a highpass, bandpass, or band-rejection *LC* ladder.

All realizations of *LC* oneports are based on consecutive *removals of poles* of the impedance or the admittance functions. An explanation of pole removal follows: If $Z_{LC}(s)$ has a pole at $s = 0$, we substract a term k_0/s where the constant k_0, the *residue* of the pole at the origin, is obtained from

$$k_0 = [sZ_{LC}(s)]_{s \to 0} \tag{2-3a}$$

i.e., the product $sZ_{LC}(s)$ evaluated at $s = 0$. Similarly, if $Z_{LC}(s)$ has a pole at infinity, a term $k_\infty s$ is extracted where the residue k_∞ is

$$k_\infty = \left[\frac{Z_{LC}(s)}{s} \right]_{s \to \infty} \tag{2-3b}$$

Finally, for an internal pole pair at $s = \pm j\omega_p$, with $0 < \omega_p < \infty$, we subtract the term $k_p s/(s^2 + \omega_p^2)$ where k_p is calculated as

$$k_p = \left[\frac{s^2 + \omega_p^2}{s} Z_{LC}(s) \right]_{s^2 \to -\omega_p^2} \tag{2-3c}$$

The remainder functions

$$Z_1(s) = Z_{LC}(s) - \frac{k_0}{s} \tag{2-4a}$$

$$Z_2(s) = Z_{LC}(s) - k_\infty s \tag{2-4b}$$

$$Z_3(s) = Z_{LC}(s) - \frac{k_p s}{s^2 + \omega_p^2} \tag{2-4c}$$

no longer have poles at the origin, at infinity, or at $s = \pm j\omega_p$, respectively, because these poles have been *removed completely*. Complete pole removal implies also that the degree of the remainder function is reduced by 1 for each pole removed. For example, $Z_3(s)$ is of order 2 less than $Z_{LC}(s)$. A completely identical development holds for admittances; we only have to replace Z by Y in Eqs. (2-3) and (2-4).

Because for each removed pole the degree of the starting immittance decreases by 1, it is apparent that the procedure described above will terminate and the realization is complete when the degree of the immittance is reduced to zero. Specifically, an immittance of order n has n poles and is, therefore, completely realized after n pole-removal steps.

Note that Eq. (2-4a),

$$Z_{LC}(s) = \frac{k_0}{s} + Z_1(s)$$

realizes $Z_{LC}(s)$ as a series connection of a capacitor with the remainder function $Z_1(s)$. Similarly, in Eqs. (2-4b) and (2-4c), $Z_{LC}(s)$ is a series connection of an inductor with $Z_2(s)$ and a parallel resonance circuit with $Z_3(s)$, respectively. Thus, the alert student will have guessed that each removed (realized) pole costs exactly one element, namely, a capacitor $C = 1/k_0$ for Eq. (2-4a), an inductor $L = k_\infty$ for Eq. (2-4b), and a parallel LC resonant circuit with $L = k_p/\omega_p^2$ and $C = 1/k_p$ for the pole pair in Eq. (2-4c). Again, the analogous (dual) statement holds for the realization of an admittance $Y_{LC}(s)$. We conclude further that this realization of an nth-order immittance with its n free parameters (the pole and zero frequencies) requires exactly n elements, which is the minimum number: the realization is said to be *canonic*.

The following examples should help to clarify the ideas just presented.

Example 2-1

Realize the fifth-order LC admittance

$$Y(s) = \frac{s(s^2 + 2)(s^2 + 4)}{(s^2 + 1)(s^2 + 3)}$$

by removing all its poles.

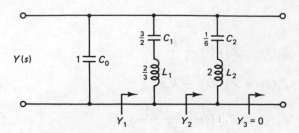

Figure 2-4   Circuit for Example 2-1.

Solution. $Y(s)$ has poles at $s = \infty$, $s = \pm j1$, and $s = \pm j\sqrt{3}$. Thus, by Eq. (2-4) we should subtract from $Y(s)$ the terms

$$k_\infty s, \qquad \frac{k_1 s}{s^2 + 1} \qquad \frac{k_2 s}{s^2 + 3}$$

to yield the *partial fraction expansion*[2] of $Y(s)$,

$$Y(s) = k_\infty s + \frac{k_1 s}{s^2 + 1} + \frac{k_2 s}{s^2 + 3} \qquad (2\text{-}5)$$

Using Eq. (2-3) with Z replaced by Y results in

$$k_\infty = \left. \frac{Y(s)}{s} \right|_{s \to \infty} = \left. \frac{(s^2 + 2)(s^2 + 4)}{(s^2 + 1)(s^2 + 3)} \right|_{s \to \infty} = 1$$

$$k_1 = \left. \frac{(s^2 + 1)Y(s)}{s} \right|_{s^2 = -1} = \left. \frac{(s^2 + 2)(s^2 + 4)}{s^2 + 3} \right|_{s^2 = -1} = \frac{3}{2}$$

and

$$k_2 = \left. \frac{(s^2 + 3)Y(s)}{s} \right|_{s^2 = -3} = \left. \frac{(s^2 + 2)(s^2 + 4)}{s^2 + 1} \right|_{s^2 = -3} = \frac{1}{2}$$

Evidently, with these constants, Eq. (2-5) leads to the network in Fig. 2-4 because the three terms are connected *in parallel*. Note that we needed *five* elements and that the two resonant branches cause short circuits at the frequencies $\omega_1 = 1/\sqrt{L_2 C_1} = 1$ and $\omega_2 = 1/\sqrt{C_2 L_2} = \sqrt{3}$ to realize the prescribed poles of $Y(s)$.

Example 2-2

Realize the *impedance* $Z(s) = 1/Y(s)$ of Example 2-1 by removing its pole at the origin, then the pole at the origin of the *remaining admittance*, next the pole at $s = 0$ of the then remaining *impedance*, and so on, until the process is completed.

Solution

$$Z(s) = \frac{1}{Y(s)} = \frac{(s^2 + 1)(s^2 + 3)}{s(s^2 + 2)(s^2 + 4)} = \frac{s^4 + 4s^2 + 3}{s^5 + 6s^3 + 8s}$$

---

[2]This realization is named after R. M. Foster; a partial fraction expansion (PFE) of $Y(s)$ gives the Foster II realization, and a PFE of $Z(s)$ results in the Foster I circuit.

First we form

$$Z_1(s) = Z(s) - \frac{k_{01}}{s} = Z(s) - \frac{0.375}{s} = \frac{0.625s^3 + 1.75s}{s^4 + 6s^2 + 8} = \frac{1}{Y_1(s)}$$

where, of course, $k_{01} = sZ(s)|_{s=0} = \frac{3}{8} = 0.375$.   Next, form

$$Y_2(s) = Y_1(s) - \frac{k_{02}}{s} = \frac{s^4 + 6s^2 + 8}{0.625s^3 + 1.75s} - \frac{4.5714}{s}$$

i.e.,

$$Y_2(s) = \frac{1}{Z_2(s)} = \frac{s^3 + 3.1429s}{0.625s^2 + 1.75}$$

Proceeding, we obtain next

$$Z_3(s) = Z_2(s) - \frac{k_{03}}{s} = \frac{0.625s^2 + 1.75}{s^3 + 3.1429s} - \frac{0.5568}{s} = \frac{0.06818s}{s^2 + 3.1429}$$

$$Y_4(s) = Y_3(s) - \frac{k_{04}}{s} = \frac{s^2 + 3.1429}{0.06818s} - \frac{46.095}{s} = \frac{s}{0.06818}$$

and

$$Y_5(s) = Y_4(s) - \frac{s}{0.06818} = 0$$

which completes the realization.   Observe that this cumbersome process results in a *continued fraction expansion*[3] of $Z(s) = N(s)/D(s)$ as

$$Z(s) = \frac{k_{01}}{s} + \cfrac{1}{\cfrac{k_{02}}{s} + \cfrac{1}{\cfrac{k_{03}}{s} + \cfrac{1}{\cfrac{k_{04}}{s} + \cfrac{1}{k_{05}/s}}}}$$

$$= \frac{0.375}{s} + \cfrac{1}{\cfrac{4.5714}{s} + \cfrac{1}{\cfrac{0.5568}{s} + \cfrac{1}{\cfrac{46.095}{s} + \cfrac{1}{0.06818/s}}}}$$

Evidently, the network is as shown in Fig. 2-5 with elements, in farads and henrys, given by $1/k_{0i}$, $i = 1, \ldots, 5$.   Note that the implementation uses again *five* components.   Note also that for $\omega \to \infty$, $Z(s)$ reduces to $1/(sC_0)$ where C_0 consists of three capacitors in series to yield $C_0 = 1.0000$ as in Example 2-1.   The two circuits in Figs. 2-4 and 2-5 are *equivalent* networks!

---

[3]This realization is named after W. Cauer.   A continued fraction expansion (CFE) around the origin $(s = 0)$ gives a Cauer II network; the topology obtained from a CFE around $s = \infty$ is called a Cauer I circuit.

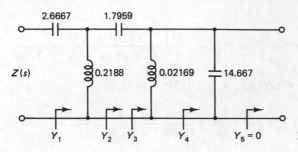

Figure 2-5   Circuit for Example 2-2.

Example 2-3

Realize the immittance of Examples 2-1 and 2-2 in the form of the circuit in Fig. 2-6.

Solution   From the given

$$Y(s) = \frac{s(s^2 + 2)(s^2 + 4)}{(s^2 + 1)(s^2 + 3)}$$

we remove first the pole at $s = \infty$, $1 \cdot s$, as in Example 2-1; this gives $C_1 = 1$ and a remainder

$$Y_1(s) = Y(s) - s = \frac{2s(s^2 + 2.5)}{(s^2 + 1)(s^2 + 3)}$$

From Fig. 2-6, the next branch required is a series arm with a parallel resonance circuit; we treat, therefore, $Z_1 = 1/Y_1$ and remove the pole pair at $s = \pm j\sqrt{2.5}$:

$$Z_1 = \frac{(s^2 + 1)(s^2 + 3)}{2s(s^2 + 2.5)} \qquad Z_2 = Z_1 - \frac{k_1 s}{s^2 + 2.5}$$

Application of Eq. (2-3c) results in $k_1 = 0.15$ and

$$Y_2(s) = \frac{1}{Z_2(s)} = \frac{2s}{s^2 + 1.2} = \frac{s/L_3}{s^2 + 1/(L_3 C_3)}$$

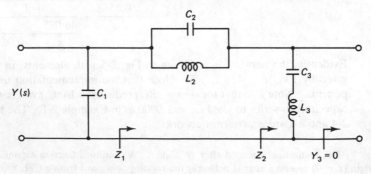

Figure 2-6   Circuit for Example 2-3.

Removing the last pole pair yields $Y_3 = 0$ and completes the process. Clearly, the *five* element values are:

$$C_1 = 1 \qquad C_2 = \frac{1}{0.15} = 6.667 \qquad L_2 = 0.06 \qquad C_3 = 1.667 \qquad L_3 = 0.5$$

Example 2-4

Realize the immittance of Example 2-3 in the topology shown in Fig. 2-7.

Solution.   Starting as in Example 2-2, we obtain

$$Y_1(s) = \frac{s^4 + 6s^2 + 8}{0.625s(s^2 + 2.8)} = \frac{0.5486s}{s^2 + 2.8} + 1.600s + \frac{4.5714}{s}$$

after removing the impedance pole at the origin (i.e., after extracting a capacitor C_1 = 1/0.375 = 2.667). Because the prescribed circuit calls for a series resonance (L_2C_2) in the shunt arm, we remove next the pole pair $s = \pm j\sqrt{2.8}$ with $k_1 = 0.5486$ found from Eq. (2-3c). The remaining function $Y_2(s)$ has a pole at the origin, $8/(1.75s)$, and one at infinity, $s/0.625$, as shown in the partial fraction expansion for $Y_1(s)$. The *five* elements of the required circuit are, therefore,

$$C_1 = 2.667 \qquad C_2 = 0.1959 \qquad L_2 = 1.8228 \qquad C_3 = 1.6 \qquad L_3 = 0.2188$$

These examples (and the problems) should be sufficient to show that there are many different *equivalent* realizations of a given immittance. Further, by comparing the example circuits with Fig. 2-1b, we note especially that all the realizations appear in the form of *ladder* networks, with series and shunt branches "resonating," i.e., having short or open circuits, at $s = 0$ (in the case of L or C), $s = \infty$ (in the case of C or L), or $s = j\omega_0$ with $0 < \omega_0 < \infty$ (in the case of series or parallel LC tank circuits with $LC = \omega_0^{-2}$). We note that, whenever in our immittance ladder realizations we desire a resonance in a *series arm*, we have to remove an *impedance pole* and, similarly, whenever a "resonance" is needed in a *shunt arm* we have to remove an *admittance pole*.

Anticipating a little the discussion in Section 2.3, we should expect from Fig. 2-1b that a parallel resonance in a series arm will cause a *transmission zero* because the series impedance $Z_i(s)$ in question would become an open circuit at the resonance frequency and thereby separate output from input. The dual state-

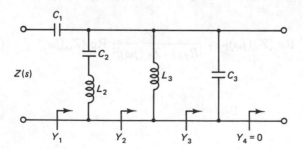

Figure 2-7   Circuit for Example 2-4.

ment applies to a series resonance in a shunt arm because at the resonance frequency $Y_j(s)$ has a pole and would appear as a short circuit across the signal path from input to output.   Because in a ladder network there is only *one* signal path from input to output, it follows further that *any* transmission zero *must* be generated by a series impedance pole or a shunt admittance pole.   Since all these immittances are *LC* functions of the form of Eq. (2-2a) with poles (and zeros) only on the $j\omega$-axis, we conclude further that *an LC ladder can have only $j\omega$-axis transmission zeros*.

Before proceeding to Section 2.3, the student is urged to reflect upon these last statements and review the example circuits in Figs. 2-4 to 2-7 with a view on the filtering problem.   If we assume our example immittances to be in the box N in Fig. 2-1b, it then becomes apparent that, for example, Fig. 2-4 is an *LC* filter with transmission zeros at $\omega = 1/\sqrt{L_1 C_1} = 1$, $1/\sqrt{L_2 C_2} = \sqrt{3}$, and ∞; Fig. 2-6 is an *LC* filter with transmission zeros at $\omega = \infty$, $\sqrt{2.5}$, and $\sqrt{1.2}$; and so forth. Once this concept is understood we have only to concern ourselves with the problem that, in *LC* filters, it is not an *LC immittance* that is prescribed but the *transfer function* of an *LC twoport* operating between two resistive terminations as shown in Fig. 2-1.   A suitable *synthesis immittance* will, therefore, first have to be derived from the prescribed transfer characteristic.

2.3 DERIVATION OF THE TWOPORT PARAMETERS OF LC FILTERS

When we discussed the transfer function $H(s)$ in Eq. (1-12) or Eq. (1-44), we did not take into consideration the terminations R_S and R_L between which the filter has to operate.   This approach is acceptable in active filters, where the source resistance can be considered a part of the filter and where the load resistor is usually of little consequence because, as will be seen, the filter output is taken from the output of an operational amplifier that can be treated as an ideal voltage source. In *LC* filters the operating conditions have to be considered more carefully, because, of course, the presence of R_S and R_L changes the character of the network from an *LC* circuit to an *RLC* circuit and thereby affects the transfer of power from source to load.

Let us go back to Fig. 2-1 and calculate the power P_1 delivered by the source to the twoport network N and the power P_2 delivered to the load R_L.   They are, respectively,

$$P_1 = |I_1(j\omega)|^2 \, \text{Re} \, \{Z_{\text{in}}(j\omega)\} = \frac{|V_S|^2}{|R_S + Z_{\text{in}}(j\omega)|^2} \, \text{Re} \, \{Z_{\text{in}}(j\omega)\} \qquad (2\text{-}6a)$$

and

$$P_2 = \frac{|V_2|^2}{R_L} \qquad (2\text{-}6b)$$

Note that $P_1 = P_2$ because N is *lossless*. Also, note that the maximum available power at the input of N occurs when $Z_{in}(j\omega) = R_S$, i.e., when N is matched to the source:

$$P_{max} = \frac{1}{4} \frac{|V_S|^2}{R_S} \tag{2-6c}$$

Let us then define the transfer function explicitly via the power transfer ratio P_2/P_{max} as

$$|H(j\omega)|^2 = \frac{4R_S}{R_L} \left| \frac{V_2}{V_S} \right|^2 \tag{2-7a}$$

or

$$H(s) = \sqrt{\frac{4R_S}{R_L}} \frac{V_2}{V_S} = \frac{N(s)}{D(s)} \tag{2-7b}$$

Clearly, since $P_2 = P_1 \leq P_{max}$, we must satisfy

$$|H(j\omega)| \leq 1 \tag{2-7c}$$

It is seen that $H(s)$ in Eq. (2-7b) differs from $H(s)$ in Eq. (1-44) by at most a multiplicative constant that depends on R_S and R_L and on how the input voltage is defined. For the remainder of the *LC* filter discussion we shall use $H(s)$ as defined in Eq. (2-7b).

As mentioned, in our *lossless* network, we have $P_1 = P_2$; thus, equating Eqs. (2-6a) and (2-6b) leads to

$$|H(j\omega)|^2 = \frac{4R_S \operatorname{Re}\{Z_{in}(j\omega)\}}{|R_S + Z_{in}(j\omega)|^2} = 1 - \left| \frac{R_S - Z_{in}(j\omega)}{R_S + Z_{in}(j\omega)} \right|^2 = 1 - |\rho(j\omega)|^2 \tag{2-8}$$

where

$$\rho(s) = \pm \frac{R_S - Z_{in}(s)}{R_S + Z_{in}(s)} \tag{2-9}$$

is recognized as the *reflection coefficient* at the input of the *LC* filter, loaded by R_L. $|\rho(j\omega)|$ is a measure of P_r, the *power reflected* from the input of N due to mismatch between R_S and $Z_{in}(s)$, as a fraction of P_{max}; that is $|\rho(j\omega)|^2 = P_r/P_{max}$. Equation (2-8), a form of *Feldtkeller's equation*, therefore means only that

$$P_1 + P_r = P_{max} \tag{2-10}$$

which makes obvious sense. The $\pm$ in Eq. (2-9) originates from the ambiguity in taking the square root of $|\rho(j\omega)|^2$ in Eq. (2-8).

Equations (2-8) and (2-9) can be used to derive the input impedance of the *LC* filter for a given transfer function $H(s)$. If $H(s) = N(s)/D(s)$, we obtain from

Eq. (2-8)

$$|\rho(j\omega)|^2 = 1 - |H(j\omega)|^2 = \frac{|D(j\omega)|^2 - |N(j\omega)|^2}{|D(j\omega)|^2}$$ (2-11a)

$$= \varepsilon^2 \frac{|F(j\omega)|^2}{|D(j\omega)|^2}$$

i.e.,

$$\rho(s) = \pm \varepsilon \frac{F(s)}{D(s)} = \pm \frac{\hat{F}(s)}{D(s)}$$ (2-11b)

where we have used the *reflection zero polynomial* $F(s)$, introduced in Eq. (1-65). Also, for convenience in later notation, we have defined the polynomial $\hat{F}(s) = \varepsilon F(s)$. Then we obtain from Eq. (2-9) $Z_{in}(s)/R_S = (1 \mp \rho)/(1 \pm \rho)$, or

$$Z_{in}(s) = R_S \frac{D(s) \mp \hat{F}(s)}{D(s) \pm \hat{F}(s)}$$ (2-12)

It is apparent that the sign ambiguity in Eqs. (2-9), (2-11b), and (2-12) only replaces Z_{in} by $1/Z_{in}$, which, as is demonstrated below, results in the dual network. In the following the upper sign will be assumed.

$Z_{in}(s)$ is the input impedance of an *LC* network terminated in a resistor R_L. If we can find such a network, our *LC* filter design problem is complete. Observe that, because of the resistive load R_L, $Z_{in}(s)$ is *not* an *LC* impedance. We can, however, separate the effect of R_L from the twoport *N* in Fig. 2-1 by expressing $Z_{in}(s)$ in terms of the *twoport parameters*, specifically, the *open-circuit impedance* parameters $z_{ij}(s)$ and/or the *short-circuit admittance* parameters $y_{ij}(s)$, $i, j = 1, 2$, defined [9, 25] as

$$\begin{pmatrix} V_1 \\ V_2 \end{pmatrix} = \begin{pmatrix} z_{11} & z_{12} \\ z_{21} & z_{22} \end{pmatrix} \begin{pmatrix} I_1 \\ I_2 \end{pmatrix} \text{ and } \begin{pmatrix} I_1 \\ I_2 \end{pmatrix} = \begin{pmatrix} y_{11} & y_{12} \\ y_{21} & y_{22} \end{pmatrix} \begin{pmatrix} V_1 \\ V_2 \end{pmatrix}$$ (2-13)

respectively. Evidently, $z_{11}(s) = (V_1/I_1)|_{I_2 = 0}$ is the open-circuit input imped-ance, $z_{12}(s) = (V_1/I_2)|_{I_1 = 0}$ is the open-circuit transfer impedance, $y_{11}(s) = (I_1/V_1)|_{V_2 = 0}$ is the short-circuit input admittance, and so forth. The student may recall that for passive *reciprocal* networks $z_{12} = z_{21}$ and $y_{12} = y_{21}$. Now observe that the *z*- and *y*-parameters are representative of the *LC* network only, separated from both R_S and R_L. Thus,

$z_{11}(s)$ and $z_{22}(s)$ are *LC impedances* and $y_{11}(s)$ and $y_{22}(s)$ are *LC admittances* of the form of Eq. (2-2a). Further, one can show that $z_{12}(s)$ and $y_{12}(s)$ are also *odd* rational functions. If the *z*-parameters are expanded into *partial fractions*, one obtains, for $i, j = 1, 2$,

$$z_{ij}(s) = \frac{k_{ij}^{(0)}}{s} + k_{ij}^{(\infty)} + \sum_{l=1}^{n} \frac{k_{ij}^{(l)} s}{s^2 + \omega_l^2}$$ (2-14)

where $k_{i,j}^{(r)}$, $r = 0, \infty, 1, \ldots, n$, are the residues obtained from Eqs. (2-3a) through (2-3c) with $Z_{LC}(s)$ replaced by $z_{ij}(s)$. One can prove [9, 25] that the important *residue condition*

$$k_{11}^{(r)} k_{22}^{(r)} - (k_{12}^{(r)})^2 \geq 0 \tag{2-15}$$

must be satisfied for all poles. It follows that both $z_{11}(s)$ and $z_{22}(s)$ must have a pole wherever $z_{12}(s)$ has a pole. Identical statements, including Eq. (2-15), are valid for the *y*-parameters.

If we express the transfer function of Eq. (2-7b) and the input impedance in terms of the *z*-parameters (Problem 2.11), the results are

$$H(s) = \frac{\sqrt{4 R_S R_L}\, z_{12}}{(z_{11} + R_S)(z_{22} + R_L) - z_{12}^2} = \frac{N(s)}{D(s)} \tag{2-16}$$

and

$$Z_{in}(s) = \frac{z_{11} R_L + z_{11} z_{22} - z_{12}^2}{z_{22} + R_L} \tag{2-17}$$

Here we note especially that for general *LC* filters $N(s)$ must be either an *even* or an *odd* polynomial because $z_{12}(s)$ is an odd rational function, and further—as pointed out earlier—that in *LC ladders* all roots of $N(s)$ (the transmission zeros) must lie on the $j\omega$-axis.

Our goal is now to extract a set of realizable *z*-parameters by equating Eq. (2-17) to Eq. (2-12), using a process called *Darlington's procedure* [11, 25]. Table 2.1 lists the resulting *z*-parameters, and the *y*-parameters derived by an analogous sequence of steps. The derivations are left to Problem 2.13. The equations in Table 2.1 show that the *z*- and *y*-parameters in any of the sets have

TABLE 2.1 THE *z*- AND *y*-PARAMETERS OF AN *LC* FILTER WITH TRANSMISSION ZERO POLYNOMIAL $N(s)$, REFLECTION ZERO POLYNOMIAL $\hat{F}(s)$, AND NATURAL FREQUENCY POLYNOMIAL $D(s)$

$N(s)$	z_{11}/R_S	z_{22}/R_L	$z_{12}/\sqrt{R_S R_L}$	$y_{11} R_S$	$y_{22} R_L$	$-y_{12}\sqrt{R_S R_L}$
Even	m_1/n_2	m_2/n_2	N/n_2	m_2/n_1	m_1/n_1	N/n_1
Odd	n_1/m_2	n_2/m_2	N/m_2	n_2/m_1	n_1/m_1	N/m_1

The even and odd polynomials $m_i(s)$ and $n_i(s)$, $i = 1, 2$, are obtained from $Z_{in}(s) = (m_1 + n_1)/(m_2 + n_2)$ and are defined as

$$m_1(s) = D_e(s) - \hat{F}_e(s) \qquad n_1(s) = D_o(s) - \hat{F}_o(s)$$

$$m_2(s) = D_e(s) + \hat{F}_e(s) \qquad n_2(s) = D_o(s) + \hat{F}_o(s)$$

$$D(s) = D_e(s) + D_o(s) \triangleq \mathrm{Ev}\,\{D(s)\} + \mathrm{Od}\,\{D(s)\}$$

$$\hat{F}(s) = \hat{F}_e(s) + \hat{F}_o(s) \triangleq \mathrm{Ev}\,\{\hat{F}(s)\} + \mathrm{Od}\,\{\hat{F}(s)\}$$

identical poles[4] and that all the functions are odd as required. By use of Eq. (2-3) it is easy to show (Problem 2.14) that the parameters satisfy the residue condition, Eq. (2-15). Specifically, defining a pole as *compact* if $k_{11}^{(r)}k_{22}^{(r)} - (k_{12}^{(r)})^2 = 0$, one finds that

All poles in $0 \leq \omega < \infty$ of the z- and y-parameters are *compact*. If $Z_{in}(s)$ has a pole at ∞, then the z-parameters have a *noncompact* pole or z_{11} has a *private*[5] pole, and the y-parameters have no pole at $s = \infty$. If $Z_{in}(s) = 0$ at $s = \infty$, then the y-parameters have a *noncompact* pole or y_{11} has a private pole, and the z-parameters have no pole at $s = \infty$.

The proof that all $z_{ii}(s)$ and $y_{ii}(s)$, $i = 1, 2$, are indeed *LC* immittances of the form of Eq. (2-2a) will be omitted [25] (see, however, Problem 2.13). Note also that the zeros of the transfer parameters $z_{12}(s)$ and $y_{12}(s)$ are the transmission zeros of the *LC* filter. Further transmission zeros of $H(s)$ can occur at *private* poles of z_{11} and/or z_{22} (and similarly, of y_{11} and/or y_{22}) and at *noncompact* poles of the z- (or y-) parameters, as is evident from Eq. (2-16).

Let us briefly recapitulate at this point where we are in our effort to realize a resistively terminated *LC* ladder filter and where we still have to go:

We saw in Section 2.2 that *LC* immittances are realized in the form of ladder structures by repeated pole removals and found that the transmission of power further down the ladder is blocked for those frequencies where the series and shunt ladder arms become open circuits and short circuits, respectively. We reasoned, therefore, that such a process should lead to an *LC* ladder design procedure if an *LC* synthesis immittance could be found that characterizes the ladder's behavior, as it is prescribed by a desired transfer function $H(s)$ and given values of R_S and R_L. We then derived a relationship between $H(s)$ and the input reflection coefficient $\rho(s)$ [Eq. (2-8)], which allowed us to calculate the *RLC* input impedance $Z_{in}(s)$ [Eq. (2-12)] of the loaded *LC* filter. Finally, we derived the *LC* z- and y-parameters of the intrinsic *LC* twoport, divorced from the effects of R_S and R_L. We shall proceed to demonstrate in Section 2.4 that the parameters $z_{11}(s)$, $z_{22}(s)$, $y_{11}(s)$, and $y_{22}(s)$ *are the desired synthesis immittances* and that a *suitable* realization of *one* of these parameters does indeed result in the correct *LC* ladder filter. First, however, let us demonstrate the process of finding the twoport parameters by an example.

Example 2-5

Find the z- and y-parameters of an *LC* lowpass filter that has transmission zeros at $f_1 = 1697.1$ Hz and $f_2 = 2078.5$ Hz; the passband characteristic is maximally flat with 0.5 dB maximum attenuation in $0 \leq f \leq 1200$ Hz. At high frequencies the attenuation should increase by 20 dB/decade.

[4]Apart from a possible pole at infinity!
[5]A *private* pole of a parameter is not shared by the other two parameters.

Also find the reflection coefficient and the input impedance $Z_{in}(s)$ of the filter if $R_S = R_L = 50 \ \Omega$.

Solution. The first step is to normalize all impedances by $R_0 = 50 \ \Omega$ and the frequency axis by $f_c = 1200$ Hz so that the filter is terminated by $R_S = R_L = 1$ and the transmission zeros are at $\omega_1 = 2\pi f_1/(2\pi f_c) = f_1/f_c = 1.4142 \simeq \sqrt{2}$ and $\omega_2 = f_2/f_c = 1.7321 \simeq \sqrt{3}$ with passband corner at $\omega = 1$. The losspole polynomial $N(s)$ therefore becomes $N(s) = k(s^2 + 2)(s^2 + 3)$ where k is a constant, and from Eq. (1-69) we find

$$|H(j\omega)|^2 = \frac{(2 - \omega^2)^2(3 - \omega^2)^2}{(2 - \omega^2)^2(3 - \omega^2)^2 + a_n\omega^{2n}}$$

The attenuation slope of -20 dB/decade for $\omega \to \infty$ requires $|H(j\omega)|$ to go to zero as $1/\omega$ for large ω; thus: $n = 5$. Further, 0.5 dB attenuation at $\omega = 1$ leads to $a_n = 4(10^{0.05} - 1) = 0.4881$ by Eq. (1-70). Note that this leads to the transfer function obtained in Example 1-6, where we found

$$H(s) = \frac{N(s)}{D(s)} = \frac{1.4314(s^2 + 2)(s^2 + 3)}{s^5 + 4.5817s^4 + 9.4717s^3 + 14.473s^2 + 11.210s + 8.5883}$$

and

$$K(s) = \frac{s^5}{1.4314(s^2 + 2)(s^2 + 3)} = \frac{\hat{F}(s)}{N(s)}$$

Further, from Eqs. (2-11) and (2-12), we find

$$\rho(s) = \frac{s^5}{s^5 + 4.5817s^4 + 9.4717s^3 + 14.473s^2 + 11.210s + 8.5883}$$

and the normalized impedance

$$Z_{in}(s) = \frac{4.5817s^4 + 9.4717s^3 + 14.473s^2 + 11.210s + 8.5883}{2s^5 + 4.5817s^4 + 9.4717s^3 + 14.473s^2 + 11.210s + 8.5883}$$

With the notation of Table 2.1, that is, $Z_{in}(s) = (m_1 + n_1)/(m_2 + n_2)$, and $N(s)$ even we find finally

$$z_{11}(s) = z_{22}(s) = \frac{4.5817s^4 + 14.473s^2 + 8.5883}{2s^5 + 9.4717s^3 + 11.210s} = \frac{4.5817(s^2 + 0.79197)(s^2 + 2.3669)}{2s(s^2 + 2.3223)(s^2 + 2.4136)}$$

$$z_{12}(s) = \frac{1.4315(s^2 + 2)(s^2 + 3)}{2s(s^2 + 2.3223)(s^2 + 2.4136)}$$

and

$$y_{11}(s) = y_{22}(s) = \frac{4.5817s^4 + 14.473s^2 + 8.5883}{9.4717s^3 + 11.210s} = \frac{4.5817(s^2 + 0.79197)(s^2 + 2.3669)}{9.4717s(s^2 + 1.1835)}$$

$$-y_{12}(s) = \frac{1.4315(s^2 + 2)(s^2 + 3)}{9.4717s(s^2 + 1.1835)}$$

Note that all the twoport parameters are *odd*, that $z_{11} = z_{22}$ and $y_{11} = y_{22}$ are *LC* immittances in contrast to z_{12} and y_{12}, whose poles and zeros do not alternate along the $j\omega$-axis. Observe further that the network can be expected to be symmetrical because $z_{11} = z_{22}$ (and $y_{11} = y_{22}$).

2.4 REALIZATION OF LC LADDERS

We derived in Section 2.3 the z- and y-parameters that characterize an *LC* filter with specified transfer function $H(s)$. We also saw that the parameters form *compatible* sets, i.e., they satisfy the residue condition [Eq. (2-15)], and that the transmission zero polynomial forms the numerator of the transfer parameter $z_{12}(s)$ or $y_{12}(s)$. Further, we stated (without proof) that the *LC* synthesis immittances are $z_{ii}(s)$ and $y_{ii}(s)$, $i = 1, 2$, of which we have to select one for the realization of the filter. Evidently, at this point we have to address three problems:

1. Which of the four parameters y_{11}, y_{22}, z_{11}, and z_{22} should be selected as the synthesis immittance?
2. How should this immittance, say, z_{11}, be realized so that the filter has the correct transmission zeros as prescribed by the numerator of z_{12}?
3. How do we make sure that the immittance at the second port, say, z_{22}, is realized *simultaneously* with z_{11} and z_{12}?

The first question is answered easily if we remember that the *LC* ladder may start and finish with a series or a shunt branch as sketched in Fig. 2-8 and that the z_{ij} are the *open-circuit* impedance and y_{ij} the *short-circuit* admittance parameters. Consequently, if the first branch of a ladder at one port is a *series* arm, it will clearly not affect the open-circuit input impedance at the second port. This imped-ance will therefore in general be of *reduced* order, i.e., its degree will be *less* than that of the prescribed transfer function. The dual statement is true for *shunt* branches at the ladder ports and the short-circuit input admittances. Because the synthesis immittance must represent the *complete* ladder, we have to select, there-fore, one of the two functions which are of higher degree. The choices are iden-tified in Fig. 2-8. In Example 2-5, for instance, we should work with the z-parameters, indicating shunt arms at the input and output of the ladder.

Turning to the second question, we recall that transmission zeros are realized via impedance poles (open circuits) in a series branch, or admittance poles (short circuits) in a shunt branch of the ladder. According to our discussion in Sec-tion 2.2, this requirement implies that the synthesis immittance at some point during the realization must have a pole or a zero which coincides with a prescribed trans-mission zero, a condition that in general is, of course, *not* satisfied automatically. The problem is solved by *zero shifting via partial pole removal* and will be explained in connection with Fig. 2-9.

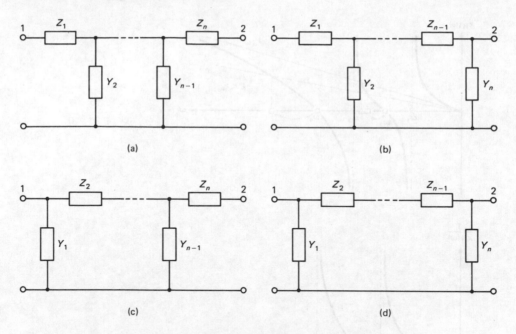

Figure 2-8 Possible input and output configurations of ladders.   (a) Use y_{11} or y_{22}; (b) use z_{11} or y_{22}; (c) use y_{11} or z_{22}; (d) use z_{11} or z_{22}.

Assume we have an *LC* synthesis impedance, say, $z_{11}(s)$ to be specific, with a zero at $s = 0$ and a pole at $s = \infty$, i.e.,

$$\lim_{s \to \infty} z_{11}(s) = sk_\infty$$

where k_∞ is the residue.   The reactance $x_{11}(\omega) = -jz_{11}(\omega)$ is sketched in Fig. 2-9a.   If we subtract this pole from z_{11}, as was discussed in Section 2.3, the remainder function $z_1 = z_{11} - k_\infty s$ no longer has a pole at $s = \infty$, because this pole is removed *completely*.   At the same time, the zero at $\omega = \omega_2$ moves to the point marked with an open triangle in Fig. 2-9a and the zero at $\omega = \omega_4$ moves all the way to infinity.   If instead we remove the pole at infinity only *partially* by subtracting a term ks with $k < k_\infty$, the function $z_1 = z_{11} - ks$ still has a pole at infinity with residue $k_1 = k_\infty - k > 0$.   In this case, the zeros of $x_{11}(\omega)$, ω_2 and ω_4, shift to the locations marked by an open square, as Fig. 2-9a clearly shows.   Note that the zero at the origin and the poles at ω_1 and ω_3, of course, do not shift.

Taking the inverse of our synthesis impedance, i.e., $1/z_{11}(s)$, simply interchanges poles and zeros.   Figure 2-9b shows a plot of the corresponding reactance, which now has a pole at $s = 0$, i.e.,

$$\lim_{s \to \infty} \frac{1}{z_{11}(s)} = \frac{k_0}{s}$$

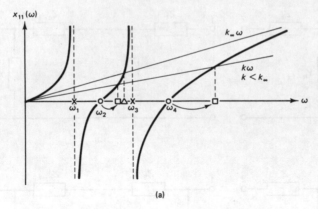

(a)

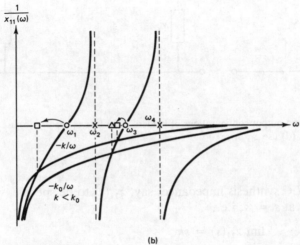

(b)

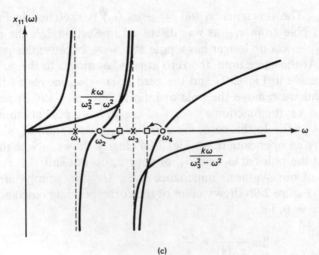

(c)

Figure 2-9 Reactance function to explain complete and partial pole removals: (a) removal of pole at ∞; (b) removal of pole at the origin; (c) removal of internal pole.

where k_0 is the residue. If this pole is subtracted from $1/z_{11}$ *completely*, the remaining admittance $y_1 = 1/z_{11} - k_0/s$ now has a zero at $s = 0$ because the zero at ω_1 moves to the origin as is illustrated in Fig. 2-9b. Also, the zero at ω_3 shifts to the point marked with an open triangle. If instead the pole is removed only *partially* by subtracting a term k/s with $k < k_0$, the remaining admittance still has a pole at the origin with residue $k_1 = k_0 - k > 0$ and the zeros at ω_1 and ω_3 move to the locations marked by an open square. We note that the zeros at infinity and the poles at ω_2 and ω_4 have not moved.

A little thought and Fig. 2-9c reveal that analogous considerations hold if an internal pole, here at $s = j\omega_3$ with residue k_3, is partially or completely removed from an immittance by subtracting a term $ks/(s^2 + \omega_3^2)$, $0 < k \le k_3$. We shall not discuss this case further, because it is used less frequently in *LC* filter design, but Example 2-8 later in this chapter will illustrate the details.

The foregoing discussion and a study of Fig. 2-9 allow us to make several observations:

1. The zeros of the immittance move *toward* the location of the partially or completely removed poles.
2. Zeros never shift across an adjacent pole.
3. The distance by which a zero shifts depends on the value k, i.e., on how much of the pole is removed.
4. The closer a zero is to the pole partially removed, the larger generally is the distance it shifts.
5. Because partial pole removal does not reduce the degree of the immittance, it costs *one additional* element (one *L or* one *C*) if a pole is partially removed ("*weakened*") at $s = 0$ or $s = \infty$, and *two additional* elements (one *L and* one *C*) if an internal pole is weakened.

It will now be apparent how a prescribed transmission zero at ω_z can be realized. By removing *partially* the pole of the synthesis immittance or of its inverse at the origin or at infinity, a zero of the immittance is shifted until it coincides with the position of the transmission zero at ω_z; the remainder function is then inverted and the new pole at ω_z removed *completely*, thus realizing a transmission zero at ω_z. Because zeros shift *toward* the pole removed, the transmission zero must lie between the zero to be shifted and the pole that is partially removed. To be specific, say that $z_{11}(s)$ with poles at $s = 0$ and $s = \infty$ is selected as the synthesis impedance. Let the transmission zero be at $\omega = \omega_z$. If a zero of z_{11} is to be shifted toward *higher* frequencies such that the remainder function is zero at $s = j\omega_z$, we weaken the pole at ∞ and find the partial residue k from

$$[z_{11}(s) - ks]_{j\omega_z} = 0 \quad \text{or} \quad k = \frac{z_{11}(j\omega_z)}{j\omega_z} \tag{2-18a}$$

If instead a zero of z_{11} is to be shifted toward *lower* frequencies, we weaken the pole at the origin and find the partial residue k from

$$\left[z_{11}(s) - \frac{k}{s} \right]_{j\omega_z} = 0 \quad \text{or} \quad k = j\omega_z z_{11}(j\omega_z) \tag{2-18b}$$

To be able to make systematic progress in this procedure, the student is *strongly* advised to sketch a set of *pole-zero diagrams* showing the relative positions of poles, zeros, and transmission zeros of the relevant z- or y-parameters along the ω-axis as is illustrated in Example 2-6 below.

The process just described realizes, of course, the synthesis immittance, say, $z_{11}(s)$, exactly. Because z_{11} and z_{12}, belonging to the same circuit, have the same poles, the denominator polynomial of $z_{12}(s)$ is also realized correctly. Finally, since we have implemented the prescribed (transmission) zeros of $z_{12}(s)$, it should be apparent also that the numerator polynomial of $z_{12}(s)$ is realized *except for an unknown multiplying gain constant K* that can be determined by analyzing the obtained network.

Example 2-6

Assume that the procedure of Section 2.3 has yielded the following y-parameters:

$$y_{11} = \frac{3s(s^2 + \frac{7}{3})}{(s^2 + 2)(s^2 + 5)} \qquad y_{22} = \frac{s(s^2 + 3)}{(s^2 + 2)(s^2 + 5)}$$

$$-y_{12} = \frac{s(s^2 + 1)}{(s^2 + 2)(s^2 + 5)}$$

(The student may want to verify that the parameters form a compatible set with all poles compact!) Realize the y-parameters as an *LC* ladder network by the zero-shifting method.

Solution. Let us choose y_{11} as our synthesis admittance and draw a set of pole-zero diagrams on an arbitrarily distorted ω-scale to show the relative positions of the poles and zeros of y_{12} and y_{11}. Figure 2-10a and b shows the result. We observe that y_{11} has no poles at $s = 0$ and $s = \infty$ and has no pole or zero that coincides with the prescribed transmission zero at $s = 1$. Therefore we invert y_{11} to get

$$z_1(s) = \frac{1}{y_{11}} = \frac{(s^2 + 2)(s^2 + 5)}{3s(s^2 + \frac{7}{3})}$$

with the pole-zero plot shown in Fig. 2-10c.

From z_1 we can now remove the pole at the origin *partially* to shift the zero at $\omega = \sqrt{2}$ to $\omega = 1$. Thus, by Eq. (2-18b),

$$k_0 = s z_1(s)|_{s=j1} = \frac{(s^2 + 2)(s^2 + 5)}{3(s^2 + \frac{7}{3})}\bigg|_{j1} = \frac{1 \cdot 4}{3 \cdot \frac{4}{3}} = 1$$

and

$$z_2(s) = z_1 - \frac{k_0}{s} = \frac{(s^2 + 1)(s^2 + 3)}{3s(s^2 + \frac{7}{3})}$$

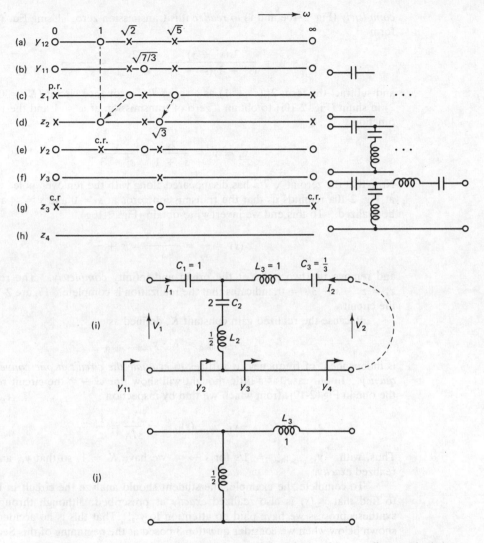

Figure 2-10   (a–h) Pole-zero diagrams of y_{12} and the initial and intermediate synthesis immittances showing partial (p.r.) and complete (c.r.) removal of poles.   (i) Realized circuit; the dashed short circuit is sketched in to emphasize that the realized y_{11} is the *short-circuit* input admittance.   (j) Asymptotic behavior of the circuit for $\omega \to \infty$.

with the pole-zero diagram shown in Fig. 2-10d.   The partially removed pole, k_0/s, results, of course, in a series capacitor $C_1 = 1/k_0 = 1$ as shown in Fig. 2-10i. $z_2(s)$ is still of degree 4, but it has now a zero at a location of the transmission zero $\omega = 1$.   Note that the zero at $\sqrt{5}$ has shifted toward the removed pole to its new location at $\omega = \sqrt{3}$. The pole at $\omega = 1$ of the admittance $y_2 = 1/z_2$ is then removed *completely* (Fig. 2-10e and f) *to realize* this transmission zero. Using Eq. (2-3c), we form

$$y_2 = \frac{3s(s^2 + \frac{7}{3})}{(s^2 + 1)(s^2 + 3)} = \frac{2s}{s^2 + 1} + \frac{s}{s^2 + 3}$$

and subtract the term $2s/(s^2 + 1)$ as a series *LC* resonance circuit ($L_2 = 0.5$, $C_2 = 2$) in shunt (Fig. 2-10i) to obtain a zero of transmission at $\omega = 1$ and the remainder function

$$y_3 = y_2 - \frac{2s}{s^2 + 1} = \frac{s}{s^2 + 3}$$

Note that the zero at $\sqrt{7/3}$ has disappeared along with the removed pole. A glance at Fig. 2-10a reminds us that the transmission zeros at $s = 0$ and $s = \infty$ are still to be realized. To this end we invert y_3 to obtain (Fig. 2-10g)

$$z_3(s) = \frac{1}{y_3} = \frac{s^2 + 3}{s} = s + \frac{3}{s}$$

and remove the two poles at the origin and infinity *completely*. The remainder, $z_4 = z_3 - s - 3/s = 0$, indicates that the realization is complete. Figure 2-10i shows the circuit.

Because the realized gain constant K, defined as

$$y_{12,\text{realized}} = K \cdot y_{12,\text{prescribed}}$$

is independent of frequency, it suffices to *evaluate the circuit at one convenient frequency*. In our case, as a little thought will show, for $\omega \to \infty$ the circuit reduces to the one in Fig. 2-10j, from which we find by inspection

$$-y_{12,\text{realized}}(s)\big|_{s \to \infty} = \frac{1}{sL_3} = \frac{1}{s}$$

Thus, with $-y_{12,\text{prescribed}} = 1/s$ for $s \to \infty$, we have $K = 1$, so that y_{11} and y_{12} are realized *exactly*.

To complete the example, the student should analyze the circuit in Fig. 2-10i to find that $y_{22}(s)$ is also realized *exactly* as prescribed, although throughout the synthesis process we have paid no attention to y_{22}! That this is no accident will be shown below when we consider question 3 posed at the beginning of this Section 2.4.

Before answering question 3, let us address a different concern that may have come to the inquisitive student in connection with the partial pole removal procedure, namely: If *complete* pole removals realize transmission zeros, do not partial pole removals realize transmission zeros, too? For instance, in Example 2-6, Fig. 2-10i, C_3, obtained by *complete* pole removal, realizes the transmission zero at $s = 0$ to block dc; but C_1, obtained by *partial* pole removal, does not. Why? For twoports with *no load* ($R_L = 0$ or $R_L = \infty$), the answer is obtained by remembering that the remainder function still has a pole at the frequency where a pole has been removed only *partially*. Thus, if the partial pole removal results in a series ladder arm, we get, intuitively, *voltage division* between two *open* circuits and therefore no transmission zero.   [In Fig. 2-10i, with $z_2(0) = \infty$, we have capacitive voltage division between C_1 and $z_2(0)$!]   Analogous reasoning shows that we have *current division* between two *short* circuits when the partially removed (admittance) pole results in a shunt arm and where the remaining admittance is still infinity at the pole frequency (current division between two parallel inductors!).   More careful reasoning is required if the twoport is *loaded* by a nonzero finite R_L.   In that case one can show [24] (Problem 2.15) that a partially removed pole in general does indeed create a transmission zero *unless the partially removed pole is later removed completely* [30] as was done in Example 2-6.

Having established that by the partial pole removal and zero-shifting procedure we can realize the synthesis immittance at one end of the *LC* ladder exactly and the transfer immittance to within a multiplying constant, we are now ready to address the remaining question 3: How can one ensure that the immittance at the other port is realized correctly?   To be specific, assume that the *prescribed* ladder parameters are $z_{11p}(s)$, $z_{22p}(s)$, and $z_{12p}(s)$ and we have *realized*

$$z_{11r}(s) = z_{11p}(s) \quad \text{and} \quad z_{12r}(s) = Kz_{12p}(s) \tag{2-19}$$

Our concern is then whether the adopted synthesis process yields the correct $z_{22p}(s)$. The solution is found in the residue condition, Eq. (2-15), repeated here for convenience,

$$k_{11}^{(r)} k_{22}^{(r)} - (k_{12}^{(r)})^2 \geq 0 \tag{2-20}$$

that the realized z- (or y-) parameters must satisfy at any of their poles ω_r, $0 \leq \omega_r \leq \infty$.   It is expressed in the following theorems [30], whose proofs are quite lengthy and shall be omitted.

Theorem 1.   If the prescribed z- (or y-) parameters have only compact poles and if each partial pole removal at a frequency ω_k in the ladder development is followed at a later stage by a complete pole removal at the same frequency ω_k, then the realized z- (or y-) parameters will have only compact poles.

One can show that, in case the complete pole removal at ω_k is *not* performed, the synthesis immittance at the other part will have a *private* pole at $\omega = \omega_k$ if ω_k *is not* a pole of the z- (or y-) parameters, and the realized parameters will have a *noncompact* pole at $\omega = \omega_k$ if ω_k *is* a pole of the z- (or y-) parameters.   By Eq. (2-16), both these cases result in a transmission zero at ω_k as mentioned earlier.

Theorem 2.   If none of the series impedances and shunt admittances of an *LC* ladder network has a pole at a pole ω_p of the z- (or y-) parameters, then the z- (or y-) parameters have a compact pole at $\omega = \omega_p$.

We remind the student that for a compact pole, Eq. (2-20) is satisfied with an equal sign and that, except possibly for a pole at infinity, all the poles of the z- (or y-) parameters obtained in Section 2.3 are compact.

Assume now that we have a set of prescribed z-parameters with compact poles and have realized $z_{11r} = z_{11p}$ and $z_{12r} = Kz_{12p}$ per Eq. (2-19) following the rules stated in Theorems 1 and/or 2. We know then that the *realized* parameter z_{22r} has compact poles, i.e., that the residues $k_{11r} = k_{11p}$, $k_{12r} = Kk_{12p}$, and k_{22r} must satisfy the residue condition [Eq. (2-20)] with the equal sign:

$$k_{11p}k_{22r} - K^2k_{12p}^2 = 0$$

Because $k_{12p}^2 = k_{11p}k_{22p}$, it follows that

$$k_{22r} = K^2k_{22p}$$

so that the prescribed parameter z_{22p} is realized to within a constant K^2 as shown in Fig. 2-11a. Equation (2-16) indicates clearly that this realization results in the wrong poles[6] for $H(s)$ unless $K = 1$. Fortunately the problem is easy to correct by connecting to the twoport output terminals 3-3' a transformer of turns ratio $n_2/n_1 = 1/K$, as is also shown in Fig. 2-11a, so that the total twoport in the dashed box has exactly the prescribed parameters. An entirely analogous procedure holds when the y-parameters are realized; only in this case the transformer has a turns ratio $n_2/n_1 = K$, as the student may verify. At this point the student should review Example 2-6, which was worked out following Theorem 2 and yielded $K = 1$. Thus, no transformer was needed!

If $z_{11}(s)$ has a *private* pole at $s = \infty$ (resulting in a transmission zero at infinity), this pole can be realized as a series inductor at the input with no effect on z_{12} and z_{22}. The remaining realization is as described above.

If the z-parameters have a *noncompact* pole at $s = \infty$, i.e., if the residue condition for that pole is

$$k_{11p}k_{22p} - k_{12p}^2 > 0$$

then we may interpret the situation by saying that k_{22p} has "excess residue," i.e., that $k_{22p} = k_{22p}' + \Delta k_{22}$ with $\Delta k_{22} > 0$ such that

$$k_{11p}k_{22}' - k_{12p}^2 = 0$$

In this case we realize as before $z_{11r} = z_{11p}$, $z_{12r} = Kz_{12p}$, and $z_{22r} = K^2z_{22}'$ and use a transformer of turns ratio $n_2/n_1 = 1/K$ to compensate for the obtained constant K. The realized z_{22} is then correct except for the missing part $\Delta k_{22}' \cdot s$ of the pole at $s = \infty$. Clearly, an inductor of value Δk_{22} in series with the output will remedy the situation as shown in Fig. 2-11b. Of course, we can also shift this inductor to the input side of the transformer, if its value is changed to $K^2 \Delta k_{22}$.

Finally, note with the help of Eq. (2-16) that the transformer can be saved *if* the value of R_L can be changed to K^2R_L, as is illustrated in Fig. 2-11c.

Let us summarize the steps involved in arriving at a resistively terminated lossless *LC* ladder filter from the initial frequency-domain magnitude specifications:

[6]It also changes the gain constant, but this is usually unimportant.

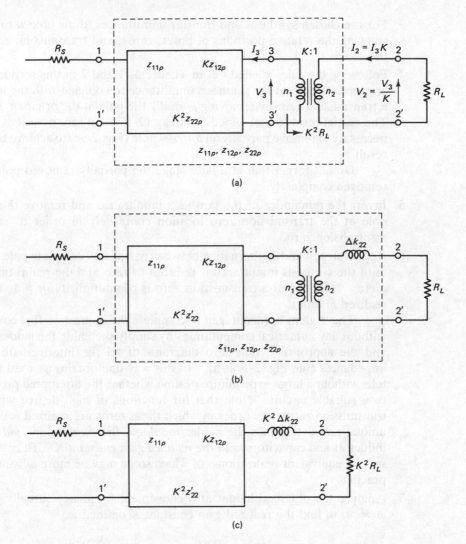

Figure 2-11 Realization of z-parameters as an LC ladder: (a) for compact poles; (b) for noncompact poles; (c) illustrating the elimination of the transformer.

1. Use the appropriate procedure of Section 1.6 to derive the transfer function $H(s) = N(s)/D(s)$.   The finite transmission zeros are the roots of $N(s)$.
2. With the help of Eqs. (2-11) and (2-12) and Table 2.1 in Section 2.3, derive the z-parameters and the y-parameters of the LC twoport.   As synthesis immittance, choose any one of the functions z_{11}, z_{22} or y_{11}, y_{22} which has the same degree as $H(s)$.   The choice will, of course, affect the final form of the ladder and the type of input and output ports as shown in Fig. 2-8.

3. For the chosen synthesis and transfer immittances, draw pole-zero diagrams showing the relative positions of poles, zeros, and transmission zeros along the ω-axis.

4. Following the rules spelled out in Theorems 1 and 2 in this section, shift an appropriate zero of the synthesis immittance to coincide with the location of a transmission zero by removing *partially* the pole at the origin or at infinity. The partial residue is obtained from Eq. (2-18). In rare occasions it may be necessary to remove partially an *internal* pole (Fig. 2-9c) to achieve the desired result.

   Do not forget that at a later stage the partially removed pole must be removed completely.

5. Invert the remainder of the synthesis immittance and remove the obtained pole at the transmission zero location *completely* in order to realize this transmission zero.

6. Repeat steps 4 and 5, alternating between partial and complete pole removals, until the synthesis immittance is reduced to zero and the realization is complete. Note that if a transmission zero is of multiplicity $m > 1$, it must be realized m times!

   The student is encouraged to complete the realization first conceptually without any numerical computations by simply sketching the ladder topology and the appropriate pole-zero diagrams of all the intermediate synthesis impedances (see Fig. 2-10a–h). In this way one obtains a "road map" that tells without a large expenditure of time whether the attempted process leads to a suitable circuit. Note that for functions of high degree with several transmission zeros, the order in which these zeros are realized is usually not unique and determines the ladder *topology*, the *number* and *values* of the inductors and capacitors, and the *realized gain constant K*. Thus, there exist several equivalent realizations, of which some may be more advantageous in practice.

7. Analyze the obtained ladder at a convenient frequency (usually $\omega = 0$ or $\omega = \infty$) to find the realized gain constant K defined as

$$K = \frac{z_{12,\text{real.}}}{z_{12,\text{prescr.}}} \quad \text{or} \quad K = \frac{y_{12,\text{real.}}}{y_{12,\text{prescr}}} \tag{2-21}$$

If $K \neq 1$, use a transformer as outlined in connection with Fig. 2-11 or change the terminating resistor to make the needed correction. Note that the termination at the port from which the design starts (R_S in the case of z_{11} or y_{11}, R_L in the case of z_{22} or y_{22}) is customarily normalized to unity.

In general, quite arbitrary filter bandpass or band-rejection magnitude requirements may be approximated and realized directly if suitable software packages such as FILTOR [11] or FILSYN [29] are available. However, many low-order and reasonably undemanding filters will be designed on the basis of a lowpass prototype obtained after suitable frequency transformation, as discussed in Section 1.6.2.   In that case the ladder obtained in step 7 is the lowpass prototype circuit that has to be transformed back into the required specification of the target filter by the following step 8:

8. Let the elements and the frequency variable in the prototype lowpass be denoted by L_{LP}, C_{LP}, and $\bar{s}$, respectively, so that the impedances and admittances are $\bar{s}L_{LP}$ and $\bar{s}C_{LP}$.

   If the target filter is an HP, the LP-to-HP transformation, Eq. (1-102), is $\bar{s} = 1/s$, and we have

$$\bar{s}L_{LP} \rightarrow \frac{1}{s} L_{LP} = \frac{1}{sC_{HP}} \tag{2-22a}$$

$$\bar{s}C_{LP} \rightarrow \frac{1}{s} C_{LP} = \frac{1}{sL_{HP}} \tag{2-22b}$$

The highpass filter is obtained by replacing each inductor L_{LP} by a capacitor $C_{HP} = 1/L_{LP}$ and each capacitor C_{LP} by an inductor $L_{HP} = 1/C_{LP}$.

   If the target filter is a BP, the LP-to-BP transformation, Eq. (1-104), is $\bar{s} = Q(s^2 + 1)/s$ and we have

$$\bar{s}L_{LP} \rightarrow QL_{LP} \frac{s^2 + 1}{s} = sQL_{LP} + \frac{QL_{LP}}{s} = sL_{BP_s} + \frac{1}{sC_{BP_s}} \tag{2-23a}$$

$$\bar{s}C_{LP} \rightarrow QC_{LP} \frac{s^2 + 1}{s} = sQC_{LP} + \frac{QC_{LP}}{s} = sC_{BP_p} + \frac{1}{sL_{BP_p}} \tag{2-23b}$$

The bandpass filter is obtained by replacing each inductor L_{LP} by a *series* combination of $L_{BP_s} = QL_{LP}$ and $C_{BP_s} = 1/(QL_{LP})$ and by replacing each capacitor C_{LP} by a *parallel* combination of $C_{BP_p} = QC_{LP}$ and $L_{BP_p} = 1/(QC_{LP})$.

   If the target filter is a BR circuit, the LP-to-BR transformation, Eq. (1-105), is $\bar{s} = 1/[Q(s^2 + 1)/s]$ and we have

$$\bar{s}L_{LP} \rightarrow \frac{sL_{LP}}{Q(s^2 + 1)} = \frac{1}{\dfrac{Q}{L_{LP}}\left(s + \dfrac{1}{s}\right)} = \frac{1}{sC_{BR_p} + \dfrac{1}{sL_{BR_p}}} \tag{2-24a}$$

$$\bar{s}C_{LP} \rightarrow \frac{sC_{LP}}{Q(s^2 + 1)} = \frac{1}{\dfrac{Q}{C_{LP}}\left(s + \dfrac{1}{s}\right)} = \frac{1}{sL_{BR_s} + \dfrac{1}{sC_{BR_s}}} \tag{2-24b}$$

The band-rejection filter is obtained by replacing each inductor L_{LP} by a *parallel* combination of $C_{BR_p} = Q/L_{LP}$ and $L_{BR_p} = L_{LP}/Q$ and each capacitor C_{LP} by a *series* combination of $L_{BR_s} = Q/C_{LP}$ and $C_{BR_s} = C_{LP}/Q$.

9. Finally, whether frequency transformation is used or not, the elements of the ladder must be denormalized as discussed in Section 1.4.

Two comments are called for at this point:

1. Although these steps will lead to a realization of the initially prescribed transfer function as a resistively terminated *LC* ladder, in practice one often finds, especially in functions of high order, that the element values are not very convenient for implementation. The problem is that the *element spread*, i.e., the difference between the largest and smallest inductor and, similarly, between the largest and smallest capacitor, is too large. In that case one may often resort to a variety of network transformations which reduce the ratios L_{max}/L_{min} and C_{max}/C_{min} to more practical values. Since these transformations are treated in a number of texts [e.g., 11 or 28], we shall not discuss them in this book. However, Problem 2.37 does provide an outline of the procedure.

2. Our second comment concerns the restrictions that must be satisfied by the transfer function in order to be realizable by the proposed *partial/complete pole removal–zero-shifting method*. The conditions pertain to the relative positions of transmission zeros and system poles; they are given, for most cases of practical interest, in reference 31 and will not be derived in this text. However, one of the conditions is implied by Theorems 1 and 2 above and will be pointed out especially. Recall that we use *partial* pole removal, usually from $s = \infty$ or $s = 0$, to *shift* an immittance zero to the location of a transfer function zero, and then a *complete* pole removal to *realize* that zero. In order to avoid obtaining unwanted transmission zeros at the partially removed poles, we then, at a later stage, must remove the partially removed pole(s) completely. Clearly, this *complete* pole removal will create a transmission zero. It follows, therefore, that for our realization process to work *the transfer function, in general, must have a transmission zero at $s = \infty$* (in the case of a lowpass or bandpass) *and/or at $s = 0$* (in the case of a highpass or bandpass). For example, an inverse Chebyshev lowpass filter of even order would not work, because its transmission at high frequencies is not zero but $|H(j\infty)|^2 = \varepsilon^2$. The same is true for an elliptic filter of even degree.

In addition to Example 2-6, we present in the following a few further examples to illustrate the details of *LC* filter design.

Example 2-7

Use frequency transformation to realize a bandpass filter to satisfy the following specifications:

Maximally flat passband with $A_p \leq 3$ dB in 900 Hz $\leq f \leq$ 1200 Hz
Transmission zero at $f_{z1} = 1582.48$ Hz

Attenuation increase at high frequencies at least 20 dB/decade

Source and load resistors: $R_S = R_L = 3 \text{ k}\Omega$

Solution. Following the development in Section 1.6.2.2, we first find the normalizing frequency $\Omega_0 = 2\pi\sqrt{900\cdot1200}/s = 2\pi \cdot 1039.23/s$ and the bandwidth $B = 2\pi(1200 - 900)/s = 2\pi \cdot 300/s$. Thus, the quality factor $Q = \Omega_0/B = 3.4641$ and the LP-to-BP transformation is

$$\bar{s} = 3.4641 \frac{s^2 + 1}{s} \tag{2-25}$$

where s is normalized with respect to Ω_0. With this transformation the two corner frequencies 900 Hz and 1200 Hz go into the frequencies $\bar{\omega} = -1$ and $\bar{\omega} = +1$, respectively, and f_{z1} is transformed into $\bar{\omega} = 3$. Note that because of the geometric symmetry of the LP-to-BP transformation, the filter will have a second transmission zero in the lower stopband at $f_{z2} = \Omega_0^2/(4\pi^2 f_{z1}) = 682.47$ Hz.

Figure 2-12a shows a sketch of the required attenuation of the normalized LP-prototype that according to Section 1.6.1.1, Eq. (1-69), results in the transfer function magnitude

$$|H_{\text{LP}}(j\bar{\omega})|^2 = \frac{(9 - \bar{\omega}^2)^2}{(9 - \bar{\omega}^2)^2 + a_n\bar{\omega}^{2n}} \tag{2-26}$$

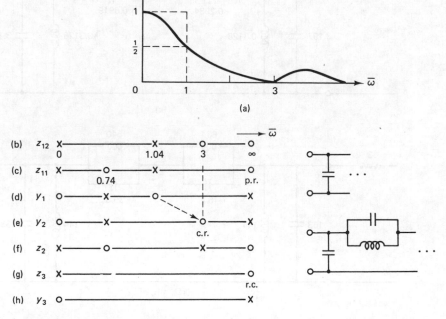

Figure 2-12 Prototype lowpass specifications, pole-zero diagrams, and realized circuits for Example 2-7. Units are $k\Omega$, mH, and nF.

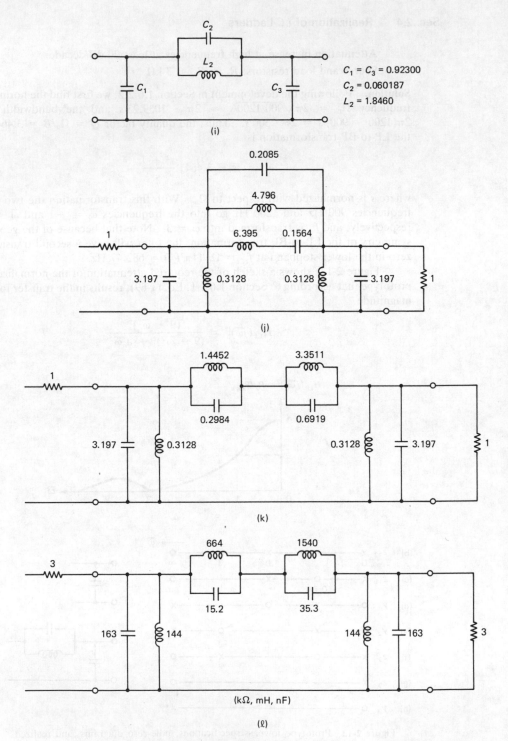

$C_1 = C_3 = 0.92300$
$C_2 = 0.060187$
$L_2 = 1.8460$

(i)

(j)

(k)

(ℓ)

Figure 2-12 (*Continued*)

Because the bandpass attenuation must increase at 20 dB/decade as $\omega \to \infty$, the transfer function must go to zero as $1/s$ as $s \to \infty$. Consequently, $n = 3$. Note that although the transformation in Eq. (2-25) converts an mth-order bandpass function into a lowpass function of order $m/2$, $H_{LP}(\bar{s})$ of Eq. (2-26) must go to zero as $1/\bar{s}$, because H_{LP} and H_{BP} have the same asymptotic behavior for large $|s|$. Finally, we need $|H_{LP}(j1)|^2 = 0.5$ to give $a_n = 64$. The magnitude squared of the prototype lowpass function is therefore

$$|H_{LP}(j\bar{\omega})|^2 = \frac{\frac{1}{64}(9 - \bar{\omega}^2)^2}{\bar{\omega}^6 + \frac{1}{64}(9 - \bar{\omega}^2)^2} \tag{2-27a}$$

and, after factoring, we obtain

$$H(\bar{s}) = \frac{\frac{1}{8}(\bar{s}^2 + 9)}{\bar{s}^3 + 2.04184\bar{s}^2 + 2.07674\bar{s} + 1.125} \tag{2-27b}$$

Also, from Eqs. (2-11) and (2-12), we find

$$|\rho(j\bar{\omega})|^2 = \frac{\bar{\omega}^6}{\bar{\omega}^6 + \frac{1}{64}(9 - \bar{\omega}^2)^2} \tag{2-28a}$$

$$\rho(\bar{s}) = \frac{\bar{s}^3}{\bar{s}^3 + 2.04184\bar{s}^2 + 2.07674\bar{s} + 1.125} \tag{2-28b}$$

and

$$Z_{in}(\bar{s}) = \frac{2.04184\bar{s}^2 + 2.07674\bar{s} + 1.125}{2\bar{s}^3 + 2.04184\bar{s}^2 + 2.07674\bar{s} + 1.125} \tag{2-29}$$

Note that the terminating resistors are normalized to $R_S = R_L = 1$. We observe that the losspole polynomial

$$N(\bar{s}) = \frac{1}{8}(\bar{s}^2 + 9)$$

is even, so that, with Eq. (2-29), from Table 2.1 we find

$$z_{11} = \frac{2.04184\bar{s}^2 + 1.125}{2\bar{s}^3 + 2.07674\bar{s}} = z_{22}; \qquad z_{12} = \frac{(\bar{s}^2 + 9)/8}{2\bar{s}^3 + 2.07674\bar{s}}$$

$$y_{11} = \frac{2.04184\bar{s}^2 + 1.125}{2.07674\bar{s}} = y_{22}; \qquad y_{12} = \frac{(\bar{s}^2 + 9)/8}{2.07674\bar{s}}$$

According to our earlier discussion, we should expect a symmetrical network, because $z_{11} = z_{22}$ (and $y_{11} = y_{22}$). Note also that $Z_{in}(\infty) = 0$, so that as predicted the z-parameters have no pole and the y-parameters have a noncompact pole at infinity. Since we need to work with the z-parameters (they are of higher order!) let us sketch the pole-zero diagrams for z_{11} and z_{12} (Fig. 2-12b, c) and contemplate a realization strategy. Clearly, z_{11} is not useful for our purpose, because there is no zero to be shifted to $\bar{\omega} = 3$ and no pole to be removed partially to effect this shift. Therefore, we invert z_{11} to get $y_1 = 1/z_{11}$ and remove the pole at infinity partially (Fig. 2-12d, e). We can then invert y_2 and remove the pole of z_2 at $\bar{\omega} = 3$ completely to realize the transmission zero (Fig. 2-12f). The remaining function z_3 is inverted once again in order to be able to realize the transmission zero at infinity by removing completely the pole of y_3 at infinity. Note also that thereby we have removed a pole completely at the frequency where previously a partial removal was performed. This last pole removal reduces the remaining immittance to zero and completes the LP synthesis.

Since our approach has succeeded, we can now proceed with the numerical calculations:

$$k_1 = \lim_{\bar{s}^2 \to -9} \frac{1}{\bar{s}} y_1(\bar{s}) = \frac{-18 + 2.07674}{-9 \cdot 2.04184 + 1.125} = 0.92300$$

$$y_2 = y_1 - k_1\bar{s} = \frac{0.11537(\bar{s}^2 + 9)\bar{s}}{2.04184\bar{s}^2 + 1.125}$$

$$z_2 = \frac{1}{y_2} = \frac{16.6147\bar{s}}{\bar{s}^2 + 9} + \frac{1}{0.92300\bar{s}}$$

$$y_3 = 0.92300\bar{s}$$

Figure 2-12i shows the circuit and the element values of the prototype lowpass filter in the $\bar{s}$-domain. Note that the circuit is symmetrical, i.e., $z_{22} = z_{11}$, so that we may deduce immediately that the realized gain constant is $K = 1$. To verify, observe that at very low frequencies $z_{12r}(\bar{s}) = 1/[\bar{s}(C_1 + C_3)] = 0.54171/\bar{s}$, which equals $z_{12p}(\bar{s})$ for $\bar{s} \to 0$.

We now have to attend to the transformation back to the bandpass (Step 8 of our summary). By use of Eq. (2-23) with $Q = 3.4641$ as found earlier, we obtain the normalized bandpass filter of Fig. 2-12j. Note that the parallel resonance circuit in Fig. 2-12i resonates at $\bar{\omega} = 3$ and that all resonance circuits in Fig. 2-12j resonate at $\omega = 1$, i.e., at the passband center of the BP filter. We know, however, that the four-element series branch of the BP ladder must yield the two prescribed transmission zeros at $\omega_{z1} = 2\pi f_{z1}/\Omega_0 = 1.52274$ and $\omega_{z2} = 2\pi f_{z2}/\Omega_0 = 0.65671$, i.e., the series branch must have parallel resonances at ω_{z1} and ω_{z2}. To make this fact more evident (and incidentally result in a more practical circuit that is easier to tune), recall that the series impedance in the LP ladder was

$$Z_L(\bar{s}) = \frac{16.6147\bar{s}}{\bar{s}^2 + 9}$$

which with Eq. (2-25) is converted to the series impedance of the BP ladder as follows:

$$Z_B(s) = \frac{16.6147Q(s^2 + 1)/s}{Q^2(s^2 + 1)^2/s^2 + 9} = \frac{(16.6147/Q)s(s^2 + 1)}{s^4 + (2 + 9/Q^2)s^2 + 1} = \frac{4.7963s(s^2 + 1)}{(s^2 + 0.6567^2)(s^2 + 1.5227^2)}$$

$$= \frac{3.3511s}{s^2 + 1.5227^2} + \frac{1.4452s}{s^2 + 0.6567^2}$$

Clearly, $Z_B(s)$ is an open circuit at ω_{z1} and at ω_{z2} and yields the circuit in Fig. 2-12k.

The design is completed by denormalizing the element values according to Eq. (1-42) with $\Omega_0 = 2\pi \cdot 1039.23/s$ and $R_0 = 3$ kΩ. We therefore multiply the resistors by 3kΩ, the inductors by $R_0/\Omega_0 = 459.44$ mH, and the capacitors by $1/(R_0\Omega_0) = 51.05$ nF to get the final circuit in Fig. 2-12l.

Because of the large computational effort that *LC* ladder design entails and because not all synthesis approaches will lead to a successful completion of an attempted design, it is of considerable benefit to go through the complete synthesis strategy with the aid of pole-zero diagrams before performing any of the numerical computations. The next example will demonstrate this fact. Also, the example will illustrate the case where an internal pole has to be partially removed as shown in Fig. 2-9c.

Example 2-8

Realize the transfer function obtained in Example 2-5.

Solution. In Example 2-5 we found the transfer function $H(s)$ and the z- and y-parameters. The z-parameters, repeated here for convenience,

$$z_{11}(s) = z_{22}(s) = \frac{4.58173s^4 + 14.4729s^2 + 8.58833}{2s^5 + 9.47170s^3 + 11.2010s}$$

$$= \frac{4.58173(s^2 + 0.79197)(s^2 + 2.36686)}{2s(s^2 + 2.32227)(s^2 + 2.41358)}$$

$$z_{12}(s) = \frac{1.43139(s^2 + 2)(s^2 + 3)}{2s(s^2 + 2.32227)(s^2 + 2.41358)}$$

are of higher order than the y-parameters, and have to be used as a starting point for the synthesis. Figure 2-13a and b contains the pole-zero diagrams of z_{12} and z_{11} on the ω^2-axis. It is apparent that z_{11} must be inverted to obtain $y_1 = 1/z_{11}$ as shown in Fig. 2-13c. There exist now different options of how to proceed:

1. Completely remove (c.r.) the pole at ∞ to realize the transmission zero at ∞; the remainder function y_2 and its inverse z_2 are shown in Fig. 2-13d and e. We can then remove partially (p.r.) the pole at ∞ to shift the zero of z_2 at $\sqrt{2.37}$ to $\sqrt{3}$, invert z_3 and remove the resulting pole completely to realize the transmission zero at $\sqrt{3}$ (Fig. 2-13f and g). From the inverse z_4 of the remainder function y_4 (Fig. 2-13h and i) we can now partially remove the pole at ∞ to shift the zero to $\sqrt{2}$ and finally remove completely the resulting pole of y_5. The structure of the obtained ladder is shown to the right of Fig. 2-13k. Unfortunately, this realization is incorrect because it realizes a *double* transmission zero at infinity contrary to our requirement. By considering the circuit, the student may be convinced that the second partial pole removal (inductor L_3) causes a transmission zero at ∞ because we have violated the rule that a complete pole removal at the same frequency must at a later stage follow any partial pole removal. Thus, we need to go back to y_1, redrawn in Fig. 2-13l.

2. Remove the pole at ∞ only partially (p.r.) to shift the zero at $\sqrt{2.41}$ to $\sqrt{3}$. Figure 2-13m and n show the corresponding diagrams for the remainder y_6 and its inverse z_6 from which the pole at $\sqrt{3}$ can be removed completely to realize the transmission zero at $\sqrt{3}$. The remainder function z_7 and its inverse y_7 are shown in Fig. 2-13o and p. At this point we have a problem. If the zero of z_6 at $\sqrt{0.79}$ shifted to a location $> \sqrt{2}$ in z_7, we could partially remove the pole at $s = 0$ to pull it back to $\sqrt{2}$, *but* this partial pole removal creates a transmission

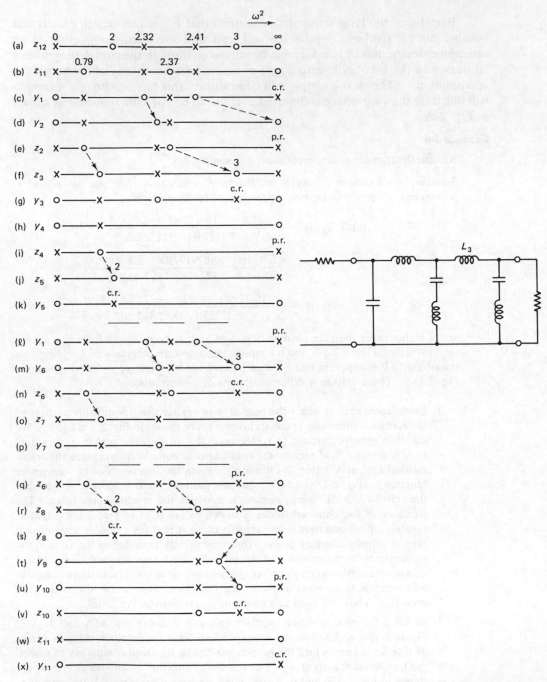

Figure 2-13 (a–x) Pole-zero diagrams for Example 2-8; (y) final *LC* filter circuit and denormalized element values.

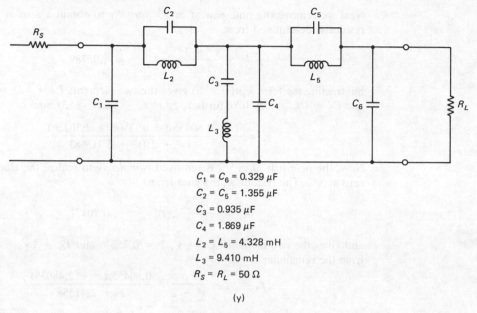

$$C_1 = C_6 = 0.329 \ \mu F$$
$$C_2 = C_5 = 1.355 \ \mu F$$
$$C_3 = 0.935 \ \mu F$$
$$C_4 = 1.869 \ \mu F$$
$$L_2 = L_5 = 4.328 \ mH$$
$$L_3 = 9.410 \ mH$$
$$R_S = R_L = 50 \ \Omega$$

(y)

Figure 2-13 (*Continued*)

zero at the origin because no complete removal at $s = 0$ will follow! If, on the other hand, the zero in z_7 is at a location $< \sqrt{2}$, we can only remove partially the internal pole to shift the zero to $\sqrt{2}$, *but* this causes again a transmission zero at the partially removed pole because a full pole removal at that frequency will not follow. The solution to this dilemma is illustrated next as option 3.

3. Return to z_6 in Fig. 2-13n and remove the pole at $\sqrt{3}$ only *partially* such that the zero at $\sqrt{0.79}$ shifts to $\sqrt{2}$ (Fig. 2-13q and r). From the inverse y_8 of the remainder function z_8 we remove the pole at $\sqrt{2}$ completely, realizing the transmission zero at $\sqrt{2}$. This results in a shift of the zero at $\sqrt{3}$ to a point $< \sqrt{3}$ in y_9 (Fig. 2-13t), but it can be pulled back to $\sqrt{3}$ by partially removing the pole at ∞ (Fig. 2-13u). Inverting y_{10}, we remove completely the zero at $\sqrt{3}$, thereby realizing the transmission zero at $\sqrt{3}$ and assuring that the previous *partial* pole removal at $\sqrt{3}$ does not create a transmission zero. Finally, we invert the remainder z_{11} and remove completely the pole at ∞ of y_{11}, thereby realizing the last prescribed transmission zero (Fig. 2-13w and x). Figure 2-13y shows the network topology, and we are now ready for the numerical calculations (rounded to five decimal places).

The first *partial* pole removal from y_1 gives a residue

$$k_1 = \left. \frac{1}{s z_{11}(s)} \right|_{s^2 = -3} = 0.12410 = C_1$$

Thus, $y_6 = y_5 - k_1 s$ and

$$z_6 = \frac{1}{y_6} = \frac{4.58173 s^4 + 14.4729 s^2 + 8.58833}{1.43143 s (s^2 + 3)(s^2 + 2.36226)}$$

Next we remove the pole pair at $\pm\sqrt{3}$ *partially* to obtain a zero at $\sqrt{2}$. The residue is determined from

$$k_2 = \frac{s^2 + 3}{s} z_6(s)\Big|_{s^2 = -2} = 1.95789$$

Subtracting the term $k_2 s/(s^2 + 3)$ gives the two elements $L_2 = k_2/3 = 0.65263$ and $C_2 = 1/k_2 = 0.51076$; further, $z_8 = z_6 - k_2 s/(s^2 + 3)$, and

$$y_8 = \frac{1}{z_8} = \frac{0.80455s(s^2 + 3)(s^2 + 2.36226)}{(s^2 + 2)(s^2 + 2.41358)}$$

Now, the pole pair at $\pm\sqrt{2}$ is removed *completely* to realize the transmission zero at $\sqrt{2}$. The residue k_3 is found from

$$k_3 = \frac{s^2 + 2}{s} y_8(s)\Big|_{s^2 = -2} = 0.70471$$

and gives the two elements $C_3 = k_3/2 = 0.35236$ and $L_3 = 1/k_3 = 1.41902$; from the remainder

$$y_9 = y_8 - \frac{k_3 s}{s^2 + 2} = \frac{0.80455s(s^2 + 2.48635)}{s^2 + 2.41358}$$

we remove *partially* the pole at infinity to shift the zero from $\sqrt{2.486}$ to $\sqrt{3}$:

$$k_4 = \frac{y_9(s)}{s}\Big|_{s^2 = -3} = 0.70471 = C_4$$

so that $y_{10} = y_9 - k_4 s$ and

$$z_{10} = \frac{1}{y_{10}} = \frac{10.01617(s^2 + 2.41358)}{s(s^2 + 3)}$$

Finally, a partial fraction expansion of z_{10} yields

$$z_{10} = \frac{1.95789s}{s^2 + 3} + \frac{8.05828}{s}$$

and the elements $L_5 = L_2$, $C_5 = C_2$, and $C_6 = C_1$! Clearly, the circuit is structurally and electrically symmetrical, from which we conclude immediately that $z_{11} = z_{22}$ and the realized gain constant $K = 1$. The student is urged to carry out the omitted numerical details in order to gain some appreciation of the complexity of the computations.

The remaining task is to denormalize the element values as discussed in Section 1.4 with $R_0 = 50 \, \Omega$ and $f_c = 1200$ Hz from Example 2-5. The resulting component values are listed in Fig. 2-13y.

Let us use one last example to demonstrate that the synthesis process becomes much simpler for *all-pole LC* lowpass functions, such as Butterworth, Bessel, or Chebyshev filters. In that case the numerator of the transfer immittance parameter z_{12} or y_{12} is a constant and all transmission zeros are at infinity, where the *LC* synthesis immittances have either a pole or a zero. Assume that there is a pole (otherwise start from the inverse!); removing this pole *completely* realizes one of the multiple transmission zeros *and* shifts a zero of the remainder immittance function to infinity. Thus, we invert this function and remove the resulting pole at ∞ *completely*, realizing thereby the second transmission zero *and* shifting a new zero of the immittance to ∞, and so on, until the realization is complete. Referring back to Example 2-2, the student will observe that this process amounts to expanding the synthesis immittance into a *continued fraction* around $s = \infty$. No partial pole removal is involved.

Example 2-9

Realize a fifth-order resistively terminated Butterworth LP filter with a 3-dB frequency at $\omega = 1$. Let $R_S = R_L = 1$.

Solution.   From Table III-1 in Appendix III we find

$$H(s) = \frac{1}{s^5 + 3.2361s^4 + 5.2361s^3 + 5.2361s^2 + 3.2361s + 1}$$

with

$$|H(j\omega)|^2 = \frac{1}{1 + \omega^{10}}$$

From Eq. (2-11) we obtain

$$|\rho(j\omega)|^2 = \frac{\omega^{10}}{1 + \omega^{10}}$$

and

$$\rho(s) = \frac{s^5}{s^5 + 3.2361s^4 + 5.2361s^3 + 5.2361s^2 + 3.2361s + 1}$$

so that, by Eq. (2-12),

$$Z_{in}(s) = \frac{3.2361(s^4 + s) + 5.2361(s^3 + s^2) + 1}{2s^5 + 3.2361s^4 + 5.2361s^3 + 5.2361s^2 + 3.2361s + 1}$$

and, from Table 2.1, with $N(s) = 1 =$ even:

$$z_{11} = z_{22} = \frac{3.2361s^4 + 5.2361s^2 + 1}{2s^5 + 5.2361s^3 + 3.2361s} \qquad z_{12} = \frac{1}{2s^5 + 5.2361s^3 + 3.2361s}$$

For completeness, Figure 2-14a contains the usual pole-zero diagrams to show the removal process. Note that there are *five* complete pole removals at ∞ to realize the transmission zero of *multiplicity* 5. Clearly, we start the process from $y_1 = 1/z_{11}$, remove the pole at $s = \infty$, invert the remainder (convert $y_2 = y_1 - k_\infty s$ into $z_2 = 1/y_2$), and so on. The resulting ladder in Fig. 2-14b shows that

$$y_1 = sC_1 + \cfrac{1}{sL_2 + \cfrac{1}{sC_3 + \cfrac{1}{sL_4 + 1/sC_5}}}$$

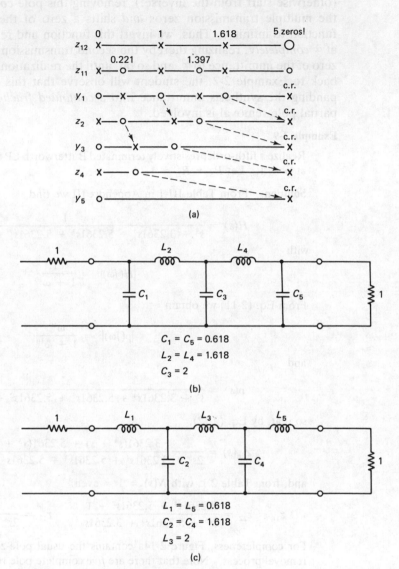

(a)

$C_1 = C_5 = 0.618$
$L_2 = L_4 = 1.618$
$C_3 = 2$

(b)

$L_1 = L_5 = 0.618$
$C_2 = C_4 = 1.618$
$L_3 = 2$

(c)

Figure 2-14   (a) Pole-zero diagrams (*note:* the scale is ω^2); (b) lowpass "minimum-inductance" ladder; (c) lowpass "minimum capacitance" ladder.

A continued fraction expansion of $1/z_{11}$ gives

$$y_1 = 0.6180s + \cfrac{1}{1.6180s + \cfrac{1}{2.0000s + \cfrac{1}{1.6180s + 1/0.6180s}}}$$

so that a comparison yields the element values

$$C_1 = C_5 = 0.618 \qquad L_2 = L_4 = 1.618 \qquad C_3 = 2.000$$

To illustrate a further point, recall that we removed the sign ambiguity in Eqs. (2-11b) and (2-12) by selecting (arbitrarily) the upper sign: in this example, $\hat{F}(s) = +s^5$. Had we chosen the lower sign, i.e., $\hat{F}(s) = -s^5$, we would have obtained

$$Z_{in}(s) = \frac{2s^5 + 3.2361(s^4 + s) + 5.2361(s^3 + s^2) + 1}{3.2361(s^4 + s) + 5.2361(s^3 + s^2) + 1}$$

i.e., we would have replaced the equation for Z_{in} by $1/Z_{in}$. Table 2.1 in this case results in

$$y_{11} = y_{22} = \frac{3.2361s^4 + 5.2361s^2 + 1}{2s^5 + 5.2361s^3 + 3.2361s} \qquad -y_{12} = \frac{1}{2s^5 + 5.2361s^3 + 3.2361s}$$

with a circuit that is the *dual* of the previous ladder (Fig. 2-14c) and is obtained by a continued fraction expansion of $z_1 = 1/y_{11}$. Thus:

$$L_1 = L_3 = 0.618 \qquad C_2 = C_4 = 1.618 \qquad L_3 = 2.000$$

As shown in Example 2-9, replacing $+\hat{F}(s)$ in Eq. (2-12) by $-\hat{F}(s)$ replaces the expression for Z_{in} by its reciprocal and interchanges the roles of the polynomials m_1, m_2 and n_1, n_2 in Table 2.1. Readers may convince themselves, therefore, that replacing $+\hat{F}(s)$ by $-\hat{F}(s)$ results, in general, in dual networks.

Finally, let us close the discussion of *LC* ladder filters by mentioning that *explicit formulas* are available for all-pole maximally flat and equiripple (Chebyshev) lowpass filters [11, 26]. Let n be the degree of the lowpass function and let ε be the ripple factor [Eq. (1-66a)]:

$$\varepsilon = \sqrt{10^{0.1A_p} - 1}$$

The elements for the *maximally flat LP minimum inductance ladder* (see Fig. 2-14b) are

$$\left.\begin{array}{l} C_k \text{ (k odd)} \\ L_k \text{ (k even)} \end{array}\right\} = 2(\varepsilon^{1/n}) \sin\left(\frac{2k - 1}{2n}\pi\right) \tag{2-30}$$

with $R_S = R_L = 1$. As was just pointed out, the minimum capacitance ladder (see Fig. 2-14c) is simply the dual circuit.

The formulas for the Chebyshev LP ladder are a little more complicated.

First, we need to compute the quantity

$$q = \frac{1}{2}\left[\left(\frac{1}{\varepsilon} + \sqrt{\frac{1}{\varepsilon^2} + 1}\right)^{1/n} - \left(\frac{1}{\varepsilon} + \sqrt{\frac{1}{\varepsilon^2} + 1}\right)^{-1/n}\right]$$

that appeared already in Eq. (1-80). Then we have for the *Chebyshev LP minimum inductance ladder*, for $k = 1, 2, \ldots, k_m$, where k_m is the largest integer less than $n/2$,

$$C_1 = \frac{2}{qR_S} \sin \frac{\pi}{2n} \tag{2-31a}$$

$$C_{2k-1}L_{2k} = \frac{4 \sin\left(\frac{4k - 1}{2n} \pi\right) \sin\left(\frac{4k - 3}{2n} \pi\right)}{q^2 + \sin^2\left(\frac{2k - 1}{n} \pi\right)} \tag{2-31b}$$

$$C_{2k+1}L_{2k} = \frac{4 \sin\left(\frac{4k - 1}{2n} \pi\right) \sin\left(\frac{4k + 1}{2n} \pi\right)}{q^2 + \sin^2\left(\frac{2k}{n} \pi\right)} \tag{2-31c}$$

Also, the end elements are given by

$$\left.\begin{array}{l} R_L C_n \ (n \text{ odd}) \\ \dfrac{1}{R_L} L_n \ (n \text{ even}) \end{array}\right\} = \frac{2}{q} \sin \frac{\pi}{2n} \tag{2-31d}$$

We observe here that the elements of the Chebyshev filter depend on R_S and R_L and we note that $R_S = R_L$ is possible in *LC* lowpass ladders *only* if there is a reflection zero at the origin, as is the case in Butterworth and in *odd-order* Chebyshev filters. To realize this, recall from Eq. (2-7) that

$$|H(j\omega)| = 2\sqrt{\frac{R_S}{R_L}}\left|\frac{V_2}{V_S}\right| \leq 1$$

so that

$$\left|\frac{V_2}{V_S}(j\omega)\right| \leq \frac{1}{2}\sqrt{\frac{R_L}{R_S}}$$

where the equality sign holds at the reflection zeros. In Butterworth and odd-order Chebyshev lowpass ladders, $s = 0$ is a point of maximum power transfer where we have, from the circuit,

$$\left|\frac{V_2}{V_S}(j0)\right| = \frac{R_L}{R_S + R_L} = \frac{1}{2}\sqrt{\frac{R_L}{R_S}}$$

and therefore $R_L = R_S$. In *even-order* Chebyshev filters, however, $s = 0$ is *not* a reflection zero. Rather, we have

$$\left|\frac{V_2}{V_S}(j0)\right| = \frac{1}{\sqrt{1 + \varepsilon^2}}\frac{1}{2}\sqrt{\frac{R_L}{R_S}}$$

which from the ladder should equal $R_L/(R_S + R_L)$. Therefore

$$\left(\frac{R_S}{R_L} + 1\right)^2 = 4\frac{R_S}{R_L}(1 + \varepsilon^2)$$

which can be solved to give

$$R_L = [1 + 2\varepsilon^2 \pm 2\varepsilon\sqrt{1 + \varepsilon^2}]R_S \qquad (2\text{-}32)$$

The two values for R_L are related by $R_{L1}R_{L2} = R_S^2$; either one may be taken, the larger value of R_L resulting in a higher voltage level according to max $|V_2/V_S| = 0.5\sqrt{R_L/R_S}$ *without* changing the power transfer.

2.5 REALIZATION OF LC ALLPASS CIRCUITS

We saw in Section 1.5.2 that the zeros of an allpass function are located in the right half *s*-plane and are mirror images of the poles, as expressed by Eq. (1-46), repeated here for convenience:

$$H(s) = \frac{D(-s)}{D(s)} \qquad (2\text{-}33)$$

We also know that the transmission zero polynomial $N(s)$ must have all its roots on the $j\omega$-axis if the transfer function is to be realizable as an *LC* ladder. We can conclude, therefore, that allpass functions cannot be realized as lossless ladder networks but that different circuit structures must be looked for. A suitable topology is the symmetrical *LC lattice* shown in Fig. 2-15, where the *LC* impedances $Z_1(s)$ and $Z_2(s)$ satisfy the relationship

$$Z_1(s)Z_2(s) = R_0^2 \qquad (2\text{-}34)$$

By finding the *y*-parameters

$$y_{11} = y_{22} = \tfrac{1}{2}(Y_1 + Y_2) \qquad y_{12} = \tfrac{1}{2}(Y_2 - Y_1)$$

and using Eq. (2-34), the student can verify that the voltage transfer function

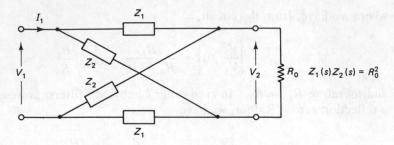

Figure 2-15   Constant-resistance lossless lattice.

V_2/V_1 equals

$$H(s) = \frac{V_2}{V_1} = \frac{1 - Z_1/R_0}{1 + Z_1/R_0} = \frac{Z_2/R_0 - 1}{Z_2/R_0 + 1} \qquad (2\text{-}35)$$

Further, the input impedance V_1/I_1 is found to be

$$Z_{in}(s) = \frac{V_1}{I_1} \equiv R_0 \qquad (2\text{-}36)$$

i.e., $Z_{in}(s)$ is constant, independent of frequency. The circuit in Fig. 2-15 is called, therefore, *a constant-resistance lattice*. Now, recall that Z_1 and Z_2 are *LC* impedances, so that, by Eq. (2.2), on the $j\omega$-axis $Z_1(j\omega)/R_0$ is purely imaginary. Consequently, from Eq. (2-35), on the $j\omega$-axis V_2/V_1 is the ratio of two conjugate complex numbers, so that

$$\left| \frac{V_2}{V_1}(j\omega) \right| \equiv 1 \qquad (2\text{-}37)$$

i.e., the constant-resistance lossless lattice is an allpass filter whose input impedance equals the load resistance for all frequencies.

The student will recognize that this is a useful property because it means that:

1. A cascade connection of several constant-resistance lattices as in Fig. 2-16a has a voltage transfer function that is simply the product of the transfer functions of the individual blocks,

$$\frac{V_o}{V_1} = \frac{V_2}{V_1} \frac{V_3}{V_2} \cdots \frac{V_n}{V_{n-1}} \frac{V_o}{V_n} \qquad (2\text{-}38)$$

because each block sees a load R_0.

2. If an *LC* ladder has been designed with a terminating resistor R_L and we have to correct its phase or delay as discussed in Section 1.5.2, we may simply replace R_L by a (cascade of) constant-resistance lattice(s) terminated in R_L as shown in Fig. 2-16b. The operation of the ladder will not be affected, because its load is not changed.

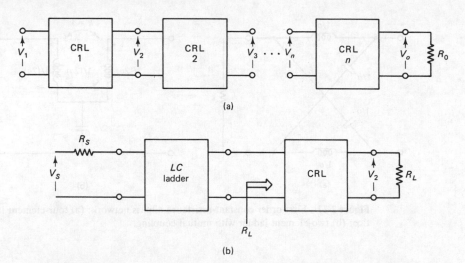

(a)

(b)

Figure 2-16   Constant-resistance lattice (CRL) blocks in cascade (a) and as phase-correcting load of an *LC* ladder (b).

Turning now to the realization of a high-order allpass function of degree n

$$H_n(s) = \frac{D_n(-s)}{D_n(s)} \tag{2-39}$$

we note first that the polynomial $D_n(s)$ may be factored into second-order terms of the form

$$s^2 + s\frac{\omega_i}{Q_i} + \omega_i^2 \tag{2-40a}$$

if n is even, and into such second-order terms and a first-order term

$$s + \sigma \tag{2-40b}$$

if n is odd.   In any event, the total function in Eq. (2-39) is factored into the product of second-order allpass functions of the form of Eq. (1-58a) with $K = 1$ and, if n is odd, a first-order function:

$$H_n(s) = \frac{\sigma - s}{\sigma + s} \cdot \prod_i \frac{s^2 - s\,\omega_i/Q_i + \omega_i^2}{s^2 + s\,\omega_i/Q_i + \omega_i^2} = H_1(s) \cdot \prod H_{2i}(s) \tag{2-41}$$

According to Eq. (2-38) and Fig. 2-16a, the individual functions H_1 and H_{2i} can then be realized separately and H_n is obtained by cascading.   Therefore, our realization problem is solved if we know how to handle first- and second-order allpass functions.   In the following discussion, let us assume that R_0 is normalized such that $R_0 = 1$.

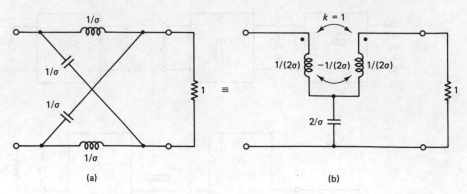

Figure 2-17 First-order constant-resistance allpass network: (a) four-element lattice; (b) two-element ladder with mutual coupling.

First-order allpass networks. To bring $H_1(s)$ into the form of Eq. (2-35), we write

$$H_1(s) = \frac{\sigma - s}{\sigma + s} = \frac{1 - s/\sigma}{1 + s/\sigma} \qquad (2\text{-}42)$$

and recognize

$$Z_1(s) = \frac{1}{Z_2(s)} = \frac{s}{\sigma} \qquad (2\text{-}43)$$

The realization is shown in Fig. 2-17a with a common-ground three-terminal equivalent in Fig. 2-17b. Routine analysis should convince the student that the circuits in Fig. 2-17a and b are indeed equivalent. Note that the allpass in Fig. 2-17b contains only two elements: one capacitor and one unity-coupled ($k = 1$) transformer with $L_1 = L_2 = -M = 1/(2\sigma)$.

Second-order allpass networks. To bring $H_{2i}(s)$ into the form of Eq. (2-35), we write

$$H_{2i}(s) = \frac{s^2 - s\omega_i/Q_i + \omega_i^2}{s^2 + s\omega_i/Q_i + \omega_i^2} = \frac{1 - \dfrac{s\omega_i/Q_i}{s^2 + \omega_i^2}}{1 + \dfrac{s\omega_i/Q_i}{s^2 + \omega_i^2}} \qquad (2\text{-}44)$$

and recognize

$$Z_1(s) = \frac{1}{Z_2(s)} = \frac{s\omega_i/Q_i}{s^2 + \omega_i^2} \qquad (2\text{-}45)$$

with the realization shown in Fig. 2-18a. Again, a common-ground, three-element (two capacitors, one transformer) equivalent circuit exists as shown in Fig. 2-18b, where the transformer "dots" are in positions ● ● or + +, depending on whether the mutual inductance

$$M = \frac{Q_i}{2\omega_i}\left(1 - \frac{1}{Q_i^2}\right) \tag{2-46a}$$

is positive or negative, respectively. Observe that the coefficient of coupling is

$$k = \frac{|M|}{L} = \frac{|Q_i^2 - 1|}{Q_i^2 + 1} < 1 \tag{2-46b}$$

i.e., the transformer is a *real* transformer with nonunity coupling.

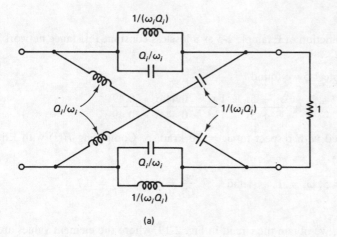

(a)

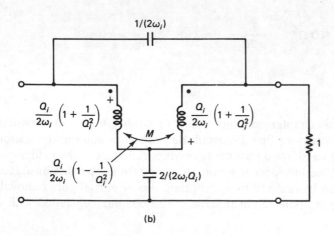

(b)

Figure 2-18 Second-order constant-resistance allpass network: (a) eight-element lattice; (b) three-element ladder with mutual coupling.

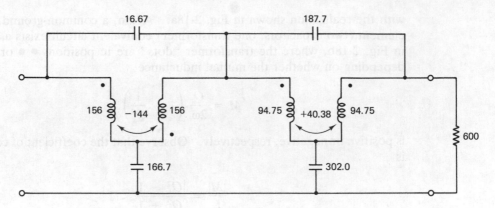

Figure 2-19   Realized circuit for Example 2-10; elements in Ω, mH, nF.

Example 2-10

Realize the allpass function of Example 1-5 as a lossless constant-resistance network. Assume $R_0 = 600 \ \Omega$.

Solution.   In Example 1-5 we found

$$H(s) = \frac{s^2 - 5s + 1}{s^2 + 5s + 1} \frac{s^2 - 0.444s + 0.49}{s^2 + 0.444s + 0.49}$$

where s is normalized with respect to $\omega_p = 10$ krad/s. Comparing $H(s)$ with Eq. (2-44), we obtain

$$\frac{\omega_1}{Q_1} = 5, \ \omega_1 = 1 \qquad \text{and} \qquad \frac{\omega_2}{Q_2} = 0.444, \ \omega_2 = 0.7$$

With these numbers, we obtain the circuit in Fig. 2-19, where the element values are denormalized by

$$R_0 = 600 \ \Omega \qquad \frac{1}{\omega_p R_0} = 166.7 \ \text{nF} \qquad \frac{R_0}{\omega_p} = 60 \ \text{mH}$$

2.6 SUMMARY

Our main concern in this chapter was the design of doubly terminated lossless ladder filters. In this context we first discussed the properties and the realization of *LC* immittances, then we showed how the twoport parameters of the ladder can be derived from a prescribed transfer function $H(s)$, and finally we demonstrated that these parameters can be realized by appropriate use of partial pole removal, zero shifting, and complete pole removal steps. The chapter concluded with a brief discussion of the implementation of lossless allpass networks for phase or delay equalization. After carefully reading this chapter, including the illustrative design examples, the student should be in a position to realize practical LC filters with quite general transmission requirements. However, we must emphasize again that the availability of suitable filter design software as well as computer-aided approximation routines becomes almost unavoidable if filter order increases beyond, say, 5 or 6, and if the classical filter functions should prove to be inappropriate.

As pointed out in Section 2.1, the treatment had to be very concise and limited to the most important aspects of LC filter design. Thus, for example, we avoided any discussion of singly terminated (only R_S or R_L) ladders [24] because, although much easier to understand, these networks are of far less practical significance.

In addition to supplying the tools necessary for the design of passive LC filters, this chapter serves the important function of providing motivation and aiding the student's insight into the methods used and the circuit topologies obtained when we discuss active simulations of passive LC ladder structures in Chapters 6, 7, and 8.

PROBLEMS

2.1 Prove that if $Z(s) = N(s)/D(s)$ is a real rational function of s (all coefficients are real), then $Z(s)$ must be an odd function if $Z(j\omega)$ is purely imaginary.

2.2 Synthesize the admittance

$$Y(s) = \frac{(s^2 + 1)(s^2 + 9)(s^2 + 25)}{s(s^2 + 4)(s^2 + 16)}$$

(a) By expanding $Y(s)$ into partial fractions (Foster II form);
(b) By expanding $Z(s) = 1/Y(s)$ into partial fractions (Foster I form);
(c) By expanding $Y(s)$ into continued fractions around $s = 0$ (Cauer II form);
(d) By expanding $Y(s)$ into continued fractions around $s = \infty$ (Cauer I form).
(e) In addition to the four canonic forms of parts (a) through (d), expand $Y(s)$ into the "fifth" canonic form shown in Fig. P2.2.

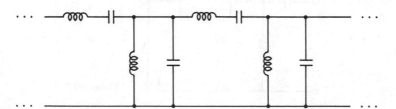

Figure P2.2

2.3 Realize an *LC* impedance to satisfy the following requirements: very small at dc; infinite at $\omega_1 = 900$, $\omega_2 = 1125$, and $\omega_3 = 1350$ rad/s; $j75 \ \Omega$ at $\omega_0 = 1000$ rad/s; $Z(s) = 1/(sC)$ at $\omega \to \infty$ where $C = 10 \ \mu F$.

2.4 Realize an impedance with the following properties:

$$Z(s) = 0 \text{ at } f_1 = 3 \text{ kHz and } f_2 = 7.5 \text{ kHz}$$
$$Z(s) = \infty \text{ at } f_3 = 2 \text{ kHz}, f_4 = 6 \text{ kHz, and } f_5 = 9 \text{ kHz}$$

At high frequencies, Z should behave like a capacitor of 200 nF. The impedance should be in such a form that a shunt capacitor appears across the input and that the zeros can be adjusted independently.

2.5 Realize the *LC* admittance

$$Y(s) = \frac{s^5 + 6s^3 + 8s}{s^4 + 4s^2 + 3}$$

in the form shown in Fig. P2.5, where $LC = 1$.

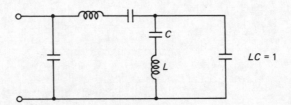

Figure P2.5

2.6 An *LC* impedance Z is to be built which has zeros at $f_1 = 800$ kHz and at ∞. It is also required that Z have a shunt capacitor of value $C_1 = 100$ pF across its input terminals, that any inductor L used satisfy $L \le 0.5$ mH, and that the largest capacitor ratio be ≤ 3. Obtain the function $Z(s)$ and the simplest possible realization.

2.7 Realize the *LC* impedance

$$Z(s) = \frac{6s^4 + 42s^2 + 48}{s^5 + 18s^3 + 48s}$$

in the form shown in Fig. P2.7.

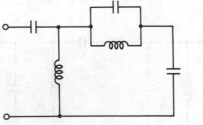

Figure P2.7

2.8 Find the z- and y-parameters of the bridged T and the twin T networks in Fig. P2.8.

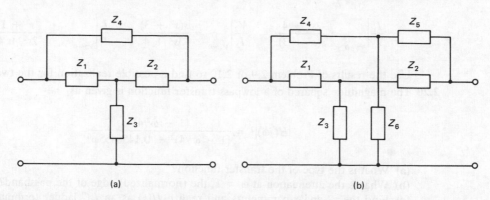

(a)   (b)

Figure P2.8   (a) Bridged T network.   (b) Twin T network.

2.9 Find the elements of the T-equivalent circuit in terms of the z-parameters.

2.10 Find the elements of the Π-equivalent circuit in terms of the y-parameters.

2.11 Derive Eqs. (2-16) and (2-17); express $H(s)$ and $Z_{in}(s)$ in terms of the y-parameters.

2.12 Using Eq. (2-8), prove that the magnitude of the reflection coefficient is very small in the filter passband.

2.13 Use the notation

$$Z(s) = \frac{m_1 + n_1}{m_2 + n_2} \cdot R_S$$

and Eq. (2-17) to derive the expressions in Table 2.1.   *Hint:* With $|z| = z_{11}z_{22} - z_{12}^2$, write Eq. (2-17) as

$$Z(s) = \frac{1 + |z|/(z_{11}R_L)}{1 + z_{22}/R_L} \cdot z_{11}$$

Bring Eq. (P2-13) into the same form, and use the fact that m_1/n_1, m_2/n_2, m_1/n_2, and m_2/n_1 are all LC immittance functions (as can be proven).   Use the same process and the result of Problem 2.11 for the y-parameter expressions.

2.14 Prove that all poles of the twoport parameters in Table 2.1 at finite frequencies are compact.   What is the residue condition for a pole at $s = \infty$?

2.15 Use the z-parameters to prove that a *partially* removed pole in a resistively terminated LC ladder realizes a transmission zero.

2.16 Find the z- and y-parameters of a symmetrial lattice (Fig. 2-15).   Derive Eqs. (2-35) and (2-36).

2.17 Show the equivalence of the circuits in Fig. 2.17a and b; repeat for Fig. 2-18a and b.

2.18 Use the results of Problem 2.16 and Eq. (2-15) to show that a symmetrical lattice can realize an arbitrary symmetrical LC twoport.

2.19 Measurements on a lossless twoport yield

$$\left.\frac{I_1}{V_1}\right|_{V_2=0} = \frac{2s^2 + 4}{s(s^2 + 3)} \qquad \left.\frac{V_2}{I_2}\right|_{V_1=0} = \frac{s(s^2 + 3)}{3s^2 + 4} \qquad \left.\frac{I_2}{I_1}\right|_{V_2=0} = -\frac{s^2 + 1}{2s^2 + 4}$$

Use the results of Problem 2.9 or 2.10 to find a possible realization for the twoport.

2.20 The magnitude squared of a lowpass transfer function is given as

$$|H(j\omega)|^2 = \frac{(1 - \omega^2/4)^2}{(1 - \omega^2/4)^2 + 0.1456455\omega^6}$$

(a) What is the type of the transfer function?
(b) What is the attenuation at $\omega = 1$, the (normalized) edge of the passband?
(c) Find the z- and y-parameters and realize $H(s)$ as an *LC* ladder terminated in $R_S = R_L = 1$.

2.21 The transfer function

$$H(\bar{s}) = \frac{0.017143(\bar{s}^2 + 6.5^2)}{\bar{s}^3 + 1.24978\bar{s}^2 + 1.53374\bar{s} + 0.72425}$$

describes a third-order prototype lowpass filter with a 0.5 dB–equiripple passband and 38 dB minimum stopband attenuation for $\bar{\omega} \geq 4$. The original bandpass has a passband in 9 kHz $\leq f \leq$ 12 kHz and a transmission zero at $f_1 = 4.5$ kHz. Source and load are 300 Ω. Find an *LC* ladder realization for the bandpass filter.

2.22 To be built is a sixth-order bandpass filter to satisfy the geometrically symmetrical requirements:
(a) Passband in 50 kHz $< f \leq$ 70 kHz with $A_p \leq 1$ dB; maximally flat magnitude
(b) Stopband in $f \geq$ 80 kHz with $A_s \geq 20$ dB
(c) One transmission zero at $f = f_0$ in the upper stopband
(d) Source and load terminations 600 Ω.

Find f_0, the minimum attenuation in the stopband, and realize the *LC* ladder.

2.23 Realize an all-pole *LC* lowpass ladder with a Chebyshev 1 dB–ripple passband in $0 \leq f \leq 3.4$ kHz and 40 dB minimum attenuation in $f > 6$ kHz. Source termination: $R_S = 1$ Ω. Find R_L for maximum power transfer.

2.24 Realize a Thomson filter for the following specifications:
(a) Source and load: 120 Ω
(b) Delay 30 μs with $\leq 1\%$ error up to 13 kHz
(c) Attenuation no larger than 2 dB up to 110 kHz

2.25 Realize the transfer function whose magnitude squared is

$$|H(j\omega)|^2 = \frac{[1 - (\omega/3.2)^2]^2}{[1 - (\omega/3.2)^2]^2 + 0.0993504\omega^6}$$

as an *LC* ladder terminated in $R_S = R_L = 1.2$ kΩ. ω is normalized with respect to $2\pi \cdot 60$ krad/s.

2.26 Realize a third-order highpass LC ladder filter to satisfy:
(a) Source and load: 600 Ω
(b) Maximally flat passband in $f \geq 32$ kHz and $A_p \leq 2$ dB
(c) Transmission zero at $f_0 = 10$ kHz
What is the minimum stopband attenuation?

2.27 Design an all-pole bandpass filter with a 1.7 dB–equiripple passband in 10 kHz $\leq f$ ≤ 18 kHz, stopband in $f \geq 26$ kHz with $A_s \geq 42$ dB, and $R_s = 750$ Ω, R_L chosen for maximum power transfer.

2.28 Design a lowpass filter to satisfy
(a) Passband: maximally flat in $0 \leq f \leq 3.3$ kHz, $A_p \leq 1.5$ dB
(b) Monotonic stopband with attenuation increase at a rate of 12 dB/octave as $f \to$ ∞
(c) transmission zero at $f_0 = 6$ kHz
(d) $R_S = R_L = 2$ kΩ

2.29 Design a doubly terminated LC ladder with $R_S = 2$ kΩ, equiripple passband in 20 kHz $\leq f \leq 30$ kHz with $A_p \leq 0.3$ dB, monotonic stopband in $f > 36$ kHz with $A_s \geq 40$ dB.   Use frequency transformation.

2.30 Design a highpass LC filter to operate between R_S and R_L both equal to 1 kΩ that meets the following specifications: passband, maximally flat in 15 kHz $< f$, $A_p \leq 2$ dB; stopband, $0 \leq f \leq 7$ kHz, $A_s \geq 40$ dB.

2.31 Design a second-order LC allpass filter whose delay $\tau(\omega)$ approximates $\tau(0) = 1$ μs in the maximally flat sense.

2.32 Design an allpass filter to meet:
(a) Maximally flat delay $\tau(0) = 90$ μs with delay error $\leq 0.8\%$ in $f \leq 9.2$ kHz
(b) Load resistor 1.3 kΩ

2.33 Realize a doubly terminated LC lowpass filter with $R_S = 2.4$ kΩ, and an elliptic magnitude response to satisfy $A_p \leq 0.9$ dB in $f \leq 3.4$ kHz and $A_s \geq 22$ dB in $f \geq 4.8$ kHz.   Find the required R_L.

2.34 An elliptic lowpass filter is to be designed to satisfy the following specifications:
(a) $A_p \leq 3$ dB in $0 \leq f \leq 9$ kHz
(b) $A_s \geq 28$ dB in $f \geq 12.4$ kHz
(c) Load resistor 1.5 kΩ

2.35 An elliptic bandpass is to be designed for the following specifications:
(a) Passband $A_p \leq 1.5$ dB in 2.4 kHz $\leq f \leq 3.8$ kHz
(b) Stopband $A_s \geq 24$ dB in $f \geq 4.14$ kHz
(c) $R_S = R_L = 3.2$ kΩ

2.36 Find a doubly terminated LC ladder to realize the transfer function of
(a) Problem 1.17   (e) Problem 1.28   (i) Problem 1.51
(b) Problem 1.18   (f) Problem 1.30   (j) Problem 1.54
(c) Problem 1.19   (g) Problem 1.36   (k) Problem 1.55
(d) Problem 1.21   (h) Problem 1.42

2.37 This problem concerns scaling the impedance level of ladder arms.
(a) Suppose that in a ladder you have a coil with n turns to make an inductor of value L and that the impedance seen to one side of it is $Z(s)$ as shown in Fig. P2.37a. Verify that this configuration is equivalent to the one in Fig. P2.37b, where the

coil is tapped at n_1 turns and the impedance is lowered by a factor $(n_1/n)^2$. Evidently, by tapping the available inductors you can scale the impedance level of ladder arms.

(b) In Fig. P2.37c and d, a T of ladder branches and its equivalent Π with two transformers are shown. Although valid in general, this transformation is normally used only when all impedances Z_i, $i = 1, 2, 3$, of the T are capacitors, i.e., when $Z_i = 1/(sC_i)$. For this case, verify that the two networks are equivalent if

$$C_a = C_1 \frac{n_1^2 C_3 + n_1(n_1 - n_2)C_2}{C_1 + C_2 + C_3} \qquad C_c = C_2 \frac{n_2^2(C_1 + C_3) - n_1 n_2 C_1}{C_1 + C_2 + C_3}$$

$$C_b = n_1 n_2 \frac{C_1 C_2}{C_1 + C_2 + C_3}$$

Verify also that

$$\frac{C_2 + C_3}{C_2} \geq \frac{n_2}{n_1} \geq \frac{C_1}{C_1 + C_3}$$

because the capacitors in the Π circuit must be positive.

One will rarely "use both transformers" in Fig. P2.37d; i.e., either n_1 or n_2 will be set equal to unity.

Suppose $n_1 = 1$; show that to *raise* the impedance Z_{11} to the right of transformer

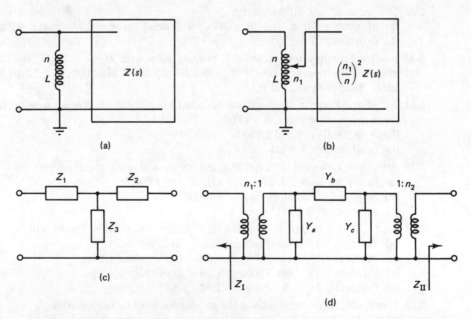

(a)   (b)

(c)   (d)

Figure P2.37 Network transformations to scale the impedance level of ladder branches. (a) Initial configuration and (b) scaling method for shunt ladder inductor in parallel with other ladder elements. (c) Ladder T section and (d) equivalent Π section with two transformers.

2, one can eliminate the transformer, set $C_2 = \infty$, and replace Z_{11} by Z_{11}/n_2^2 with

$$1 \geq n_2 \geq \frac{1}{1 + C_3/C_1}$$

Similarly, to lower Z_{11}, you set $C_1 = \infty$ so that $1 \leq n_2 \leq 1 + C_3/C_2$.

Analogously, show that to *lower* the impedance Z_1 to the left of transformer 1, one eliminates that transformer, sets $C_2 = \infty$, $n_2 = 1$ and replaces Z_1 by Z_1/n_1^2 with

$$1 \leq n_1 \leq 1 + \frac{C_3}{C_1}$$

Of course, one can raise Z_1 by setting $C_1 = \infty$ so that $1 \geq n_1 \geq 1/(1 + C_3/C_2)$. Evidently, then, one can raise or lower the impedances of ladder arms by appropriately transforming a capacitive T. Note that only two of the three arms of the T must be available, because either Z_1 or Z_2 is always set to zero.

Use this procedure to scale the ladder elements in the circuit realized in Problem 2.27 such that $R_S = R_L$.

Sensitivity

THREE

3.1 INTRODUCTION

In general, a desired filter function is realized by interconnecting electrical components of carefully chosen nominal values in a network of suitable topology. In Chapter 2, we illustrated this process for passive LC ladder filters, and in Chapters 5 through 8 we shall discuss a number of synthesis procedures for active RC and switched-capacitor circuits. *Network synthesis* implies, therefore, (1) the selection of an appropriate network configuration, for which, especially in active filter design, there exist many possibilities, and (2) calculation of the nominal element values. The designer now faces the difficult task of how to select among the numerous available filter configurations a "best" circuit, given the fact that, in practice, real components will deviate from their nominal design values. Not only will circuit elements be initially inaccurate due to fabrication tolerances, but also the components will not maintain their initial values but will drift due to environmental effects such as temperature and humidity variations, and chemical changes which occur as the circuit ages. Furthermore, inaccuracies in modeling the passive and active devices, such as those resulting from nonideal op amp characteristics (to be discussed in Chapter 4) and parasitic capacitor losses, will also cause practical components to deviate from their ideal behavior. Because all coefficients, and therefore the poles and zeros of a transfer function $H(s)$, depend on the circuit elements [see, for example, Eq. (1-18) in Example 1-1], it is to be expected that a filter will deviate from its specified ideal behavior. The size of the error depends on how large the component tolerances are and how *sensitive* the circuit's performance is to these tolerances.

The concept of *sensitivity* is one of the most important criteria for comparing circuit configurations and for establishing their practical utility in meeting desired requirements. If applied carefully, sensitivity calculations not only allow the designer to select the better circuits from the many available ones in the active filter literature, but they also permit him or her to conclude, given the available component tolerances and their expected long-term stability, whether a chosen filter circuit satisfies and will keep satisfying the given specifications.

Given a component x, then in general any performance criterion P, such as the quality factor or a pole or zero frequency, will depend on x; that is, $P = P(x)$. If the performance measure P stands for the transfer function $H(s)$ or its magnitude or phase, then P is also a function of frequency, so that we may generally write $P = P(s, x)$. An intuitively appealing method for determining mathematically the deviation in P caused by an error $dx = x - x_0$ of the element x is provided by a Taylor expansion of $P(s, x)$ around the nominal value x_0 of the component:

$$P(s, x) = P(s, x_0) + \left.\frac{\partial P(s, x)}{\partial x}\right|_{x_0} dx + \frac{1}{2}\left.\frac{\partial^2 P(s, x)}{\partial x^2}\right|_{x_0} (dx)^2 + \ldots \quad (3\text{-}1)$$

If we assume that $(dx/x_0) \ll 1$ and that the curvature of $P(s, x)$ in the vicinity of x_0 is "not too large," we can neglect the second and higher derivative terms in Eq. (3-1) to obtain

$$\Delta P(s, x_0) = P(s, x_0 + dx) - P(s, x_0) \simeq \left.\frac{\partial P(s, x)}{\partial x}\right|_{x_0} dx \quad (3\text{-}2)$$

In most situations, the designer is less interested in the absolute changes ΔP caused by absolute changes dx and more interested in the relative changes. Therefore, it is customary to normalize Eq. (3-2) to yield

$$\frac{\Delta P(s, x_0)}{P(s, x_0)} \simeq \frac{x_0}{P(s, x_0)} \left.\frac{\partial P(s, x)}{\partial x}\right|_{x_0} \frac{dx}{x_0} \quad (3\text{-}3)$$

where the quantity

$$S_x^P = \frac{x_0}{P(s, x_0)} \left.\frac{\partial P(s, x)}{\partial x}\right|_{x_0} = \left.\frac{\partial P/P}{\partial x/x}\right|_{x_0} = \left.\frac{d(\ln P)}{d(\ln x)}\right|_{x_0} \quad (3\text{-}4)$$

is the *sensitivity* to a *small change* in a single parameter.

The small-change sensitivity S_x^P as defined in Eq. (3-4) forms the basis of most work in sensitivity analysis of electronic circuits; by Eq. (3-4), the "number" S_x^P tells us immediately that the *relative change*, also referred to as *variability*, of a performance measure P,

$$\frac{\Delta P}{P} \simeq S_x^P \frac{dx}{x} \quad (3\text{-}5)$$

is S_x^P times as large as the *relative change* of the circuit parameter x on which P

depends. It is evident, therefore, and intuitively obvious, that "good" circuits should have "small" sensitivities to their elements: for acceptable performance deviations (variabilities) $\Delta P/P$, they can be assembled from elements having larger *tolerances dx/x* and they are, therefore, more economical to build. Also, if their components vary during operation, they are less likely to drift out of the acceptable range of specification (the acceptability region).

Once this simple physical interpretation of sensitivity has been understood, the student should not forget how S_x^P was derived and which mathematical restrictions underlie its meaning. Even at this early stage in our discussion of sensitivity, it is therefore not too soon to emphasize a few important points that the reader should always keep in mind in any future work with sensitivity problems:

1. The fact that the Taylor expansion was truncated after the linear term implies that we replaced the function $P(s, x)$ by its *slope* at $x = x_0$, the *nominal* element value, and at a given frequency s. In general, this linearization means that S_x^P can give useful information only if dx/x is a "small" change. Whether a component tolerance is indeed small may depend on the situation and should be investigated from case to case.

2. Note that if the performance measure P of interest depends on frequency, then also the "number" S_x^P is really a function of frequency as is indicated clearly in Eq. (3-4). Thus, when calculating sensitivities, the relevant frequency range should be used in the evaluation. The sensitivity of a transfer function to the gain of an operational amplifier at dc, for example, may be of very little significance to the filter's operation at 10 kHz.

3. Although only one parameter, $x = x_0$, was shown explicitly in Eqs. (3-1) to (3-5), note that P depends on more than one and, possibly, on all elements in a circuit. Therefore, clearly, S_x^P depends on all the remaining parameters at their nominal points. If these nominal parameter values are altered, then S_x^P will change and the circuit's sensitivity performance will have to be re-evaluated.[1] It also means that if any free parameters are available (as is usually the case in active filters), then they can often be chosen to optimize (reduce) sensitivities for an overall better circuit performance.

4. An important practical point to keep in mind is that in the final analysis it is not sensitivity but *variability* that is important for a circuit's performance. Thus, by Eq. (3-5), large sensitivity to a very stable parameter can be quite acceptable whereas even moderate sensitivities to an element with large tolerances may well render a design useless.

Before treating sensitivity calculations in more detail in the following sections, let us give an example, which the student should study carefully because it illustrates these important points and concepts.

---

[1]Multiparameter sensitivity measures will be discussed in Section 3.3.

Example 3-1

The circuit in Fig. 3-1 is to realize a second-order bandpass function with center frequency $f_0 = 3$ kHz and a quality factor $Q = 20$. Find a set of element values and calculate the sensitivities $S_x^{\omega_0}$, S_K^Q and $S_K^{|H|}$, where x stands for all the passive components. "K" is a voltage amplifier of positive gain K. The midband gain of the filter is unimportant in this example.

Solution. Writing the node equations for the nodes marked V' and $V'' = V_2/K$ yields, with $G_i = 1/R_i$ and $C_1 = C_2 = C$,

$$V'(G_1 + 2sC) = G_1 V_1 + sC\left(1 + \frac{1}{K}\right)V_2$$

$$V'sC = \frac{V_2}{K}(sC + G_2) - V_2 G_2$$

from which V' may be eliminated to give the transfer function

$$H(s) = \frac{V_2}{V_1} = -\frac{K}{K-1}\frac{s/\tau_1}{s^2 + s\left(\dfrac{2}{\tau_2} - \dfrac{1}{\tau_1}\dfrac{1}{K-1}\right) + \dfrac{1}{\tau_1\tau_2}} \tag{3-6}$$

where $\tau_1 = CR_1$ and $\tau_2 = CR_2$. Comparing this expression with the notation introduced in Eq. (1-56) shows that the center frequency

$$\omega_0 = \frac{1}{\sqrt{\tau_1\tau_2}} = \frac{1}{\sqrt{C^2 R_1 R_2}} \tag{3-7}$$

and the quality factor

$$Q = \frac{r}{2 - r^2/(K-1)} \tag{3-8}$$

where the resistor ratio was defined as $r^2 = R_2/R_1$. Now observe that ω_0 is independent of the amplifier gain K, i.e.,

$$S_K^{\omega_0} \equiv 0$$

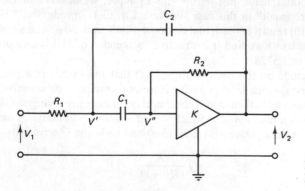

Figure 3-1 Second-order bandpass for Example 3-1.

which is a very desirable situation, and that, using Eq. (3-4),

$$S_{R_i}^{\omega_0} = \frac{R_i}{\omega_0} \frac{\partial \omega_0}{\partial R_i} = -\frac{1}{2} \qquad S_C^{\omega_0} = \frac{C}{\omega_0} \frac{\partial \omega_0}{\partial C} = -1$$

Note that $S_C^{\omega_0} = -1$ only because we have assumed that the two capacitors are equal! If the two capacitors are different $(C_1 \neq C_2)$, then

$$\omega_0 = \frac{1}{\sqrt{C_1 R_1 C_2 R_2}} \tag{3-9}$$

and $S_{C_i}^{\omega_0} = -0.5$, as the student should verify. It is also useful to observe in this context that from *dimensional* considerations alone (ω_0 is a frequency with units of s^{-1}), ω_0 must be inversely proportional to the square root of the product of two time constants as in Eq. (3-9). Therefore, 0.5 is the minimum value that the magnitude of the sensitivity of ω_0 to passive components in *any* active filter can take on. All second-order active filter circuits should be designed with this goal in mind, in addition to making ω_0 independent of the active parameters as is the case for the circuit in Fig. 3-1.

Returning to Eq. (3-8), we note that we have two parameters, r and K, to set the value of $Q = 20$. Either r or K can be chosen arbitrarily, subject to the restriction $K > 0.5r^2 + 1$ because Q must be positive. Choosing for convenience $r = 1$, i.e., equal resistors, yields

$$Q = \frac{K - 1}{2K - 3} \tag{3-10}$$

and $K = K_0 = 1.5128$ for $Q = 20$. Then, using Eqs. (3-4) and (3-10) gives

$$S_K^Q = \frac{K}{Q} \frac{dQ}{dK} = \frac{-K}{(K - 1)(2K - 3)} \bigg|_{K_0} = -115 \tag{3-11}$$

so that even a change of the amplifier gain as small as 0.25% results in a 28% change (!) in the value of Q. Because accuracy and stability of active device parameters of 0.25% are hardly possible in practice, this situation would be quite unacceptable; it is caused, of course, by the steep slope of Q in the vicinity of its pole at $K = 1.5$. Note especially that a mere 0.85% decrease in the nominal value K_0 would result in oscillations because Q will be negative for $K < 1.5$!

Referring to point 1 made just before this example, we observe that even a $\pm 0.5\%$ change is *not* "small" in this case, because Eq. (3-11) predicts a $\mp 57\%$ Q error whereas Eq. (3-10) results in the actual errors of -36% and $+140\%$, a significant discrepancy. The student may find it instructive to sketch Eq. (3-10) and observe the slope at $K = K_0 = 1.5128$.

Before we conclude on the basis of Eq. (3-11) that this circuit is impractical, let us recall the observation made in point 3 before our example: the sensitivity is a function of all elements and will change if their nominal values are changed. In our case, by Eq. (3-8), Q is a function of r, which we set arbitrarily to unity. Leaving r as a free parameter, we recalculate with Eqs. (3-8) and (3-4) the Q sensitivity,

$$S_K^Q = -rQ \frac{K}{(K - 1)^2} \tag{3-12}$$

where we note that K and r are related by Eq. (3-8):

$$\frac{1}{K-1} = \frac{2}{r^2} - \frac{1}{rQ} \tag{3-13}$$

Inserting Eq. (3-13) into Eq. (3-12) results, after a little algebra, in

$$S_K^Q = 1 + \frac{4}{r^2} - \frac{1}{rQ} - \frac{2Q}{r}\left(1 + \frac{2}{r^2}\right) \tag{3-14a}$$

We see now that the small value $r = 1$, giving $S_K^Q = -115$ for $Q = 20$, was a very unfortunate nominal point because S_K^Q decreases with increasing r. A much better choice would have been a larger value for the resistor ratio r^2, for example, $r. = 6$, which yields $S_K^Q \simeq -5.9$, a dramatic improvement of Q sensitivity to amplifier gain obtained for the price of different component values and, especially, a *larger* resistor *spread*. Note also that the Q sensitivity with respect to r decreases with increasing r:

$$S_r^Q = \frac{4Q}{r} - 1 \tag{3-14b}$$

We may conclude, therefore, that the circuit in Fig. 3-1 is after all a good low-sensitivity bandpass section *if the elements are chosen apropriately*. Recall point 3!

To complete the design, we choose $r = 6$ to give, from Eqs. (3-13) and (3-7) with $f_0 = 3$ kHz and a choice of $C = 5$ nF:

$$R_1 = \frac{1}{12\pi f_0 C} = 1.768 \text{ k}\Omega \qquad R_2 = r^2 R_1 = 63.66 \text{ k}\Omega \qquad K = 22.18$$

Note that the gain has increased over our previous value K_0, but it is still relatively small and can be realized easily.

Finally, let us calculate $S_K^{|H|}$. Using Eqs. (3-4), (3-6), and (3-14a), the student may show as an exercise that

$$S_K^{|H|} = \frac{-1}{K-1} + \frac{(\omega\omega_0/Q)^2}{(\omega_0^2 - \omega^2)^2 + (\omega\omega_0/Q)^2} S_K^Q \tag{3-15}$$

where the first term is due to the constant multiplier $K/(K-1)$ in Eq. (3-6) and the second, more important term is due to the dependence of Q on K. Because $H(s)$ is a bandpass function with $|H(j\omega)|^2$ as sketched in Fig. 1-18c, it is evident that $S_K^{|H|}$ is a strong function of frequency as alluded to in our point 2 above. It must be evaluated in the frequency range of interest, i.e., in the passband, where $S_K^{|H|}$ is a maximum. Again, a simple sketch of Eq. (3-15) should help the student's appreciation of the frequency dependence of $S_K^{|H|}$.

The purpose of this introduction has been to convey the concept of sensitivity and its importance to designers of filters, especially active filters. It should be clear that a good understanding of sensitivity is essential for the successful realization of practical active filters. In the succeeding sections of this chapter we turn our attention to definitions and to computations of different sensitivity measures. We begin with a detailed look at the somewhat simplistic, but nevertheless very useful, single-parameter sensitivity, defined in Eq. (3-4), and conclude with a treatment of multiparameter sensitivity concepts and related topics.

3.2 SINGLE-PARAMETER SENSITIVITY

It was mentioned in Section 3.1 that the single-parameter sensitivity S_x^P is somewhat simplistic because it neglects the dependence of the performance measure P on all the other parameters. Nevertheless, if interpreted carefully, much useful insight can be gained from S_x^P. Often, the smallest and largest single-parameter sensitivities are of practical importance because they identify the least and most critical circuit components. Clearly, one would use the most precise (and hence expensive) components where sensitivities are large in magnitude, whereas one can afford to use cheaper components having larger tolerances when sensitivities are small. Also, the treatment of single-parameter sensitivity is important because many multiparameter sensitivity measures are conveniently expressed in terms of single-parameter sensitivities.

With this in mind, let us concentrate in this section on the single-parameter sensitivity measure defined in Eq. (3-4),

$$S_x^{P(x)} = \frac{x}{P}\frac{dP}{dx} = \frac{d(\ln P)}{d(\ln x)} \tag{3-16}$$

and on the *semi*relative sensitivity measure[2]

$$Q_x^{P(x)} = x\frac{dP}{dx} \tag{3-17}$$

which provides more useful insight in some circumstances. For example, if S_x^P is evaluated at or near a nominal point where $P = 0$, then $S_x^P \rightarrow \infty$, which is not a usable result. Q_x^P, however, stays finite and lends itself to meaningful interpretations. A case in point might be the sensitivity of a transfer function at or near a transmission zero. Also, in some situations, the designer is indeed interested in the absolute rather than the relative change of P, most notably when P is the location of a pole or a zero, where the absolute shift in the s-plane is important regardless of the initial pole or zero position.

Although S_x^P can always be calculated by applying directly the definition in Eq. (3-16), there are a number of convenient relationships that simplify the computations significantly. Some of the more important ones are

$$S_x^{P_1 P_2} = S_x^{P_1} + S_x^{P_2}; \tag{3-18a}$$

$$S_x^{P_1/P_2} = S_x^{P_1} - S_x^{P_2} \tag{3-18b}$$

---

[2]Various notations exist for the semirelative sensitivity; our choice of Q_x^P should cause no confusion with the quality factor, because sensitivity always has a subscript and a superscript.

for the case where $P(x) = P_1(x)P_2(x)$ or $P(x) = P_1(x)/P_2(x)$, and

$$S_x^{P(y(x))} = S_y^P S_x^y \qquad (3\text{-}19)$$

where $P = P(y)$ and $y = y(x)$. [These equations were used to derive Eq. (3-15).] For example, if the zero frequency ω_z of a notch filter depends on a resistor R and a capacitor C which in turn depend on temperature T, then the sensitivity of ω_z to temperature is

$$S_T^{\omega_z} = S_R^{\omega_z} S_T^R + S_C^{\omega_z} S_T^C$$

which is a convenient result. The student is asked to prove Eqs. (3-18) and (3-19), as well as a number of additional useful sensitivity equations, in Problem 3.2.

3.2.1 Sensitivity Invariants

A number of interesting and useful relationships can be established between the single-parameter sensitivities $S_{x_i}^P$ of a performance function $P(x_1, x_2, \ldots, x_n)$ that satisfies the so-called *homogeneity condition*. This condition means that if all parameters are multiplied by a constant, say k, then

$$P(kx_1, kx_2, \ldots, kx_n) = k^\lambda P(x_1, x_2, \ldots, x_n) \qquad (3\text{-}20)$$

where λ is an integer constant. Differentiating Eq. (3-20) with respect to k gives

$$\sum_{i=1}^n x_i \frac{\partial P}{\partial (kx_i)} = \lambda k^{\lambda-1} P$$

Dividing both sides by P and setting $k = 1$ results in Euler's formula

$$\sum_{i=1}^n S_{x_i}^P = \lambda \qquad (3\text{-}21)$$

For our applications, if P is a voltage transfer function $H(s)$ and the parameters are the passive elements R_i, C_i, and L_i in addition to voltage amplifiers of gain μ_i, current amplifiers of gain α_i, transconductances g_{mi}, and transresistances r_{mi},[3] then

$$H(s) = H(s, R_i, C_i, L_i, \mu_i, \alpha_i, g_{mi}, r_{mi}) \qquad (3\text{-}22)$$

Because we know that H is not changed by an impedance scaling factor $1/R_0$, we have, with $\lambda = 0$,

$$H\left(s, \frac{R_i}{R_0}, R_0 C_i, \frac{L_i}{R_0}, R_0 g_{mi}, \frac{r_{mi}}{R_0}\right) = H(s)$$

and application of Eq. (3-21) yields

$$\sum_i S_{R_i}^H + \sum_i S_{L_i}^H - \sum_i S_{C_i}^H - \sum_i S_{g_{mi}}^H + \sum_i S_{r_{mi}}^H = 0 \qquad (3\text{-}23a)$$

---

[3] These active device parameters are, of course, defined as: $\mu = V_{out}/V_{in}$, $\alpha = I_{out}/I_{in}$, $g_m = I_{out}/V_{in}$, and $r_m = V_{out}/I_{in}$.

Similarly, applying frequency scaling by a factor $1/\omega_n$ to the original function in Eq. (3-22) simply scales the elements L_i and C_i:

$$H\left(s, R_i, \frac{L_i}{\omega_n}, \frac{C_i}{\omega_n}, \mu_i, \alpha_i, g_{mi}, r_{mi}\right) = H\left(\frac{s}{\omega_n}\right)$$

Differentiating with respect to $1/\omega_n$, setting $\omega_n = 1$, and dividing by H yields

$$\sum_i S_{L_i}^H + \sum_i S_{C_i}^H = S_s^H = \frac{d(\ln H)}{d(\ln s)} \tag{3-23b}$$

Note that in Eqs. (3-23a) and (3-23b) the sums are to be taken over *all* the respective elements.

The important consequence of Eqs. (3-23a) and (3-23b) is that the sensitivities are not independent; therefore, if by clever circuit design the sensitivity to one circuit element is reduced, the sensitivities to one or more of the remaining elements have to increase, because the sums must remain constant. Note that the results in the two equations are independent of the method of realization of the function $H(s)$.

For example, in an *LC* filter we have

$$\sum_i S_{L_i}^H - \sum_j S_{C_j}^H = - \sum_{k=1}^{2} S_{R_k}^H \tag{3-24a}$$

$$\sum_i S_{L_i}^H + \sum_j S_{C_j}^H = \frac{s}{H} \frac{\partial H(s)}{\partial s} \tag{3-24b}$$

where R_1 and R_2 are the source and load resistors. It is apparent that the sum of all sensitivities to the inductors and capacitors of the filter is predetermined by the *slope* of the transfer function versus s. Equation (3-24b) provides a ready explanation for the fact that in practice the sensitivities usually have their maxima at the passband edges, where filters have steep transition regions.

Considering active *RC* filters, which are mostly built using voltage amplifiers (μ), we have from Eq. (3-23)

$$\sum_i S_{R_i}^H - \sum_j S_{C_j}^H = 0 \tag{3-25a}$$

$$\sum_j S_{C_j}^H = S_s^H \tag{3-25b}$$

Equations (3-25a) and (3-25b) imply that the *sum* of *all* sensitivities to the capacitors is equal to the *sum* of all sensitivities to the resistors and they are both given by the transfer function slope, independently of the type of circuit chosen for the realization.

The concept expressed in Eqs. (3-20) and (3-21) can be used to derive several further relationships that are of relevance to our study of filters. For example, note that a pole p_l (or for that matter *any* frequency parameter) of an active filter transfer function is independent of impedance scaling, i.e.,

$$p_l(R_i, C_j, \mu_k) = p_l\left(\frac{R_i}{R_0}, C_j R_0, \mu_k\right)$$

Therefore, we have by Eq. (3-21) with $\lambda = 0$

$$\sum_i S_{R_i}^{p_l} = \sum_j S_{C_j}^{p_l} \tag{3-26a}$$

(see Example 3-1).   Similarly, for frequency scaling we find

$$p_l\left(R_i, \frac{C_j}{\omega_0}, \mu_k\right) = \omega_0 p_l(R_i, C_j, \mu_k)$$

and, therefore, from Eq. (3-21) with $\lambda = -1$

$$\sum_j S_{C_j}^{p_l} = -1 \tag{3-26b}$$

Also, for the quality factor $Q = Q(R_i, C_j, \mu_k)$ of a pole, we find

$$\sum_i S_{R_i}^{Q} = \sum_j S_{C_j}^{Q} = 0 \tag{3-26c}$$

because Q is a dimensionless parameter that depends only on *ratios* of resistors and capacitors.   The student should verify the correctness of Eqs. (3-26a) through (3-26c) for the circuit of Example 3-1 (Problem 3.3).

Example 3-2

An engineer has to construct an active filter in thin- or thick-film technology.   The design should be such that the pole frequency is insensitive to temperature variations. Find the required relationship between the temperature coefficients, α_R of the resistors and α_C of the capacitors.

Solution.   In general, for $\omega_0 = \omega_0(R_i, C_j)$, with Eq. (3-19),

$$S_T^{\omega_0} = \sum_i S_{R_i}^{\omega_0} S_T^{R_i} + \sum_j S_{C_j}^{\omega_0} S_T^{C_j} \tag{3-27}$$

For the required technology, with all resistors on a common substrate and fabricated with the same material, we can assume that temperature dependence is the same for all resistors:

$$R_i = R_{i0}[1 + \alpha_R(T - T_0)]$$

and similarly for all capacitors:

$$C_i = C_{i0}[1 + \alpha_C(T - T_0)]$$

where T_0 is the nominal operating temperature.   Therefore, we have

$$S_T^{R_i} = \left.\frac{T}{R_i}\frac{dR_i}{dT}\right|_{T_0} \approx \alpha_R T_0 \qquad \text{and} \qquad S_T^{C_i} = \alpha_C T_0$$

for all i, and Eq. (3-27) becomes

$$S_T^{\omega_0} = \alpha_R T_0 \sum_i S_{R_i}^{\omega_0} + \alpha_C T_0 \sum_j S_{C_j}^{\omega_0} \tag{3-28}$$

Now, according to Eqs. (3-26a) and (3-26b), both sensitivity sums are equal to -1, so that Eq. (3-28) reduces to

$$d\omega_0 = -\omega_0(\alpha_R + \alpha_C)\, dT \tag{3-29}$$

Clearly, therefore, to make ω_0 insensitive to temperature we need to choose thin- or thick-film materials with temperature coefficients such that $\alpha_R = -\alpha_C$, a result that has intuitive appeal because it implies that RC products are constant to the first order.

3.2.2 The Gain-Sensitivity Product

The most commonly used active device in active filter design is the operational amplifier (op amp), whose properties and operation will be treated in detail in Chapter 4. For now it suffices to know that an op amp is a voltage amplifier of large differential gain A, defined by the relation

$$V_o = A(V^+ - V^-) \tag{3-30}$$

as shown in Fig. 3-2a. Because the *open-loop* gain A is very large and has significant variability dA/A, op amps are always used with feedback to achieve a reduced *closed-loop* gain μ of smaller variability $d\mu/\mu$. Two examples are shown in Figs. 3-2b and c. For Fig. 3-2b, we calculate

$$V_2 = A\left(V_1 - \frac{R_1}{R_1 + (K-1)R_1} V_2\right) = A\left(V_1 - \frac{1}{K} V_2\right)$$

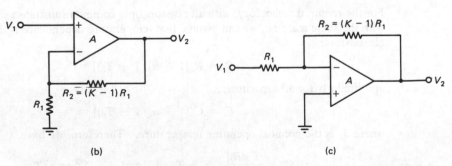

Figure 3-2 Operational amplifier circuits: (a) op amp symbol; (b–c) feedback configurations for positive or noninverting (b) and negative or inverting (c) gains.

so that

$$\mu_1 = \frac{V_2}{V_1} = \frac{K}{1 + K/A}\bigg|_{K<<A} \simeq K \tag{3-31a}$$

Similarly, for Fig. 3-2c, we find

$$\mu_2 = \frac{V_2}{V_1} = -\frac{K-1}{1 + K/A}\bigg|_{K<<A} \simeq -(K-1) \tag{3-31b}$$

The sensitivities of the closed-loop gains μ_1 and μ_2 to the open-loop gain A are obtained via Eq. (3-4) as

$$S_A^{\mu_1} = \frac{K/A}{1 + K/A} = \frac{\mu_1}{A} \tag{3-32a}$$

and

$$S_A^{\mu_2} = \frac{+K/A}{1 + K/A} \simeq \frac{-\mu_2}{A} \qquad \text{if } K \gg 1 \tag{3-32b}$$

Observe that the sensitivity of μ_i to A is proportional to μ_i, i.e., it increases with increasing closed-loop gain. Also note that the variability of μ_i is significantly reduced from that of A:

$$\frac{d\mu_i}{\mu_i} = S_A^{\mu_i}\frac{dA}{A} \simeq \frac{\mu_i}{A}\frac{dA}{A} \tag{3-33}$$

because in practice $\mu_i \ll A$; in fact, the variability of μ_i is approximately equal to that of the resistor ratio K, as Eqs. (3-31a) and (3-31b) show.

Assume now, that a filter parameter P depends on the closed-loop gain μ of an op amp [see, for example, Eq. (3-8)]; then, by Eqs. (3-19) and (3-32),

$$S_A^P = S_\mu^P S_A^\mu = \frac{\mu}{A} S_\mu^P$$

and the variability becomes

$$\frac{dP}{P} = (\mu S_\mu^P)\frac{dA}{A^2} = \Gamma_\mu^P \frac{dA}{A^2} \tag{3-34}$$

i.e., dP/P is proportional to the *product* of gain and sensitivity (Γ_μ^P) and to a term dA/A^2 that depends only on the particular op amp used. We observe, therefore, that for a given op amp the variability of a performance parameter P depends not only on the sensitivity but on the gain-sensitivity product (GSP) Γ_μ^P; that is, dP/P depends not only on how sensitive P is to μ, but also on how large a closed-loop gain μ is needed for the implementation. The GSP has therefore been proposed [32] and used as a more representative measure for comparing different designs. For example, assume that an op amp is available with $A = 10^4$ and

$dA/A = 60\%$; design 1 has $S_\mu^{P_1} = 6$ and needs $\mu = 95$, and design 2 has a larger sensitivity, $S_\mu^{P_2} = 38$, but needs $\mu = 4$. For these numbers we find

$$\frac{dP_1}{P_1} = 95 \cdot 6 \cdot 0.6 \cdot 10^{-4} = 3.4\%$$

$$\frac{dP_2}{P_2} = 4 \cdot 38 \cdot 0.6 \cdot 10^{-4} = 0.91\%$$

indicating that design 2 is preferable in spite of the more than 6 times higher sensitivity.

Finally, we wish to point out that the GSP for the open-loop gain A is equal to the GSP for the closed-loop gain μ (Problem 3.6):

$$\Gamma_A^P = \Gamma_\mu^P \tag{3-35}$$

and that Eqs. (3-31) to (3-34) lead to the intuitively expected and important conclusion that the designer of an active filter should choose op amps whose gain A in the frequency range of interest is as large as possible in order to reduce the term dA/A^2 in Eq. (3-34).

Example 3-3

The finite-gain amplifier of Example 3-1 is implemented as shown in Fig. 3-3. Assume that an op amp with gain $A = 10^4$ and variability $dA/A = 40\%$ is used in the design. Calculate the variability $\Delta Q/Q$ for both choices of r, $r = 1$ and $r = 6$, used in Example 3-1.

Solution. From Fig. 3-3 we find

$$\mu = \frac{V_2}{V^-} = \frac{K}{1 - K/A}\bigg|_{K \ll A} \simeq K$$

and

$$S_A^\mu = -\frac{K/A}{1 - K/A} = -\frac{\mu}{A} \simeq -\frac{K}{A}$$

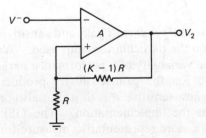

Figure 3-3 Finite gain noninverting amplifier for Example 3-3.

Thus, with Eqs. (3-12) and (3-13),

$$\frac{\Delta Q}{Q} = S_\mu^Q S_A^\mu \frac{dA}{A} = rQ\left(\frac{K}{K-1}\right)^2 \frac{dA}{A^2} = rQ\left(1 + \frac{2}{r^2} - \frac{1}{rQ}\right)^2 \frac{dA}{A^2}$$

For the given numbers, we find

$$\frac{\Delta Q}{Q} \simeq \begin{cases} 0.7\% & \text{for } r = 1 \\ 0.5\% & \text{for } r = 6 \end{cases}$$

Note that in spite of the large reduction in S_K^Q from -115 to -5.9 in Example 3-1, $\Delta Q/Q$ has not improved much, because of the required increase in closed-loop gain from 1.51 to 22.2.   Nevertheless, increasing r to 6 is still worthwhile because of the reduction in sensitivities to the resistor ratios r and K in Eqs. (3-14a) and (3-14b). The student may wish to verify that the GSP in this example,

$$\Gamma_\mu^Q = rQ\left(1 + \frac{2}{r^2} - \frac{1}{rQ}\right)^2$$

has a minimum at the value $r^2 \simeq 6$.

3.2.3 Transfer Function Sensitivity

The transfer function of a filter was introduced in Chapter 1, Eq. (1-44), as

$$H(s) = \frac{N(s)}{D(s)} = \frac{a_m s^m + \ldots + a_1 s + a_0}{s^n + \ldots + b_1 s + b_0} \tag{3-36}$$

If we assume that both N and D depend on the circuit element x, so that they are in effect the functions $N(s, x)$ and $D(s, x)$, then the sensitivity of H to x, using Eq. (3-18b), is the difference between the numerator and denominator sensitivities:

$$S_x^H = S_x^N - S_x^D = x\left(\frac{1}{N}\frac{\partial N}{\partial x} - \frac{1}{D}\frac{\partial D}{\partial x}\right) \tag{3-37}$$

If, for example, only the coefficient b_i depends on x, then, from Eqs. (3-37) and (3-19),

$$S_x^H = -S_x^D = -S_{b_i}^D S_x^{b_i} \tag{3-38}$$

where we note that

$$S_{b_i}^D = \frac{1}{D(s)} b_i s^i \tag{3-39}$$

with

$$\sum_{i=0}^{n} S_{b_i}^D = 1 \tag{3-40}$$

If more than one coefficient depends on x, we can use Eqs. (3-37) through (3-39) to calculate

$$S_x^H = \frac{x}{N} \sum_j \frac{\partial a_j}{\partial x} s^j - \frac{x}{D} \sum_i \frac{\partial b_i}{\partial x} s^i \tag{3-41}$$

where the summations extend over those coefficients a_j and b_i which depend on x.

Example 3-4

Calculate the sensitivity of the transfer function $H(s)$ in Eq. (3-6) to the resistor R_1 and to the amplifier gain K.

Solution. $H(s)$ is of the form of Eq. (3-36) with

$$b_2 = 1 \qquad b_1 = \frac{2}{\tau_2} - \frac{1}{C(K-1)} \frac{1}{R_1} \qquad b_0 = \frac{1}{C^2 R_2} \frac{1}{R_1}$$

$$a_1 = -\frac{K}{C(K-1)} \frac{1}{R_1}$$

where we have shown the explicit dependence of the coefficients on R_1 and K. Thus, with Eq. (3-41),

$$S_{R_1}^H = \frac{R_1}{a_1 s} s \frac{-a_1}{R_1} - \frac{R_1}{D(s)} \left[\frac{-1}{C(K-1)} \frac{-1}{R_1^2} s + \frac{1}{C^2 R_2} \frac{-1}{R_1^2} \right]$$

$$= -1 - \frac{1}{D(s)} \left[\frac{1}{\tau_1(K-1)} s - \frac{1}{\tau_1 \tau_2} \right]$$

$$S_K^H = \frac{-1}{K-1} - \frac{s/\tau_1}{D(s)} \frac{K}{(K-1)^2} = \frac{-1}{K-1} [1 - H(s)] \tag{3-42}$$

A very interesting relationship is obtained by finding the sensitivity S_x^H when H is first expressed as

$$H(j\omega, x) = |H(j\omega, x)| e^{j\phi(\omega, x)}$$

Using Eq. (3-18a) or the part of Eq. (3-16) that involves logarithms, we obtain

$$S_x^{H(j\omega)} = S_x^{|H(j\omega)|} + jQ_x^{\phi(\omega)} \tag{3-43}$$

which says that

The real part of the transfer function sensitivity is the magnitude sensitivity:

$$\text{Re}\{S_x^{H(j\omega)}\} = S_x^{|H(j\omega)|} \tag{3-44a}$$

and the imaginary part of the transfer function sensitivity is the *semi*relative phase sensitivity:

$$\text{Im}\{S_x^{H(j\omega)}\} = Q_x^{\phi(\omega)} = x \frac{\partial \phi(\omega, x)}{\partial x} \tag{3-44b}$$

Example 3-5

Use Eq. (3-42) to find magnitude and phase sensitivities of the transfer function [Eq. (3-6)] in Example 3-1.

Solution.   Using the notation

$$D(s) = s^2 + s\frac{\omega_0}{Q} + \omega_0^2$$

with ω_0 and Q as defined in Eqs. (3-7) and (3-8), we find from Eq. (3-42):

$$S_K^{H(j\omega)} = \frac{-1}{K-1} - \frac{K}{(K-1)^2}\frac{j\omega/\tau_1}{\omega_0^2 - \omega^2 + j\omega\omega_0/Q}$$

$$= \frac{-1}{K-1} - \frac{K}{(K-1)^2}\frac{\omega^2\omega_0/(\tau_1 Q) + j(\omega_0^2 - \omega^2)\omega/\tau_1}{(\omega_0^2 - \omega^2)^2 + (\omega\omega_0/Q)^2}$$

Thus, we have

$$\mathrm{Im}\ \{S_K^{H(j\omega)}\} = Q_K^{\phi(\omega)} = -\frac{K}{(K-1)^2}\frac{\omega}{\tau_1}\frac{\omega_0^2 - \omega^2}{|D(j\omega)|^2} = \frac{1}{rK}\left(\frac{\omega}{\omega_0} - \frac{\omega_0}{\omega}\right)|H(j\omega)|^2 \quad (3\text{-}45)$$

where $r^2 = R_2/R_1 = 1/(\omega_0\tau_1)$ was defined in Example 3-1, and

$$\mathrm{Re}\ \{S_K^{H(j\omega)}\} = S_K^{|H(j\omega)|} = \frac{-1}{K-1} + \frac{(\omega\omega_0/Q)^2}{|D(j\omega)|^2}\left[-\frac{Q}{\omega_0\tau_1}\frac{K}{(K-1)^2}\right]$$

which is recognized as the result in Eq. (3-15) because the term in brackets equals S_K^Q by Eq. (3-12).

We have seen that Eq. (3-41) provides a useful result that permits us to calculate the transfer function sensitivity when the polynomial coefficients, defined in Eq. (3-36), change due to element tolerances. Clearly, altering the coefficients will have the effect of shifting the roots of the polynomials, i.e., the poles and zeros of $H(s)$. We should, therefore, ask what consequences the shifting of poles and zeros has on the behavior of $H(s)$. The question is answered most easily by taking the logarithm of $H(s)$ in its factored form[4] as given in Eq. (1-14):

$$\ln H = \ln K + \sum_{i=1}^{m} \ln (s - z_i) - \sum_{i=1}^{n} \ln (s - p_i) \quad (3\text{-}46)$$

where $K = a_m/b_n$ and where K, z_i, and p_i are assumed to be functions of the element x. Taking the derivative of Eq. (3-46) and multiplying by x yields the desired equation,

$$S_x^H = S_x^K - \sum_{i=1}^{m} \frac{Q_x^{z_i}}{s - z_i} + \sum_{i=1}^{n} \frac{Q_x^{p_i}}{s - p_i} \quad (3\text{-}47)$$

---

[4]We assume only simple roots; see Problem 3.10 for the modifications necessary if the roots are multiple.

which shows the not unexpected result that any pole or zero shift influences the transfer function most strongly in the neighborhood of that pole or zero (where $|s - z_i|$ and $|s - p_i|$ are small). By expressing Eq. (3-47) in terms of its common denominator $N(s)D(s)$, we observe that S_x^H has poles at all poles *and* zeros of $H(s)$, a result that is also apparent from Eq. (3-37). Equation (3-47) is the *partial fraction expansion* of S_x^H, so that the pole and zero sensitivities $Q_x^{p_i}$ and $-Q_x^{z_i}$ are the residues of the poles p_i and z_i of S_x^H (Problem 3.11). An additional insight to be gained from Eq. (3-47) is that the transfer function sensitivity will in general have a large magnitude at all frequencies which are close to a pole p_i or a zero z_i of $H(s)$. Specifically, for physical frequencies $s = j\omega$, we see that $S_x^H \to \infty$ at a $j\omega$-axis transmission zero $z_i = j\omega_{z_i}$ as mentioned earlier! Also, for frequencies $s = j\omega$ in the neighborhood of any pole pair with large quality factor Q, we have to expect high sensitivities unless special measures are taken: note that a conjugate complex pole pair arising from a denominator factor $s^2 + s\omega_0/Q + \omega_0^2$ is located very close to the $j\omega$-axis if Q is large:

$$p_{1,2} = -\frac{\omega_0}{2Q} + j\omega_0 \sqrt{1 - \frac{1}{4Q^2}} \tag{3-48}$$

so that $|j\omega - p_1|$ becomes small. We shall return to Eq. (3-47) and the related comments in Section 3.2.6.

3.2.4 Root Sensitivity

We saw in Eq. (3-47) that the transfer function deviation can be calculated as soon as we know the sensitivities of the poles and zeros,

$$Q_x^{p_i} = x \frac{\partial p_i}{\partial x} \quad and \quad Q_x^{z_i} = x \frac{\partial z_i}{\partial x}$$

However, for computing these sensitivities we need to obtain the functions $z_i(x)$ and $p_i(x)$, which poses a seemingly insurmountable problem because a symbolic factorization of a polynomial of order 4 or higher is in general impossible. Therefore, the expression showing the explicit dependence of p_i or z_i on a circuit element x cannot be found. A way out of this dilemma lies in the fact that, as mentioned, Eq. (3-47) is a partial fraction expansion of S_x^H in Eq. (3-37), so that $-Q_x^{z_i}$ and $Q_x^{p_i}$ are the residues. Thus, if we equate Eqs. (3-47) and (3-37) and evaluate the result as $s \to p_i$, we obtain

$$\lim_{s \to p_i} x \left(\frac{1}{N} \frac{\partial N}{\partial x} - \frac{1}{D} \frac{\partial D}{\partial x} \right) = \lim_{s \to p_i} \frac{Q_x^{p_i}}{s - p_i} \tag{3-49}$$

where we used the fact that $Q_x^{p_i}/(s - p_i)$ dominates the remaining terms in Eq. (3-47). If we recall further that p_i is a pole of the transfer function $H(s)$, i.e., a zero of $D(s)$, then we recognize that $(1/D)\, \partial D/\partial x$ dominates $(1/N)\partial N/\partial x$ and Eq.

(3-49) can be brought into the form

$$\lim_{s \to p_i} \left(-\frac{x}{D} \frac{\partial D}{\partial x} - \frac{Q_x^{p_i}}{s - p_i} \right) = 0$$

or

$$Q_x^{p_i} = -\lim_{s \to p_i} \left[x(s - p_i) \frac{1}{D(s, x)} \frac{\partial D(s, x)}{\partial x} \right] \tag{3-50a}$$

In some situations, the root-sensitivity calculations can be simplified further by noting that

$$\lim_{s \to p_i} \frac{D(s, x)}{s - p_i} = \lim_{s \to p_i} \frac{\partial D(s, x)}{\partial s} \tag{3-51}$$

so that, for example, Eq. (3-50a) becomes

$$Q_x^{p_i} = -\lim_{s \to p_i} \left[x \frac{\partial D(s, x)/\partial x}{\partial D(s, x)/\partial s} \right] \tag{3-50b}$$

Similarly, we find for the zero sensitivity

$$Q_x^{z_i} = -\lim_{s \to z_i} \left[x(s - z_i) \frac{1}{N(s, x)} \frac{\partial N(s, x)}{\partial x} \right] = -\lim_{s \to z_i} \left[x \frac{\partial N(s, x)/\partial x}{\partial N(s, x)/\partial s} \right] \tag{3-50c}$$

A useful special case of Eq. (3-50a) is obtained when x is one of the coefficients of $D(s)$, $x = b_j$. In this case $\partial D/\partial b_j = s^j$ and we have

$$Q_{b_j}^{p_i} = -\lim_{s \to p_i} \left[b_j s^j \frac{s - p_i}{D(s, b_j)} \right] = -\lim_{s \to p_i} \frac{b_j s^j}{\partial D(s, b_j)/\partial s} \tag{3-52a}$$

and, similarly,

$$Q_{a_j}^{z_i} = -\lim_{s \to z_i} \left[a_j s^j \frac{s - z_i}{N(s, a_j)} \right] = -\lim_{s \to z_i} \frac{a_j s^j}{\partial N(s, a_j)/\partial s} \tag{3-52b}$$

The advantage of Eqs. (3-50) and (3-52) is that the root sensitivities can be found without the need to know $p_i(x)$ or $z_i(x)$ explicitly, and, for that matter, without having to factor the polynomials (beyond the specific root under consideration). Note that Eqs. (3-50) and (3-52) are valid only for *simple* roots. Because *multiple* roots occur very rarely in filter functions (except at $s = 0$ and $s = \infty$), this case will not be considered [26, 33] (Problem 3.13).

Once the sensitivity is known, the approximate position of the shifted root can be calculated.   To be specific, assume that $Q_x^{p_0}$ has been determined; then the new location p_1 is obtained from the nominal pole p_0 as

$$p_1 = p_0 + dp_0 \simeq p_0 + Q_x^{p_0} \frac{dx}{x} \tag{3-53}$$

Note that $Q_x^{p_0}$ is a complex number.   The following example will demonstrate the process.

Example 3-6

In Example 1-6 we found the lowpass function

$$H(s) = \frac{1.4314(s^2 + 2)(s^2 + 3)}{s^5 + 4.5817s^4 + 9.4717s^3 + 14.473s^2 + 11.210s + 8.5883}$$

with poles at

$$p_i = -2.63759 \qquad p_{2,3} = -0.184484 \pm j1.14676$$

$$p_{4,5} = -0.787586 \pm j1.33914$$

Assume that in the realization of $H(s)$ as an active filter, the denominator coefficients of s^4 and s^2 depend on a parameter K as

$$4.5817 = 23.6721\left(1 - \frac{K}{1.24}\right)$$

and

$$14.473 = 38.08684(1 - 0.62K)$$

respectively, with $K = K_0 = 1$ nominally.   Normally, the poles closest to the $j\omega$-axis [those with the highest pole Q; see Eq. (3-48)] have the largest sensitivities.   Find, therefore, $Q_K^{P_2}$ and give the new locations of p_2 when K varies by $\pm 3\%$ from its nominal value $K_0 = 1$.

Solution.   Using Eq. (3-50b), we compute

$$\frac{\partial D(s, K)}{\partial K} = -\frac{23.6721}{1.24} s^4 - 38.08684 \cdot 0.62s^2$$

and

$$\frac{\partial D(s, K_0)}{\partial s} = 5s^4 + 18.3268s^3 + 28.4151s^2 + 28.946s + 11.210$$

Evaluating these two expressions at $s = p_{20}$ where $p_{20} = -0.18448 + j1.14676$ is the nominal pole yields, after some complex arithmetic,

$$\left.\frac{\partial D(p_{20}, K)}{\partial K}\right|_{K_0} = 2.33984 - j10.7035$$

$$\left.\frac{\partial D(s, K_0)}{\partial s}\right|_{p_{20}} = -9.99696 + j1.09942$$

so that

$$Q_K^{P_2} = -1.0894e^{-j251.4°} = 0.3476 - j1.0325$$

The new pole locations are obtained from Eq. (3-53); i.e.,

$$p_2 = p_{20} + dp_2 \approx p_{20} + Q_K^{P_2} \frac{dK}{K}$$

Thus, for $dK/K = +0.03$ and -0.03, respectively, we find

$$p_2 = -0.174 + j1.116$$

$$p_2 = -0.195 + j1.178$$

The exact new locations (to five decimal places), calculated by a numerical root finder, are $p_2 = -0.17150 + j1.12208$ for $K = 1.03$ and $p_2 = -0.18688 + j1.18383$ for $K = 0.97$. The agreement is seen to be reasonably good.

The student is cautioned at this point to use Eqs. (3-50) to (3-53) only for determining the "trend" of a root shift but not for any accurate calculation of the new root location *unless* the parameter or coefficient deviations are indeed incremental. The roots of a polynomial are very sensitive to the exact values of the coefficients, so that the linearization involved in sensitivity calculations cannot be expected or relied upon to give accurate results in this case. For example, already if dK/K is as large as -10%, the exact pole location is $p_2 = -0.12967 + j1.23814$, whereas sensitivity calculations yield $p_2 = -0.219 + j1.250$, giving a significant discrepancy.

3.2.5 Second-Order Active Filters

The most important practical building block in the design of active filters is the second-order, or *biquadratic*, section (the 'biquad') introduced in Chapter 1, Eq. (1-49), and repeated here for convenience:

$$H_2(s) = \frac{a_2(s - z_1)(s - z_2)}{(s - p_1)(s - p_2)} = \frac{a_2 s^2 + a_1 s + a_0}{s^2 + s\omega_p/Q_p + \omega_p^2} \qquad (3\text{-}54)$$

In the only case of real interest, the poles are complex ($Q_p > 0.5$), so that p_1 and p_2 are conjugate, i.e., $p_2 = p_1^*$, with

$$p_1 = -\omega_p\left(\frac{1}{2Q_p} - j\sqrt{1 - \frac{1}{4Q_p^2}}\right) \qquad (3\text{-}55)$$

as in Eq. (3-48). Because, as mentioned in Section 1.6, the poles dominate the passband behavior of filters, accurate pole positions are most important, and p_1 should be insensitive to variations in any parameter x. Assuming that Q_p and ω_p are functions of x, we calculate from Eq. (3-55) with Eq. (3-17)

$$Q_x^{p_1} = p_1\left(S_x^{\omega_p} - j\frac{S_x^{Q_p}}{\sqrt{4Q_p^2 - 1}}\right) \qquad (3\text{-}56)$$

and $Q_x^{p_2} = (Q_x^{p_1})^*$, as the student should verify. From Eq. (3-56) we observe that the location of a pole is $\sqrt{4Q_p^2 - 1} \approx 2Q_p$ times more sensitive to variations in ω_p than it is to variations in Q_p. For the actual shift dp_1 of the pole p_1, we find from

Eq. (3-56)

$$dp_1 = p_1\left(\frac{d\omega_p}{\omega_p} - j\,\frac{1}{\sqrt{4Q_p^2 - 1}}\,\frac{dQ_p}{Q_p}\right) \qquad (3\text{-}57)$$

which means that for $dQ_p = 0$ the pole moves radially away from or toward the origin:

$$dp_1 = \frac{d\omega_p}{\omega_p}\,p_1 \qquad Q_p = \text{const} \qquad (3\text{-}58a)$$

and for $d\omega_p = 0$ the pole moves tangentially to a circle of radius ω_p around the origin:

$$dp_1 = -j\,\frac{dQ_p/Q_p}{\sqrt{Q_p^2 - 1}}\,p_1 \qquad \omega_p = \text{const} \qquad (3\text{-}58b)$$

Figure 3-4 shows a graphic representation of these facts for positive $d\omega_p$ and dQ_p.

Having established an expression for the pole sensitivity $Q_x^{p_1}$ in Eq. (3-56), we are now ready to investigate its effect on the passband of the biquadratic transfer function $H_2(s)$ in Eq. (3-54). To this end we evaluate Eq. (3-47) for $H_2(s)$ under the assumption that the parameter x does not affect the zeros and the gain constant:

$$S_x^{H_2(s)} = \frac{Q_x^{p_1}}{s - p_1} + \frac{(Q_x^{p_1})^*}{s - p_1^*} \qquad (3\text{-}59)$$

If we express the right-hand side in terms of its common denominator $s^2 +$

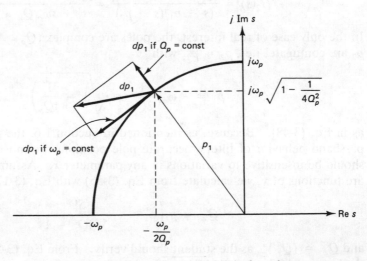

Figure 3-4 Illustration of pole shift in the s-plane.

$s\omega_p/Q_p + \omega_p^2$ and use Eq. (3-56), we obtain

$$S_x^{H_2(s)} = -\frac{\left(2\omega_p^2 + \dfrac{s\omega_p}{Q_p}\right)S_x^{\omega_p} - \dfrac{s\omega_p}{Q_p}S_x^{Q_p}}{s^2 + s\omega_p/Q_p + \omega_p^2} \tag{3-60}$$

Finally, if we make use of Eq. (3-44a) we can derive the magnitude sensitivity

$$S_x^{|H_2(j\omega)|} = -\frac{2(1 - \omega_n^2) + \omega_n^2/Q_p^2}{(1 - \omega_n^2)^2 + \omega_n^2/Q_p^2}S_x^{\omega_p} + \frac{\omega_n^2/Q_p^2}{(1 - \omega_n^2)^2 + \omega_n^2/Q_p^2}S_x^{Q_p} \tag{3-61}$$

$$= S_{\omega_p}^{|H_2|}S_x^{\omega_p} + S_{Q_p}^{|H_2|}S_x^{Q_p}$$

where $\omega_n = \omega/\omega_p$ is a normalized frequency and

$$S_{\omega_p}^{|H_2|} = -\frac{2(1 - \omega_n^2) + \omega_n^2/Q_p^2}{(1 - \omega_n^2)^2 + \omega_n^2/Q_p^2} \tag{3-62a}$$

$$S_{Q_p}^{|H_2|} = \frac{\omega_n^2/Q_p^2}{(1 - \omega_n^2)^2 + \omega_n^2/Q_p^2} \tag{3-62b}$$

A better appreciation of the meaning of Eq. (3-62) is obtained from a plot of $S_{\omega_p}^{|H_2|}$ and $S_{Q_p}^{|H_2|}$; as shown in Fig. 3-5, both $S_{\omega_p}^{|H_2|}$ and $S_{Q_p}^{|H_2|}$ are strong functions of frequency with

$$\max\{S_{Q_p}^{|H_2|}\} = 1 \quad \text{at} \quad \omega = \omega_p \tag{3-63a}$$

and, for large Q_p,

$$\max\{S_{\omega_p}^{|H_2|}\} \simeq \frac{Q_p}{1 + 1/Q_p} \quad \text{at} \quad \omega \simeq \omega_p\left(1 + \frac{1}{2Q_p}\right) \tag{3-63b}$$

$$\min\{S_{\omega_p}^{|H_2|}\} \simeq -\frac{Q_p}{1 - 1/Q_p} \quad \text{at} \quad \omega \simeq \omega_p\left(1 - \frac{1}{2Q_p}\right) \tag{3-63c}$$

as the student should verify from Eq. (3-62). Note that the extreme values of $S_{\omega_p}^{|H_2|}$ occur approximately at the 3-dB frequencies of the high-Q second-order functions and that the ratio

$$\frac{S_{\omega_p}^{|H_2|}}{S_{Q_p}^{|H_2|}} = 2Q_p^2\left(1 - \frac{1}{\omega_n^2}\right) - 1 \tag{3-64}$$

increases rapidly with the distance from the pole frequency $\omega = \omega_p$. Specifically, at the critical "bandedge" frequencies, $\omega \simeq \omega_p[1 \pm 1/(2Q_p)]$, we find

$$S_{\omega_p}^{|H_2|} \simeq \pm 2Q_pS_{Q_p}^{|H_2|} \tag{3-65}$$

All these considerations lead to the important conclusion that in a search for "good, practical" second-order sections with high values of Q_p it is more important to pay attention to low values of $S_x^{\omega_p}$ than to small values of $S_x^{Q_p}$. If the variabilities of

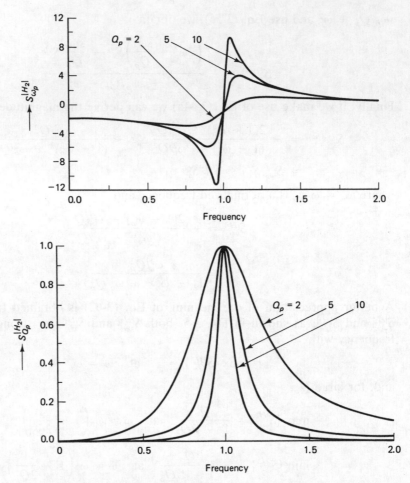

Figure 3-5 Transfer function magnitude sensitivities to pole frequency ω_p and pole quality factor Q_p.

w_p and Q_p are to have comparable effects on the transfer function, then $d\omega_p/\omega_p$ should be approximately Q_p times smaller than dQ_p/Q_p, because, by Eq. (3-61),

$$\frac{d|H_2(j\omega)|}{|H_2(j\omega)|} = S_{\omega_p}^{|H_2|} \frac{d\omega_p}{\omega_p} + S_{Q_p}^{|H_2|} \frac{dQ_p}{Q_p} \qquad (3\text{-}66)$$

This fact is further illustrated in Fig. 3-6, where for ease of comparison the two sensitivities are plotted for $Q_p = 10$ on the same scale. Finally, we want to emphasize that the results expressed in Eqs. (3-59) to (3-66) and Figs. 3-5 and 3-6 are valid for *any* second-order filter section, *independent* of its realization. Thus, the importance of small pole-frequency variabilities $d\omega_p/\omega_p$, as expressed in

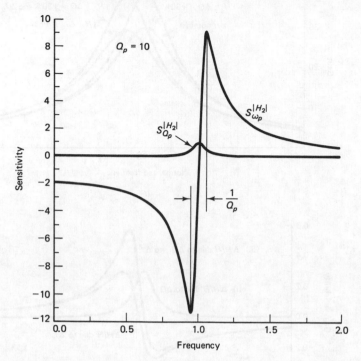

Figure 3-6 $S_{\omega_p}^{|H_2|}$ and $S_{Q_p}^{|H_2|}$ plotted on the same scale to show different magnitude.

Eq. (3-66), should be foremost in the designer's mind when selecting second-order active filter building blocks.

The following example illustrates some of these considerations:

Example 3-7

Consider a bandpass function $H(s)$ with center frequency $f_0 = 1$ kHz, pole quality factor $Q_p = 10$, and midband gain of 30. Assume that f_0 and Q_p depend on varying circuit parameters that cause variabilities of $\Delta f_0/f_0 = 3\%$ and $\Delta Q_p/Q_p = 30\%$. Find the resulting variabilities of the transfer function $H(s)$.

Solution. For the given requirements, the nominal transfer function is

$$H(s) = \frac{3 \cdot 2\pi f_p s}{s^2 + s\,(2\pi f_p/Q_p) + (2\pi f_p)^2}$$

where $f_p = f_0 = 1$ kHz nominally. Normalizing the frequency, i.e., taking $s_n = s/(2\pi f_0)$, yields

$$H(s_n) = \frac{3as_n}{s_n^2 + s_n a/Q_p + a^2}$$

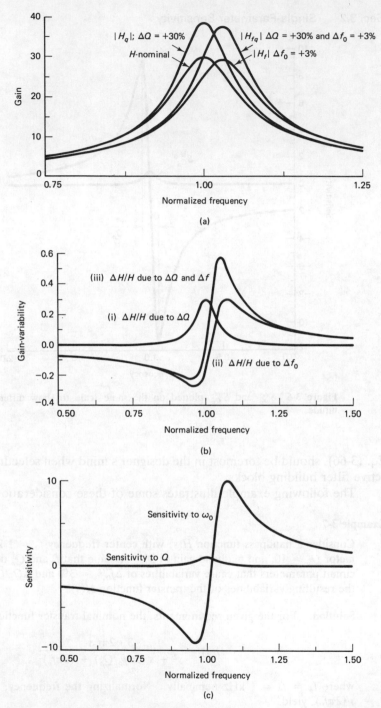

Figure 3-7 Expected gain curves (a), gain variabilities (b), and sensitivity comparison (c) in Example 3-7 for a 30% Q_p change and a 3% ω_p change.

with $a = f_p/f_0$ so that the magnitude becomes

$$|H(j\omega)| = \frac{3a\omega_n}{\sqrt{(a^2 - \omega_n^2)^2 + (\omega_n a/Q_p)^2}}$$

Figure 3-7a contains plots of $|H_{nom}|$ for $a = 1$, $Q_p = 10$; $|H_q|$ for $a = 1$, $Q_p = 13$; $|H_f|$ for $a = 1.03$, $Q_p = 10$; and $|H_{fq}|$ for $a = 1.03$, $Q_p = 13$. The three plots show the expected absolute deviations of the gain for the different cases. From these curves we can calculate

$$\frac{\Delta H}{H} = \left|\frac{H_q}{H_{nom}}\right| - 1 \qquad \text{due to the } Q_p \text{ tolerances } \Delta Q_p$$

$$\frac{\Delta H}{H} = \left|\frac{H_f}{H_{nom}}\right| - 1 \qquad \text{due to the } f_0 \text{ tolerances } \Delta f_0$$

and

$$\frac{\Delta H}{H} = \left|\frac{H_{fq}}{H_{nom}}\right| - 1 \qquad \text{due to both } \Delta f_0 \text{ and } \Delta Q_p$$

The resulting curves as a function of frequency are seen in Fig. 3-7b. Note that curves (i) and (ii) resemble those of Fig. 3-5 but that in this example curve (iii) is *not* simply the addition of curves (i) and (ii) as suggested by Eq. (3-66), because the Q_p and f_0 variances of 30% and 3%, respectively, are *not small* changes[5] as was assumed in the derivation of Eq. (3-66). Observe also that the extrema of curves (i) and (ii) are approximately the same, as should have been expected from our earlier discussion because we chose $\Delta Q_p/Q_p = Q_p \cdot \Delta f_0/f_0 = 0.3$. Nevertheless

$$\max S_{\omega_0}^{|H|} \simeq Q_p \max S_{Q_p}^{|H|} = 10 \max S_{Q_p}^{|H|}$$

as can be seen from Fig. 3-7c, where $S_{Q_p}^{|H|}$ and $S_{\omega_0}^{|H|}$ were obtained by dividing curves (*i*) and (*ii*) of Fig. 3-7b by $\Delta Q_p/Q_p = 0.3$ and $\Delta f_0/f_0 = 0.03$, respectively.

3.2.6 High-Order Active Filters

After having derived some important criteria for the selection of good second-order active filters, we shall in this section use sensitivity considerations to establish guidelines for the design of active filters of higher than second order. The transfer function is given by Eq. (3-36) and can be written in factored form as

$$H(s) = \frac{N(s)}{D(s)} = \frac{\displaystyle\prod_{k=1}^{m/2} (\alpha_{2k}s^2 + \alpha_{1k}s + \alpha_{0k})}{\displaystyle\prod_{i=1}^{n/2} (s^2 + s\omega_{pi}/Q_{pi} + \omega_{pi})} \tag{3-67}$$

---

[5] $\Delta Q/Q = 0.3$ is certainly not small, and even $\Delta f_0/f_0 = 0.03$ is large when one considers that max $S_{\omega_0}^{|H|} \simeq Q$, so that, by Eq. (3-66), $Q \Delta f_0/f_0 = \Delta f_0/B$ affects the variability of $|H(j\omega)|$. The important change is, therefore, Δf_0 *as a fraction of the bandwidth* $B = f_0/Q$, which in our example is also 30%! This consideration often limits the maximum achievable value of Q in active filters.

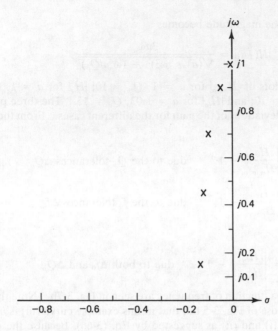

Figure 3-8 Pole locations for tenth-order 1 dB–ripple Chebyshev filter.

where we have assumed that n is even.[6] If n is odd, the denominator also contains a factor $(s + \sigma)$.

According to Eqs. (3-37) and (3-47), the single-parameter small-change sensitivity of $H(s)$ to x equals

$$S_x^H = \frac{x}{N(s)D(s)} \left[D(s) \frac{\partial N(s, x)}{\partial x} - N(s) \frac{\partial D(s, x)}{\partial x} \right] \tag{3-68}$$

$$= S_x^K - \sum_{i=1}^{m} \frac{Q_x^{z_i}}{s - z_i} + \sum_{i=1}^{n} \frac{Q_x^{p_i}}{s - p_i}$$

As mentioned earlier, this equation tells us that the transfer function sensitivity to any circuit parameter x will be quite large in the neighborhood of any pole or zero because S_x^H has poles at all poles and zeros of $H(s)$. Consequently, in a high-order highly selective filter, the sensitivity must be expected to be large throughout the passband and especially near the passband edges, because many or all of the filter poles are distributed over the passband close to the $j\omega$-axis, i.e., they have large Q factors and $D(s)$ becomes small. As an illustration, Fig. 3-8 shows the upper half-plane poles of a tenth-order 1 dB–ripple Chebyshev filter. We may conclude from Eq. (3-68), therefore, that a *direct realization* of a high-order transfer function will in general *not* lead to practically useful designs, because, for the

---

[6]If the degree m of $N(s)$ is odd, one of the coefficients α_{2k} is zero.

achievable finite component variabilities dx/x due to fabrication tolerances or drifts, the transfer function variability dH/H will be unacceptably large.

In a direct realization, in general all poles and zeros of $H(s)$ will depend on each element, so that none of the sensitivities $Q_x^{z_i}$ and $Q_x^{p_i}$ are zero. If instead we can devise a design topology where any component x affects only *one* pole pair or zero pair, then all the remaining sensitivities are zero and Eq. (3-68) reduces to Eq. (3-59) with much-improved results. This approach leads to the *cascade design* to be discussed in Section 3.2.6.1.

Alternatively, if one can devise a circuit topology for which the magnitude of the polynomial in brackets in Eq. (3-68), $[D \, \partial N/\partial x - N \, \partial D/\partial x]$, is very small in the frequency range of interest, then, of course, the passband sensitivities can be low even if the filter function $H(s)$ contains many high-Q poles. The method of *LC ladder simulation* discussed in Section 3.2.6.2 leads to the appropriate design technique.

3.2.6.1 Cascade design.

As pointed out above, our goal is to have the element x influence only *one* pole pair and/or zero pair so that the effects of any variations Δx are isolated from all other critical frequencies. Evidently, this implies that the coefficients of only a single biquadratic function of the form of Eqs. (1-49) and (1-50), i.e.,

$$H_k(s) = \frac{\alpha_{2k}s^2 + \alpha_{1k}s + \alpha_{0k}}{s^2 + s\omega_{pk}/Q_{pk} + \omega_{pk}^2} \tag{3-69}$$

may depend on x. It requires that $H(s)$ of Eq. (3-67) be factored into

$$H(s) = \prod_{k=1}^{n/2} H_k(s) = \prod_{k=1}^{n/2} \frac{\alpha_{2k}s^2 + \alpha_{1k}s + \alpha_{0k}}{s^2 + s\omega_{pk}/Q_{pk} + \omega_{pk}^2} \tag{3-70}$$

with each function $H_k(s)$ realized as a second-order section (a *biquad*) with no interaction between sections. In this case the circuit structure appears as in Fig. 3-9, with

$$H(s) = \frac{V_{out}}{V_{in}} = \frac{V_1}{V_{in}} \frac{V_2}{V_1} \frac{V_3}{V_2} \cdots \frac{V_{n/2-1}}{V_{n/2-2}} \frac{V_{out}}{V_{n/2-1}}$$

$$= H_1 H_2 H_3 \cdots H_{n/2-1} H_{n/2}$$

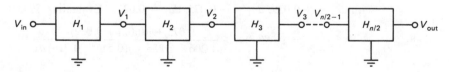

Figure 3-9   Cascade connection.

as required by Eq. (3-70), and the sensitivities of the transfer function $H(s)$ are

$$S_{H_j(s)}^{H(s)} = 1 \qquad (3\text{-}71a)$$

$$S_x^{H(s)} = S_x^{H_j(s)} \qquad (3\text{-}71b)$$

where we have assumed that the element of interest, x, is in biquad j. Equations (3-71a) and (3-71b) indicate that the sensitivity of the high-order transfer function to the element x and its variability in relation to x are as large as those of the second-order building block which contains the element. It follows that any cascade realization of $H(s)$ should use the "best possible" biquads, i.e., those of lowest sensitivity, according to the criteria established in Section 3.2.5.

Example 3-8

To be designed is a fourth-order Chebyshev lowpass filter with 2 dB passband ripple. From Appendix III, Table III-2c, the frequency-normalized transfer function is

$$H(s) = \frac{N(s)}{D(s)} = \frac{0.163}{s^4 + 0.716s^3 + 1.256s^2 + 0.517s + 0.206}$$

$$= \frac{0.163}{(s^2 + 0.210s + 0.928)(s^2 + 0.506s + 0.221)}$$

Assume that $H(s)$ is realized in two forms:

(a) Directly as one fourth-order block where a variation $\Delta x/x = 0.1$ of an element x is found to cause the following changes in the coefficients:
$a_3 = 0.716$: -4%
$a_2 = 1.256$: -6%
$a_1 = 0.517$: $+4\%$

(b) As a cascade of two second-order lowpass sections where $\Delta x/x = 0.1$ causes the coefficient $b_1 = 0.506$ to increase by 12%

Calculate the variability $\Delta|H|/|H|$ and $\Delta\phi$ for the two cases at the passband edge, $\omega = 1$. ϕ is the phase of $H(j\omega)$.

Solution.   According to Eq. (3-37),

$$S_x^H = -\frac{x}{D}\frac{\partial D}{\partial x} = -\frac{x}{D}\left(s^3\frac{da_3}{dx} + s^2\frac{da_2}{dx} + s\frac{da_1}{dx}\right)$$

$$= -\frac{s^3\,da_3 + s^2\,da_2 + s\,da_1}{D(s)} \cdot \frac{x}{dx}$$

with $da_3 = -0.029$, $da_2 = -0.075$, $da_1 = 0.021$. Evaluating S_x^H at $s = j1$ yields

$$S_x^{H(j1)} = -\frac{0.075 + j(0.021 + 0.029)}{1.206 - 1.256 + j(0.517 - 0.716)}\frac{x}{dx}$$

$$= (0.325 - j0.295)\frac{x}{dx}$$

Because, by Eq. (3-43),

$$S_x^{H(j1)} = \frac{x}{|H(j1)|} \frac{d|H(j1)|}{dx} + jx \frac{d\phi(j1)}{dx}$$

we have at $\omega = 1$:

$$\frac{\Delta|H|}{|H|} \simeq 0.33 = 33\% \qquad \Delta\phi \simeq -0.30 \text{ rad} \simeq -17.2°$$

For the cascade realization, we find similarly

$$S_x^H = \frac{0.061s}{s^2 + 0.506s + 0.221} \frac{x}{dx}$$

and

$$S_x^{H(j1)} = \frac{-(0.055 + j0.036)}{0.1}$$

so that

$$\frac{\Delta|H|}{|H|} \simeq -0.055 = -5.5\% \qquad \Delta\phi \simeq -0.036 \text{ rad} \simeq -2.1°$$

Note that even in this simple fourth-order case with low values of pole quality factors, the magnitude variability of the direct design is 6 times larger than that of the cascade implementation. This is so in spite of the fact that the coefficient error of b_1 is twice as large as the largest of the a_i errors. Also observe that the phase error is 8 times larger in the direct implementation.

We have established that, on the basis of sensitivity considerations alone, realizing a high-order transfer function $H(s)$ as a cascade of second-order sections [possibly in addition to a first-order section if the degree of $H(s)$ is odd] is in general far superior to a direct realization as a single high-order block. Also, a cascade implementation is more modular, using largely identical building blocks; is easier to tune because the blocks do not interact; and, as we shall see, is generally easier to design. Therefore, it is worthwhile to consider the realization of biquadratic transfer functions as good low-sensitivity second-order filter sections. Chapter 5 is devoted to this problem; further details of cascade design are discussed in Section 6.2.

3.2.6.2 Simulation of *LC* ladders.

We pointed out earlier that the sensitivity equation [Eq. (3-68)] depends only on the transfer function $H(s)$ and is valid regardless of the method of realization and the type of network. The equation holds, therefore, also for the transfer function

$$H(j\omega, x) = \exp\left[-\alpha(\omega, x) + j\phi(\omega, x)\right] \qquad (3-72)$$

of an LC filter, as defined via Eq. (2-7):

$$|H(j\omega, x)|^2 = \frac{P_2}{P_{max}} = \left|\frac{N(j\omega, x)}{D(j\omega, x)}\right|^2 \tag{3-73}$$

From Eq. (3-72) we can calculate the attenuation α, in nepers:

$$\alpha(\omega, x) = -\ln|H(j\omega, x)| \tag{3-74}$$

and if the circuit parameter x varies, the attenuation sensitivity is obtained from the *real part* of Eq. (3-68):

$$x\frac{\partial\alpha}{\partial x} = -\frac{x}{|H(j\omega, x)|}\frac{\partial|H(j\omega, x)|}{\partial x} = -S_x^{|H(j\omega,x)|} = -\operatorname{Re} S_x^{H(j\omega,x)} \tag{3-75}$$

Now, if the network is a *passive lossless* twoport designed for *maximum power transfer* as the LC ladders in Chapter 2, then, by Eq. (2-7c), we have $|H(j\omega, x)| \leq 1$ with $|H(j\omega, x)| = 1$ for $x = x_{nominal}$ at the reflection zeros ω_{ri} [see Eqs. (2-8) and (2-11b)]. Consequently, $\alpha(\omega, x) = 0$ at these points of perfect transmission and, for a change in frequency or in *any* component, the attenuation α of the *passive* network can only *increase*. This implies, of course, that α not only has double zeros as a function of ω, but it also has zeros of even multiplicity as a function of x, so that $\partial\alpha/\partial x = 0$ at $x = x_{nom}$ and $\omega = \omega_{ri}$ (Fig. 3-10) [34]. We note, therefore, from Eq. (3-75) that for passive lossless filters designed for maximum power transfer the magnitude sensitivity $S_x^{|H(j\omega,x)|}$ to any element x is zero in the passband at all the reflection zeros ω_{ri}.

A more careful look at the LC transfer function shows that at passband frequencies other than reflection zeros, the transfer function sensitivities are still very small, although not zero [11, 24, 35]. To this end, consider the voltage transfer function V_2/V_S, Eq. (2-16):

$$H(s) = \frac{2\sqrt{R_S R_L}z_{12}}{(z_{11} + R_S)(z_{22} + R_L) - z_{12}^2} \tag{3-76}$$

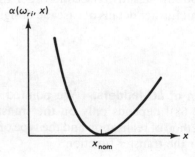

Figure 3-10 Quadratic dependence of attenuation of LC filter on any of the elements x at a reflection zero ω_{ri}.

From this equation it is easy to show (Problem 3.26) that

$$S_{R_S}^{H(s)} = -\frac{1}{2}\frac{R_S - Z_{in}(s)}{R_S + Z_{in}(s)} = -\frac{1}{2}\rho_1(s) \tag{3-77a}$$

$$S_{R_L}^{H(s)} = -\frac{1}{2}\frac{R_L - Z_{out}(s)}{R_L + Z_{out}(s)} = -\frac{1}{2}\rho_2(s) \tag{3-77b}$$

where the input reflection coefficient $\rho_1(s)$ was defined in Eq. (2-9) and the output reflection coefficient $\rho_2(s)$ is defined similarly, with $Z_{out}(s)$ being the impedance "seen into" the output of the LC filter when terminated at the input by R_S. Further, making use of Eq. (3-24), we find for the sensitivities to the reactive elements L_i and C_j

$$\sum_i S_{L_i}^{H(s)} = \frac{1}{2}\left\{\frac{s}{H(s)}\frac{\partial H(s)}{\partial s} + \frac{1}{2}[\rho_1(s) + \rho_2(s)]\right\} \tag{3-78a}$$

$$\sum_j S_{C_j}^{H(s)} = \frac{1}{2}\left\{\frac{s}{H(s)}\frac{\partial H(s)}{\partial s} - \frac{1}{2}[\rho_1(s) + \rho_2(s)]\right\} \tag{3-78b}$$

Finally, from Eqs. (3-77) and (3-44) we obtain

$$R_i\frac{\partial\alpha(\omega)}{\partial R_i} = \frac{1}{2}\,\text{Re}\,\{\rho_i(j\omega)\} \qquad i = 1, 2 \tag{3-79a}$$

and, from Eqs. (3-78) and (3-72),

$$\sum_i L_i\frac{\partial\alpha(\omega)}{\partial L_i} = \frac{1}{2}\left\{\omega\frac{\partial\alpha(\omega)}{\partial\omega} - \frac{1}{2}\text{Re}\,[\rho_1(j\omega) + \rho_2(j\omega)]\right\} \tag{3-79b}$$

$$\sum_j C_j\frac{\partial\alpha(\omega)}{\partial C_j} = \frac{1}{2}\left\{\omega\frac{\partial\alpha(\omega)}{\partial\omega} + \frac{1}{2}\text{Re}\,[\rho_1(j\omega) + \rho_2(j\omega)]\right\} \tag{3-79c}$$

Recall now that in lossless filters designed for maximum power transfer, ρ_1, ρ_2, and $\partial\alpha/\partial\omega$ are zero in the passband at $\omega = \omega_{ri}$, which confirms that all element sensitivities are zero at these frequencies. In the remainder of the passband, we have $|\rho_i(j\omega)| \ll 1$, $i = 1, 2$; and, by Eq. (2-8),

$$\alpha(\omega) = -\tfrac{1}{2}\ln\,(1 - |\rho_1(j\omega)|^2) \ll 1$$

so that the element sensitivities are seen to be very small throughout the passband. Note, however, that in the stopband and transition band(s) the sensitivities to L_i and C_j are usually large because of the steep slope of the attenuation $\alpha(\omega)$ (compare Fig. 1-22).

Equations (3-78a) and (3-78b) can be used to identify the deviations in LC filter performance that must be expected when the components are lossy [93–95]. Let us assume that the inductor losses can be represented by a series loss resistor

R_L and that analogously the capacitor losses are modeled by a parallel conductance G_C; then we have for the inductive and capacitive immittances, respectively,

$$j\omega L \rightarrow j\omega L_0 + R_L = j\omega L_0\left(1 - j\frac{R_L}{\omega L_0}\right)$$

$$= j\omega L_0\left(1 - j\frac{1}{Q_L}\right) = j\omega(L_0 + dL) \quad (3\text{-}80a)$$

$$j\omega C \rightarrow j\omega C_0 + G_C = j\omega C_0\left(1 - j\frac{G_C}{\omega C_0}\right)$$

$$= j\omega C_0\left(1 - j\frac{1}{Q_C}\right) = j\omega(C_0 + dC) \quad (3\text{-}80b)$$

Q_L and Q_C are the components' quality factors, and the "component errors" dL and dC are defined as follows:

$$dL = -j\frac{R_L}{\omega} = -j\frac{L_0}{Q_L} \quad \text{and} \quad dC = -j\frac{G_C}{\omega} = -j\frac{C_0}{Q_C} \quad (3\text{-}81)$$

With these definitions, the variability of the transfer function $H(s, L_i, C_i, R_k)$ of the LC network is given by

$$\frac{dH}{H} = \sum_i S_{L_i}^H \frac{dL_i}{L_{0i}} + \sum_j S_{C_j}^H \frac{dC_j}{C_{0j}} = -j\left[\sum_i S_{L_i}^H \frac{1}{Q_{Li}} + \sum_j S_{C_j}^H \frac{1}{Q_{Cj}}\right] \quad (3\text{-}82)$$

Making now the realistic assumption that all inductors and capacitors, respectively, in the filter have the same quality factors and using Eq. (3-78) for the sensitivity sums, we can rewrite Eq. (3-82) as

$$\frac{dH}{H} = -j\left\{\frac{1}{2Q_L}\left[S_s^H + \frac{1}{2}(\rho_1 + \rho_2)\right] + \frac{1}{2Q_C}\left[S_s^H - \frac{1}{2}(\rho_1 + \rho_2)\right]\right\}$$

$$= -j\left[\frac{1}{2}\left(\frac{1}{Q_L} + \frac{1}{Q_C}\right)S_s^H + \frac{1}{4}\left(\frac{1}{Q_L} - \frac{1}{Q_C}\right)(\rho_1 + \rho_2)\right] \quad (3\text{-}83)$$

More insight can be gained if we evaluate Eq. (3-83) on the $j\omega$-axis and express H in terms of the filter's loss $\alpha(\omega)$ and its phase $\phi(\omega)$ as in Eq. (3-72). The result is

$$-d\alpha(\omega) + j\, d\phi(\omega) = -\frac{j}{2}\left[\left(\frac{1}{Q_L} + \frac{1}{Q_C}\right)\left(-\omega\frac{d\alpha}{d\omega} + j\omega\frac{d\phi}{d\omega}\right)\right.$$

$$\left. + \frac{1}{2}\left(\frac{1}{Q_L} - \frac{1}{Q_C}\right)(\rho_1 + \rho_2)\right] \quad (3\text{-}84)$$

Finally, we note that $-d\phi/d\omega = \tau(\omega)$, the filter delay, and we remember that $\rho_i(j\omega)$ is a complex number. Therefore, separating real and imaginary parts of

Eq. (3-84) results in the expressions

$$da(\omega) = \frac{1}{2}\left(\frac{1}{Q_L} + \frac{1}{Q_C}\right)\omega\tau(\omega) - \frac{1}{4}\left(\frac{1}{Q_L} - \frac{1}{Q_C}\right)\text{Im}\{\rho_1(j\omega) + \rho_2(j\omega)\} \qquad (3\text{-}85a)$$

$$d\phi(\omega) = \frac{1}{2}\left(\frac{1}{Q_L} + \frac{1}{Q_C}\right)\omega\frac{da(\omega)}{d\omega} - \frac{1}{4}\left(\frac{1}{Q_L} - \frac{1}{Q_C}\right)\text{Re}\{\rho_1(j\omega) + \rho_2(j\omega)\} \qquad (3\text{-}85b)$$

We saw earlier that in the passband of the *LC* filter, $|\rho_1(\omega) + \rho_2(\omega)| \ll 1$, so that the second terms in Eq. (3-85) will mostly be negligible.[7] Also note that in the passband, $|\alpha(\omega)| \ll 1$, so that the first term in Eq. (3-85b) will be very small. Consequently, the dominant effects of finite inductor and capacitor losses are given by

$$d\alpha(\omega) \simeq \frac{1}{2}\left(\frac{1}{Q_L} + \frac{1}{Q_C}\right)\omega\tau(\omega) = \frac{1}{2}\left(\frac{r_L}{L_0} + \frac{g_C}{C_0}\right)\tau(\omega) \qquad (3\text{-}86)$$

This expression shows that for *positive* values of Q_L and Q_C the attenuation $\alpha(\omega)$ *increases* with the same frequency dependence as that of the delay $\tau(\omega)$. The largest errors can, therefore, be expected at the bandedge (compare Fig. 1-28). Similarly, if $(1/Q_L + 1/Q_C) < 0$, a situation that will be seen in later chapters to occur relatively frequently in *active* simulations of *LC* ladders, the total attenuation $\alpha(\omega)$ will *decrease*. Figure 3-11 illustrates the typical behavior for these cases. In

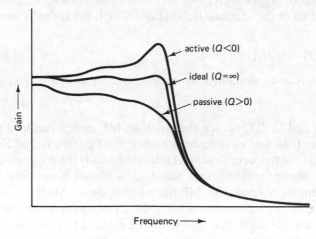

Figure 3-11 Curves indicating the typical performance (exaggerated) of an *LC* filter with no losses (ideal), with positive losses ($Q > 0$), and of an active simulation when losses are negative ($Q < 0$).

active filters, the gain enhancement caused by negative quality factors usually requires that compensation measures be taken [96].

After the excellent sensitivity properties of *LC* filters with maximum power transfer were recognized [34], considerable effort went into the search for *active*

[7]An exception may occur in lowpass filters, where the two terms can be comparable in magnitude [95] because $\omega\tau(\omega)$ also is small.

RC filter designs that would somehow "inherit" these properties. The resulting methods use either operational or topological simulations of *LC* ladders. The details will be discussed in Chapter 6.

3.2.6.3 Multiple-feedback topologies.

Among high-order active filters, those that simulate the behavior of *LC* ladders have been found to have the lowest sensitivity to component variations. Cascade realizations, although much less sensitive to element tolerances than direct implementations (other than via *LC* ladder simulations), often—especially for high-*Q* narrowband filters—have transfer function variabilities that are quite unacceptable in practice. On the other hand, their ease of design and the modularity achieved by interconnecting second-order building blocks are often perceived as practical advantages. As a result, several circuit configurations have been proposed where second-order sections are interconnected not just by simple cascading, as in Fig. 3-9, but instead in various *multiple-feedback* (MF) topologies with the goal of reducing sensitivities by means of carefully chosen coupling between sections. Here, an MF topology is understood to mean a network with a single forward path from input to output through unilateral low-order sections (e.g., biquads), which in turn are embedded in a (usually) resistive feedback configuration. Quite a number of such topologies have been found useful in practice; see reference 36 and Chapter 6. Typical examples, the so-called *follow-the-leader feedback* (FLF) and *leapfrog* (LF) configurations, are shown in Fig. 3-12a and b, respectively. In MF filters, the transfer function *H*(*s*) is not simply the product of the sections $H_k(s)$ [Eq. (3-70)], but rather a more general function

$$H(s) = f\{H_k(s)\} \qquad k = 1, \ldots, n \qquad (3\text{-}87)$$

so that the sensitivity of *H*(*s*) to an element *x* in section *j* becomes, by Eq. (3-19),

$$S_x^H = S_{H_j}^f S_x^{H_j} \qquad (3\text{-}88)$$

By comparing Eqs. (3-88) and (3-71), we see that such an MF design *can* lead to a sensitivity behavior better than that of cascade circuits *if* the function *f*{ } of Eq. (3-87) is such that $|S_{H_j}^f| < 1$ in the frequency range of interest (usually the passband).

On the basis of the above considerations alone, it is almost impossible to arrive at generally valid sensitivity results for MF filter topologies. Also, it is not our intention here to delve deeply into the sensitivity theory of multiple-feedback filters. Instead, we shall use the particular cases, the FLF and LF topologies of Fig. 3-12, to convey to the reader some intuitive understanding of the potential and the merit of the MF concept. To this end we analyze the block diagrams of Fig. 3-12 to yield (Problem 3.29)

$$H_{\text{FLF}}(s) = \frac{V_{\text{out}}}{V_{\text{in}}} = \frac{\displaystyle\prod_{i=1}^{n} H_i(s)}{1 + \displaystyle\sum_{k=1}^{n} F_k \prod_{j=1}^{k} H_j(s)} \qquad (3\text{-}89\text{a})$$

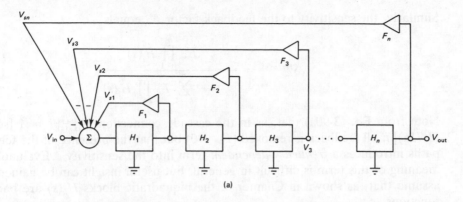

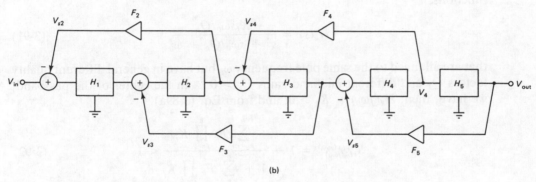

Figure 3-12 "Follow-the-leader feedback" (FLF) (a) and "leapfrog" (LF) (b) topologies of high-order active filters.

and

$$H_{LF}(s) = \cfrac{\displaystyle\prod_{j=1}^{5} H_j(s)}{\begin{array}{c} 1 + F_2 H_1 H_2 + F_3 H_2 H_3 + F_4 H_3 H_4 + F_5 H_4 H_5 + F_2 F_4 H_1 H_2 H_3 H_4 \\[2mm] + F_2 F_5 H_1 H_2 H_4 H_5 + F_3 F_5 H_2 H_3 H_4 H_5 \end{array}} \tag{3-89b}$$

Equations (3-89a) and (3-89b) define the function $f\{H_k\}$ introduced in Eq. (3-87); the $H_i(s)$ are the transfer functions of second-order sections, and the feedback factors F_i are defined via $F_i = V_{si}/V_i$ as indicated in Fig. 3-12. We observe from Eq. (3-89) that both $H_{FLF}(s)$ and $H_{LF}(s)$ reduce to a cascade transfer function, Eq. (3-70), if all the feedback factors are set to zero. Because sensitivity behavior is our concern here, we evaluate $S_{Hm(s)}^{H_{FLF}(s)}$, $1 \le m \le n$, from Eq. (3-89a) to obtain

$$S_{Hm(s)}^{H_{FLF}(s)} = 1 - \cfrac{\displaystyle\sum_{k=m}^{n} F_k \prod_{j=1}^{k} H_j(s)}{1 + \displaystyle\sum_{k=1}^{n} F_k \prod_{j=1}^{k} H_j(s)} \tag{3-90a}$$

Similarly, the sensitivity to the feedback factor F_m equals

$$S_{F_m}^{H_{FLF}(s)} = - \frac{F_m \prod\limits_{j=1}^{m} H_j(s)}{1 + \sum\limits_{k=1}^{n} F_k \prod\limits_{j=1}^{k} H_j(s)} \tag{3-90b}$$

Note from Eq. (3-90a) that, as in the cascade connection, $S_{H_m(s)}^{H_{FLF}(s)} = 1$ [see Eq. (3-71a)] *if* all feedback is removed ($F_k = 0$) but that the presence of the feedback paths introduces a *frequency-dependent* term into the sensitivity. Evaluating the meaning of this term is difficult in general, but useful insight can be gained if we assume that, as shown in Chapter 6, the biquadratic blocks $H_k(s)$ are bandpass functions,

$$H_k(s) = \frac{K_k s \omega_0 / Q_k}{s^2 + s \omega_0 / Q_k + \omega_0^2} \tag{3-91}$$

that are all tuned to the same pole frequency ω_0 but have in general different quality factors Q_k and different midband gains $K_k > 0$.[8] In the center of the passband, we have, then, $H_k(j\omega_0) = K_k > 0$; and from Eq. (3-89a),

$$S_{H_m(j\omega_0)}^{H_{FLF}(j\omega_0)} = 1 - \frac{\sum\limits_{k=m}^{n} F_k \prod\limits_{j=1}^{k} K_j}{1 + \sum\limits_{k=1}^{n} F_k \prod\limits_{j=1}^{k} K_j} < 1 \tag{3-92}$$

verifying that, at least at $\omega = \omega_0$, the FLF sensitivity is reduced below the cascade sensitivity, i.e., with Eq. (3-88),

$$|S_x^{H_{FLF}(j\omega_0)}| < |S_x^{H_m(j\omega_0)}|$$

Note that the sensitivity in Eq. (3-92) is *real*, i.e., Equation (3-92) gives us directly the magnitude sensitivity at $\omega = \omega_0$. Similar reasoning shows that $|S_{F_m}^{H_{FLF}(j\omega_0)}| < 1$, from Eq. (3-90b).

Turning now to the lower and upper stopbands, i.e., $|s| \ll \omega_0$ and $|s| \gg \omega_0$, respectively, we find from Eq. (3-91)

$$H_k(s) \simeq \frac{\omega_0}{s} \frac{K_k}{Q_k} \to 0 \qquad |s| \gg \omega_0 \tag{3-93a}$$

and

$$H_k(s) \simeq \frac{s}{\omega_0} \frac{K_k}{Q_k} \to 0 \qquad |s| \ll \omega_0 \tag{3-93b}$$

---

[8]The simple FLF topology of Fig. 3-12a is especially useful for geometrically symmetrical all-pole bandpass filters where ω_0 is the passband center frequency.

Consequently, in both stopbands, the sensitivity

$$S_{H_{m(s)}}^{H_{\mathrm{FLF}}(s)} \to 1 \tag{3-94a}$$

as in a cascade connection and the sensitivities to F_m reduce to zero:

$$S_{F_m}^{H_{\mathrm{FLF}}(s)} \to 0 \tag{3-94b}$$

Again, the results in Eqs. (3-94a) and (3-94b) are intuitively appealing because, for $H_k(s) \to 0$, Eq. (3-89) reduces to the transfer function of a cascade filter. In Problem 3.31 the student is asked to show by the same reasoning that analogous results hold for the LF topology.

Although our discussion has dealt only with a special case, by simple extrapolation we are led to the following conclusions about multiple feedback topologies.

Feedback paths around low-order sections in an MF filter topology *can* lead to passband sensitivities lower than those of an equivalent cascade design. The sensitivity improvement is usually largest in the center and becomes less pronounced toward the edges of the passband. In the stopbands, where the feedback paths lose their effectiveness, MF and cascade sensitivities are approximately of the same magnitude.

To further support these conclusions, we show in Fig. 3-13 the standard deviations in dB of the gain variation $|\Delta|H||$ as a function of frequency as obtained by a Monte Carlo simulation (see Section 3.3.3) of a sixth-order Butterworth bandpass filter, realized in cascade, FLF, and LF topologies and as a passive *LC* ladder. The comparison shows that MF deviations are located between those of the cascade design at the upper end and those of the resistively terminated *LC* ladder at the lower end. We note that in all active realizations, the stopband deviations are comparable. The overall lower deviations of the *LC* ladder (by approximately a factor of 2) can be attributed to the fact that frequency parameters in *LC* filters are set by only two components, via *LC* resonance, whereas in active *RC* circuits at least four elements (two *R*'s and two *C*'s) are needed. Thus, more components contribute to the variance in active filters. For practical active filter realizations, this component-count disadvantage exists regardless of the synthesis approach.

3.3 MULTIPARAMETER SENSITIVITY

In Sections 3.1 and 3.2, we demonstrated that a significant amount of information on circuit performance can be gained by careful interpretation of single-parameter sensitivities. However, to gain a more realistic and complete picture of filter behavior under the influence of element tolerances, we have to take into consideration that the network function depends on many parameters or elements x_i, $i = 1, \ldots, k$, all of which will simultaneously be subject to tolerances (compare Example 3-1). Although the resulting *multiparameter* sensitivity can be understood and discussed via simple extensions of the concept of *single-parameter* sensitivity,

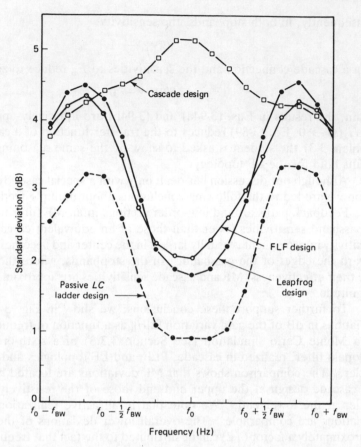

Figure 3-13 Simulated variations of the sixth-order Butterworth bandpass filter (1.732% passive component tolerances) for different realizations.

any actual evaluation in all but the simplest cases will involve computer-aided routines because of the large volume of computations in a multidimensional element space. In this section we shall discuss some of the theoretical and computational aspects and definitions of multiparameter sensitivity, first with the assumption that element tolerances are deterministic and later for the more realistic case of statistical parameter variations.

3.3.1 Deterministic Case

If a performance measure P depends on the k parameters x_i, for convenience gathered into the vector

$$\mathbf{x} = \{x_1 \quad x_2 \ldots x_k\}^t \tag{3-95}$$

where t denotes *transpose*, then by simple extension of Eqs. (3-1) to (3-3) we can use a truncated Taylor expansion to express the deviation of $P(s, \mathbf{x})$ as

$$dP \simeq \sum_{i=1}^{k} \left.\frac{\partial P}{\partial x_i}\right|_{x_{i0}} \cdot dx_i = P \sum_{i=1}^{k} \left(\frac{x_i}{P}\frac{\partial P}{\partial x_i}\right)_{x_{i0}} \frac{dx_i}{x_{i0}} \tag{3-96a}$$

or

$$\frac{dP}{P} \simeq \sum_{i=1}^{k} S_{x_i}^{P} \frac{dx_i}{x_{i0}} \tag{3-96b}$$

As always, sensitivities are evaluated at the nominal point

$$\mathbf{x}_0 = \{x_{10} \quad x_{20} \cdots x_{k0}\}^t$$

Defining sensitivity and element variability vectors, respectively, as

$$S_{\mathbf{x}}^{P} = \{S_{x_1}^{P} \quad S_{x_2}^{P} \dots S_{x_k}^{P}\}^t \tag{3-97}$$

and

$$d\hat{\mathbf{x}} = \left\{\frac{dx_1}{x_{10}} \quad \frac{dx_2}{x_{20}} \dots \frac{dx_k}{x_{k0}}\right\}^t \tag{3-98}$$

we may rewrite Eq. (3-96b) in compact form as

$$\frac{dP}{P} \simeq (S_{\mathbf{x}}^{P})^t \cdot d\hat{\mathbf{x}} \tag{3-99}$$

Evidently, Eq. (3-3) is a special case of Eq. (3-99).

As an example, consider the case where $P(s, \mathbf{x})$ is a transfer function $H(s)$ that depends on the m coefficients c_j, $j = 1, \ldots, m$, which in turn are functions of the k elements x_i, $i = 1, \ldots, k$. Then we may write, using Eqs. (3-19) and (3-96),

$$\frac{dH}{H} \simeq \sum_{i=1}^{k} \left[\sum_{j=1}^{m} \frac{c_j}{H}\frac{\partial H}{\partial c_j}\frac{x_{i0}}{c_j}\frac{\partial c_j}{\partial x_i}\frac{dx_i}{x_{i0}}\right]$$
$$= \sum_{i=1}^{k} \sum_{j=1}^{m} S_{c_j}^{H} S_{x_i}^{c_j} \frac{dx_i}{x_{i0}} \tag{3-100}$$

Using the more compact matrix notation, Eq. (3-100) becomes

$$\frac{dH}{H} \simeq (S_{\mathbf{c}}^{H})^t \, S_{\mathbf{x}}^{\mathbf{c}} \, d\hat{\mathbf{x}} \tag{3-101}$$

where the matrix $\mathbf{S}_x^c$ is defined as follows:

$$\mathbf{S}_x^c = \begin{pmatrix} S_{x_1}^{c_1} & \cdots & S_{x_k}^{c_1} \\ \cdot & & \cdot \\ \cdot & & \cdot \\ \cdot & & \cdot \\ S_{x_1}^{c_m} & \cdots & S_{x_k}^{c_m} \end{pmatrix} \tag{3-102}$$

and the vectors $\mathbf{S}_c^H$, $d\hat{\mathbf{x}}$, and $\mathbf{c}$ are defined through Eqs. (3-95), (3-97), and (3-98).

Observe that Eq. (3-101) provides significant insight into the causes contributing to the variability of $H(s, \mathbf{c}(\mathbf{x}))$. dH/H has been expressed in terms of three factors: the vector $\mathbf{S}_c^H$, which depends only on the *type* of transfer function $H(s, \mathbf{c})$; the matrix $\mathbf{S}_x^c$, which is determined solely by the dependence of the transfer function coefficients on the elements $\mathbf{c}(\mathbf{x})$, i.e., by network topology; and finally, the vector $d\hat{\mathbf{x}}$, which depends on the technology chosen to implement the network. This result supports our intuitive reasoning that the final performance of a network should depend on the choice of transfer function (see Section 1.6), on the circuit topology selected for the synthesis, and, of course, on the technology used for realizing the circuit components.

For optimization or for comparing the sensitivity performance of different networks, the sensitivity vector $\mathbf{S}_x^P$ of Eq. (3-97) is not suitable. Rather, a scalar quantity must be available. For these purposes a convenient multiparameter sensitivity measure is

$$S = \sum_{i=1}^{k} |S_{x_i}^P|^2 = (\mathbf{S}_x^P)^{t*} \mathbf{S}_x^P \tag{3-103}$$

as defined by Schoeffler [37]. As introduced earlier, t stands for *transpose* and * means complex conjugate. The square root of Eq. (3-103), i.e., the length of the vector $\mathbf{S}_x^P$, is also used by some authors. Because the different multiparameter sensitivity measures in Eqs. (3-97), (3-102), and (3-103) are based on the single-parameter sensitivity as defined in Eq. (3-4), it should be apparent to the student that all the comments and warnings made in the beginning of Section 3.1 also hold for multiparameter sensitivity calculations. Specifically, we note that $\mathbf{S}_x^P$ is in general a function of frequency if P is frequency-dependent and therefore must be evaluated in the appropriate frequency range. Further, the components of $\mathbf{S}_x^P$ are calculated at the *nominal point* $\mathbf{x}_0$, so that the sensitivity measures change if $\mathbf{x}_0$ is changed; this fact can be used for sensitivity optimization just as in Example 3-1. Finally, the student should remember that the transfer function sensitivity has a pole at a zero of $H(s)$ [Eq. (3-37)]. Therefore, if the multiparameter sensitivity of filters at or near a transmission zero is to be calculated, all computations should be based on the semirelative sensitivity measure defined in Eq. (3-17).

Multiparameter sensitivities provide a fairly easy method for obtaining an estimate of the *worst-case* variabilities dP/P of a performance characteristic $P(\mathbf{x})$: in the worst case, all element tolerances simultaneously take on values such that their effects on $dP(\mathbf{x})$ add; i.e., the worst-case deviations can be obtained from Eq. (3-96b) by taking the magnitudes

$$\max \left| \frac{dP}{P} \right| = \sum_{i=1}^{k} |S_{x_i}^P| \left| \frac{dx_i}{x_{i0}} \right| = |\mathbf{S}_{\mathbf{x}}^P|^t \, |d\hat{\mathbf{x}}| \tag{3-104}$$

With Eq. (3-104), the maximum and minimum values of $P(\mathbf{x})$ may be estimated from

$$\max P(\mathbf{x}) = P(\mathbf{x}_0)[1 + |\mathbf{S}_{\mathbf{x}}^P|^t \, |d\hat{\mathbf{x}}|]$$

$$= P(\mathbf{x}_0) + \sum_{i=1}^{k} \left| \frac{\partial P}{\partial x_i} \right|_0 |dx_i| \tag{3-105a}$$

$$\min P(\mathbf{x}) = P(\mathbf{x}_0) - \sum_{i=1}^{k} \left| \frac{\partial P}{\partial x_i} \right|_0 |dx_i| \tag{3-105b}$$

In most all cases of practical interest, these relationships give reasonable estimates of the maximum and minimum values of $P(\mathbf{x})$; they fail only when $P(\mathbf{x})$ has a quadratic dependence on all x_i (so that $\partial P/\partial x_i = 0$) as in doubly terminated lossless filters at a point of maximum power transfer (see Section 3.2.6.2). In such cases, the "linearized Taylor expansion" in Eq. (3-105) must be extended to include second derivatives.

In practice, especially if k is large, the estimates in Eqs. (3-104) and (3-105) are usually very pessimistic, because it is not likely due to the *statistical* nature of the tolerances that all elements cause additive variations in $P(\mathbf{x})$. Therefore, a further refinement of multiparameter sensitivity measures must take into account that circuit elements generally follow a statistical distribution, and that the values of different circuit parameters may be correlated. For example, in integrated circuits, a change in temperature will change all resistors by the same percentage and all capacitors by another percentage. Similarly, mask errors or changes in material parameters tend to cause strongly correlated errors in electrically similar elements. Because such situations cannot, in general, be handled deterministically without leading to very misleading results, a different treatment is needed. In the following section, we discuss some methods of statistical sensitivity analysis that yield a more realistic representation of expected circuit behavior.

3.3.2 Statistical Case[9]

In practice, the circuit designer generally does not know the exact value of a component x_i, but only the *probability* that its value lies in the range dx_i with $dx_i \to 0$. The probability is defined[10] as $f(x_i) \, dx_i$, where $f(x_i)$ is the *marginal*

---

[9]See references 41 to 43 for background and introductions to probability and statistics.

[10]If the components $\mathbf{x}$ are not independent, the probability that $\mathbf{x}$ lies in the range dx_i, $i = 1$, $\ldots, k$, is $f(\mathbf{x}) \, dx_1 \, dx_2 \ldots \, dx_k$, where $f(\mathbf{x}) = f(x_1, x_2, \ldots, x_k)$ is the *joint* probability density function.

probability density function (pdf). The pdf contains the information about the statistical distribution of the component values as obtained in a manufacturing process. Frequently used pdf's are the *uniform distribution* (Fig. 3-14a),

$$f(x) = 0 \qquad\qquad x \leq x_m, x \geq x_M$$

$$f(x) = \frac{1}{x_M - x_m} \qquad x_m \leq x \leq x_M \qquad\qquad (3\text{-}106)$$

which assumes that all values of x in the range $x_{\min} = x_m \leq x \leq x_{\max} = x_M$ are equally likely; and the *normal* (or *Gaussian*) distribution

$$f(x) = \frac{1}{\sqrt{2\pi}\sigma} \exp \frac{(x - \bar{x})^2}{2\sigma^2} \qquad\qquad (3\text{-}107)$$

shown in Fig. 3-14b. Note that the pdf must be defined such that

$$\int_{-\infty}^{\infty} f(x)\, dx = 1 \qquad\qquad (3\text{-}108)$$

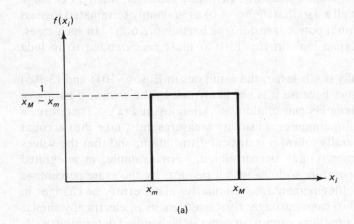

(a)

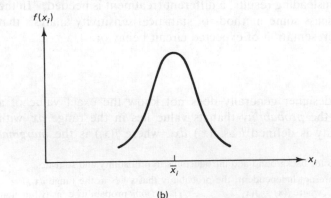

(b)

Figure 3-14 Typical statistical distributions: (a) uniform, (b) normal.

In Eq. (3-107), $\bar{x}$ and σ are the *mean* and the *standard deviation*, respectively.   The mean $\bar{x}$ or expected value $E\{x\}$ is defined as

$$\bar{x} = E\{x\} = \int_{-\infty}^{\infty} xf(x) \, dx \qquad (3\text{-}109)$$

It is the center around which the random values of x cluster.   Information about how broad the cluster is or how widely the random values are dispersed around $\bar{x}$ is obtained from the *variance*, defined as

$$\text{Var}\{x\} = E\{(x - \bar{x})^2\} = \int_{-\infty}^{\infty} (x - \bar{x})^2 f(x) \, dx \qquad (3\text{-}110)$$

For the normal distribution [Eq. (3-107)] one can show that

$$\sigma^2 = \text{Var}\{x\} \qquad (3\text{-}111)$$

i.e., the standard deviation is the square root of the variance.   Often only the mean $\bar{x}_i$ and the variance σ_i^2 are known for the distributions for each component x_i.

Because our applications deal with functions of more than one statistically varying parameter, further useful concepts for the analysis are the *correlation* between two random variables,

$$E\{x_i x_j\} = \int_{-\infty}^{\infty} \int_{-\infty}^{\infty} x_i x_j f(x_i, x_j) \, dx_i \, dx_j \qquad (3\text{-}112)$$

the *covariance*

$$\text{Cov}\{x_i, x_j\} = E\{(x_i - \bar{x}_i)(x_j - \bar{x}_j)\} = E\{x_i x_j\} - E\{x_i\}E\{x_j\} \qquad (3\text{-}113)$$

and the *correlation coefficient*

$$\rho_{ij} = \frac{\text{Cov}\{x_i, x_j\}}{\sigma_i \sigma_j} \qquad (3\text{-}114a)$$

which can be shown to obey

$$-1 \le \rho_{ij} \le +1 \qquad (3\text{-}114b)$$

The variables are said to be uncorrelated if $\text{Cov}\{x_i, x_j\} = 0$ or $\rho_{ij} = 0$.

If a performance characteristic, say, the transfer function, is a function of the statistically varying elements x_i, that is, takes the form $H(s, \mathbf{x})$, then clearly also H is a random variable and we need to determine the statistical properties of H from those of $\mathbf{x}$.   Because this is in general an impossible problem to solve analytically, the task can be approached only numerically (via Monte Carlo methods; see Section 3.3.3) or by appropriate approximations: assuming, as for Eq. (3-99), that the tolerances are small enough to permit linearization, we obtain, with $dH \triangleq H(s, \mathbf{x}) - H(s, \bar{\mathbf{x}})$ and $H \triangleq H(s, \bar{\mathbf{x}})$,

$$\frac{dH}{H} \simeq (\mathbf{S}_{\mathbf{x}}^H)^t \, d\hat{\mathbf{x}}$$

and

$$\left|\frac{dH}{H}\right|^2 \simeq [(\mathbf{S_x^H})^t \, d\hat{\mathbf{x}}]^* \, [d\hat{\mathbf{x}}^t \, \mathbf{S_x^H}] \tag{3-115}$$

Note that the sensitivity expressions $S_{x_i}^H$ are evaluated at the mean values $\bar{x}_i$ and, therefore, are not random values. Thus, taking the expected value of Eq. (3-115) results in

$$S_T \triangleq E\left\{\left|\frac{dH}{H}\right|^2\right\} \simeq (\mathbf{S_x^H})^{t*} \, \mathbf{C} \, \mathbf{S_x^H} \tag{3-116a}$$

where

$$\mathbf{C} = E\{d\hat{\mathbf{x}} d\hat{\mathbf{x}}^t\} \tag{3-117}$$

is the $k \times k$ covariance matrix with matrix elements

$$c_{ij} = \text{Cov}\left\{\frac{dx_i}{x_i}, \frac{dx_j}{x_j}\right\} \tag{3-118}$$

as defined in Eq. (3-113). If the off-diagonal terms are zero, i.e., if there is no correlation between x_i and x_j, $i \neq j$, then

$$S_T \simeq \sum_{i=1}^k |S_{x_i}^H|^2 \hat{\sigma}_i^2 \tag{3-116b}$$

where we have used the fact that, by Eqs. (3-110) and (3-113), $c_{ii} = \hat{\sigma}_i^2$, the variance of (dx_i/x_i).

We note that the results in Eqs. (3-115) and (3-116) are frequency-dependent. If, for example, for comparison purposes a single number is more convenient, we may integrate over some designer-specified frequency band $\omega_1 \leq \omega \leq \omega_2$, such as the filter passband, to obtain

$$S_I \triangleq E\left\{\int_{\omega_1}^{\omega_2} \left|\frac{dH}{H}\right|^2 d\omega\right\} \simeq \int_{\omega_1}^{\omega_2} [(\mathbf{S_x^H})^{t*} \, \mathbf{C} \, \mathbf{S_x^H}] \, d\omega \tag{3-119a}$$

which for $c_{ij} = 0$, $i \neq j$, becomes

$$S_I \simeq \sum_{i=1}^k \int_{\omega_1}^{\omega_2} [|S_{x_i}^H|^2 \, \hat{\sigma}_i^2] \, d\omega \tag{3-119b}$$

Note that the deterministic measure in Eq. (3-103) can be obtained as a special case of Eq. (3-116b) or (3-119b) for $\hat{\sigma}_i = 1$.

The sensitivity measure of Eq. (3-116) or (3-119) was introduced in reference 39; its advantages for the designer are that:

1. Multiparameter statistical element variations are included.
2. Correlated element variations are easily considered.

3. Sensitivity may be averaged [Eq. (3-119)] over a frequency range (expressed as an integral) or it may be evaluated as a function of frequency (shown as an integrand) [Eq. (3-116)].

4. It can easily be optimized; i.e., sensitivity minimization can be incorporated into the design.

5. It gives good agreement with Monte Carlo methods [47] (see Section 3.3.3).

When evaluating the integrand measure S_T in Eq. (3-116), one obtains just a number or possibly a scalar function of frequency that must be interpreted by the designer; thus, it is important not to lose track of its physical meaning: S_T is the mean value of the square of the relative deviations (the variabilities) of the transfer function $H(s, \mathbf{x})$ at each frequency point, caused by the statistical tolerances in the parameter vector $\mathbf{x}$. Figure 3-15 shows an appropriate diagram containing a *tolerance tube* (the shaded area) bounded by the curves

$$M(\omega) \triangleq \max \left| \frac{dH}{H} \right|^2$$

and

$$m(\omega) \triangleq \min \left| \frac{dH}{H} \right|^2$$

calculated as a function of frequency from Eq. (3-115). S_T is then the mean of the statistical distribution of the values of $|dH/H|^2$ in the tolerance tube at each frequency point. For the frequent case where the deviations are *zero-mean* random

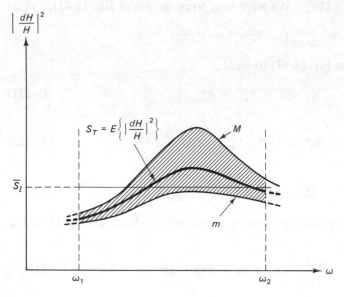

Figure 3-15   Tolerance tube to show the graphic interpretation of S_I and $\bar{S}_I$.

variables, an alternative useful practical interpretation of Eq. (3-116a) can be obtained by noting from Eqs. (3-110), (3-111) and (3-113) that

$$E\left\{\left|\frac{\Delta H}{H}\right|^2\right\} = \left[E\left\{\left|\frac{\Delta H}{H}\right|\right\}\right]^2 + \text{Var}\left\{\left|\frac{\Delta H}{H}\right|\right\}$$

Thus, S_T in Eq. (3-116a) can be interpreted as the variance of $\Delta H/H$,

$$S_T = E\left\{\left|\frac{\Delta H}{H}\right|^2\right\} \simeq \text{Var}\left\{\left|\frac{\Delta H}{H}\right|\right\} = \sigma^2$$

provided that the mean of the deviation, $E\left\{\left|\dfrac{\Delta H}{H}\right|\right\}$, is negligibly small.   The

integral measure S_I of Eq. (3-119) is, of course, simply the area under the curve S_T in the frequency interval $\omega_1 \leq \omega \leq \omega_2$.   A possibly more meaningful interpretation can be obtained by dividing S_I by the frequency interval $\omega_2 - \omega_1$, i.e.,

$$\overline{S}_I = \frac{S_I}{\omega_2 - \omega_1} \tag{3-120}$$

$\overline{S}_I$ is then the mean of the statistical distribution of the squared variability

$$\left|\frac{dH(j\omega, \mathbf{x})}{H(j\omega, \mathbf{x})}\right|^2$$

averaged over the designer-specified frequency band $\omega_1 \leq \omega \leq \omega_2$.

It is a relatively easy matter to obtain the mean gain and phase deviations from Eqs. (3-116) and (3-119).   We have only to make use of Eq. (3-43),

$$S_x^{H(j\omega)} = S_x^{|H(j\omega)|} + jQ_x^{\phi(\omega)}$$

and insert this result into Eq. (3-97) to yield

$$\mathbf{S}_\mathbf{x}^H = \mathbf{S}_\mathbf{x}^{|H|} + j\mathbf{Q}_\mathbf{x}^\phi \tag{3-121}$$

where

$$\mathbf{S}_\mathbf{x}^{|H|} = \{S_{x1}^{|H|} \ \ S_{x2}^{|H|} \ldots S_{xk}^{|H|}\}^t \tag{3-122a}$$

and

$$\mathbf{Q}_\mathbf{x}^\phi = \{Q_{x1}^\phi \ Q_{x2}^\phi \ldots Q_{xk}^\phi\}^t \tag{3-122b}$$

are the gain and phase sensitivity vectors, respectively.   With Eq. (3-121), Eq. (3-99) becomes

$$\frac{dH}{H} \simeq (\mathbf{S}_\mathbf{x}^{|H|})^t \, d\hat{\mathbf{x}} + j(\mathbf{Q}_\mathbf{x}^\phi)^t \, d\hat{\mathbf{x}} \tag{3-123}$$

and we obtain with Eq. (3-115)

$$\left|\frac{dH}{H}\right|^2 \simeq (\mathbf{S}_x^{|H|})^t \ d\hat{\mathbf{x}} \ d\hat{\mathbf{x}}^t \ \mathbf{S}_x^{|H|} + (\mathbf{Q}_x^\phi)^t \ d\hat{\mathbf{x}} \ d\hat{\mathbf{x}}^t \ \mathbf{Q}_x^\phi \tag{3-124}$$

From Eq. (3-124), the mean integrand and integral gain and phase deviations can be calculated in the usual manner. For example, using Eq. (3-119a) with Eq. (3-120), the frequency-averaged mean gain and phase deviations become

$$\overline{S}_I^{|H|} \simeq \frac{1}{\omega_2 - \omega_1} \int_{\omega_1}^{\omega_2} [(\mathbf{S}_x^{|H|})^t \ \mathbf{C} \ \mathbf{S}_x^{|H|}] \ d\omega \tag{3-125a}$$

$$\overline{Q}_I^\phi \simeq \frac{1}{\omega_2 - \omega_1} \int_{\omega_1}^{\omega_2} [(\mathbf{Q}_x^\phi)^t \ \mathbf{C} \ \mathbf{Q}_x^\phi] \ d\omega \tag{3-125b}$$

where $\mathbf{C}$ is the covariance matrix defined in Eq. (3-117). If there is no correlation between the values of x_i and x_j, $i \neq j$, then $c_{ij} = 0$, $i \neq j$, and as in Eq. (3-119b) we obtain from Eq. (3-125)

$$\overline{S}_I^{|H|} \simeq \frac{1}{\omega_2 - \omega_1} \sum_{i=1}^{k} \hat{\sigma}_i^2 \int_{\omega_1}^{\omega_2} |S_{x_i}^{|H|}|^2 \ d\omega \tag{3-126a}$$

$$\overline{Q}_I^\phi \simeq \frac{1}{\omega_2 - \omega_1} \sum_{i=1}^{k} \hat{\sigma}_i^2 \int_{\omega_1}^{\omega_2} |Q_{x_i}^\phi|^2 \ d\omega \tag{3-126b}$$

where $\hat{\sigma}_i^2 = Var\{dx_i/x_i\}$.

It should be apparent that statistical sensitivity calculations must make extensive use of computers. However, to demonstrate some of the details involved, Example 3-9 will illustrate the process for an extremely simple circuit. Further discussions of these measures and their evaluations can be found in references 44 to 46.

Example 3-9

Calculate magnitude and phase deviations for the first-order lowpass filter in Fig. 3-16. The nominal cutoff frequency is to be 1 krad/s; the elements have a normal distribution, their variations are uncorrelated, and their variance is $\sigma_e^2 = 10^{-4}$ (i.e., 1% variation in each element).

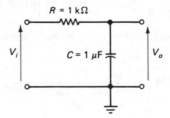

Figure 3-16 Simple RC circuit for Example 3-9.

Solution.   Clearly, the transfer function is

$$H(s) = \frac{1/RC}{s + 1/RC} \tag{3-127}$$

and nominal design values of $R_0 = 1$ kΩ and $C_0 = 1$ μF give the desired cutoff frequency of 10^3 rad/s.   According to Eq. (3-123), we need

$$\mathbf{x} = [R \quad C]^t \tag{3-128a}$$

$$\mathbf{d\hat{x}} = \left[\frac{dR}{R_0} \quad \frac{dC}{C_0}\right]^t \triangleq [dr \quad dc]^t \tag{3-128b}$$

and the vector

$$\mathbf{S}_{\mathbf{x}}^{H(j\omega)} = \mathbf{S}_{\mathbf{x}}^{|H(j\omega)|} + j\mathbf{Q}_{\mathbf{x}}^{\phi(\omega)}$$

The definitions of dr and dc should be apparent.   With $\sigma_0 \triangleq 1/(R_0 C_0)$, we find, from Eq. (3-127),

$$H(j\omega) = \sqrt{\frac{\sigma_0^2}{\sigma_0^2 + \omega^2}} \exp\left(-j \tan^{-1}\frac{\omega}{\sigma_0}\right)$$

and, making use of Eq. (3-19),

$$\mathbf{S}_{\mathbf{x}}^{|H(j\omega)|} = \frac{\omega^2}{\sigma_0^2 + \omega^2}[-1 \quad -1]^t \tag{3-129a}$$

$$\mathbf{Q}_{\mathbf{x}}^{\phi(\omega)} = -\frac{\sigma_0\omega}{\sigma_0^2 + \omega^2}[-1 \quad -1]^t \tag{3-129b}$$

Inserting Eqs. (3-128) and (3-129) into Eq. (3-124) yields

$$\left|\frac{dH}{H}\right|^2 \simeq \left(\frac{-\omega^2}{\omega^2 + \sigma_0^2}\right)^2 (1 \quad 1)\binom{dr}{dc}(dr \; dc)\binom{1}{1}$$

$$+ \left(\frac{-\sigma_0\omega}{\omega^2 + \sigma_0^2}\right)^2 (1 \quad 1)\binom{dr}{dc}(dr \; dc)\binom{1}{1} \tag{3-130a}$$

$$= \left[\frac{\omega^4}{(\omega^2 + \sigma_0^2)^2} + \frac{\sigma_0^2\omega^2}{(\omega^2 + \sigma_0^2)^2}\right](1 \quad 1)\begin{pmatrix}(dr)^2 & dr \; dc \\ dr \; dc & (dc)^2\end{pmatrix}\binom{1}{1}$$

and finally

$$\left|\frac{dH}{H}\right|^2 = \left[\frac{\omega^4}{(\omega^2 + \sigma_0^2)^2} + \frac{\sigma_0^2\omega^2}{(\omega^2 + \sigma_0^2)^2}\right][(dr)^2 + 2 \; dr \; dc + (dc)^2] \tag{3-130b}$$

By Eq. (3-116b), the expected value of $|dH/H|^2$, with no correlation between R and C, results in

$$S_T = 2\hat{\sigma}_e^2\left[\frac{\omega^4}{(\omega^2 + \sigma_0^2)^2} + \frac{\sigma_0^2\omega^2}{(\omega^2 + \sigma_0^2)^2}\right] = E\left\{\left|\frac{dH}{H}\right|^2\right\}$$

where we set

$$E\left\{\left(\frac{dR}{R_0}\right)^2\right\} = E\left\{\left(\frac{dC}{C_0}\right)^2\right\} = \hat{\sigma}_e^2 = 10^{-4}$$

per assumption. S_T can now be evaluated at any desired frequency; e.g., at the passband edge $\omega = \sigma_0$, we find $S_T = \hat{\sigma}_e^2 = 10^{-4}$. Similarly, we find for the gain and phase deviations, using Eqs. (3-124) and (3-129),

$$S^{|H|} = 2\hat{\sigma}_e^2 \left.\frac{\omega^4}{(\omega^2 + \sigma_0^2)^2}\right|_{\omega = \sigma_0} = 0.5 \cdot 10^{-4}$$

$$Q^\phi = 2\hat{\sigma}_e^2 \left.\frac{\sigma_0^2\omega^2}{(\omega^2 + \sigma_0^2)^2}\right|_{\omega = \sigma_0} = 0.5 \cdot 10^{-4}$$

with $S_T = S^{|H|} + Q^\phi$. These numbers mean that at $\omega = \sigma_0$, the 1% element variations can be expected to cause a *mean* error in H, $|H|$, and ϕ, respectively, of 10^{-2} (1%), $\sqrt{0.5} \cdot 10^{-2}$ (0.7%), and 0.7%.

To obtain the frequency-averaged mean deviations in Eqs. (3-125a) and (3-125b), we need to integrate the gain and phase sensitivity vectors. In general, this step will require numerical integration routines, but in our simple case, Eqs. (3-129) or (3-130), closed-form solutions are possible. Assuming we are interested in the range $0 \le \omega \le \sigma_0$, we obtain, from Eq. (3-126), with $\lambda = \omega/\sigma_0$,

$$\overline{S}_I^{|H|} = \frac{1}{\sigma_0} 2\sigma_e^2 \int_0^{\sigma_0} \frac{\omega^4}{(\omega^2 + \sigma_0^2)^2} \, d\omega = 2\sigma_e^2 \int_0^1 \frac{\lambda^4}{(\lambda^2 + 1)^2} \, d\lambda$$

$$= \sigma_e^2 \left.\left(3\lambda - 3\tan^{-1}\lambda - \frac{\lambda^3}{1 + \lambda^2}\right)\right|_0^1 = 0.1438 \cdot 10^{-4}$$

$$\overline{Q}_I^\phi = \frac{1}{\sigma_0} 2\sigma_e^2 \int_0^{\sigma_0} \frac{\sigma_0^2\omega^2}{(\omega^2 + \sigma_0^2)^2} \, d\omega = 2\sigma_e^2 \int_0^1 \frac{\lambda^2}{(\lambda^2 + 1)^2} \, d\lambda$$

$$= \sigma_e^2 \left.\left(\tan^{-1}\lambda - \frac{\lambda}{\lambda^2 + 1}\right)\right|_0^1 = 0.2854 \cdot 10^{-4}$$

and

$$\overline{S}_I = \overline{S}_I^{|H|} + \overline{Q}_I^\phi = 0.4292 \cdot 10^{-4}$$

This means that over the range $0 \le \omega \le \sigma_0$, the *average mean* deviations of H, $|H|$, and ϕ caused by the statistically varying elements with variance of 10^{-4} are $\sqrt{0.4292} \cdot 10^{-2}$ (0.66%), $\sqrt{0.1438} \cdot 10^{-2}$ (0.38%), and $\sqrt{0.2854} \cdot 10^{-2}$ (0.53%), respectively.

Observe that the measures in Eqs. (3-116), (3-119), (3-125), and (3-126) are *not* sensitivities. Rather, they are variabilities; i.e., the numbers obtained in Example 3-9 give the actual mean percentage deviations for the assumed element tolerances. If sensitivities are needed, e.g., for the purpose of comparing different circuits, we must make use of the sensitivity vector in Eq. (3-121) [Eq. (3-129), in the example]. Often, a convenient sensitivity measure may be obtained by dividing the numbers S_I, $\bar{S}_I$, and so forth, by the variance of the elements (assuming it is reasonably constant for all components). Thus, in Example 3-9, with $\hat{\sigma}_R^2 = \hat{\sigma}_C^2 = \hat{\sigma}_e^2$, we find, for example, the mean gain *sensitivity at* $\omega = \sigma_0$ to be $\sqrt{S^{|H|}/\hat{\sigma}_e^2} \simeq 0.7$ and the frequency-averaged mean gain sensitivity to be $\sqrt{\bar{S}_I^{|H|}/\hat{\sigma}_e^2} \simeq 0.38$.

3.3.3 Monte Carlo Analysis [48]

We pointed out earlier that it is in general not possible to establish analytically the statistical properties of a performance measure $P(\mathbf{x})$ when the parameters x_i in the vector $\mathbf{x}$ are subject to statistical variation. In Section 3.3.2 we therefore used approximate methods, based on linearized Taylor expansions, to arrive at the expected value of $P(\mathbf{x})$. These approximate evaluations are reasonably accurate if carefully used and interpreted; their primary advantage is that they permit statistical network variations to be analyzed with only a moderate amount of computer time. This computational efficiency is particularly important when sensitivities are to be minimized (see Section 3.3.4).

If a more complete or more accurate picture beyond the capabilities of the approximate methods is required, the network behavior must be analyzed using numerical techniques. The most common one is a "brute-force" method known as *Monte Carlo simulation*, which can be made arbitrarily accurate at the cost of a significant increase in computer time. In a Monte Carlo simulation, in effect, each randomly varying parameter or component x_i in a circuit is replaced by a random number generator[11] that produces (*pseudo*) random values r_i with the *same* statistical properties (pdf, mean, variance, and so forth; see Fig. 3-14) as those of x_i. For each parameter value so generated, the performance measure P is evaluated and the process is repeated N times. Here, N must be chosen large enough (usually $100 < N < 10,000$) to obtain results which are statistically significant. The object of this exercise is, of course, to obtain a sufficiently large number of values for the measure $P(\omega, \mathbf{r})$ such that its statistical properties may be estimated. Once a set of N random values $P(\omega, \mathbf{r})$ is known, approximate values for the mean and standard deviations, for example, can readily be obtained at each frequency point from

$$\bar{P}(\omega) \simeq \frac{1}{N} \sum_{i=1}^{N} P_i(\omega) \tag{3-131}$$

$$\sigma_P^2(\omega) \simeq \frac{1}{N} \sum_{i=1}^{N} P_i^2(\omega) - \bar{P}^2(\omega) \tag{3-132}$$

where $P_i(\omega)$ is the value of the sample $P(\omega, x_i)$. Equations (3-131) and (3-132) are analogous to Eqs. (3-109) and (3-110). If required, e.g., for purposes of

---

[11]Available in most computer libraries.

comparing the performance of different networks, Eqs. (3-131) and (3-132) can be integrated or averaged over the relevant frequency band by use of a numerical integration routine in a manner similar to the way Eqs. (3-119) and (3-120) were derived.   Note that determining the sample value $P_i(\omega)$ requires, in general, a complete circuit analysis.   Thus, depending on the desired accuracy, the network must be simulated hundreds or thousands of times, a fact that is indicative of the need for significant amounts of computer time.

In the foregoing discussion, the specific meaning of the "performance measure" $P(\omega, \mathbf{x})$ was on purpose left undefined in order to keep the treatment general. In practice, $P(\omega, \mathbf{x})$ can be any characteristic that is important for a circuit's proper functioning, such as a frequency parameter, the selectivity expressed via Q, or the complete transfer function $H(j\omega, \mathbf{x})$.   In the latter case, the circuit simulation provides us with real and imaginary parts that can be used in the usual way [Eqs. (3-43), (3-123)–(3-126)] to obtain information about gain and phase sensitivities or variabilities.   For example, the statistical mean relative gain error at a frequency point ω_0 is simply

$$E\left\{\frac{d|H|}{|H|}\right\} = \frac{1}{N} \sum_{i=1}^{N} \text{Re} \left\{1 - \frac{H(j\omega_0, \mathbf{x}_i)}{H(j\omega_0, \mathbf{x}_n)}\right\} \tag{3-133a}$$

where $\mathbf{x}_i$ is the ith sample, $1 \leq i \leq N$, and $\mathbf{x}_n$ is the nominal value (the mean) of the component vector $\mathbf{x}$.   Similarly, the mean phase deviation can be obtained from

$$E\{d\phi\} \simeq \frac{1}{N} \sum_{i=1}^{N} \text{Im} \left\{1 - \frac{H(j\omega_0, \mathbf{x}_i)}{H(j\omega_0, \mathbf{x}_n)}\right\} \tag{3-133b}$$

Further, as before in Eqs. (3-125a) and (3-125b), integrating Eqs. (3-133a) and (3-133b) over the frequency band of interest results in the corresponding frequency-averaged mean values.

Monte Carlo simulation can give the engineer quite a realistic picture of practical system performance *if* the statistical element variations are modeled correctly by the samples r_i obtained from the random number generator.   This means that the samples r_i must have the same statistical properties and distributions as the circuit parameters x_i they represent.   Evidently, a good statistical model requires detailed knowledge of the fabrication or process environment and the availability of extensive manufacturing data from which the statistical properties can be determined.   The difficult modeling job is often facilitated by assuming that the elements x_i are uncorrelated, i.e., that they are independent random variables. However, correlation is an important statistical property of many filter implementation technologies, such as thin or thick film, or monolithic integration, where electrically similar elements tend to have highly correlated variations.   For example, although absolute values of integrated resistors and capacitors have large ($\geq 20\%$) tolerances, ratios of like components are usually implemented very accurately [e.g., $\Delta(C_i/C_j)/(C_i/C_j) < 0.5\%$].   To account for such correlation, one may either specify the element ratio as the statistical variable instead of the individual elements, or one can identify the common parameter that causes the correlated variations, such as the oxide thickness for MOS capacitors, and use it as the random variable.

If the statistical modeling of the manufacturing process is handled with care and a sufficiently large number of samples is used, Monte Carlo analysis does provide a realistic representation of filter performance. It can be used advantageously to simulate the final design and provides a useful estimate of the *manufacturing yield*,[12] i.e., of the percentage of circuits that will meet the design specifications when the components vary because of fabrication tolerances or drift during operation. Apart from the extensive computer time required, the only weakness of Monte Carlo analysis is the lack of information about the circuit's sensitivity to individual elements x_i; i.e., the engineer obtains no clues about which parameters are the most critical for reliable circuit performance and so there is little feedback given to help in a redesign for a more optimal circuit with improved yield.

The questions of sensitivity and yield optimization will be addressed briefly in the following sections.

---

[12]The manufacturing yield $Y(\mathbf{x}_n)$ is simply the probability that a performance measure $P(\mathbf{x})$ meets the desired specifications when $\mathbf{x}$ takes on all likely combinations of values. To calculate $Y(\mathbf{x}_n)$ for a design with nominal component values $\mathbf{x}_n$, we need to evaluate the multidimensional integral

$$Y(\mathbf{x}_n) = \int_{G_k} F(\mathbf{x})p(\mathbf{x}_n, \mathbf{x}) \, d\mathbf{x}$$

where $p(\mathbf{x}_n, \mathbf{x})$ is the probability density function of the components $\mathbf{x}$ in multidimensional element space G_k and $F(\mathbf{x})$ is a function that takes on the value 1 if the circuit passes the test and the value 0 if the circuit fails. Because the dimensionality of $\mathbf{x}$ is very high in realistic circuits, the (numerical) evaluation of this integral is not practicable. Monte Carlo simulation provides a feasible alternative that estimates the yield from

$$Y(\mathbf{x}_n) \simeq \frac{1}{N} \sum_{i=1}^{N} F(\mathbf{x}_i)$$

[43, 48] and the standard deviation as

$$\sigma\{Y(\mathbf{x}_n)\} \simeq \sqrt{\frac{Y(\mathbf{x}_n)[1 - Y(\mathbf{x}_n)]}{N}}$$

where N is the number of samples. Therefore, a larger N reduces σ and increases the confidence we can have in the validity of $Y(\mathbf{x}_n)$. The probability that $Y(\mathbf{x}_n)$ is in $Y(\mathbf{x}_n) \pm 2\sigma\{Y(\mathbf{x}_n)\}$ is 95%. For example, if $Y(\mathbf{x}_n)$ was calculated as 80% using 1000 samples, then $\sigma \simeq 0.0126$ and we can only say with 95% certainty that the yield is in $77.5\% < Y(\mathbf{x}_n) < 82.5\%$. For $N = 100$ samples, the 95% confidence interval is in $(72\%, 88\%)$. The importance of a large sample size is thus apparent.

3.3.4 Sensitivity Optimization

Active filters generally have more elements or parameters than are needed to satisfy the design equations. The remaining components, the *degrees of freedom*, can therefore be determined for the convenience of the engineer, such as to simplify the design equations, to get preferred component values, or to ease the manufacturing process. It was pointed out several times in this chapter that the sensitivity really represents nothing but the *slope* of the design criterion (e.g., quality factor or transfer function) in multidimensional element space at the nominal point. Consequently, if the nominal design point is changed by arbitrarily selecting the free elements or parameters, then the sensitivities are also altered and must be reevaluated. Conversely, the engineer may, of course, attempt to select the free parameters such that a point of minimum sensitivity (slope) is reached. It is this concept—reduction of the slope of the performance measure—that forms the basis of sensitivity optimization. It was used in Section 3.1, Example 3-1, to reduce the sensitivity of a bandpass quality factor to the amplifier gain K by an appropriate selection of the single free parameter, the resistor ratio r^2. The student is urged to review this example at this point. Example 3-10 below will further illustrate sensitivity optimization in line with our more recent discussion and notation.

Example 3-10

In the lowpass circuit of Fig. 3-17, determine any free parameters such that the gain variability S_T of Eq. (3-116a) is minimized at the pole frequency $\omega_0 = 1000$ rad/s. The pole quality factor is $Q = 4$ and the dc gain is arbitrary. Assume that the relevant statistical properties of the circuit components are known.

Solution. The transfer function of the circuit is

$$H(s) = \frac{V_2}{V_1} = \frac{N(s)}{D(s)} = \frac{K/(\tau_1\tau_2)}{s^2 + s\left(\dfrac{1 + \rho}{\tau_1} + \dfrac{1 - K}{\tau_2}\right) + \dfrac{1}{\tau_1\tau_2}} \tag{3-134}$$

where $\tau_1 = C_1R_1$, $\tau_2 = C_2R_2$, and $\rho = R_1/R_2$. Evidently, $H(0) = K = $ arbitrary, and

$$\omega_0 = \frac{1}{\sqrt{\tau_1\tau_2}} \qquad \frac{1}{Q} = \sqrt{\frac{\tau_2}{\tau_1}}(1 + \rho) + \sqrt{\frac{\tau_1}{\tau_2}}(1 - K) \tag{3-135}$$

From Eq. (3-134), it is clear that it is not the individual element values that are critical for the performance, but instead the time constants τ_1 and τ_2, the resistor ratio ρ, and the amplifier gain K. With Eq. (3-135), we can find

$$\tau_2 = \frac{1}{\tau_1\omega_0^2} \tag{3-136a}$$

and

$$K = 1 + \frac{1 + \rho}{\omega_0^2\tau_1^2} - \frac{1}{\omega_0\tau_1 Q} \tag{3-136b}$$

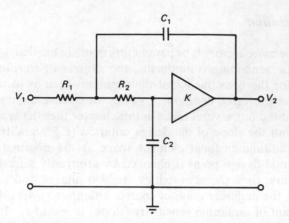

Figure 3-17 Simple active lowpass filter
for Example 3-10.

We have, therefore, two free parameters, ρ and τ_1, that can be used for the minimization of S_T. From Eqs. (3-116a) and (3-117) we have:

$$S_T \triangleq E\left\{\left|\frac{dH}{H}\right|^2\right\} \simeq (\mathbf{S_x^H})^{t*}\, \mathbf{C}\, \mathbf{S_x^H}$$

$$\mathbf{C} = E\{\mathbf{d\hat{x}d\hat{x}^t}\} \tag{3-137}$$

The vector $\mathbf{d\hat{x}}$ in this case is

$$\mathbf{d\hat{x}} = \left\{\frac{d\tau_1}{\tau_{10}} \quad \frac{d\tau_2}{\tau_{20}} \quad \frac{d\rho}{\rho_0} \quad \frac{dK}{K_0}\right\}^t$$

$$\triangleq \{t_1 \quad t_2 \quad r \quad k\}^t$$

with the definitions of t_i, r, and k evident. The covariance matrix $\mathbf{C}$ is then

$$\mathbf{C} = E\left\{\begin{pmatrix} t_1^2 & t_1 t_2 & t_1 r & t_1 k \\ t_1 t_2 & t_2^2 & t_2 r & t_2 k \\ rt_1 & rt_2 & r^2 & rk \\ kt_1 & kt_2 & kr & k^2 \end{pmatrix}\right\}$$

Let us assume for the further development a hybrid circuit realization with thin-film resistors and discrete capacitors, and assume that all electrically dissimilar elements and also the capacitors C_1 and C_2 are statistically independent variables. Consequently, there will be no correlation between t_1, t_2, r, and k, so that

$$\mathbf{C} = \begin{pmatrix} \sigma_{t1}^2 & 0 & 0 & 0 \\ 0 & \sigma_{t2}^2 & 0 & 0 \\ 0 & 0 & \sigma_r^2 & 0 \\ 0 & 0 & 0 & \sigma_k^2 \end{pmatrix} \tag{3-138}$$

Further,

$$\mathbf{S_x^H} = \{S_{\tau 1}^H \quad S_{\tau 2}^H \quad S_\rho^H \quad S_K^H\}^t \tag{3-139}$$

where

$$S_{\tau_1}^H = -1 + \frac{\dfrac{1 + \rho}{\tau_1} s + \dfrac{1}{\tau_1 \tau_2}}{D(s)} \tag{3-140a}$$

$$S_{\tau_2}^H = -1 + \frac{\dfrac{1 - K}{\tau_2} s + \dfrac{1}{\tau_1 \tau_2}}{D(s)} \tag{3-140b}$$

$$S_\rho^H = - \frac{s\rho/\tau_1}{D(s)} \tag{3-140c}$$

$$S_K^H = 1 + \frac{sK/\tau_2}{D(s)} \tag{3-140d}$$

With Eqs. (3-138) and (3-139), S_T in Eq. (3-137) becomes

$$S_T \simeq |S_{\tau_1}^H|^2 \sigma_{t_1}^2 + |S_{\tau_2}^H|^2 \sigma_{t_2}^2 + |S_\rho^H|^2 \sigma_r^2 + |S_K^H|^2 \sigma_k^2 \tag{3-141}$$

where the sensitivities are given in Eq. (3-140).

Evidently, S_T can now be evaluated as a function of frequency or averaged over frequency [Eq. (3-119a)] as soon as the design parameters (ω_0, Q), the circuit components (R_i, C_i, K), and the technology (σ_r, σ_t, σ_k) are known. In our example, S_T is to be minimized at $\omega = \omega_0 = 1/\sqrt{\tau_1 \tau_2}$, with the constraint $Q = 4$, i.e.,

$$\frac{1 + \rho}{\omega_0 \tau_1} + (1 - K) \omega_0 \tau_1 = \frac{1}{Q} = 0.25 \tag{3-142}$$

Using this notation in Eq. (3-140) yields

$$S_{\tau_1}^H(j\omega_0) = -Q[(1 - K)\omega_0\tau_1 + j] \qquad S_{\tau_2}^H(j\omega_0) = -1 + Q[\omega_0\tau_1(1 - K) - j]$$

$$S_\rho^H(j\omega_0) = -Q \frac{\rho}{\omega_0 \tau_1} \qquad\qquad\qquad S_K^H(j\omega_0) = 1 + QK\omega_0\tau_1$$

so that we obtain for S_T from Eq. (3-141)

$$S_T(j\omega_0) \simeq [(1 - K)^2\omega_0^2\tau_1^2 + 1]Q^2\sigma_{t_1}^2 + \{[-Q^{-1} + (1 - K)\omega_0\tau_1]^2 + 1\}Q^2\sigma_{t_2}^2 \tag{3-143}$$

$$+ \left(Q\frac{\rho}{\omega_0\tau_1}\right)^2 \sigma_r^2 + (Q^{-1} + K\omega_0\tau_1)^2 Q^2\sigma_k^2$$

Note that S_T is proportional to Q^2.

At this point we have to concern ourselves with the statistical properties of the elements. Let us assume that the resistor *ratio* $\rho = R_1/R_2$ can be implemented accurately so that the term involving σ_r^2 is negligible. Further, for convenient implementation let us choose $C_1 = C_2 = C$, which implies that $\rho = \omega_0^2\tau_1^2$ is fixed because of Eq. (3-136a) as soon as τ_1 is chosen. If the gain K is not "too large" and is realized with an op amp and a resistor ratio K [see Eq. (3-31a) and Fig. 3.2], we can assume also that the σ_k^2 term is small, so that the variabilities $\sigma_{t_1}^2$ and $\sigma_{t_2}^2$ dominate. For a final simplification, let us set $\sigma_{t_1}^2 = \sigma_{t_2}^2 = \sigma_t^2$ so that Eq. (3-143), with $x = \sqrt{\rho} \triangleq$

$\omega_0\tau_1$, becomes

$$S_T \approx 2Q^2\left[(1 - K)^2x^2 - \frac{1 - K}{Q}x + 1 + \frac{1}{2Q^2}\right]\sigma_t^2 \tag{3-144}$$

To find a minimum for this expression, we must remember that K by Eq. (3-136b) is a function of x:

$$K = 1 + \frac{1 + x^2}{x^2} - \frac{1}{xQ} = 2 + \frac{1}{x^2} - \frac{1}{xQ} \tag{3-145}$$

Calling now for $y \triangleq (1 - K)x$, $dS_T/dx = 0$ from Eq. (3-144) yields

$$2y\frac{dy}{dx} - \frac{1}{Q}\frac{dy}{dx} = 0 \tag{3-146a}$$

with

$$\frac{dy}{dx} = \frac{1}{x^2} - 1 \tag{3-146b}$$

Equation (3-146b) yields $x = \omega_0\tau_1 = 1$; and Eq. (3-146a), for $dy/dx \neq 0$, gives $y = (1 - K)x = 1/(2Q)$. The second value for x, with $\rho \triangleq x^2$ and Eq. (3-142), results in $\rho = 4Q^2(1 + \rho)^2$, which has no positive solution for ρ and is therefore useless. Thus, the minimum of S_T with $x = 1$ and $K = 3 - 1/Q$ is found from Eq. (3-144) to be

$$\min S_T = (10Q^2 - 4Q + 1)\sigma_t^2 = 145\sigma_t^2$$

If we assume $\sigma_t^2 = 0.0009$ (corresponding to a time-constant variation of 3%), then the mean variation of the relative gain at $\omega = \omega_0$ equals $\Delta H/H = \sqrt{0.1305} = 0.36$. Note that this result assumes $\sigma_k^2 \approx \sigma_r^2 \approx 0$. The design is completed by choosing, for example, $C = 25$ nF and calculating, with $\rho = \omega_0\tau_1 = 1$,

$$R_1 = \frac{1}{\omega_0 C} = 40 \text{ k}\Omega \qquad R_2 = \frac{R_1}{\rho} = R_1 = 40 \text{ k}\Omega$$

$$K = 3 - \frac{1}{Q} = 2.75$$

By carefully working through Example 3-10 and Problems 3.35 to 3.38, the student should begin to understand the process of sensitivity minimization. We need to bear in mind that in general all circuit components vary simultaneously, so that the optimization has to be carried out in multidimensional element space, subject to performance and fabrication constraints. Thus, it should become apparent that the computations can quickly become overwhelming and computer aids must be used for any but the most trivial problems, even if a number of assumptions are used, as in our example, that simplify the complexity significantly. The student should note especially that the "optimal solution" obtained depends in general quite critically on the assumptions made and on the statistical model used. Extreme care should, therefore, be taken when formulating the optimization problem and the constraints and conditions. The references [49–53] provide further details about this important aspect of the process of designing practical filter circuits.

Sensitivity calculations have one weakness, even if optimization is used: in a practical design environment, filter requirements are usually not prescribed exactly but are only specified with certain tolerances or limits. For example, system requirements may call for a lowpass filter to have *no more than* A_p (in dB) attenuation in the passband and *at least* A_s (in dB) attenuation in the stopband. Let us assume that the appropriate calculations indicate that a Butterworth function of a certain degree just meets the specifications and that the designer selects an active filter topology with acceptable sensitivity performance. · All information about potential deviations that is available at this point is based on the *slope* of the *nominal Butterworth function* with respect to the circuit elements at the *nominal design values*. Even if multiparameter statistical sensitivity measures are used, the designer can arrive only at an estimate of the *expected mean deviations* caused by a set of components with known tolerances. However, two questions that are more important for the manufacturing engineer cannot be answered with any degree of confidence:

1. Given a fabrication process, what is the expected manufacturing *yield;* i.e., out of the total production run, how many circuits will meet the design specifications?
2. Assuming that the yield is less than 100%, how can it be improved without going to a more expensive, low-tolerance process? In other words, instead of the chosen Butterworth filter, is there possibly a *better nominal design* which still, of course, meets the imprecise passband and stopband requirements of $\leq A_p$ (in dB) and $\geq A_s$ (in dB) and at the same time results in a lower number of rejected samples?

Both these questions are addressed and can often be answered by a statistical circuit design method generally referred to as *design centering*. The principles of this technique will be explained in the next subsection, and one of the available computer algorithms will be discussed briefly.

3.3.5 Design Centering

The chief object of *design centering* is the following:

> Find a *nominal* circuit *design* that is *centered* in the *acceptability region* sufficiently far away from its borders that the prescribed specifications are met by the maximum number of circuits assembled with any set of components which are in the *tolerance range* of the available process.

Design centering is in general so computation-intensive that the availability of computer algorithms is essential. However, the concepts can best be explained by an example, which is kept simple enough to permit the whole process to be completed by hand.

Example 3-11

Design a passive first-order lowpass filter to meet the specifications given in Fig. 3-18a. The available components are resistors with ±10% and capacitors with ±20% tolerances. The yield should be maximized.

Solution. A suitable circuit is shown in Fig. 3-18b; its transfer function is

$$H(s) = \frac{R_2}{R_1 + R_2} \frac{1}{1 + sCR_1 R_2/(R_1 + R_2)} \triangleq a \frac{1}{1 + s\tau a}$$

Because the impedance level is clearly irrelevant, the three available variables R_1, R_2, and C can be reduced to two:

$$a = \frac{R_2}{R_1 + R_2} = \frac{1}{1 + R_1/R_2} \quad \text{and} \quad \tau = CR_1$$

Unconcerned with (or unaware of) yield problems, the designer may proceed to meet the specification $0.6 \le H(j0) = a \le 0.75$ by choosing $a = 0.65$, i.e., $R_1/R_2 = 0.5385$, and, for example,

$$R_1 = 3 \text{ k}\Omega \qquad R_2 = 5.57 \text{ k}\Omega$$

Further, a little trial and error yields, for example, $\tau = 0.3$ ms, i.e., $C = 0.1$ μF, to

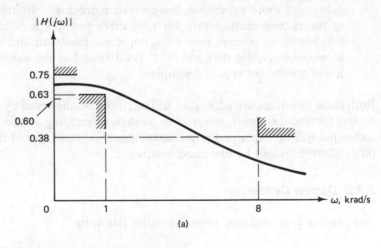

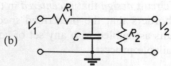

Figure 3-18 Design specifications (a) and passive lowpass circuit (b) for Example 3-11.

satisfy the requirement that

$$|H(j\omega)| = \frac{a}{\sqrt{1 + (\omega\tau a)^2}} \tag{3-147}$$

is larger than 0.63 at $\omega \leq \omega_p = 1000$ rad/s and less than 0.38 at $\omega \geq \omega_s = 8000$ rad/s.

The given tolerances, $r = \Delta R/R = \pm 0.1$ and $c = \Delta C/C = \pm 0.2$, result in

$$\tau_0(1 + \Delta\tau) = R_1 C(1 + r)(1 + c) = \tau_0(1 \pm 0.32)$$

and

$$a_0(1 + \Delta a) = \frac{1}{1 + \dfrac{R_{10}}{R_{20}} \dfrac{1 + r_1}{1 + r_2}} \simeq a_0 (1 \pm 0.07)$$

A quick check confirms that at the pair of extremes ($\Delta\tau = +0.32$, $\Delta a = +0.07$), all specifications are met, but that the design fails at the other three points of extreme tolerance values: $(\Delta\tau, \Delta a) = (+0.32, -0.07)$, $(-0.32, +0.07)$, and $(-0.32, -0.07)$. Consequently, the production yield Y is certainly less than 100%. As a matter of fact, Y is only $\simeq 45\%$, as we shall demonstrate presently.

To understand the nature of the problem and recognize methods for improving this design, we need to consider more carefully the limits imposed on a and τ by Fig. 3-18a. If we note the passband and stopband limits by $a_p = 0.63$ and $a_s = 0.38$, respectively, Eq. (3-147) results in

$$\tau \leq \frac{1}{\omega_p} \sqrt{\frac{1}{a_p^2} - \frac{1}{a^2}} \tag{I}$$

$$\tau \geq \frac{1}{\omega_s} \sqrt{\frac{1}{a_s^2} - \frac{1}{a^2}} \tag{II}$$

in addition to the obvious constraints

$$a \geq 0.6 \tag{III}$$

$$a \leq 0.75 \tag{IV}$$

Limits (I) to (IV) are drawn in Fig. 3-19; the inside of the bold "sail-shaped" boundary constitutes the *acceptability region* (AR); i.e., any set of two values τ and a inside AR leads to an acceptable circuit. Specifically, our previous design choice, marked N_1 (i.e., $a_0 = 0.65$, $\tau_0 = 0.3$ ms), is, of course, inside AR; and a sketch of the tolerance region $a_0 \pm \Delta a = 0.65(1 \pm 0.07)$, $\tau_0(1 \pm \Delta\tau) = 0.3(1 \pm 0.32)$ ms, shows immediately why the yield $\simeq 45\%$: only approximately 45% of the total area of the tolerance rectangle lies inside AR. With the problem recognized, the solution is now apparent: if we use a better "centered" point, such as $a \simeq 0.705$, $\tau \simeq 0.38$ (marked N_2 in Fig. 3-19), then the tolerance region lies completely in AR and the yield is 100%. The algorithm to be discussed below gives

$$R_1 = 2.66 \text{ k}\Omega \qquad R_2 = 6.35 \text{ k}\Omega \qquad C = 0.145 \text{ }\mu\text{F}$$

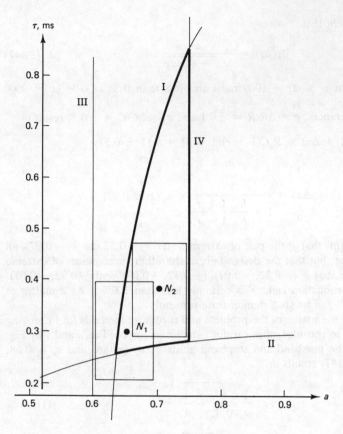

Figure 3-19 Acceptability region for the circuit elements and specifications of Fig. 3-18, Example 3-11.

that is,

$$a = 0.704 \qquad \tau = 0.385 \text{ ms}$$

for 100% final yield from an initial yield of 46%, based on 1000 Monte Carlo iterations and assuming uniform distribution (Fig. 3-14a) and $\Delta R/R = 0.1$, $\Delta C/C = 0.2$. If we instead want to go with the standard R and C series of *preferred* values, then we can use

$$R_1 = 2.7 \text{ k}\Omega \pm 10\% \qquad R_1 = 6.8 \text{ k}\Omega \pm 10\%$$

and

$$C = 0.15 \text{ }\mu\text{F} \pm 20\%$$

which results in $Y = 99.5\%$.

Example 3-11 should help to explain the concept of design centering. In particular it should now have become apparent that the choice of a nominal design point is not arbitrary if, as is mostly the case, yield considerations are important. The student may rightly object that the nominal design N_1 in Example 3-11 was, at least in hindsight, "obviously" a poor choice; a better one could have been picked intuitively by choosing an initial curve $|H(j\omega)|$, such as the sample sketched in Fig. 3-18a, that is "farther away" from the critical corners of the attenuation specifications. Let us point out, however, that this example was on purpose made almost trivially simple in order to make the process as transparent as possible, and that even in this case the *optimal* point N_2 could not easily have been found "by inspection." In general, the situation is far more complicated: if a circuit has n varying components, then the acceptability region AR is a volume in n dimensional space. Furthermore, this volume may have "holes" (embedded forbidden regions) and consist of disconnected parts. Considering that in such a case it is a near impossible task even to describe the boundary of AR, the student will appreciate that it is not a trivial matter to locate a nominal point such that the also n-dimensional tolerance range is as completely as possible within AR. Several computer algorithms have been proposed in the literature [54–64] which attempt to solve the design centering (yield optimization) problems. One method that is relatively easy to understand and that at the same time operates with the least number of assumptions about the (in general unknown) acceptability region will be described in the following [59].

Let us assume that we have a circuit with a set of n initial nominal element values $\mathbf{x}_n$. It is unimportant for the following discussion whether the set $\mathbf{x}_n$ meets the design criteria or not, i.e., whether or not $\mathbf{x}_n$ lies within the acceptability region, $\mathbf{x}_n \in$ AR. To be specific, however, let $\mathbf{x}_n \in$ AR, as shown in two dimensions[13] in Fig. 3-20. From our earlier discussions and Fig. 3-20, it is apparent that the nominal point $\mathbf{x}_n$ with tolerances $\Delta\mathbf{x}$ does not result in 100% yield, because a portion of the tolerance region TR_0 falls outside AR. Because the boundaries of AR are generally not known, the difficulty we have to face is to decide which part of TR lies within AR. The problem is addressed by randomly varying the components $\mathbf{x}$ around $\mathbf{x}_n$ inside TR_0 and checking the circuit's performance for each set $\mathbf{x}$ (Monte Carlo analysis, Section 3.3.3). If after a sufficiently large statistically significant number of samples all tests have passed, the yield can be estimated to be 100%, which, we assume, is our goal. As soon, however, as one point fails, we interrupt the process and expand around this point—p_1 in Fig. 3-20—a new tolerance region, TR_1. For the further development of the algorithm, it is now important to realize that TR_1 is a *forbidden region*, because *any nominal* point within TR_1 will include in its tolerance region the point p_1, a known failure, and consequently will result in less than 100% yield. Therefore, TR_1 is *cut* out of the region of possible nominal points, and we proceed to test the circuit's performance at a new *nominal* point, say, p_2 in Fig. 3-20. If, as assumed in our example, this point results in a circuit failure, the corresponding tolerance region TR_2 is cut out, i.e., it is *added* to the forbidden region, thereby further narrowing the possible locations of acceptable

---

[13]To keep the discussion lucid, we assume that AR is known. The reader should bear in mind, however, that in general the shape of AR is unknown and is n-dimensional, so that Fig. 3-20 represents only a two-dimensional cross section through AR, with all parameter values except x_1 and x_2 fixed.

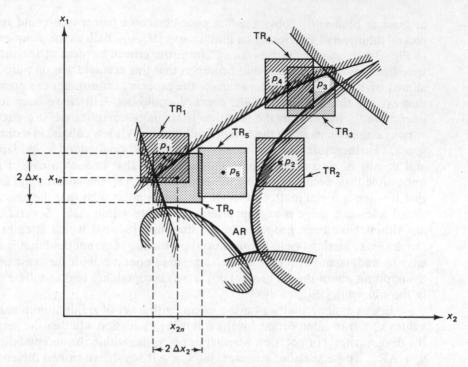

Figure 3-20 Two-dimensional acceptability region AR and tolerance regions (TR_i) to illustrate design centering algorithm.

parameter sets **x**. The process then continues in the same manner; if the next choice of a nominal point, say, p_3, lies within AR, i.e., the circuit passes, a new Monte Carlo analysis is initiated in TR_3 around p_3. As soon as a failing parameter set p_4 is encountered, the region TR_4 is *added* to the forbidden region and a new nominal point, say, p_5, is tested. If p_5 and all Monte Carlo simulations in TR_5 around p_5 pass, as is assumed in Fig. 3-20, we have found one nominal parameter point that for the given tolerances results in 100% yield.

In addition to being easy to understand and implement, this algorithm has a number of important advantages:

1. Any narrow "tails" of AR, where very tight component tolerances are required for meeting circuit specifications, tend to be cut out by the algorithm.

2. No *a priori* knowledge of or assumptions about the shape of AR are necessary.

3. The algorithm is efficient because the union of all forbidden regions, the *cut region*, quickly limits the choice of available nominal design points.

4. If a solution region of 100% yield exists, the algorithm is guaranteed to find it, i.e., convergence can be proved [59].

As described, this design centering algorithm attempts to locate a nominal design point in AR for which the yield is 100%. Although this will clearly be the most desirable goal in manufacturing, it is quite possible in practice that no 100% yield solution exists for the filter under consideration with a given fabrication process. Under these circumstances the algorithm will not converge, but the designer is still interested in determining which yield ($< 100\%$) is obtainable, i.e., they are interested in a suboptimal solution. Fortunately, the design centering algorithm can easily be modified to account for this case. Assume that a yield of Y_a percent is acceptable. Then, instead of one failure, we need only to permit $n_F = N(1 - Y_a/100)$ failures to occur before we break off the N-sample Monte Carlo analysis to look for a new nominal point. Of course, each failing point is still surrounded by a cut region, and the union of all cut regions forms a forbidden region with no allowed nominal points. In this case, the cut region is not the full tolerance region TR but only that part[14] of TR in which statistically Y_a percent of the samples can be expected to lie. Clearly, the size of the cut region depends on the assumed statistical distribution (see Fig. 3-14). Example 3-12 illustrates some of the relevant points.

Example 3-12

Repeat the design of Example 3-11, but assume that the dc gain must lie between 0.64 and 0.70.

Solution. Following Example 3-11, we find a nominal point N_1 at $a = 0.65, \tau = 0.3$ ms that meets specifications. The acceptability region and the tolerance region around N_1 are shown in Fig. 3-21, with $R_1 = 3$ kΩ, $R_2 = 5.57$ kΩ, $C = 0.1$ μF, and uniform distribution, as before, the yield can be computed to be $Y = 42\%$ (area of TR in AR, divided by TR). The unmodified algorithm does not converge to 100% yield, because TR does not fit into AR. If the acceptable yield is reduced, say, $Y_a = 75\%$, and the algorithm modified correspondingly, a new nominal design point is found at $R_1 = 2586$ Ω, $R_2 = 5522$ Ω, and $C = 138.6$ nF to give point N_2 at $a = 0.681, \tau = 0.358$ ms, for a yield of 76%.

The designer now has several options: either accept these values, or use further iterations starting from different initial points, or go to more expensive, low-tolerance elements. Suppose we take the latter choice and select 5% resistors and 10% capacitors. A centered design with 100% yield is then point N_3 at $a = 0.677$ and $\tau = 0.336$ ms, with $R_1 = 2655$ Ω, $R_2 = 5552$ Ω, and $C = 126.5$ nF.

At this point the designer may also try to influence circuit cost by staying with 20% capacitors but reducing only the resistor tolerances to 5%. (An "obvious" route,

---

[14]The reader should bear in mind, however, that for a significant reduction in the size of the individual cut regions the cuts become increasingly less important for large dimensionality n. Assume that the full tolerance region is TR and each dimension is reduced, due to $Y_a < 100\%$, by $p\%$. Then the cut "volume," TR $(1 - p/100)^n$, may become so small that its being hit by a new sample point becomes increasingly unlikely and the cut routine may as well be bypassed. For example, for $Y_a = 0.85$ and uniform distribution, 85% of the samples are in 0.85 TR. If there are $n = 35$ parameters, the cut volume is only 0.85^{35} TR $= 0.0038$ TR.

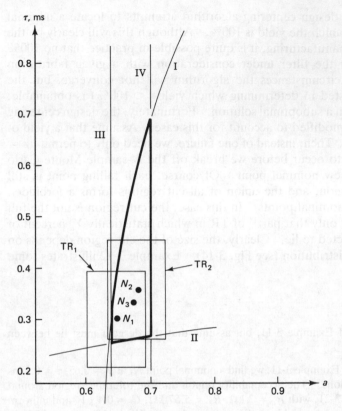

Figure 3-21 Acceptability region for Example 3-12.

since in this case we know the shape of AR!). This attempt leads to $R_1 = 2782 \, \Omega$, $R_2 = 5827 \, \Omega$, and $C = 127.6$ nF for $a = 0.677$, $\tau = 0.355$ ms (close to point N_2), and 99.7% yield.

All calculations were performed with the cut algorithm described in this section [59]. However, because the example was kept sufficiently simple, most computations can readily be repeated by the student with pencil, paper, and a calculator.

We hope that this short subsection has given the reader some appreciation of the concept, the problems, and the possibilities of design centering for yield optimization with practical manufacturing constraints. Further details of and insights into these heavily computer-based procedures can be found in the references [54–64].

3.4 LARGE-CHANGE SENSITIVITY

Apart from Monte Carlo–based methods (Sections 3.3.3 and 3.3.5) that use direct evaluations of the filter characteristics for the actual element values, all sensitivity-related calculations in this chapter assume *small changes* in component values because they are based on the truncated (linearized) Taylor series expansions in Eqs. (3-2) and (3-96). If the expected variation dx of the component x is large,[15] Eq. (3-2) may give a very poor estimate of the deviation of the performance measure $P(x + dx)$ because its curvature, $\partial^2 P/\partial x^2$, has been neglected. Under these circumstances, the second derivatives must be included in the calculations according to Eq. (3-1):

$$\frac{P(s, x)}{P(s, x_0)} \simeq 1 + \left.\frac{\partial[P(s, x)/P(s, x_0)]}{\partial x/x_0}\right|_{x_0} \frac{dx}{x_0}$$
$$+ \frac{1}{2} \left.\frac{\partial^2[P(s, x)/P(s, x_0)]}{\partial(x/x_0)^2}\right|_{x_0} \left(\frac{dx}{x_0}\right)^2 \tag{3-148}$$

Similarly, if multiparameter variations are to be considered, Eq. (3-96a) must be extended to include second-order derivatives:

$$dP(\mathbf{x}) \simeq \sum_{i=1}^{k} \left.\frac{\partial P}{\partial x_i}\right|_{x_{i0}} dx_i + \frac{1}{2} \sum_{i=1}^{k} \sum_{j=1}^{k} \left.\frac{\partial^2 P}{\partial x_i\, \partial x_j}\right|_{x_{i0}x_{j0}} dx_i\, dx_j \tag{3-149}$$

Finally, if the element variations cannot be considered deterministic but must be assumed to be statistical, as in Section 3.3.2, an appropriate interpretation of Eq. (3-149) is

$$dP(\mathbf{x}) = P(\mathbf{x}) - P(\overline{\mathbf{x}}) \simeq \sum_{i=1}^{k} (x_i - \overline{x}_i) \left.\frac{\partial P}{\partial x_i}\right|_{\overline{\mathbf{x}}}$$
$$+ \frac{1}{2} \sum_{i=1}^{k} \sum_{j=1}^{k} (x_i - \overline{x}_i)(x_j - \overline{x}_j) \left.\frac{\partial^2 P}{\partial x_i\, \partial x_j}\right|_{\overline{\mathbf{x}}} \tag{3-150}$$

where $\overline{\mathbf{x}}$ is the k-dimensional vector of the mean element values as defined in Eq. (3-109). To find the expected value of $P(\mathbf{x})$, we use Eqs. (3-109) through (3-113) to obtain from Eq. (3-150)

$$E\{P(\mathbf{x})\} = P(\overline{\mathbf{x}}) + \frac{1}{2} \sum_{i=1}^{k} \sigma_i^2 \left.\frac{\partial^2 P}{\partial x_i^2}\right|_{\overline{\mathbf{x}}} + \sum_{i<j}^{k} \sum_{j=1}^{k} \text{Cov}\{x_i, x_j\} \left.\frac{\partial^2 P}{\partial x_i\, \partial x_j}\right|_{\overline{\mathbf{x}}} \tag{3-151}$$

The interpretation of Eq. (3-151) is obvious; note especially that in general $E\{P(\mathbf{x})\} \neq P(\overline{\mathbf{x}})$ if the second derivatives must be taken into account, but that the expected value of $P(\mathbf{x})$ depends on the variances σ_i^2 and the covariances $\text{Cov}\{x_i, x_j\}$ of the components. Clearly, if the elements can be assumed to be statistically independent, the last term in Eq. (3-151) is zero.

Evidently, the evaluation of large-change deviations by means of Eqs. (3-148) to (3-151) becomes quite involved, so that use of computer aids will become necessary. Therefore, if large element changes are to be expected, it is probably

---

[15]As mentioned earlier, whether a given dx is large or small depends on the situation and must be evaluated carefully in each case.

simpler and usually more accurate to evaluate (by computer) the circuit's performance directly at the extreme values $\mathbf{x} \pm d\mathbf{x}$ of the components. A direct computation of the performance variabilities caused by large changes in many circuit components via Eqs. (3-149) to (3-151) does not, of course, result in *sensitivity* information; i.e., no direct feedback is provided to the designer for determining the most critical components. However, the advantage is that no first- and second-order derivatives have to be found, and existing circuit analysis programs can be used directly to obtain a reliable picture of the circuit's expected performance.

3.5 SUMMARY

This chapter, along with the problems and references, should have given the reader a good understanding of the concept and the importance of sensitivity. We have presented methods for calculating both single-parameter and multiparameter sensitivities, and we have discussed ways of dealing with this problem in situations where the component changes can be assumed to be deterministic and also with the more realistic case of statistical variations. Because component values are rarely precise and usually do not even maintain their initial values due to manufacturing tolerances and environmental effects, it is extremely important in the design of electrical filters to determine how sensitive a circuit's performance is to changes in network elements. As a matter of fact, sensitivity has been established as the primary criterion for separating "good" circuits from those which will not perform satisfactorily in a practical environment. Clearly, even an apparently most attractive "paper" design will be useless in practice if its performance drifts out of the prescribed specification range because of small unavoidable component tolerances. We cannot emphasize too much, therefore, that before accepting a design, the engineer *must* evaluate the sensitivities of a proposed circuit and its variabilities for realistic component tolerances.

Of course, the best way to arrive at a low-sensitivity structure is to avoid the *causes* of large sensitivities from the beginning. The most serious contributor to sensitivity that is found in a large number of designs is the *difference effect* by which a small number is set through the difference of two relatively larger numbers. An example is the determination of the quality factor Q via an expression $1/Q = a - b$. In a high-Q circuit, $1/Q$ is a small number that is often set by the difference of some functions of components, a and b, with $1/Q \ll (a, b)$. Clearly, even small variations in a and b will give rise to large changes in the value of Q; i.e., Q is very sensitive to a and b. Another frequently overlooked sensitivity problem may arise when a circuit's performance relies on the cancellation of a pole and a zero. Because pole and zero positions depend on circuit elements, the cancellation will be destroyed in practice due to element tolerances, and the resulting "doublet" may cause serious performance deviations—i.e., the circuit is very sensitive to the components which determine the relevant pole and zero.

We mentioned before that sensitivities are used to separate "good" designs from "unsatisfactory" ones.   Evidently, this selection process implies a *comparison* of different circuits that must be handled with considerable thought and caution if erroneous conclusions are to be avoided.   Reference 33 contains a detailed discussion of this point.   The most frequently encountered errors in sensitivity comparison are the following:

1. *Forgetting that sensitivity is a small-change concept.*   Sensitivity is a measure of the *slope at the nominal point* of the performance criterion.   If component changes are not incremental, the sensitivity evaluation may be quite meaningless.   Even if the sensitivity is zero, it usually only means that the dependence on the element is quadratic and that second-derivative effects will dominate the performance.

2. *Using the wrong frequency range.*   We have pointed out repeatedly that sensitivities are functions of frequency; therefore, the circuits should be evaluated in the frequency *range* of interest.   Just a few *test* frequencies may be insufficient, because sensitivities do have zeros and $S_x^P(j\omega_j) = 0$ does not necessarily say much about the performance over the frequency *range*.   Also, calculating only the ω_0 and Q sensitivities may be insufficient unless interpreted carefully.

3. *Comparing optimized with unoptimized circuits.*   This point appears to be quite obvious but should nevertheless be mentioned.   We discussed earlier that most circuits have more components than are necessary for satisfying the design constraints, and that these degrees of freedom can and should be used to optimize the circuit's behavior.   Therefore, before comparing two circuits, the engineer should make sure that *both* have been optimized for the *same* performance criterion.

4. *Considering only sensitivities instead of the more important variabilities.*   In the final analysis, it is not sensitivity but the *variability* that determines practical circuit performance.   Large sensitivities to very accurate and stable components are not necessarily bad, but even moderate sensitivities to elements with large tolerances may well render a design impractical.

5. *Relying too much on single-parameter sensitivities.*   Unless the designer has made certain that a particular component is the most critical one (a very difficult task in practice), single-parameter sensitivities can lead to very misleading conclusions, because the variabilities caused by all the other components may well dominate the performance.   Furthermore, we remind the reader again that sensitivities are only a measure of the *slope* of the performance criterion *at the nominal point*.   If the nominal point is altered because of a different choice of elements, the whole sensitivity picture changes and must be reevaluated.   Also note the comments made in item 1, above.

Considering the importance of sensitivity concepts, we have devoted a full chapter to this topic and have discussed in some detail even difficult but practically important subjects that are not often encountered in a textbook on filter design, namely, multiparameter and statistical sensitivity, and design centering. Although the aid of computer programs is necessary for effectively performing many of the actual computations, any serious designer of analog signal filters should neverthe-less attempt to gain a good theoretical understanding of sensitivity and variability if he or she expects the circuits to perform reliably in practice. We shall find frequent occasions to make use of sensitivity considerations in subsequent chapters.

PROBLEMS

3.1 As a further illustration of the sensitivity improvement in Example 3-1 when r is increased from 1 to 6, calculate the pole locations and the Q values for both $r = 1$ and $r = 10$ when $\Delta K/K_0 = \pm 0.5\%$. For $r = 10$, what percentage change in K will cause instability? Compare with the value for $r = 1$.

3.2 Using Eq. (3-16), prove Eqs. (3-18) and (3-19) and the expressions below. In all cases, the P_i are functions of x, and n and k are constants.

$$S_x^{1/P} = S_{1/x}^P = -S_x^P \qquad S_x^{kP} = S_x^P$$

$$S_x^{P^n} = nS_x^P \qquad S_x^{kx^n} = n$$

$$S_x^{k+P} = \frac{P}{k+P} S_x^P \qquad S_x^{\Sigma P_i} = \frac{1}{\Sigma P_i} \sum (P_i S_x^{P_i})$$

3.3 In Example 3-1, assume that $C_1 \neq C_2$ and verify Eqs. (3-25) and (3-26).

3.4 Realize the transfer function

$$H_2(s) = \frac{s^2 + \omega_0^2}{s^2 + s\omega_0/Q + \omega_0^2}$$

with $f_0 = 2.4$ kHz and $Q = 8$, using the circuit of Fig. P3.4.

(a) Find $H_2(s) = V_o/V_i$ and find the appropriate element values. Assume that the amplifier gain A is large.

(b) Use the element values obtained in part (a) to calculate Q when $K = A/(A + 1)$ changes from $K_{\text{nom}} = 1$ by -5%. Assume K is real. Repeat, but calculate Q using sensitivities. Explain any large discrepancies between the two results.

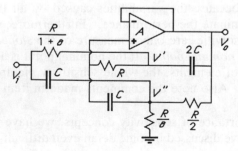

Figure P3.4

3.5 Consider a second-order active filter built with resistors R_i and capacitors C_j (and an amplifier, of course). Assume that due to varying temperature all resistors and capacitors change their values by r percent and c percent, respectively. Show that the value of the quality factor does not change and that the pole frequency changes by $-(r + c)$ percent.

3.6 Prove Eq. (3-35).

3.7 (a) For the circuit in Fig. P3.7, find the transfer function $H(s) = V_o/V_i$ and determine the elements for given values of ω_0 and Q. What is $|H(j\omega_0)|$?

    (b) Find the sensitivities S_x^Q, $S_x^{\omega_0}$, and $S_x^{H(j\omega)}$, where x stands for all the active and passive elements. Assume that the amplifier K is realized as in Fig. 3-3. Try to express the sensitivities in terms of ω_0 and Q.

    (c) Find the gain-sensitivity product Γ_K^Q. For which value of K is it a minimum? What is the percentage change of Q when A varies by 50%?

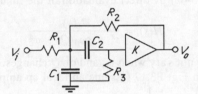

Figure P3.7

3.8 For the circuit in Fig. P3.7, find and sketch magnitude and phase sensitivities. Assume $f_0 = 1000$ Hz, $Q = 25$, and K chosen for minimum Γ_K^Q as in Problem 3.7.

3.9 Calculate the phase of $H(j\omega)$ in Eq. (3-6) and verify by direct computation that Eq. (3-45) is the phase sensitivity Q_K^ϕ.

3.10 Derive an expression analogous to Eqs. (3-46) and (3-47) for the case that poles and zeros of $H(s)$ are multiple.

3.11 Prove that $Q_x^{z_i}$ in Eq. (3-47) is the residue of the zero at $s = z_i$ of S_x^H. The analogous statement is true, of course, for $Q_x^{p_i}$.

3.12 One can prove [9] that a network function $H(s)$ is a bilinear function of any of its circuit components x, i.e.,

$$H(s, x) = \frac{N(s, x)}{D(s, x)} = \frac{a(s) + xb(s)}{c(s) + xd(s)}$$

where, as indicated, a, b, c, and d are functions of s but *not* of x. For the situation of interest in this text, x stands for any of the elements R, L, or C, or the gain parameter of a controlled source. The above equation indicates that the relevant network polynomials are linear functions of x; e.g.,

$$D(s, x) = c(s) + xd(s)$$

Use this fact to derive an expression for $Q_x^{p_i}$ where $D(p_i) = 0$, that is, p_i is a root of $D(s)$.

    Hint: For $x \rightarrow x_0 + \Delta x$, we have $p_1 = p_0 + \Delta p_0$ where $s = p_0$ is the nominal root if $x = x_0$. Use these relationships and expand $D(s, x)$ into a Taylor series up to the first order.

3.13 Assume that $D(s)$ has a root p_0 of multiplicity $m > 1$. Use the results of Problem 3.12 to show that the root p_0 is converted into m simple roots by changing the parameter x to $x + \Delta x$.

 Hint: Bring $D(s, x) = c(s) + (x + \Delta x)d(s) = 0$ into the form $1 + (\Delta x/x) \cdot [x \, d(s)]/D(s, x) = 0$ and expand $x \, d(s)/D(s)$ into partial fractions.

3.14 Use the results of Problem 3.12 to determine the approximate roots of

$$D(s) = s^3 + a_2 s^2 + a_1 s + a_0 = s^3 + 3s^2 + 12s + 10$$

$$= (s + 1)(s^2 + 2s + 10)$$

when the coefficient $a_2 = 3$ changes by -5%.

3.15 Use Eq. (3-52) to determine the shift of the pole with highest quality factor of a seventh-order 1 dB–ripple Chebyshev filter when the coefficient of the s^2 term changes by 5%.

3.16 Derive Eqs. (3-60) to (3-62) by direct evaluation of the function

$$H(s) = \frac{N(s)}{s^2 + s\omega_0/Q + \omega_0^2}$$

Assume that $N(s)$ does not vary when ω_0 and/or Q changes.

3.17 For the circuit shown in Fig. P3.17 (assume an ideal op amp):

(a) Show that
$$\frac{V_o}{V_i} = \frac{-1}{R_1 C} \frac{s}{s^2 + 2s/(R_2 C) + 1/(R_1 R_2 C^2)}$$

(b) Determine the various sensitivity functions $S_{x_i}^{\omega_0}$ and $S_{x_i}^Q$, where the x_i are the R's and the C's.

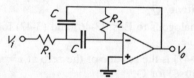

Figure P3.17

3.18 For the circuit shown in Fig. P3.18, calculate the following:

(a) All $S_{x_i}^H$, where the x_i are the RLC components.

(b) The sensitivities of ω_0 and Q relative to R, C, and L.

(c) For a design value of $\omega_0 = 2\pi 10^5$ rad/s and $Q = 10$, determine the changes in ω_0 and ω_{3dB} if:

(1) C changes by $+5\%$.

(2) R changes by $+5\%$.

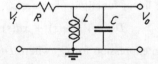

Figure P3.18

3.19 For the filter in Fig. P3.17, verify Eqs. (3-23) and (3-25).

3.20 For the filter in Fig. P3.17, calculate the poles as functions of the elements and verify Eqs. (3-26a) to (3-26c).

3.21 For the filter in Fig. P3.17, calculate the element values for $\omega_0 = 5$ krad/s, $Q = 12$ and determine the pole displacement if: (a) all resistor and capacitor values increase by 3%; (b) all resistor values increase and all capacitors decrease by 3%.

3.22 The circuit in Fig. P3.22 realizes

$$H(s) = \frac{V_o}{V_i} = \frac{-s/R_1C_1}{s^2 + \dfrac{s}{C_2}\left(\dfrac{1}{R_1} + \dfrac{1}{R_2}\right) + \dfrac{1}{R_1R_2C_1C_2}}$$

For this function/filter, complete
(a) Problem 3.19
(b) Problem 3.20
(c) Problem 3.21

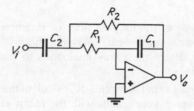

Figure P3.22

3.23 For the twin T circuit of Problem 1.1, calculate the sensitivity of the zero frequency ω_z to all circuit elements R_i and C_i, $i = 1, 2, 3$.

3.24 Use Eq. (3-66) to express the gain variability $\Delta|H_2|/|H_2|$ as a function of the ω_p and Q_p variabilities:
(a) At the pole frequency $\omega = \omega_p$
(b) At the bandedge $\omega = \omega_p(1 \pm 1/(2Q_p))$

3.25 In an active RC network, the pole frequency is given by $\omega_0 = 1/\sqrt{R_1R_2C_1C_2}$. Assume that the temperature coefficients of the elements are

$$\alpha_R = (100 \pm 30) \text{ ppm/°C} \qquad \alpha_C = -(50 \pm 10) \text{ ppm/°C}$$

and you have tracking components, i.e., α_R and α_C are the same for all resistors and capacitors, respectively. If the temperature rises by 75°C, what are the bounds between which ω_0 can be expected to vary?

3.26 Prove Eq. (3-77).

3.27 Derive equations corresponding to Eq. (3-79) for the sensitivities of the phase $\phi(\omega)$.

3.28 Use Eq. (3-79) to find the total variance $d[\ln H(s)]$ and assume that the variances dR_i/R_i, dL_i/L_i, and dC_i/C_i are the same for all i. Under these conditions, derive expressions for the total variations of $\alpha(\omega)$ and $\phi(\omega)$.

3.29 Derive Eqs. (3-89a) and (3-89b).

3.30 Derive Eqs. (3-90a) and (3-90b).

3.31 Show that in LF filters (Fig. 3-12b), the passband sensitivites are reduced below those of cascade filters and the stopband sensitivities are comparable in size to those of a cascade connection. Assume $n = 5$ sections as in Eq. (3-89b).

3.32 For the circuit in Fig. 3-18, obtain the expression for the deterministic multiparameter sensitivity given by Eq. (3-107).

3.33 Find the deterministic multiparameter sensitivity [Eq. (3-103)] of the circuit in Fig. P3.18 and evaluate it at:

(a) $\omega = \omega_0$

(b) $\omega = \omega_{3dB}$

3.34 Determine the multiparameter sensitivity measures $S^{|H|}$, Q^ϕ, and S_T over a frequency range $\omega_1 = 2\pi(0.95 \times 10^5)$ to $\omega_2 = 2\pi(1.05) \times 10^5$ rad/s for the circuit in Fig. P3.18 with the design values $\omega_0 = 2\pi(10^5)$ rad/s and $Q = 10$. Assume no correlation and a uniform distribution with variance $\sigma_{x_i}^2 = 10^{-4}$.

3.35 For the optimized design of the lowpass filter in Example 3-10, what is the contribution of the terms involving σ_k^2 and σ_r^2 (that were neglected in the text) to the value of S_T?

3.36 In Example 3-10, assume that the passive parameters τ_1, τ_2, and ρ have negligible variances, and that the σ_k^2 term dominates. Optimize S_T for this situation. What is S_T, from Eq. (3-144), for this optimization? Compare your results with those for Problem 3.35.

3.37 For ease of manual tuning of the lowpass filter of Example 3-10, the resistor R_1 is chosen as a discrete external component. Assume that σ_{R1}^2 dominates all other variances. Optimize your design, in a way analogous to the procedure in Example 3-10, so that S_T is minimized.

3.38 Using the same assumptions as in Example 3-10, calculate the maximum and minimum values of $|\Delta H/H|^2$. Compare your results with the mean value provided in the text example. (Use linearization for your error calculations!)

3.39 Rework Example 3-10 but assume that also the capacitors are realized as film capacitors on the substrate. Thus, the time constants τ_1 and τ_2 will now be correlated. For simplicity, assume perfect correlation, i.e., $E\{t_i\ t_j\} = \sigma_t^2$ for $i, j = 1, 2$. As before, let k, ρ, and t_i be uncorrelated.

3.40 Design the circuit of Fig. P3.18 so that it meets the specifications shown in Fig. P3.40. Center the design to maximize the yield. Estimate the yield for tolerances of $\Delta R/R = 1\%$, $\Delta C/C = 3\%$, and $\Delta L/L = 5\%$.

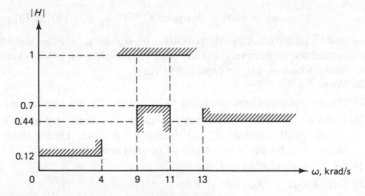

Figure P3.40

Operational Amplifiers and Fundamental Active Building Blocks

FOUR

4.1 INTRODUCTION

We have seen in Chapter 2 that *LC resonance* effects are necessary for the creation of complex (high-Q) poles in the realization of electrical signal filters with steep transfer function skirts. However, the requirement of inductors presents a problem in any attempt to miniaturize the design, because inductors tend to be large and bulky, especially at low frequencies. On the other hand, although *RC* networks can have complex *zeros* (compare Problem 3.23), it is not hard to show [9, 10, 24, 25] that the *poles of RC* networks are simple and restricted to the negative real axis. Consequently, steep slopes are very difficult to realize with *RC* networks unless the network is of very high order. Fortunately, there is a solution to this problem: as we have seen, e.g., in Examples 3-1 and 3-10 in Chapter 3, combining an *RC* network with a *gain* element can lead to complex poles with high Q factors. An elementary more formal illustration of this fact is contained in Fig. 4-1, which shows an *RC* network and an amplifier of gain $-K$ in a feedback configuration. If we label the *RC* transfer functions V_3/V_i by

$$T_{3i}(s) = \frac{V_3}{V_i} = \frac{N_{3i}(s)}{D(s)} \qquad i = 1, 2 \tag{4-1}$$

then the transfer function of the complete feedback network is

$$H(s) = \frac{V_2}{V_1} = -\frac{T_{31}(s)}{T_{32}(s) + 1/K} = -\frac{N_{31}(s)}{N_{32}(s) + D(s)/K} \tag{4-2}$$

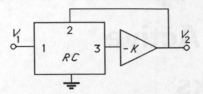

Figure 4-1 Active *RC* feedback network.

If the amplifier gain is very large, the function further simplifies to $H(s) \simeq N_{31}(s)/N_{32}(s)$ and evidently can have both complex poles and complex zeros because, as mentioned before, the transfer functions of the passive *RC* network can have complex zeros.

This simple but quite general feedback circuit indicates clearly the need for *inexpensive gain devices* if filters with sharp cutoff behavior are to be built in microelectronic form without the use of inductors. The *operational amplifier*, often referred to as *op amp*, fills this need. Indeed, without the availability of inexpensive high-performance op amps, much of the modern development in analog signal-processing circuitry would not have been possible. Although some recent work in *active RC filters* aimed at fully integrated implementations uses other active elements, most notably *voltage-to-current converters* (*transconductances;* see Section 4.2.3) [65], the large majority of active filters are built with op amps. Therefore, in this chapter we shall primarily discuss op amps and their properties in some detail, along with a number of active building blocks that are fundamental to the design of active filters. Transconductance amplifiers and their applications will be treated in Sections 4.2.3 and 4.4.

Basically, op amps are direct-coupled differential amplifiers with very high low-frequency gain A_0 (typically $A_0 > 10^5$); generally, they are used with external feedback to control gain and bandwidth and to be able to stabilize the applications against the extremely large variability of the op amp parameters. Simple examples of such feedback schemes were encountered in Fig. 3-2; refer also to Eq. (3-33). Currently, op amps are available in all semiconductor integrated-circuit processing techniques (bipolar, NMOS, PMOS, CMOS GaAs), so that circuits can be designed which are compatible with the technology of choice.

Fortunately, for most work in active filter design it is sufficient to consider op amps as circuit elements described by their input-output terminal relationships. Detailed knowledge of the internal configuration is rarely necessary. We shall discuss, therefore, in this chapter only the input-output operation, the properties, and the terminology of op amps as far as they are germane to the design of active filters. In addition, a number of simple circuits will be presented that form op amp building blocks fundamental to the design of various different kinds of active filters. Those readers who are interested in details of op amp design are referred to the literature [66–71].

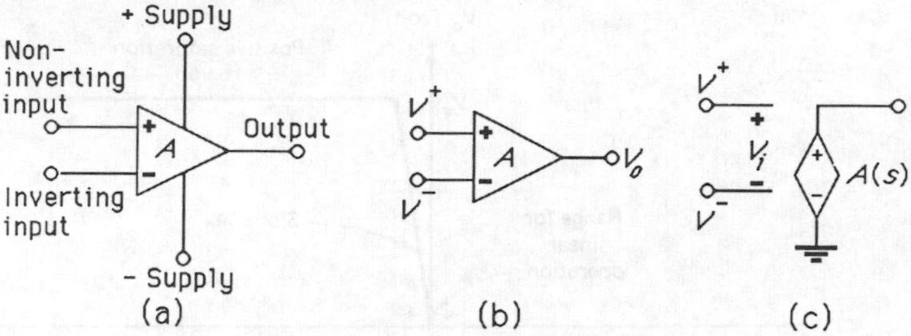

Figure 4-2   (a) Operational amplifier symbol showing main terminals; (b) customary circuit symbol; (c) small-signal circuit.

4.2 OP AMP NOTATION AND CHARACTERISTICS

4.2.1 The Ideal Operational Amplifier

The ideal operational amplifier was introduced briefly in Section 3.2.2; its circuit symbol is shown in Fig. 3-2 and is repeated for convenience in Fig. 4-2.   In addition to the *signal* terminals V^+, V^-, and V_o shown in Fig. 4-2, an op amp has positive and negative *power supply* terminals and may also have a number of further terminals where ground, offset adjustment, or external frequency compensation can be provided.   The operation of an op amp is described by the equation

$$V_o(s) = A(s)[V^+(s) - V^-(s)] \tag{4-3}$$

where $A(s)$ is the gain of the op amp and, as indicated, is in general a function of frequency. V^+ and V^- are the noninverting and inverting input voltages, respectively.   At low frequencies, the magnitude $A_0 = |A(j\omega)|$ of the op amp gain is very large, typically $A_0 > 10^5$ (100 dB) in bipolar op amps and $A_0 > 10^4$ (80 dB) for MOS op amps.   Because the output voltage is limited to a value somewhat less than the power supply, it follows that for linear operation the differential input voltage

$$V_i = V^+ - V^- = \frac{V_o}{A_0} \leq \frac{V_s}{A_0} \tag{4-4}$$

must be very small in practical op amps (usually measured in mV or μV). V_s is the output saturation level identified in Fig. 4-3, which illustrates the op amp's input-output transfer characteristic.

In the first-order analysis of linear circuits it is most convenient to assume the operational amplifier to be *ideal*.   The ideal op amp is characterized by an infinite open loop gain, $A_0 = \infty$, in addition to infinite input impedance, $R_i = \infty$, and zero output impedance, $R_o = 0$.   As shown in Fig. 4-2c, this means that the amplifier output is an ideal voltage source and that both currents into the signal

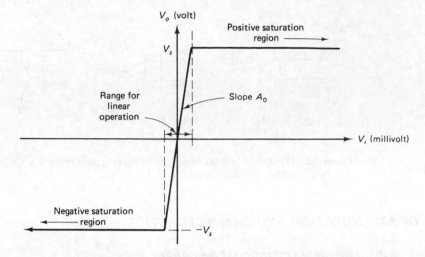

Figure 4-3 Typical op amp transfer characteristic.

input terminals and, by Eq. (4-4), also the differential input voltage are zero:

$$I^+ = I^- = 0 \tag{4-5}$$

$$V_i = V^+ - V^- = 0 \tag{4-6}$$

If, as is often the case, the noninverting input terminal is grounded, then $V^- = 0$ and the inverting input terminal is said to be at *virtual ground.*

These ideal conditions represent a good first-order approximation to a real op amp and greatly simplify the analysis and synthesis of active filters in undemanding applications at very low frequencies. Whether the approximation is acceptable in practice depends on a number of factors, such as the circuit's impedance level and, most importantly, the frequency range. In exacting applications, the design must be based on a more accurate op amp model that permits the imperfections of real op amps to be taken into consideration. The more important nonidealities and a suitably accurate op amp model will be discussed below along with simple analysis techniques that provide insight into the effects of real op amps on the performance of active filters. Of course, when computer aids are available, an appropriately accurate op amp model can be used to predict a circuit's performance reliably.

Before discussing the behavior of real operational amplifiers, let us illustrate the analysis of circuits with ideal op amps on a simple example. The student may also wish to review the earlier examples of op amp circuits.

Example 4-1

Determine the function of the circuit in Fig. 4-4. The op amp is ideal.

Solution. Because the noninverting op amp input terminal is grounded, we have, by Eq. (4-6), $v^-(t) = 0$; therefore, a node equation at the inverting input terminal, with

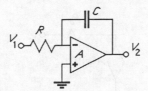

Figure 4-4   Active *RC* integrator.

$i^-(t) = 0$, results in

$$\frac{v_1(t) - v^-(t)}{R} = C \frac{d[v^-(t) - v_2(t)]}{dt}$$

or

$$v_2(t) = -\frac{1}{RC} \int_{-\infty}^{t} v_1(\lambda)\, d\lambda$$

Alternatively, going directly into the frequency domain, we could have written, of course,

$$\frac{V_1}{R} = -sCV_2$$

i.e.,   (4-7)

$$V_2 = -\frac{1}{sRC} V_1$$

Thus, the circuit is an ideal integrator built with an *ideal* op amp. Integrators are extremely useful and important building blocks for active filters; we shall need to investigate, therefore, the performance of these circuits when they are constructed with *real* op amps. In Section 4.2.2 we shall see that the integrator becomes lossy if a real op amp is used.

4.2.2 Real Operational Amplifiers

In practice, op amps are electronic circuits, composed of many transistors, resistors, and capacitors. Consequently, their input and output impedances cannot in general be expected to be infinite or zero, respectively, and, especially, their gain and bandwidth must be finite and not infinite as assumed for the ideal model. Because an exact representation not the electronic op amp circuit would be too complicated to handle, it is still necessary to develop an approximate model, but one that contains all relevant op amp imperfections with sufficient accuracy to permit realistic predictions of op amp and active filter behavior. A suitable small-signal circuit is shown in Fig. 4-5, and typical parameters are listed in Table 4-1.

4.2.2.1 Frequency-dependent gain.   For applications of relevance in this text, it has been found that the op amp's *finite gain* and *bandwidth* are the most important limiting factors for filter operation, and much work has been done to investigate their effects and to develop suitable models [6, 11, 14, 33, 52, 66, 71]. In general, the frequency response of the electronic op amp circuit is determined

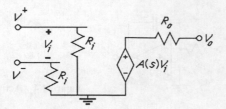

Figure 4-5

by many poles and zeros [52, 66, 71]; however, in order to assure stability[1] in closed-loop feedback configurations [72, 73], most op amps are designed to have a dominant negative real pole at $s = -\sigma$ so that a suitably accurate op amp model is

$$A(s) = \frac{\omega_t}{s + \sigma} \simeq \frac{\omega_t}{s} \tag{4-8}$$

where ω_t is the *gain-bandwidth product* defined as

$$\omega_t = A_0 \sigma \tag{4-9}$$

and σ is, of course, the 3-dB frequency of the op amp gain. As is evident from Eqs. (4-8) and (4-9), A_0 is the dc gain that was defined in connection with Eq. (4-4). Figure 4-6 shows plots of magnitude and phase of $A(j\omega)$ which shows that $\omega_t = A_0 \sigma$ is the *unity-gain (0-dB) bandwidth*. For frequencies below the *unity-gain frequency* ω_t, the magnitude and phase of a real op amp,

$$|A(j\omega)| = \frac{\omega_t}{\sqrt{\omega^2 + \sigma^2}} \simeq \frac{\omega_t}{\omega} \tag{4-10a}$$

$$\phi(\omega) = -\tan^{-1} \frac{\omega}{\sigma} \tag{4-10b}$$

are represented quite well by Eq. (4-8). We note from Table 4-1 that the 3-dB frequency σ is very small so that in most situations $\omega \gg \sigma$. Therefore, most active filter work can be based on the approximations indicated in Eqs. (4-8) and

TABLE 4-1 TYPICAL OP AMP PARAMETERS

$R_i \geq 10^6 \ \Omega$	$R_o \leq 500 \ \Omega$
$A_0 > 10^4 \ (80 \ \text{dB})$	
$5 \ \text{Hz} \leq \sigma/2\pi \leq 100 \ \text{Hz}$	$\omega_t/2\pi \geq 10^6 \ \text{Hz}$

[1]Stability is determined by factoring the denominator polynomial of the output-input transfer function. If all roots, the *poles of the system*, have a *negative* real part, the system is said to be stable. A number of methods are available that permit stability checks without explicitly factoring the system polynomial, such as the *Routh-Hurwitz test*, *root-locus techniques*, or the *Nyquist criterion*. Because numerical root finders are now widely available not only on mainframes but also on mini- and micro-computers, we shall in this text not discuss any of these methods but assume that the relevant polynomials can actually be factored. The interested student is referred to the literature for details on any of the stability tests [e.g., 72].

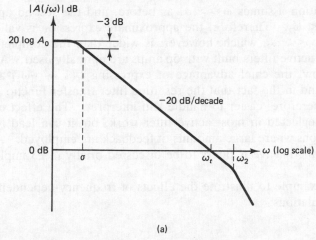

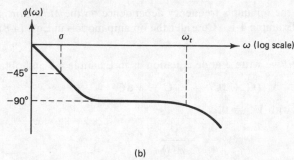

(b)

Figure 4-6 Typical frequency response of real op amps: (a) magnitude response (Bode plot); (b) phase response. The deviations for frequencies around and above ω_t lead to the model in Eq. (4-11).

(4-10a); i.e., the effect of the 3-dB frequency is neglected and the op amp is assumed to act as an ideal integrator. However, as sketched in Fig. 4-6, for frequencies close to and above ω_t further deviations do occur; most important, the phase decreases to values below $-90°$. Therefore, at higher frequencies, a more accurate model should be used that takes into account the second-order effects of additional poles and zeros which were neglected in the model of Eq. (4-8). Although a two-pole–one-zero model is sometimes used advantageously [52], mostly all parasitic effects are modeled by a pole at a frequency ω_2 that is approximately 2.5 to 4 times as large as ω_t. The model then becomes[2]

$$A(s) = \frac{\omega_t}{(s + \sigma)(1 + s/\omega_2)} \simeq \frac{\omega_t}{s(1 + s/\omega_2)} \simeq \frac{\omega_t}{s}\left(1 - \frac{s}{\omega_2}\right) \qquad (4\text{-}11)$$

[2]Because the magnitude response of real op amps is represented very well by Eq. (4-10a) but phase deviations occur already below the unity-gain frequency ω_t, the alternative model

$$A(s) = \frac{\omega_t}{s + \sigma}\, e^{-s\tau}$$

which changes only the phase of $A(s)$ is sometimes used. τ is an extra delay, and $\Delta\phi(\omega) = -\omega\tau$ is termed the *excess phase*. Clearly, all three expressions are equivalent because, for $\omega\tau \ll 1$, $e^{-s\tau} \simeq 1/(1 + s\tau) \simeq 1 - s\tau$. τ corresponds to $1/\omega_2$ in Eq. (4-11).

where the first approximation assumes $\omega \gg \sigma$, as before, and the second approximation assumes $\omega \ll \omega_2$. Therefore, the approximate expression is valid only in the range $\sigma \ll \omega \ll \omega_2$, which, however, is wider than the complete frequency range for which active filters built with op amps are normally used. As will be demonstrated below, the chief advantage of expressing $1/(1 + s/\omega_2)$ as $1 - s/\omega_2, |s| \ll \omega_2$, is found in the fact that the resulting filter transfer functions are of lower order and, therefore, easier to handle and interpret. The effect of this second pole can be neglected in most active filter work, but it can lead to instabilities in some situations where large amounts of feedback are employed. A prime example is the so-called *active R* filters to be discussed briefly in Example 4-5 and in Chapter 5.

Let us consider an example to illustrate the effects of frequency-dependent gain and the relevant calculations:

Example 4-2

Determine the effect of the op amp's frequency dependence on the *Miller integrator* in Fig. 4-4, analyzed in Example 4-1. Consider the op amp models in Eqs. (4-8) and (4-11).

Solution. With $G = 1/R$ we write a node equation as in Example 4-1 and obtain

$$V^-(G + sC) = V_1 G + V_2 sC$$

Further, from Eq. (4-3) with $V^+ = 0$,

$$V^- = -\frac{V_2}{A(s)}$$

so that for arbitrary $A(s)$ the circuit's function is described by

$$\frac{V_2}{V_1} = -\frac{1}{sCR + (1 + sCR)/A(s)} \tag{4-12}$$

[The reader should compare Eq. (4-12) with Eq. (4-2); evidently, for the integrator we have $T_{32}(s) = sCR/(1 + sCR)$ and $T_{31}(s) = 1/(1 + sCR)$]. Clearly, for $A = \infty$ the transfer function reduces to that of Eq. (4-7). If $A(s)$ is modeled as in Eq. (4-8), Eq. (4-12) results in

$$\frac{V_2}{V_1} = -\frac{1}{sCR + \dfrac{s + \sigma}{\omega_t}(1 + sCR)} = -\frac{1}{s^2\dfrac{CR}{\omega_t} + s\left(CR + \dfrac{1}{\omega_t}\right) + \dfrac{\sigma}{\omega_t}} \tag{4-13a}$$

$$\simeq -\frac{1}{\dfrac{CR}{\omega_t}\left(s + \dfrac{1}{A_0 CR}\right)(s + \omega_t)}$$

Note that the "integrator" transfer function now has two negative real poles (at $\simeq -\omega_t$ and at $\simeq -\sigma/(\omega_t CR)$; see Problem 4.5) so that performance deviations must be expected. Had we approximated $A(s)$ as suggested in Eq. (4-8) by further neglecting σ, then the term $\sigma/\omega_t = 1/A_0$ in Eq. (4-13a) would have been absent. In that case, the integrator function has one pole at the origin as desired and one parasitic pole at $s \simeq -\omega_t$:

$$\frac{V_2}{V_1} \simeq \frac{1}{sCR(1 + s/\omega_t)} \tag{4-13b}$$

where we have assumed that in practice $\omega_t CR \gg 1$. One the $j\omega$-axis, Eq. (4-13a) becomes

$$\frac{V_2}{V_1} \simeq -\frac{1}{j\omega CR - \omega^2 CR/\omega_t + \sigma/\omega_t}$$

which shows that the real integrator is lossy and has a phase

$$\phi(\omega) = -\frac{\pi}{2} - \tan^{-1}\left(\frac{\omega}{\omega_t} - \frac{1}{\omega CRA_0}\right) \tag{4-14a}$$

which is different from the ideal integrator phase $-\pi/2$. For applications at very low frequencies, the phase *error* becomes $+\tan^{-1}[1/(\omega CRA_0)]$ caused by the finite low-frequency op amp pole at $-\sigma$; at frequencies $\omega \gg \sigma$, where active filters normally operate, the integrator phase *error* is

$$\Delta\phi \simeq -\tan^{-1}\frac{\omega}{\omega_t} \simeq -\tan^{-1}\frac{1}{|A(j\omega)|} \tag{4-14b}$$

i.e., *it is determined by the gain of the op amp at the frequency of operation.* Evidently, then, large op amp gain throughout the filter's operating range is very important. The two-pole–one-zero frequency compensation discussed in reference 52 was introduced precisely with this goal in mind.

If the frequency is so large that the effect of the second pole at ω_2 must be considered (and σ can, of course, be neglected), we insert Eq. (4-11) into Eq. (4-12) to obtain the third-order function

$$\frac{V_2}{V_1} = -\frac{1}{sCR + (1 + sCR)\dfrac{s(1 + s/\omega_2)}{\omega_t}}$$

Because lower-order functions are easier to handle, we use instead the approximation in Eq. (4-11), where ω_2 is represented by a parasitic zero, and obtain the second-order expression

$$\frac{V_2}{V_1} = -\frac{1 - s/\omega_2}{sCR(1 - s/\omega_2) + s(1 + sCR)/\omega_t} \tag{4-15}$$

$$= -\frac{1 - s/\omega_2}{s\left(CR + \dfrac{1}{\omega_t}\right) + \dfrac{s^2 CR}{\omega_t}\left(1 - \dfrac{\omega_t}{\omega_2}\right)}$$

A comparison of Eqs. (4-13a) and (4-15) shows that the integrator function has acquired a high-frequency parasitic zero[3] at $+\omega_2$ and that the real part in the denominator

[3]With the approximations made, this zero can equivalently be represented by a high-frequency pole at $-\omega_2$!

of Eq. (4-13), $-\omega^2 CR/\omega_t$, is reduced by the factor $(1 - \omega_t/\omega_2)$. The phase *error* for this case is

$$\Delta\phi(\omega) \simeq -\tan^{-1}\frac{\omega}{\omega_2} - \tan^{-1}\frac{1 - \omega_t/\omega_2}{|A(j\omega)|} \tag{4-16}$$

We have demonstrated in this section that the op amp's frequency-dependent gain is likely to cause deviations in circuit performance and have shown how these effects can be analyzed. Specifically, as shown in Example 4-2, Eq. (4-13a), the designer has to expect a *small shift in the dominant poles* of the active network and, in addition, *the creation of parasitic high-frequency poles* if the op amp gain is finite and frequency-dependent as modeled in Eq. (4-8). These facts can further be illustrated by referring to the feedback network of Fig. 4-1 and Eq. (4-2): replacing the amplifier of gain $-K$ in Fig. 4-1 by an op amp of gain $A(s)$ as defined in Eq. (4-8) results instead of Eq. (4-2) in the expression

$$H(s) = -\frac{N_{31}(s)}{N_{32}(s) + \dfrac{s + \sigma}{\omega_t} D(s)}.$$

If Fig. 4-1 is to realize a *second-order* section (the usual case), then the *RC* network must evidently be chosen such that $N_{32}(s)$ and $D(s)$ are second-order polynomials and $N_{31}(s)$ is at most of second order. For ideal op amps with $\omega_t = \infty$, the nominal transfer function is $H(s) = N_{31}(s)/N_{32}(s)$, but for finite ω_t the denominator of $H(s)$ is clearly of third order with an additional high-frequency root. Further, as a little thought will show, the nominal (dominant) poles of $H(s)$, i.e., the roots of $N_{32}(s)$, have shifted due to the presence of the term $D(s)/A(s)$ (Problem 4.11).

For the most part, the parasitic high-frequency poles will not be important for the operation of an active filter[4] but the effect of shifting the dominant poles must be carefully examined. Particularly in filters with large Q values, deviations caused by dominant pole shifts can become so large that *passive or active compensation methods* are employed to eliminate the effects of the op amp's frequency dependence. Details will be discussed in Section 4.3 and in Chapter 5.

For most active filter work, the ideal integrator representation of Eq. (4-8), i.e., $A(s) = \omega_t/s$, is the appropriate and adequate op amp model to use. Only at relatively high frequencies or in cases that show unexpected Q *enhancement* or even oscillations, the designer may have to consider the effects of the second op amp pole and the resulting *excess phase* (below $-90°$) on his or her circuit.

4.2.2.2 Other Op Amp Nonidealities

Input and Output Impedances. Although the input and output impedances in general are frequency-dependent, they are usually treated as purely resistive. We shall stay with this approximation and convention in this book.

---

[4]The designer should be aware, though, of the small phase errors caused by these high-frequency poles and/or zeros. In feedback systems they may cause serious filter errors!

The input resistance of modern operational amplifiers, especially those with FET input stages, is so large compared to the impedance level of other elements in the circuit that it may be considered an open circuit for almost all applications. Thus, in this respect the ideal amplifier approximation $R_i = \infty$ is very accurate and will be used throughout the text.

Because the output resistance of practical op amps is not equal to zero but is located in the range $50\ \Omega \le R_o \le 1000\ \Omega$, its effects on the amplifier's feedback network and on the remainder of the active filter system cannot necessarily be neglected. Rather, the influence of R_o must be evaluated from case to case if there is any concern that its effect may be important. Problem 4.12 provides examples for the relevant calculations and shows that R_o can cause the output impedance of a feedback network to be strongly frequency-dependent. Usually, no difficulties arise if the circuit's impedance level in the frequency range of interest is much larger than R_o.

Recall from Chapter 3 that high-order active filters for sensitivity reasons are frequently realized by cascading second-order sections which are assumed to be noninteracting (see Fig. 3-9), i.e., each section must have a high input and/or low output impedance. We mentioned at that time that for this reason active filter sections are usually constructed such that the amplifier output forms also the section output. Clearly, if the op amp's output resistance is not zero, then also the section's output impedance is finite and, as just mentioned (see Problem 4.13), strongly frequency-dependent. Consequently, the independence of operation of cascaded sections may be questionable, and loading effects may well affect the transfer function.

Slew-Rate Limitations—Maximum Undistorted Output Signals.   Any op amp has specified by the manufacturer a maximum possible slewing rate (SR) which refers to the maximum rate of change of its output voltage:[5]

$$\text{SR} = \left.\frac{dv_o(t)}{dt}\right|_{\text{max}} \tag{4-17}$$

Slew-rate limitations are caused mainly by the finite current drive capabilities of the amplifier stages, which have to charge and discharge the internal or external compensation or load capacitors. A good discussion can be found in reference 66. Violating the SR limit will result in distorted signals, as shown in Fig. 4-7a, and must be avoided in linear circuit applications.[6] Because we are mainly concerned with sinusoidal signals of the form $v(t) = V_o \sin \omega t$ whose maximum slope

---

[5]Note that SR may be different for inverting and noninverting gain configurations.

[6]A potentially more serious difficulty than a distorted signal is the often grossly misshapen transfer function magnitude shown in Fig. 4-7c that arises if the limit in Eq. (4-18) is violated and in the extreme leads to the so-called *jump phenomenon*. The jump phenomenon, also called *jump resonance*, is caused by the *nonlinear* behavior of the slewing amplifier; its treatment is beyond the scope of this text. The interested reader is referred to the literature [74, 75].

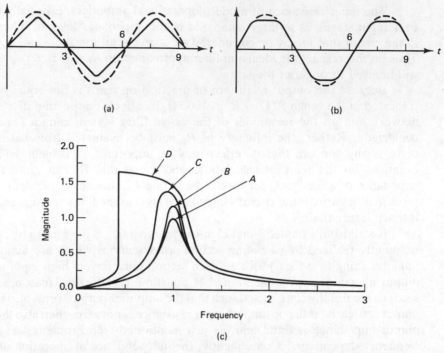

Figure 4-7 Signal distortion caused by (a) slew-rate limiting [Eq. (4-18)] and (b) clipping [Eq. (4-19)]; (c) shape of the transfer function magnitude of a second-order bandpass filter with $Q = 10$ for nominal design (no slewing: A) and for continuously increasing signal amplitudes (B, C, D). Curve C shows marked deviations caused by nonlinear op amp operation due to slewing, and curve D shows the *jump phenomenon* observed when frequency is varied.

is ωV_o, evidently, the limit to be obeyed for a given frequency range is

$$V_o < \frac{\text{SR}}{\omega_{\text{max}}} \qquad (4\text{-}18)$$

For example, if SR $= 0.7$ V/μs and $f_{\text{max}} = 50$ kHz, the undistorted signal amplitude is limited by $V_o < 2.2$ V. Note that the limit SR implies an already slewing signal, which in turn means some distortion; thus, it is normally advisable to stay well below the limit in Eq. (4-18).

A more obvious signal level limitation is obtained from the fact that op amps are transistor amplifiers whose output voltages must be less than the power supply voltages applied by the designer:

$$V_o < |V_{\text{power supply}}| \qquad (4\text{-}19)$$

Because the maximum signal levels are limited by Eqs. (4-18) and (4-19) to prevent distortion by *triangularization* (Fig. 4-7a) or *clipping* (Fig. 4-7b), respectively, of the signals, these two relationships are important for determining the *dynamic range* of a circuit. Dynamic range is defined as the ratio of the largest undistorted output signal and the *noise* floor. Electrical noise will be discussed briefly in the next section.

Noise. We can classify as *noise* all signals which appear at the output of a filter but are unrelated to the applied input signals. The noise sources may originate in the op amp, which is our concern in this subsection, or they may arise from the (resistors in the) feedback network, from the power supply, or even from sources external to the filter circuit. Noise at the filter output is calculated by modeling the physical noise sources as voltage and/or current generators, $|e_{ni}|$ and $|i_{ni}|$, respectively, at the appropriate locations in the circuit and computing their individual contributions, $|e_{o,ni}|$, to the total output noise signal as

$$|e_{o,ni}| = |e_{ni}| \, |H_{ni}(j\omega)| \qquad (4\text{-}20)$$

where $H_{ni}(s)$ is the noise transfer function from the location of the ith noise source, $i = 1, \ldots, m$, to the filter output. Because noise sources can be considered random and uncorrelated, noise is additive and the total rms noise, $|e_{on}|$, is the square root of the sum of the squares of the individual noise generator outputs:

$$|e_{on}| = \sqrt{\sum_{i=1}^{m} \left[|e_{ni}|^2 \, |H_{ni}(j\omega)|^2 \right]} \qquad (4\text{-}21)$$

The smallest signal that can then be effectively processed by the filter is limited by this *noise floor*, and *dynamic range* can be defined as

$$DR = 20 \log \left| \frac{\text{maximum undistorted rms signal}}{e_{on}} \right| \quad dB \qquad (4\text{-}22)$$

which lies between 70 and <100 dB for well-designed active filters.

Excluding noise from the power supply lines and from external sources, we need to consider in active *RC* filters the thermal noise

$$|e_{Rn}| = \sqrt{4kTR \, \Delta f} \qquad V_{rms} \qquad (4\text{-}23)$$

contributed by resistors, and the amplifier noise. In Eq. (4-23), k is Boltzmann's constant ($8.63 \cdot 10^{-5}$ eV/K), T is the absolute temperature (K), R is the value of the resistor (Ω), and Δf is the bandwidth (Hz). Two convenient models are shown in Fig. 4-8. Note that the op amp's noise performance is modeled by *three* noise

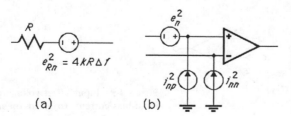

(a)   (b)

Figure 4-8 Resistor and op amp noise models: (a) noise-free resistor R in series with rms noise voltage source; (b) noise-free op amp with two rms noise currents and an rms noise voltage source.

sources (as specified by the manufacturer) and that the "remaining" circuit elements in the models are *noiseless*.

Although conceptually not difficult, noise computations are in general quite lengthy, and a thorough treatment is beyond the scope of this text. The interested reader is referred to the literature for an in-depth discussion and for typical noise calculations [76–78]. Equations (4-20) to (4-23) provide an indication of the steps involved in computing noise performance; Problems 4.15 and 4.16 give further details and should help with gaining understanding and insight into the noise analysis of active filters.

Offset Voltage, Offset Current, and DC Bias Currents. Ideally, operational amplifiers have zero input currents, and their output voltage is zero if no input signal is applied. In practice, however, imperfect elements and imbalances in the op amp circuitry cause a nonzero dc output voltage to appear even if no input signal is applied. Usually this dc output offset voltage is represented as an equivalent input voltage at the noninverting input terminal as shown in Fig. 4-9a and is referred to as *input offset voltage V_{os}*. Furthermore, op amps with bipolar input stages require, for correct transistor operation, input dc bias currents which must flow through the external circuitry at the inverting and the noninverting input terminals, as shown in Fig. 4-9b. *Offset current* is defined as the difference between the two input bias currents, $I_{os} = I_{Bn} - I_{Bp}$.

Clearly, these two phenomena are of little concern in active filters as long as there is no transmission at dc or the dc level is not important. The designer should bear in mind, though, that op amps are electronic circuits with transistor output stages. If these experience a significant dc offset voltage, the positive and negative signal swings may be very unequal and significant distortion may result at much lower signal levels than anticipated. We shall assume for the remainder of this text that no dc offset problems are present and refer the reader to the literature for more detailed discussions [71]. However, some of the potential difficulties, the relevant calculations, and suitable simple compensation methods are outlined in Problem 4.17.

Common Mode Rejection. A further op amp imperfection that the designer of active filters should be aware of although it is of little concern in most applications is *common mode gain* and the *common mode rejection ratio* (CMRR). Recall that in our definition of an op amp's operation in Eq. (4-3) it was implied that the gain from the noninverting input to the output, $A^+(s)$, and from the inverting input to

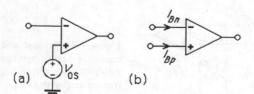

(a)   (b)

Figure 4-9 Input offset voltage (a) and input bias currents (b) of an op amp.

the output, $A^-(s)$, are equal; i.e., we assumed

$$A^+(s) = A^-(s) = A(s)$$

If, due to component imperfections and circuit imbalances, $A^+(s)$ and $A^-(s)$ are not equal, Eq. (4-3) becomes

$$V_o(s) = A^+(s)V^+(s) - A^-(s)V^-(s) \qquad (4\text{-}24)$$

so that, in contrast to what is implied by the ideal situation in Eq. (4-3), the output voltage

$$V_o(s) = [A^+(s) - A^-(s)]V_c(s)$$

is not zero if a *common mode signal* $V_c(s) = V^+ = V^-$ is applied to the op amp inputs. The term $[A^+ - A^-]$ is referred to as *common mode gain*. Similarly, the output becomes

$$V_o(s) = [A^+(s) + A^-(s)] \frac{V_d(s)}{2} = \frac{A^+(s) + A^-(s)}{2} V_d(s)$$

if a *differential mode signal* $V_d = V^+ - V^-$, i.e., $0.5V_d(s) = V^+ = -V^-$, is applied. $(A^+ + A^-)/2$ is called the *differential mode gain*, which ideally is equal to $A(s)$. The ratio of the common mode and differential mode gains,

$$\text{CMRR} = \frac{\frac{1}{2}[A^+(s) + A^-(s)]}{A^+(s) - A^-(s)} \qquad (4\text{-}25)$$

is called the *common mode rejection ratio*. If we define $V^+ = V_c + 0.5V_d$ and $V^- = V_c - 0.5V_d$ as in Fig. 4-10, Eq. (4-24) can be brought into the form

$$V_o = \frac{A^+(s) + A^-(s)}{2}\left(V_d + \frac{V_c}{\text{CMRR}}\right) \qquad (4\text{-}26)$$

which indicates that CMRR is a measure of the degree of rejection of the usually undesired common mode signal V_c compared to the desired differential mode signal V_d.

As Eq. (4-25) shows, A^+ and A^- and therefore also CMRR are functions of frequency. In practice, CMRR is found to decrease rapidly for increasing frequencies from a low-frequency value > 10,000 (80 dB). Nevertheless, in most active filter applications common mode effects are negligible, because the term V_c/CMRR in Eq. (4-26) is small enough to be ignored; also, many active filter structures do not use the op amp's differential input features.

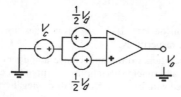

Figure 4-10 Op amp showing definitions of differential and common mode signals.

4.2.3 Transconductance Amplifiers

Although currently a large majority of active filters are built with voltage-controlled voltage sources, specifically, the conventional operational amplifier described by Eq. (4-3), it has become increasingly apparent that filters based on op amps are very restricted in their applications. Most important, the frequency-dependent gain of op amps (Section 4.2.2.1) tends to impose quite serious deviations on filter behavior and thereby precludes active filter applications significantly above the audio range. Further, apart from a few uncritical applications, designers so far have generally not been successful in developing op amp–based active RC analog filters in monolithic form; i.e., no suitable method has been found for integrating demanding op amp–RC filters along with other analog or digital circuitry on the same silicon chip.

Mainly to combat the latter shortcoming, a certain amount of effort is being directed lately toward designing active filters with "more natural"[7] gain devices, namely, *transconductance amplifiers*, where the output *current* is controlled by an applied input voltage signal. Transconductance amplifiers generally have significantly higher bandwidths[8] than op amps; also, they provide generally simpler circuitry for integration and easy methods for electronic tuning by changing a bias current. Finally, as we shall see later in this book, analog filters built with transconductances usually turn out to require fewer components than their op amp counterparts.

The circuit symbol and the equivalent circuit of an *ideal* operational transconductance amplifier (OTA) are shown in Fig. 4-11. An ideal OTA is a voltage-controlled current source described by

$$I_o = g_m(V^+ - V^-) \tag{4-27}$$

whose input and output impedances are both infinite, as illustrated in the diagram of Fig. 4-11b. In many designs, the transconductance g_m is variable by setting a control bias current, I_{cntl}, such that g_m is proportional to I_{cntl}, or $g_m = kI_{cntl}$. Quite a number of useful circuits have been developed based on the ideal transconductance model; an illustration follows in Example 4-3, below. However, when developing filters based on OTAs, the designer must take into consideration that real transconductances have finite input and output impedances that can be modeled as shown in Fig. 4-12. In many applications it is not the intrinsic and always present frequency dependence of g_m that imposes limits on transconductance operation but rather the two time constants set by the input and output impedances in addition to any possible load effects.

---

[7]The intrinsic electronic gain devices, BJT and MOS transistors, are current-controlled and voltage-controlled current sources, respectively.

[8]Commercially available operational transconductance amplifiers (OTAs) have bandwidths in the low megahertz region [79, 81], and simple special-purpose monolithic transconductances for use in integrated filters have been designed with bandwidths of tens of megahertz up to more than 100 MHz.

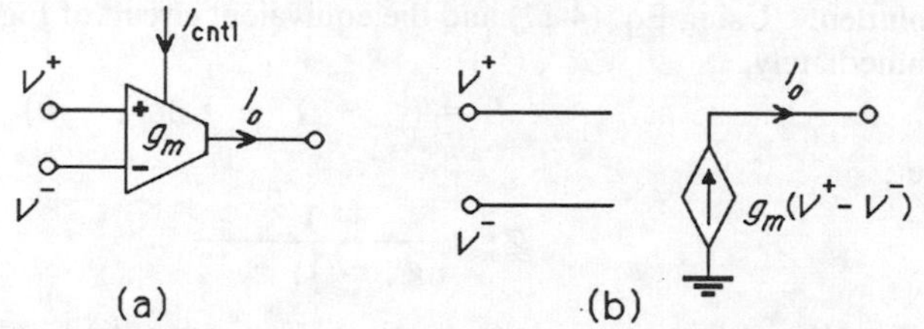

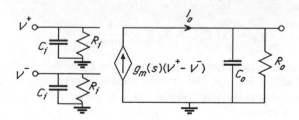

Figure 4-11   Circuit symbol (a) and small-signal equivalent circuit (b) of an OTA. In most designs the transconductance g_m is variable by means of a control bias current I_{cntl}.

When using commercially available transconductance amplifiers, there are a number of practical limitations to be observed that are best obtained from the manufacturer's data sheets and application notes [79–81]. Most important among these restrictions currently is the limited range of the input signal (≤ 20 mV) that must be obeyed if the operation is to be linear. Naturally, many of the observations made in connection with op amps apply also here, such as limited output signal swing and the requirement of a dc path at the input terminals for bias currents (in bipolar OTAs), but the most serious limitation on op amp operation, the frequency-dependent gain, is removed because of the much larger bandwidth of g_m.

The examples below should provide some indication of the capabilities of and design methods for g_m-RC circuits. Further examples are contained in Problems 4.19 to 4.22 and in Section 4.4.

Example 4-3

Analyze the configuration in Fig. 4-13 to find the input impedance. What happens if the g_m input terminals are interchanged?

Assume that the circuit is to be used in an integrated audio-frequency filter with a pole frequency of 10 kHz. If the maximum capacitor value is 50 pF, find the necessary value of g_m. For reasonable values of the parasitic capacitors, justify that the transconductance can be assumed to be ideal. Use the model of Fig. 4-12.

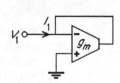

Figure 4-12   OTA model with frequency-dependent g_m and finite input and output impedances. The impedances from the inverting and the noninverting input terminals to ground are assumed to be equal, and any impedance *between* the input terminals is neglected.

Figure 4-13   Circuit for simulating a grounded resistor.

Solution.   Using Eq. (4-27) and the equivalent circuit of Fig. 4-12 with $V^+ = 0$ yields immediately

$$I_1 = (g_m + Y_i + Y_o)(V_1 - 0)$$

i.e.,   (4-28a)

$$Z_{in} = \frac{1}{g_m + Y_i + Y_o}$$

where $Y_i = sC_i + G_i$ and $Y_o = sC_o + G_o$ as identified in Fig. 4-12.   Evidently, for ideal OTAs $(Y_i = Y_o = 0)$ the circuit simulates a grounded resistor of value

$$R_{in} = \frac{1}{g_m} \tag{4-29}$$

If the g_m input terminals are interchanged, Equation (4-28a) shows that g_m is replaced by $-g_m$.   (The same principle holds for any other circuit that uses OTAs.)   Therefore, both positive and negative resistors are implemented quite easily.

It will be seen in Chapter 7 that the configuration of Fig. 4-13 is very useful because it enables us to design monolithic g_m-RC filters with only transconductances and capacitors (i.e., g_m-C filters)!   Note also that very large resistors can be implemented very efficiently (in a small area of silicon) simply via small transconductances. If the OTA is not ideal, then per Eq. (4-28a) the input impedance becomes capacitive:

$$Z_{in}(s) = \frac{1}{g_m + G_i + G_o + s(C_i + C_o)} \tag{4-28b}$$

with a 3-dB frequency

$$\omega_c = \frac{g_m + G_i + G_o}{C_i + C_o} \simeq \frac{g_m}{C_i + C_o} \tag{4-30}$$

The approximation in Eq. (4-30) assumes, of course, that in realistic applications $(G_i + G_o) \ll g_m$.

Let the pole of the audio-frequency filter be at $f_p = 10$ kHz; g_m is then calculated from

$$\omega_p = \frac{g_m}{C_{max}}$$

i.e.,

$$g_m = 2\pi f_p C_{max} \simeq 3 \ \mu S \simeq \frac{1}{320 \ k\Omega}$$

If we further assume the total parasitic capacitance $C_i + C_o$ to be 1.5 pF, f_c by Eq. (4-30) becomes 330 kHz.   Thus, ω_c and also the intrinsic 3-dB frequency of g_m are much higher than ω_p, so that the transconductance may indeed be assumed ideal in the frequency range of interest.

Example 4-4

Show that for ideal transconductances the g_m-C filter of Fig. 4-14 realizes a second-order bandpass function.   Design the circuit for a pole frequency of $\omega_0 = 10^5$ rad/s and a quality factor of value $Q_0 = 8$.   Assume that the transconductances are identical

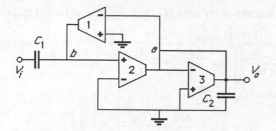

Figure 4-14 Transconductance-C bandpass filter.

and that the largest realizable capacitor is 250 pF. Analyze the problems and deviations that may occur in the filter when the transconductances are implemented in MOS technology with $G_i = 0$, $C_i = 0.5$ pF, $R_o = 20$ MΩ, and $C_o = 0.1$ pF. The frequency dependence of g_m can be neglected.

Solution. Let us analyze the circuit directly for the case where the finite input and output admittances $Y_a = Y_{o2} + Y_{o3} + Y_{i1} + Y_{i3}$ and $Y_b = Y_{o1} + Y_{i2}$ are connected to nodes a and b, respectively. The two node equations are

$$V_a(sC_2 + g_{m3} + Y_a) = g_{m2}V_b$$

$$V_b(sC_1 + Y_b) - V_i sC_1 = -g_{m1}V_o$$

Also note that $V_o = V_a$. Eliminating V_b from these equations results in the bandpass function

$$\frac{V_o}{V_i} = \frac{sC_1 g_{m2}}{s^2 C_1 C_2 + s[C_1(g_{m3} + Y_a) + C_2 Y_b] + g_{m1}g_{m2} + Y_b(g_{m3} + Y_a)} \qquad (4\text{-}31)$$

Assuming for now the parasitic elements Y_a and Y_b to be zero, the pole frequency and the pole quality factor become, with $g_{m1} = g_{m2} = g_{m3} = g_m$,

$$\omega_0 = \frac{g_m}{\sqrt{C_1 C_2}} \qquad Q_0 = \sqrt{\frac{C_2}{C_1}}$$

respectively. We select, therefore,

$$C_2 = C_{max} = Q^2 C_1 = 250 \text{ pF} \quad \text{and} \quad C_1 = \frac{250 \text{ pF}}{64} = 3.9 \text{ pF}$$

in order to keep the smallest capacitor well above the level of the parasitics. (Problem 4.23 addresses the errors arising when the parasitic capacitors are not small compared to the actual circuit capacitors C_1 and C_2.) With these values, g_m becomes

$$g_m = \omega_0 \sqrt{C_1 C_2} = 3.12 \ \mu\text{S}$$

If the input and output admittances are not zero, the denominator polynomial of Eq. (4-31) becomes

$$s^2(C_1 + C_o + C_i)[C_2 + 2(C_o + C_i)]$$

$$+ s\{(C_1 + C_o + C_i)(g_m + 2G_o) + [C_2 + 2(C_o + C_i)]G_o\} \qquad (4\text{-}32)$$

$$+ g_m^2 + G_o(g_m + 2G_o)$$

Neglecting the small second-order term $(G_o/g_m)^2$ we obtain from Eq. (4-32)

$$\omega_0^2 \approx \frac{g_m^2}{C_1 C_2} \frac{1 + G_o/g_m}{\left(1 + \dfrac{C_o + C_i}{C_1}\right)\left[1 + \dfrac{2(C_o + C_i)}{C_2}\right]} \approx \frac{g_m^2}{C_1 C_2}\left(1 + \frac{G_o}{g_m}\right) \approx (100.8 \text{ krad/s})^2$$

and for the actual value, Q_a, of Q,

$$Q_a \approx Q_0 \frac{\sqrt{1 + G_o/g_m}}{1 + (2 + Q_0^2)G_o/g_m} \approx 3.92$$

We observe, therefore, that the parasitic admittances of OTAs may in general not be neglected; typically, they give rise to "small" deviations in pole frequency and to significant errors in quality factor. Note that even a relatively large output resistor of value 20 MΩ in our example caused a Q error $\approx 50\%$!

4.3 ACTIVE BUILDING BLOCKS USING OP AMPS

After our introduction to op amps and their behavior in Sections 4.1 and 4.2, we are now ready to discuss the fundamental op amp building blocks that are used to construct active filters. Although it is sometimes difficult to identify the function of the op amp building block in terms of the elementary operations to be treated below, in many cases it is readily apparent that the op amp is used to *amplify* signals, to *sum* signals, or to *simulate* the behavior of inductors and/or capacitors. The latter is accomplished either by *integrating* a signal, i.e., by an *operational simulation* of the function of L's and C's in the sense illustrated in Fig. 4-15, or by building a circuit whose impedance "looks" like that of an inductor as shown in Fig. 4-16.

In Fig. 4-15a we show a series inductor and a shunt capacitor described, respectively, by

$$IR = \frac{V_1 - V_2}{sL/R} \quad \text{and} \quad V = \frac{I_1 R - I_2 R}{sCR} \tag{4-33}$$

where we have converted the currents into *voltage signals* through multiplication

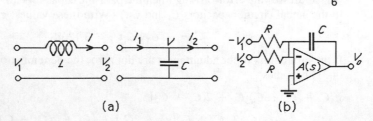

(a)   (b)

Figure 4-15   (a) Series L and shunt C filter elements performing "integration"; (b) op amp equivalent.

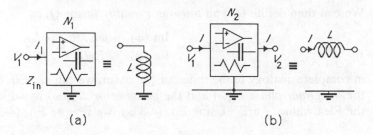

Figure 4-16 Illustration of the simulation (a) of a grounded inductor and (b) of a floating inductor via op amp–RC circuits N_1 and N_2, respectively.

by a normalizing resistor R. The same integrating function, i.e., division by s, evidently is performed by the op amp circuit in Fig. 4-15b, which realizes, for $|A(j\omega)| \gg 1$,

$$V_o = \frac{V_1 - V_2}{(sCR + 2)/A(s) + sCR} \simeq \frac{V_1 - V_2}{sCR} \tag{4-34}$$

Inductor simulation on an element-by-element basis is accomplished as shown in Fig. 4-16: one attempts to construct an op amp–RC circuit whose impedance approximates a grounded or floating inductor in the frequency range of interest as closely as possible. Thus, for a grounded inductor we need a circuit N_1 whose input impedance equals $Z_{in}(s) = sL$, and simulation of a floating inductor requires a twoport N_2 whose admittance matrix equals

$$\begin{pmatrix} I_1 \\ I_2 \end{pmatrix} = \frac{1}{sL} \begin{pmatrix} 1 & -1 \\ -1 & 1 \end{pmatrix} \begin{pmatrix} V_1 \\ V_2 \end{pmatrix} \tag{4-35}$$

where $I_1 = -I_2 = I$, as indicated in Fig. 4-16b.

Because of such factors as component tolerances and op amp imperfections, the simulations will not be perfect and will be valid only over a limited frequency range. The designer must, therefore, expect not only that the realized "inductor" will have an incorrect "element value" L_r and be affected by parasitics but, especially, that it will be lossy, i.e., the "inductor" will be nonideal:

$$j\omega L \rightarrow j\omega L_r + R_L = j\omega L_r \left(1 - j\frac{R_L}{\omega L_r} \right) = j\omega L_r \left(1 - j\frac{1}{Q_L(\omega)} \right) \tag{4-36}$$

where $Q_L(\omega)$ is the inductor quality factor evaluated at some appropriate frequency. The concept of *lossy inductors* is easy to visualize in the component simulations illustrated in Fig. 4-16; it is not as obvious how this effect expresses itself in the operational simulations of Fig. 4-15. However, if we label the inductor-integrator function in Eq. (4-33) as $T_L(s) = R/(sL)$ and replace $j\omega L$ by $j\omega L_r + R_L$, we obtain

$$T_L(j\omega) = \frac{R}{j\omega L_r + R_L} = \frac{1}{j\, \text{Im}\,(\omega) + \text{Re}\,(\omega)} \tag{4-37}$$

We can then define [11] an *integrator quality factor* Q_I as

$$Q_I(\omega) = \frac{\text{Im } (\omega)}{\text{Re } (\omega)} = \frac{\omega L_r/R}{R_L/R} = \frac{\omega L_r}{R_L} \qquad (4\text{-}38a)$$

in complete analogy to the inductor Q. Also, we can use Eq. (4-37) to investigate the *integrator phase* $\phi_I(\omega)$ and the *phase error* $\Delta\phi_I(\omega)$ by which ϕ_I deviates from the ideal value $-\pi/2$. Using Eq. (4-38a), we rewrite Eq. (4-37) in the form

$$T_L(\omega) = \frac{1}{j \text{ Im } (\omega) + \text{Re } (\omega)} = \frac{1}{j \text{ Im } (\omega) \left[1 - j \dfrac{1}{Q_I(\omega)} \right]} \qquad (4\text{-}39)$$

which results in

$$\phi_I(\omega) = -\frac{\pi}{2} + \tan^{-1}\frac{1}{Q_I(\omega)} = -\frac{\pi}{2} + \Delta\phi_I(\omega) \qquad (4\text{-}40)$$

As should have been expected, the integrator phase error is determined by the integrator's quality factor.

The corresponding development holds, of course, for the integrator Q factor of a capacitor integrator where

$$Q_I(\omega) = \frac{\text{Im } (\omega)}{\text{Re } (\omega)} = \frac{\omega C_r R}{G_C R} = \omega C_r R_C \qquad (4\text{-}38b)$$

R_C would be the parallel loss resistor of an actual capacitor.

The concepts of a *lossy integrator*, the *integrator Q factor*, or the *integrator phase error* will prove to be extremely convenient and useful in the design of active filters. Not only do they provide insight into the circuits' behavior and give clues for possible correction or compensation of errors, but they also allow us to use many of the results that were developed earlier for passive *LC* filters.

4.3.1 Summers and Related Circuits

During the design of active filters one often finds that a signal must be amplified ("scaled") or that the scaled versions of two or more signals must be added. The corresponding circuits are all based on the *inverting* and *noninverting* amplifiers shown in Fig. 4-17a and b. Their gains are, respectively,

$$\frac{V_o}{V_1} = -\frac{R_F}{R_1} \frac{1}{1 + \dfrac{1}{A(s)}\left(1 + \dfrac{R_F}{R_1}\right)} \qquad (4\text{-}41a)$$

and

$$\frac{V_o}{V_1} = \left(1 + \frac{R_F}{R_1}\right) \frac{1}{1 + \dfrac{1}{A(s)}\left(1 + \dfrac{R_F}{R_1}\right)} \qquad (4\text{-}41b)$$

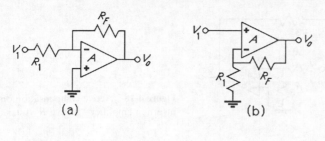

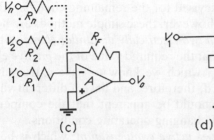

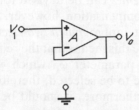

Figure 4-17 Elementary op amp circuits: (a) inverting amplifier; (b) noninverting amplifier; (c) inverting summer; (d) unity-gain buffer.

If $|A(j\omega)| \to \infty$, the two expressions become

$$\frac{V_o}{V_1} = -K_0 \quad \text{and} \quad \frac{V_o}{V_1} = 1 + K_0 \qquad (4\text{-}41c)$$

where $K_o = R_F/R_1$ is the ideal *nominal gain* factor. Note, however, that these amplifiers in practice have finite *frequency-dependent* gain because $A(s)$ is finite. If we model $A(s)$ as $A(s) = \omega_t/(s + \sigma) \simeq \omega_t/s$ as introduced in Eq. (4-8), both finite-gain amplifiers are seen to have the reduced bandwidth

$$\omega_K \simeq \frac{\omega_t}{1 + K_0} \qquad (4\text{-}42)$$

which depends on K_0, the finite gain realized. As a matter of fact, we find that the *closed-loop gain-bandwidth product*, $(1 + K_0)\omega_K$, is approximately equal to the *open-loop gain-bandwidth product*, ω_t. If the bandwidth ω_K is too small, the designer can improve the situation by a number of minor or major circuit modifications; for example, placing a small capacitor C_c across the input resistor R_1 of Fig. 4-17a changes Eq. (4-41a) into

$$\frac{V_o}{V_1} \simeq -K_0 \frac{1 + sC_cR_1}{1 + (1 + K_0)s/\omega_t} \qquad (4\text{-}43)$$

That is, a capacitor of value

$$C_c = \frac{1}{\omega_K R_1} = \frac{1 + K_0}{\omega_t R_1} \qquad (4\text{-}44)$$

will cancel the parasitic pole at ω_K by creating a zero at the same frequency. Similar

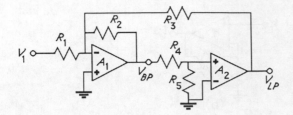

Figure 4-18 Active compensation on an inverting amplifier–"active R" filter.

passive compensation schemes can be devised for the remaining circuits. Beyond approximate first-order compensation, however, these simple methods are not very practical, because they attempt *to match two electrically dissimilar elements* to each other. In the present case this means that the required value of the *passive* element C_c must track the *active* parameter ω_t, which we know to have extremely wide tolerances; C_c would have to be selected, therefore, and have a different value for each op amp used. Furthermore, it should be apparent that the compensation will break down if ω_t varies because of changing operating conditions.

A far more practical approach uses *active compensation*, which is based on the idea of making the "resistive" *passive* feedback network *active* by employing a second *matched* op amp.[9] Many such designs are available in the literature [e.g., 11, 82–84]; their common concepts are either to structure the circuits such that the matched and tracking frequency dependence of the op amps cancels as far as possible or to shape the resulting higher-order transfer functions in some desirable way. Example 4-5 illustrates this approach for an inverting amplifier.

Example 4-5

Investigate the circuit of Fig. 4-18 and determine the element values such that the magnitude is "flat" with the widest possible bandwidth or that the phase is as linear as possible.

Solution. Assuming that the two op amps are matched, i.e., that $A_1 = A_2 = A$, routine analysis yields

$$\frac{V_{LP}}{V_1} = \frac{-ad}{1/A(s)^2 + b/A(s) + cd}$$

where we have used the abbreviations

$$a = \frac{G_1}{G} \qquad b = \frac{G_2}{G} \qquad c = \frac{G_3}{G} \qquad d = \frac{G_4}{G_4 + G_5} \qquad G = G_1 + G_2 + G_3$$

in order to make the following derivations less cumbersome. Note that the resistor ratios a, b, c, and d are separately adjustable, but their values are, of course, less than or equal to 1. If we now model the op amp gain as ω_t/s, we obtain the amplifier

---

[9]And even a third or fourth op amp [85, 86]; dual, triple, and even quad matched op amps on the same IC chip are available. Of course, these approaches come with additional costs: more components, increased power consumption, and higher noise.

gain as a second-order function of frequency:

$$\frac{V_{LP}}{V_1} = -\frac{ad}{S^2 + bS + cd} = -\frac{ad}{S^2 + \sqrt{cd}\,(b/\sqrt{cd})S + cd} \tag{4-45}$$

where $S = s/\omega_t$ is a normalized frequency. The equation was written in this form to facilitate our analysis, because we recognize in Eq. (4-45) a "lowpass" filter[10] with dc gain, quality factor, and pole frequency, respectively, given by

$$K = \frac{a}{c} = \frac{R_3}{R_1} \qquad Q = \frac{\sqrt{cd}}{b} \qquad \omega_p = \omega_t \sqrt{cd} \tag{4-46}$$

We are now in a position to choose the resistor ratios to achieve the desired objectives. In order to make the gain as "flat" as possible, i.e., to obtain the widest possible bandwidth without "peaking," we recall from Chapter 1 that for *maximally flat magnitude* our function must be a Butterworth lowpass with $Q = \sqrt{cd}/b = 1/\sqrt{2}$ and 3-dB bandwidth $\omega_c = \omega_p = \omega_t\sqrt{cd}$. The student may show (Problem 4.24) that the bandwidth is improved over the value ω_K of the single–op amp inverting amplifier [Eq. (4-42)] for $K \gg 1$ by the potentially substantial factor $\sqrt{1 + K}$; i.e.,

$$\omega_c \simeq \omega_K \sqrt{1 + K}$$

Our second objective, to design an amplifier with minimal phase distortion, means simply that the "lowpass" filter of Eq. (4-45) must have *maximally flat delay*, i.e., we need to realize a Bessel filter. According to our discussion in Chapter 1 and Table III in Appendix III, we obtain $b = cd = 3$; the resulting amplifier has a "constant" delay of value $\tau(0) = 1/\omega_t$; delay and magnitude errors can be estimated from Fig. 1-29.

We hope that this example has conveyed to the reader the principal idea behind active frequency compensation: one has to construct an op amp circuit, using one or more op amps, and then use the available parameters (element values) to shape the transfer function in a way that is appropriate for the application at hand. Note that this does *not* necessarily imply that one should build the most ideal op amp building block (summer, amplifier, or integrator, for example) but

---

[10]The astute reader will have noticed that in Fig. 4-18 we have actually built a lowpass filter (for the output V_{LP}) and simultaneously, as can be shown by simple analysis, a bandpass filter (at the output V_{BP}), described by

$$\frac{V_{BP}}{V_1} = -\frac{aS}{S^2 + bS + cd} = -\frac{\dfrac{a}{b}\sqrt{cd}\,\dfrac{b}{\sqrt{cd}}S}{S^2 + \sqrt{cd}\,(b/\sqrt{cd})\,S + cd}$$

with midband gain $K_B = a/b = R_2/R_1$, and the remaining relevant parameters defined as in Eq. (4-46). We note also that these filter functions are obtained *without using any* (external) *capacitors*. These and many other related circuits have been labeled, therefore, *active R filters* [87]; they were introduced primarily in an effort to obtain more predictable high-frequency performance by including a more realistic op amp model directly in the design. A more detailed discussion of active R filters will be found in Chapter 5.

rather that the performance of the op amp building block should be optimized for best *overall* filter performance [84].

We return now to the remaining two circuits in Fig. 4-17, the summer (Fig. 4-17c) and the *unity-gain buffer* (Fig. 4-17d). The inverting amplifier of Fig. 4-17a has a convenient virtual ground node at the inverting input terminal of the op amp to which further inputs can be connected as shown in Fig. 4-17c. Elementary analysis of the resulting summing circuit yields

$$V_Q = - \frac{1}{1 + \frac{1}{A(s)} \frac{G}{G_F}} \sum_{i=1}^{n} \frac{G_i}{G_F} V_i \qquad (4\text{-}47)$$

where we set $G = G_F + \Sigma (G_i)$. Evidently, for $A(s) \approx \omega_t/s$, all summing coefficients have the same frequency dependence as the amplifiers discussed above [Eq. (4-41)] and are, therefore, amenable to the same active or passive compensation techniques. In the ideal case, $|A(j\omega)| \rightarrow \infty$, Equation (4-47) reduces to

$$V_o = - \sum_{i=1}^{n} \frac{G_i}{G_F} V_i \qquad (4\text{-}48)$$

which shows that all summing coefficients can be set independently, a convenience for adjusting and tuning an active filter. Note that this feature of noninteractive tuning is available only if the inverting op amp input is used as the summing node. Addition of signals can, of course, also be obtained at the noninverting input, but tuning the coefficients is no longer independent (Problem 4.25). We also wish to emphasize at this point that the summing operation can be performed wherever a virtual ground node is available in an op amp circuit; often there is no need to construct a separate summer [see Fig. 4-15b and Eq. (4-34)].

Finally, the buffer circuit in Fig. 4-17d is, of course, nothing but a special case of the noninverting amplifier in Fig. 4-17b with $R_F = 0$ and $R_1 = \infty$, so that from Eq. (4-41b) we obtain

$$\frac{V_o}{V_1} = \frac{A(s)}{1 + A(s)} = \left. \frac{1}{1 + 1/A(s)} \right|_{s=j\omega} \simeq \frac{1}{1 + s/\omega_t} \qquad (4\text{-}49)$$

Observe that the 3-dB bandwidth of this buffer equals ω_t and that $|V_o/V_1| \simeq 1$ for all reasonable active filter frequencies, $\omega \ll \omega_t$. The main reason for using this *unity-gain* amplifier is its high input impedance and its extremely low output impedance:

$$Z_{\text{in}} \simeq R_i \quad \text{and} \quad Z_{\text{out}} \simeq \frac{R_o}{1 + A(s)} \simeq 0$$

Thus, the circuit is a *buffer* that can drive large loads without loading the active filter itself, because (almost) no current is drawn from the filter.

4.3.2 Integrators

We saw in the beginning of Section 4.3 that integrators perform an important function in active filter work; not only can they simulate the performance of inductors and capacitors, but, as illustrated in Fig. 4-19, they are also used to build second-order filter sections by connecting two integrators in a loop. With each integrator described by $1/s$, simple analysis of the circuit in Fig. 4-19 yields

$$H(s) = \frac{V_2}{V_1} = \frac{Ks}{s^2 + as + 1} \tag{4-50}$$

i.e., by combining one *noninverting* and one *inverting* integrator, we obtain a second-order transfer function. The inverting integrator is needed, of course, for stability reasons to make the loop gain negative. The student should compare the circuit in Fig. 4-19 with the active R filter in Fig. 4-18, where the op amp itself was used as an integrator!

The foregoing discussion shows that the availability of good inverting and noninverting integrators is essential and that we need to investigate the behavior and practical design of such circuits.

4.3.2.1 Inverting integrators. An inverting integrator circuit was introduced in Example 4-1, Fig. 4-4. This basic *Miller integrator*, repeated in Fig. 4-20, can be obtained from the inverting amplifier of Fig. 4-17a by replacing R_F by $1/(sC)$; thus, from Eq. (4-41a), we have

$$\frac{V_2}{V_1} = -\frac{1}{sCR}\frac{1}{1 + \dfrac{1}{A(s)}\left(1 + \dfrac{1}{sCR}\right)} \tag{4-51}$$

which, with $A(s) = \omega_t/s$ and $\omega_t CR \gg 1$, becomes

$$\frac{V_2}{V_1} = -\frac{1}{sCR\left(1 + \dfrac{s}{\omega_t} + \dfrac{1}{\omega_t CR}\right)} \approx -\frac{1}{sCR}\frac{1}{1 + \dfrac{s}{\omega_t}}\bigg|_{s=j\omega}$$

$$= -\frac{1}{j\omega CR - \omega^2 CR/\omega_t} \tag{4-52}$$

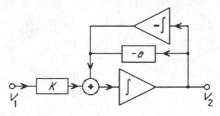

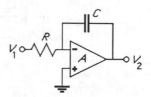

Figure 4-19 Two-integrator loop to realize second-order filter.

Figure 4-20 Miller integrator, $Q_t = -|A(j\omega)|$.

Comparing this expression with Eqs. (4-37) through (4-40) shows that the Miller integrator has a phase error and quality factor, respectively, of

$$\Delta\phi \simeq -\tan^{-1}\frac{\omega}{\omega_t} = -\tan^{-1}\frac{1}{|A(j\omega)|} \quad \text{and} \quad Q_I \simeq -\frac{\omega_t}{\omega} = -|A(j\omega)| \tag{4-53}$$

Note that the integrator Q is *negative* and is given by the magnitude of the op amp gain at the frequency of interest. Consequently, large gains are important, as was pointed out earlier.

If the quality factor of Eq. (4-53) is too low, improvements via passive or active compensation can be sought just as for the devices discussed in Section 4.3.1. Two methods for passive compensation are shown in Fig. 4-21. We leave it to the reader (Problem 4.26) to investigate the operation of Fig. 4-21a; here we concentrate on the circuit in Fig. 4-21b, which proves to be preferable. Analyzing the integrator of Fig. 4-21b yields, after a few steps,

$$\frac{V_2}{V_1} = \frac{-(1 + sCR_c)}{(1 + sCR)\left(\dfrac{s}{\omega_t} + \dfrac{sCR_1}{1 + sCR}\right)} = -\frac{1}{sCR}\frac{1 + sCR_c}{\dfrac{s}{\omega_t} + \dfrac{1}{\omega_t CR} + \dfrac{R_1}{R}}$$

$$= -\frac{1}{sCR}\frac{s + 1/(CR_c)}{\dfrac{1}{\omega_t CR_c}\left(s + \dfrac{1}{CR} + \omega_t\dfrac{R_1}{R}\right)} \tag{4-54}$$

where $R = R_c + R_1$. Thus, if we make

$$\frac{1}{CR_c} = \frac{1}{CR}(1 + \omega_t CR_1) \tag{4-55}$$

i.e.,

$$R_c = \frac{1}{\omega_t C}$$

the transfer function equals

$$\frac{V_2}{V_1} = -\frac{1}{sCR}$$

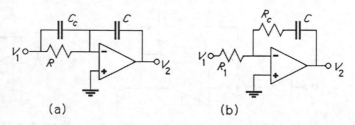

| (a) | (b) |

Figure 4-21 Passive compensation methods for Miller integrator.

i.e., apart from the mentioned matching or tracking problems the compensated circuit realizes an ideal inverting integrator. In practice, the two resistors can be realized via one potentiometer of value R, tapped such that $R_1 = aR$ and $R_c = (1 - a)R$.

If passive compensation proves unsatisfactory in practice, active compensation schemes must be used. An excellent one is shown in the circuit [11, 92] in Fig. 4-22, whose transfer function can be derived to equal

$$\frac{V_2}{V_1} = -\frac{1}{\dfrac{1}{1 + 1/A_2}sCR + \dfrac{1 + sCR}{A_1}} \tag{4-56}$$

To judge the performance of the circuit, we bring Eq. (4-56) into the form of Eq. (4-37) and evaluate Q_I. Setting as always $A(s) = \omega_t/s$ and approximating, for $|s/\omega_t| \ll 1$, $1/[1 + 1/A(s)]$ by $1 - s/\omega_t + (s/\omega_t)^2 - (s/\omega_t)^3$, we obtain for $s = j\omega$

$$\frac{V_2}{V_1} = \frac{1}{j\omega CR\left(1 + \dfrac{1}{\omega_{t1}CR} - \dfrac{\omega^2}{\omega_{t2}^2}\right) + \dfrac{\omega^2 CR}{\omega_{t2}}\left(1 - \dfrac{\omega_{t2}}{\omega_{t1}} - \dfrac{\omega^2}{\omega_{t2}^2}\right)}$$

$$= -\frac{1}{j\,\mathrm{Im}\,(\omega) + \mathrm{Re}\,(\omega)}$$

i.e., by Eq. (4-38a), the quality factor of the circuit of Fig. 4-22 equals

$$Q_I = \frac{\mathrm{Im}\,(\omega)}{\mathrm{Re}\,(\omega)} = \frac{\omega_{t2}}{\omega}\frac{1 + \dfrac{1}{\omega_{t1}CR} - \dfrac{\omega^2}{\omega_{t2}^2}}{1 - \dfrac{\omega_{t2}}{\omega_{t1}} - \dfrac{\omega^2}{\omega_{t2}^2}} \simeq \frac{\omega_{t2}}{\omega}\frac{1}{1 - \dfrac{\omega_{t2}}{\omega_{t1}} - \dfrac{\omega^2}{\omega_{t2}^2}} \tag{4-57}$$

For the approximation, we have assumed that $\omega_{t1}CR \gg 1$ and $\omega/\omega_{t2} \ll 1$ in the numerator of Eq. (4-57); for the denominator expression we must be more careful: if the two op amps A_1 and A_2 are perfectly matched, i.e., $\omega_{t1} = \omega_{t2} = \omega_t$, Fig. 4-22 results in a *high-Q integrator* with

$$Q_I \simeq -\left(\frac{\omega_t}{\omega}\right)^3 = -|A(j\omega)|^3 \tag{4-58a}$$

For the more realistic case of a slight mismatch, even between two op amps on

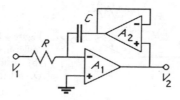

4-22 Active compensation of Miller integrator, $Q_I \simeq -|A(j\omega)|^3$.

the same chip, we can neglect $(\omega/\omega_{t2})^2$ in the denominator of Eq. (4-57) and obtain

$$Q_I \simeq \frac{\omega_{t2}}{\omega} \frac{1}{1 - \omega_{t2}/\omega_{t1}} \simeq \frac{|A_2(j\omega)|}{1 - \omega_{t2}/\omega_{t1}} \tag{4-58b}$$

Consider, for example, a pair of op amps on the same chip with $f_t = 1$ MHz and 1% mismatch between their f_t values. Let the frequency of operation be 10 kHz. According to Eq. (4-58b), the integrator Q factor for this case becomes $|Q_I| = 1000$, whereas for perfect matching Eq. (4-58a) results in $Q_I = 10^6$. Normally, mismatch between the two op amps will be the dominant effect. Note also that Q_I can be positive or negative depending on whether ω_{t2} is smaller or larger than ω_{t1}.

4.3.2.2 Noninverting integrators.   We pointed out in connection with Fig. 4-19 that in general both inverting and noninverting integrators are needed for the realization of active filters. The simplest method for achieving noninverting integration is to connect an inverter (Fig. 4-17a) and a Miller integrator (Fig. 4-20) in cascade as shown in Fig. 4-23a. The student may show (Problem 4.28) that for $A_1(j\omega) = A_2(j\omega) = A(j\omega)$ the additional op amp reduces the integrator Q_I of the Miller-inverter cascade to

$$Q_I \simeq -\frac{1}{3}\frac{\omega_t}{\omega} = -\frac{1}{3}|A(j\omega)|$$

Fortunately, a simple wiring change [11, 92], shown in Fig. 4-23b, results in the integrator function

$$\frac{V_2}{V_1} = \frac{1}{sCR} \frac{1}{1 + \frac{1}{A(s)}\left(1 + \frac{1}{sCR}\right)} \frac{1 + 2/A_1(s)}{1 + 2/A_2(s)} \tag{4-59}$$

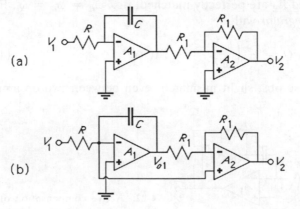

(a)

(b)

Figure 4-23 Noninverting integrators: (a) Miller-inverter cascade, $Q_I \simeq -|A(j\omega)|/3$; (b) modified Miller-inverter cascade, $Q_I = |A(j\omega)|$.

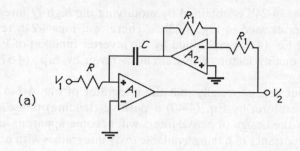

(a)

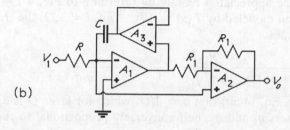

(b)

Figure 4-24 Noninverting integrators: (a) positive $Q_I \approx +|A(j\omega)|$ (phase-lead); (b) modified high-Q_I, $Q_I \approx -|A(j\omega)|^3$.

which is the same as in Eq. (4-51) but multiplied by the gain

$$\frac{V_o}{V_{o1}} = -\frac{1 + 2/A_1(s)}{1 + 2/A_2(s)} \tag{4-60}$$

of the inverter in the connection shown in Fig. 4-23b. Therefore, the modified Miller-inverter cascade has the same quality factor, $Q_I = -|A(j\omega)|$, as the simple Miller integrator of Eq. (4-52).

There exist several other noninverting integrators [11, 92], which have been found useful in active filter work; two popular ones are shown in Fig. 4-24. The circuit in Fig. 4-24a is obtained[11] by placing an inverting unity-gain amplifier into the feedback loop of the Miller integrator after interchanging the input terminals of $A_1(s)$ to maintain stability. Simple analysis yields, for Fig. 2-24a,

$$\frac{V_2}{V_1} = \frac{1}{sCR\left[\dfrac{1}{1 + 2/A_2(s)} + \dfrac{1}{A_1(s)} + \dfrac{1}{sCRA_1(s)}\right]} \tag{4-61}$$

from which we can calculate for $A_1(s) = A_2(s) = A(s)$ with the same methods as were used above

$$Q_I \approx +\frac{\omega_t}{\omega} = +|A(j\omega)| \tag{4-62}$$

---

[11]The phase-lead integrator can become unstable for certain types of op amps, depending on their parameters. See reference 123 and Problem 4.40 for the modifications necessary to assure stability.

Similarly, the circuit in Fig. 4-24b is obtained by modifying the high-Q integrator of Fig. 4-22 with the inverter used in Fig. 4-23b. Therefore, Fig. 4-24b realizes the integrator function of Eq. (4-56) multiplied by the inverter function of Eq. (4-60). Consequently, the quality factor of this circuit is given by Eqs. (4-57) and (4-58).

We wish to point out here especially that the integrator of Fig. 4-24a has a *positive Q* factor and, therefore, by Eq. (4-40) a positive (leading) phase error. The utility of this fact for the design of active filters will become apparent in later sections; however, the importance of having available *lossy* integrators with positive and with negative Q_I can be appreciated readily by returning to Fig. 4-19: if the two integrators are lossy and modeled by $T_i(s) = 1/(s + q_i)$, $i = 1, 2$, the transfer function [Eq. (4-50)] becomes

$$\frac{V_2}{V_1} = \frac{K(s + q_2)}{s^2 + s(a + q_1 + q_2) + 1 + aq_2 + q_1q_2} \tag{4-63}$$

It should be apparent from Eq. (4-50) that $a = 1/Q$, which for large Q is a small number; also, the parameters q_1 and q_2, being inversely proportional to the corresponding integrator Q_I factors, are small but not necessarily small compared to a. Thus, as a result of lossy integrators we can expect the *realized Q_r*,

$$Q_r = \frac{Q}{1 + (q_1 + q_2)/a} \tag{4-64}$$

to be *enhanced* or *reduced* depending on whether the q_i are negative or positive, respectively. If, however, integrators with positive and with negative Q_I are available such that $q_1 \simeq -q_2$, as from the circuits in Figs. 4-20 and 4-24a, the Q errors just cancel! The second potential problem that is evident from Eq. (4-63) is pole-frequency deviation; fortunately, however, in most applications we have $aq_2 + q_1q_2 \ll 1$, so that this error turns out to be negligible.

4.3.3 Gyrators and Immittance Converters

Let us turn our attention now to the simulation of inductors illustrated in Fig. 4-16a, i.e., to constructing an active RC circuit N_1 such that the input impedance "looks" inductive.

The best-known method for inductor simulation is probably that of using a *gyrator*. A gyrator is a twoport whose input impedance equals

$$Z_{in}(s) = \frac{r^2}{Z_{load}(s)} = r^2 Y_{load}(s) \tag{4-65}$$

where r is the gyration resistance. Thus, if $Y_{load} = sC$, Z_{in} will be an inductor. The student may show that in terms of the y-parameters of the twoport, the input impedance equals

$$Z_{in}(s) = \frac{y_{22} + Y_{load}}{y_{11}Y_{load} + \Delta y}$$

where Δy is the determinant of the y matrix. Therefore, in order for Z_{in} to be proportional to Y_{load}, we must build a twoport with $y_{11} = y_{22} = 0$. The customary circuit symbol for a gyrator is shown in Fig. 4-25a; it is a twoport described by the equations

$$I_1 = y_{12}V_2 \qquad I_2 = y_{21}V_1 \qquad\qquad (4\text{-}66a)$$

and

$$Z_{in} = \frac{V_1}{I_1} = \frac{-1}{y_{12}y_{21}} \frac{-I_2}{V_2} = -\frac{Y_{load}}{y_{12}y_{21}} \qquad\qquad (4\text{-}66b)$$

Normally, one attempts to construct the twoport such that the gyrator does not introduce any (parasitic) frequency dependence of its own; i.e., one chooses

$$y_{12}(s) = g_2 \quad \text{and} \quad y_{21}(s) = -g_1$$

which results in the generic equivalent circuit in Fig. 4-25b. Further, to obtain the relationship of Eq. (4-65), frequently the gyrator is designed such that $g_1 = g_2 = g = 1/r$.

Because most electronic gyrator designs have a common ground connection between input and output ports as shown in Fig. 4-26a, the simulated inductor will be grounded. A so-called *floating* gyrator (with no common ground connection) is difficult to build; therefore, to simulate a floating inductor one usually resorts to cascading two grounded gyrators and a capacitor as illustrated in Fig. 4-26b. The student may show that for two equal gyrators, each with gyration resistance r, Fig. 4-26b does indeed realize a floating inductor of value $L = r^2C$ (Problem 4.30).

It is evident from Eq. (4-66a) and Fig. 4-25b that gyrator operation is inherently based on transconductances, i.e., the design of such devices is accomplished much more easily by use of voltage-controlled *current* sources than via op amps, which are voltage-controlled *voltage* sources. We shall defer, therefore, a detailed discussion of gyrator implementation to Section 4.4 and Chapter 7. However, for

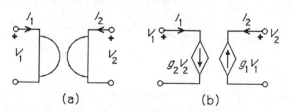

(a)   (b)

Figure 4-25 (a) Gyrator symbol; (b) generic small-signal equivalent circuit.

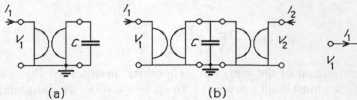

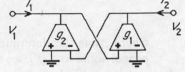

Figure 4-26 Gyrator simulation of (a) grounded inductor; (b) floating inductor.

Figure 4-27 Gyrator realization using transconductance amplifiers.

illustration we show in Fig. 4-27 an easy realization that uses the generic equivalent circuit of Fig. 4-25b and the transconductance amplifiers introduced in Section 4.2.3.

Inductance simulation with op amp circuits can be based on the configuration shown in Fig. 4-28, which consists of an as yet undetermined twoport N and a feedback resistor $R = 1/G$. To make our analysis simple, let us further assume that the twoport has infinite input and zero output impedances. Because $I_1 = G(V_1 - V_2)$ and we wish to have

$$\frac{V_1}{I_1} = Z_{in}(s) = sL$$

we must clearly build a twoport N such that

$$\frac{I_1}{V_1} = \frac{1}{sL} = G\left(1 - \frac{V_2}{V_1}\right)$$

i.e.,

$$\frac{V_2}{V_1} = 1 - \frac{R}{sL} \tag{4-67}$$

The realization is fairly easy if we recognize that V_2/V_1 is obtained according to Eq. (4-67) by subtracting an *integration* from a *constant gain*. The necessary two

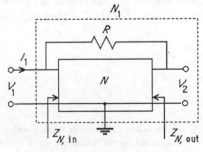

Figure 4-28 Generic inductance-simulation circuit.

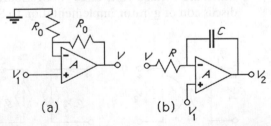

Figure 4-29 Subcircuits for achieving inductance simulation: (a) noninverting amplifier of gain $K_0 = 2$ and (b) Miller integrator with ground terminal used as additional input.

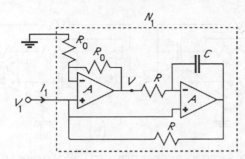

Figure 4-30 Riordan inductor-simulation circuit [88].

subcircuits are shown in Fig. 4-29; the Miller integrator realizes

$$\frac{V_2}{V_1} = \frac{1 + sCR}{sCR} - \frac{1}{sCR}\frac{V}{V_1} = 1 + \frac{1}{sCR}\left(1 - \frac{V}{V_1}\right) \qquad (4\text{-}68)$$

Therefore, with $V/V_1 = 2$ as implemented by Fig. 4-29a, the interconnection of the two subcircuits along with the feedback resistor R between V_1 and V_2 (Fig. 4-30) simulates an inductor of value $L = CR^2$.

A number of such inductance-simulation circuits exist in the literature, all based on the generic configuration in Fig. 4-28 but different realizations of the twoport N. A particularly popular approach implements the *difference* $[1 - 1/(s\tau)]$ needed in Eq. (4-67) directly by means of a Miller integrator with V^+ removed from ground in order to be able to make use of the *differential* op amp input. This circuit, shown framed in Fig. 4-31, realizes

$$\frac{V_2}{V} = \frac{V^+}{V} - \frac{1}{sCR}\left(1 - \frac{V^+}{V}\right) \qquad (4\text{-}69)$$

Therefore, to bring Eq. (4-69) into the form of Eq. (4-67), we need $V^+/V = \frac{1}{2}$, which gives

$$\frac{V_2}{V_1} = \frac{V_2}{V}\frac{V}{V_1} = \frac{1}{2}\left(1 - \frac{1}{sCR}\right)\frac{V}{V_1} \qquad (4\text{-}70)$$

and is implemented by the resistive voltage divider connected in dashed lines in Fig. 4-31. Finally, to remove the factor $\frac{1}{2}$, we need to implement $V/V_1 = 2$ as done at a cost of one op amp and two resistors by the noninverting amplifier in Fig. 4-31. The two resistors can be saved by recognizing that we can make $V/2$

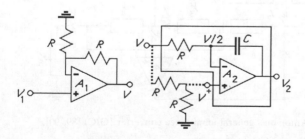

Figure 4-31 Circuit blocks showing the development of the Antoniou inductance simulator.

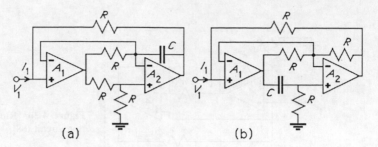

Figure 4-32 Antoniou's general impedance converters (GICs) realizing $L = CR^2$: (a) type I GIC; (b) type II GIC [89, 90].

$= V_1$ by connecting the inverting input of op amp A_1 directly to the node labeled $V/2$ in Fig. 4-31. The resulting circuit, together with the feedback resistor R between V_1 and V_2, is one type of Antoniou's *general impedance converter (GIC)* shown in Fig. 4-32a.

In order to be able to concentrate on the basic derivation of the circuits without obscuring the issues with op amp nonidealities, all our development so far is based on the assumption of ideal op amps. But, because GICs are extremely important devices and widely used for the design of active filters, we need to develop a clear understanding of the circuits' behavior with *real* op amps. To this end we analyze the GIC with general admittances in the redrawn form in Fig. 4-33 that is usually found in the literature. To start, let us assume again that $A_1 = A_2 = \infty$; then we have $V_1 = V_3 = V_5$, so that

$$I_4 = V_1 Y_5 \qquad V_4 = V_1\left(1 + \frac{Y_5}{Y_4}\right) \qquad I_3 = I_2 = -\frac{Y_3 Y_5}{Y_4} V_1$$

$$V_2 = V_1\left(1 - \frac{Y_3 Y_5}{Y_2 Y_4}\right) \qquad I_1 = \frac{Y_1 Y_3 Y_5}{Y_2 Y_4} V_1$$

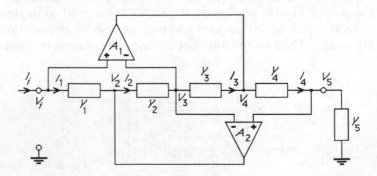

Figure 4-33 Antoniou's general immittance converter (GIC) [89, 90].

The GIC's input impedance is, therefore, ideally

$$Z_{in}(s) = \frac{Z_1(s)Z_3(s)Z_5(s)}{Z_2(s)Z_4(s)} \tag{4-71}$$

which, for $Z_2 = 1/(sC)$ and $Z_1 = Z_3 = Z_4 = Z_5 = R$ yields $Z_{in} = sL = sCR^2$ as derived earlier. We also see now that the same inductive input impedance can be obtained by interchanging Z_2 and Z_4, i.e., setting $Z_4 = 1/(sC)$ and $Z_1 = Z_2 = Z_3 = Z_5 = R$. This latter choice results in the so-called type II GIC of Fig. 4-32b. In order to investigate the effects of op amp imperfections on GIC performance, we need to reevaluate the circuit without the assumption $A_1 = A_2 = \infty$; routine but somewhat lengthy analysis (Problem 4.35) yields

$$Y_{in}(s) = \frac{Y_1 Y_3 Y_5}{Y_2 Y_4} \tag{4-72}$$

$$\cdot \frac{1 + \dfrac{1}{A_2}\left(1 + \dfrac{Y_4}{Y_5}\right) + \dfrac{1}{A_1}\dfrac{Y_2}{Y_3}\left(1 + \dfrac{Y_4}{Y_5}\right) + \dfrac{1}{A_1 A_2}\left(1 + \dfrac{Y_2}{Y_3}\right)\left(1 + \dfrac{Y_4}{Y_5}\right)}{1 + \dfrac{1}{A_1}\left(1 + \dfrac{Y_5}{Y_4}\right) + \dfrac{1}{A_2}\dfrac{Y_3}{Y_2}\left(1 + \dfrac{Y_5}{Y_4}\right) + \dfrac{1}{A_1 A_2}\left(1 + \dfrac{Y_3}{Y_2}\right)\left(1 + \dfrac{Y_5}{Y_4}\right)}$$

Evidently, for $A_1 = A_2 = \infty$, the input impedance reduces to the ideal form given by Eq. (4-71). For the following analysis, let us neglect terms divided by $A_1 A_2$, because, for reasonable frequencies,

$$\left| \frac{1}{A_1(j\omega)A_2(j\omega)} \right| \simeq \frac{\omega^2}{\omega_{t1}\omega_{t2}} \ll 1$$

Under these conditions we have from Eq. (4-72) for a type II GIC, i.e., $Y_4 = sC_4$ and $Y_i = G_i$, $i = 1, 2, 3, 5$,

$$Z_{in}(s) \simeq s\frac{C_4 G_2}{G_1 G_3 G_5} \frac{1 + \left(\dfrac{s}{\omega_{t1}} + \dfrac{s}{\omega_{t2}}\dfrac{G_3}{G_2}\right)\left(1 + \dfrac{G_5}{sC_4}\right)}{1 + \left(\dfrac{s}{\omega_{t1}}\dfrac{G_2}{G_3} + \dfrac{s}{\omega_{t2}}\right)\left(1 + \dfrac{sC_4}{G_5}\right)}$$

$$\simeq sL_0 \left[1 + \left(\frac{s}{\omega_{t1}} + \frac{s}{\omega_{t2}}\frac{G_3}{G_2}\right)\left(1 + \frac{G_5}{sC_4}\right)\right] \tag{4-73}$$

$$\left[1 - \left(\frac{s}{\omega_{t1}}\frac{G_2}{G_3} + \frac{s}{\omega_{t2}}\right)\left(1 + \frac{sC_4}{G_5}\right)\right]$$

where

$$L_0 = C_4 \frac{G_2}{G_1 G_3 G_5} \tag{4-74}$$

is the nominal value of the realized inductor and where we have used the approximation $1/(1 + x) \simeq 1 - x$ for $x \ll 1$.

On the $j\omega$-axis, Z_{in} becomes

$$Z_{in}(j\omega) = j\omega L_0[a(\omega) + jb(\omega)] = j\omega L_0 a(\omega)\left[1 - j\frac{1}{Q_L(\omega)}\right] \qquad (4\text{-}75)$$

where $Q_L = -a(\omega)/b(\omega)$, with

$$a(\omega) = 1 + \left(\frac{G_3}{G_2} + \frac{\omega_{t2}}{\omega_{t1}}\right)\frac{G_5}{\omega_{t2}C_4} + \left(1 + \frac{G_2}{G_3}\frac{\omega_{t2}}{\omega_{t1}}\right)\frac{\omega}{\omega_{t2}}\frac{\omega C_4}{G_5} \qquad (4\text{-}76a)$$

and

$$b(\omega) = \left(1 - \frac{G_3}{G_2}\right)\left(\frac{\omega}{\omega_{t2}} + \frac{\omega}{\omega_{t1}}\frac{G_2}{G_3}\right) \qquad (4\text{-}76b)$$

The interesting observation here is that in a type II GIC,

$$Q_L = \infty \qquad \text{for} \qquad R_2 = R_3$$

because b goes to zero. Therefore, a type II GIC enables us to simulate a high-quality inductor simply by trimming the resistor ratio R_2/R_3. Realistically, assuming a small remaining matching error δ_g such that

$$G_3 = G_2(1 - \delta_g) \qquad \delta_g \ll 1$$

we find from Eqs. (4-74) to (4-76) with $\omega_{t1} \simeq \omega_{t2} = \omega_t$ for the realized inductor value L_r

$$L_r = L_0 a(\omega) = L_0 + \Delta L \simeq \frac{C_4}{G_1 G_5} + \left[\frac{C_4}{G_1 G_5}2\frac{\omega}{\omega_t}\left(\frac{G_5}{\omega C_4} + \frac{\omega C_4}{G_5}\right)\right]$$

The error ΔL is minimized by choosing $G_5 = \omega_c C_4$ at some critical frequency ω_c, e.g., the filter's bandedge. In that case we have

$$L_r \simeq L_0\left(1 + \frac{4}{|A(j\omega_c)|}\right) \qquad (4\text{-}77a)$$

with a realized quality factor Q_{Lr}:

$$Q_{Lr} = -\frac{a(\omega)}{b(\omega)} \simeq \frac{1 + 4/|A(j\omega_c)|}{(1 - G_3/G_2)2/|A(j\omega_c)|} \simeq \frac{|A(j\omega_c)|}{2\delta_g} \qquad (4\text{-}77b)$$

Finally, the inductor value is set by choosing

$$R_1 = \omega_c L_0$$

In passing, we note from Eq. (4-77b) that Q_{Lr} can be made positive or negative by trimming the ratio G_3/G_2 to be less than or larger than 1, respectively. As was pointed out earlier, Q values of both signs can be very useful for the design of active filters.

Completely analogous treatment for the type I GIC (Problem 4.36) results in $R_1 = \omega_c L_0$, $G_3 = \omega_c C_2$ to minimize the inductor error ΔL and, to maximize the quality factor,

$$R_4 = R_5 \frac{\omega_{t1}}{\omega_{t2}} \qquad (4\text{-}78)$$

Note that for best performance the type I GIC requires *matched* op amps, a small disadvantage. Thus, generally, type II GICs are the preferred devices for inductance simulation. However, we shall find in Chapter 5 that the type I GIC is used as the basic building block for one of the best-performing second-order active filter sections. Finally, we need to return to the general immittance converter in Chapter 6 because, for the realization of some high-order filters in the form of ladder simulations, certain transformations turn out to be convenient which require GICs for their implementation.

4.4 ACTIVE BUILDING BLOCKS USING TRANSCONDUCTANCES

Because operational amplifiers are widely available, highly developed, and inexpensive, op amps have become the primary active component in the design of (discrete) active filters. Therefore, this chapter is mainly devoted to these devices and their use for building useful subcircuits for analog filters. However, the increasing demand for analog filter functions that can be integrated together with other signal-processing circuitry on one silicon chip has prompted designers to look for more suitable gain devices. As mentioned in Section 4.2.3, *transconductance amplifiers* appear to be the appropriate components because they are easy to implement in monolithic form, can be tuned electronically, and, as we shall see, lead to intriguingly simple filter circuitry. Therefore, operational transconductance amplifiers (OTAs) are likely to play an increasingly important role in analog filter design so that a discussion of their properties and a special section on methods for developing useful filter building blocks appears appropriate.

The fundamental operation of OTAs was introduced in Section 4.2.3, and Examples 4-3 and 4-4 gave an indication of the simplicity of their use in filter design. Further building blocks, analogous to the ones developed for integration and element simulation using op amps, will be presented in this section [65]. In all cases, we show circuits using *only* OTAs and capacitors, because not only are these two elements sufficient in general, but they also lead to circuits that are simpler to integrate.

Consider first the circuit in Fig. 4-13 for simulating a *grounded* resistor. If instead we need a *floating* resistor, we must "lift the circuit off ground" and devise a method for the current I_1 to flow out of the second terminal. Figure 4-34 shows the result, from which we calculate

$$I_1 = g_{m1}(V_1 - V_2) \quad \text{and} \quad I_2 = g_{m2}(V_1 - V_2) \qquad (4\text{-}79)$$

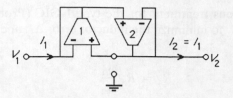

Figure 4-34 Floating resistor $R = 1/g_m$ for $g_{m1} = g_{m2} = g_m$.

Thus, for matched OTAs, i.e., $g_{m1} = g_{m2} = g_m$, we have a *floating* resistor of value $R = 1/g_m$.

To build a *summer*, we require one g_m element for each signal to be added, as shown in Fig. 4-35a; the summing current, I_s, can flow into *any* impedance (we show a positive "resistor" in Fig. 4-35a). Clearly, we have

$$V_s = \frac{g_{m1}}{g_{ms}} V_1 + \frac{g_{m2}}{g_{ms}} V_2 \tag{4-80}$$

i.e., V_s is the sum of two *scaled* (compare Problem 4.21) voltages. Extensions to more than two signals are obvious; we also remind the student (Example 4-3) that interchanging the input terminals of any feed-in OTA will change the sign of the corresponding summing coefficient. Thus, it is easy to form the *difference* of two signals.

Figure 4-35b shows a *lossy* (differential) *integrator* which realizes

$$V_o = \frac{g_{m3}}{sC + g_{m4}} (V_i^+ - V_i^-) \tag{4-81}$$

Clearly, the integrator can be made lossless (for *ideal* transconductances) if g_{m4} is removed. If we connect V_s of Fig. 4-35a to either V_i^+ or V_i^- of Fig. 4-35b with the other integrator input grounded, we will have constructed a *summing* noninverting or inverting integrator, respectively.

Suppose, now, that we wish to design a grounded impedance Z_{in} that is inversely proportional to a second impedance Z_L, i.e., $Z_{in} = r^2/Z_L$. Using OTAs, we accomplish this task readily by noting that the g_m output current is proportional to a voltage. If we make this voltage proportional to Z_L and the g_m output current the impedance input current, we have an admittance proportional to Z_L, i.e., our

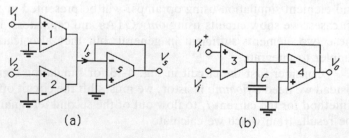

(a)   (b)

Figure 4-35 (a) g_m-C summer; (b) lossy differential g_m-C integrator.

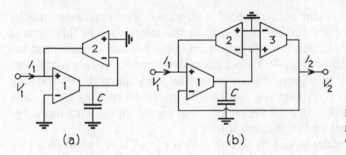

Figure 4-36 g_m-C simulation of (a) a grounded inductor and (b) a floating inductor for $g_{m2} = g_{m3} = g_m$.

job is complete.   Figure 4-36a shows the resulting circuit developed with the goal of simulating a *grounded inductance* by choosing $Z_L = 1/(sC)$.   Elementary analysis yields

$$V_C = \frac{g_{m1}}{sC} V_1 \quad \text{and} \quad I_1 = g_{m2} V_C = \frac{g_{m1} g_{m2}}{sC} V_1$$

Therefore, as predicted, the input impedance of Fig. 4-36a[12] is that of an inductor:

$$Z_{\text{in}}(s) = \frac{V_1}{I_1} = s \frac{C}{g_{m1} g_{m2}} \tag{4-82}$$

which is electronically tunable by varying g_{mi}.

The student will recall that we found *floating* inductors difficult to simulate with op amp circuits; using transconductances, however, makes the task quite trivial because we need only to lift the inverting g_{m1} terminal in Fig. 4-36a off ground and devise a circuit that generates $I_2 = I_1$ at the second impedance terminal.   Figure 4-36b shows a simple method that needs only *one* additional transconductance: analysis of Fig. 4-36b yields

$$V_C = \frac{g_{m1}}{sC} (V_1 - V_2)$$

and

$$I_1 = g_{m2} V_C = \frac{g_{m1} g_{m2}}{sC} (V_1 - V_2)$$

$$I_2 = g_{m3} V_C = \frac{g_{m1} g_{m3}}{sC} (V_1 - V_2)$$

Evidently, for $g_{m2} = g_{m3} = g_m$ we have simulated a floating inductor of value $L = C/(g_m g_{m1})$.

We have shown in the foregoing brief treatment that all the essential building blocks required for the design of active filters can be implemented with only transconductances and capacitors.   Once these blocks are available, second-order sec-

---

[12]The circuit in Fig. 4-36a is the transconductance equivalent of the GIC in Fig. 4-33.

tions or ladder simulations can be assembled with ease. The problems and the material in later chapters will further demonstrate the methods. In this context the student may find it also instructive to revisit Example 4-4 and observe that the second-order bandpass filter of Fig. 4-14 was obtained by connecting two integrators of the form of Fig. 4-35b in a loop: one noninverting lossy integrator (consisting of g_{m2}, g_{m3}, and C_2) and one inverting lossless integrator (consisting of g_{m1} and C_1). The general method is that of Fig. 4-19. For signal input, we simply disconnected C_1 from ground and connected it to V_1.

In order not to be sidetracked by details which are of less importance in a first derivation of potentially useful circuits, we found it appropriate to assume all OTAs as *ideal*. Of course, the careful designer will have to investigate how good these assumptions are and how much the imperfections of a *real* OTA might possibly affect the circuit's performance. We have given some indication of the necessary calculations in Examples 4-3 and 4-4; Example 4-6, below, will further illustrate that care must be exercised when dealing with real components.

Example 4-6

Find a circuit model for the floating inductor simulated by Fig. 4-36b. Use Fig. 4-12 to model the OTAs. Assume that the three OTAs are sufficiently well matched that differences in their input and output impedances are negligible; any possible effects of g_m mismatch should be investigated. Neglect the frequency dependence of g_m.

Solution. The small-signal model for Fig. 4-36b, with the OTAs represented by Fig. 4-12, is shown in Fig. 4-37. With this model established, the analysis is easy. We find

$$V_C = \frac{g_{m1}}{sC + Y_2}(V_1 - V_2)$$

and

$$I_1 = Y_1 V_1 + g_{m2} V_C = Y_1 V_1 + \frac{g_{m1} g_{m2}}{sC + Y_2}(V_1 - V_2) \qquad (4\text{-}83a)$$

$$I_2 = Y_3 V_2 - g_{m3} V_C = Y_3 V_2 + \frac{g_{m1} g_{m3}}{sC + Y_2}(V_2 - V_1) \qquad (4\text{-}83b)$$

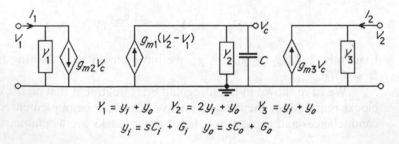

$$Y_1 = y_i + y_o \qquad Y_2 = 2y_i + y_o \qquad Y_3 = y_i + y_o$$
$$y_i = sC_i + G_i \qquad y_o = sC_o + G_o$$

Figure 4-37 Small-signal model for the g_m-C floating inductor of Fig. 4-36b.

If we label $g_{m3} = g_{m2} + \Delta g_m$ in order to investigate the effect of g_m mismatch, Eq. (4-83b) becomes

$$I_2 = Y_3 V_2 + \frac{g_{m1} g_{m2}}{sC + Y_2} (V_2 - V_1) + \frac{g_{m1} \Delta g_m}{sC + Y_2} (V_2 - V_1) \qquad (4\text{-}83\text{c})$$

With Eqs. (4-83a) through (4-83c), it is easy to establish a circuit model for the floating inductor as simulated by Fig. 4-36b. The model, Fig. 4-38, shows that in addition to the inductor of value $L = (C + 2C_i + C_o)/(g_{m1} g_{m2}) = L_0 + \Delta L$, we have to accept inductor losses $R_L = (2G_i + G_o)/(g_{m1} g_{m2})$ and two parasitic shunt admittances $Y = s(C_i + C_o) + G_i + G_o$. Finally, if the two transconductances g_{m2} and g_{m3} have a matching error Δg_m, the model contains a controlled source with the frequency-dependent transadmittance

$$y_m(s) = \frac{g_m \Delta g_m/(C + 2C_i + C_o)}{s + (G_o + 2G_i)/(C + 2C_i + C_o)} \qquad (4\text{-}84)$$

This example shows clearly that the "inductor" is far from ideal if real OTAs are used in the design. Whether the deviations caused by the inductor error, the losses, and the various parasitic elements are acceptable or not depends on the application and must be evaluated in each particular case (Problems 4.32 and 4.39).

A final comment concerning applications of OTA circuits is called for. The designer should keep in mind that *real* transconductances are not ideal current sources (with zero output admittance) and that many of the g_m building blocks have nonzero output admittances. Their behavior will, therefore, be extremely susceptible to loading effects, which can *completely alter* the intended circuit function. Observe, for example, that a load admittance Y_L on the summing circuit of Fig. 4-35a will change g_{ms} into $g_{ms} + Y_L$ in Eq. (4-80). Similarly, a load Y_L will cause g_{m4} to be replaced by $g_{m4} + Y_L$ in the integrator transfer function of Eq. (4-81). Thus, transconductance circuits generally should be designed so that the building blocks drive high-impedance nodes, such as the inputs of other OTAs. If large loads must be driven, the OTA circuit must be buffered by, for example, a unity-gain op amp (Fig. 4-17d) or, preferably, to keep with the all-g_m-C strategy, by the scheme illustrated in Fig. 4-39, which can be used by itself or in combination with other subcircuits. To determine the circuit's behavior, recall that a parasitic admittance $y_p = y_i + y_o$ (Fig. 4-37) is found at each connecting node between two

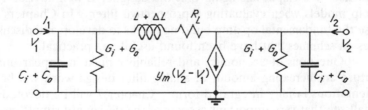

Figure 4-38 Twoport model for a floating inductor, simulatd by Fig. 4-36b with real OTAs.

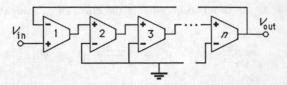

Figure 4-39 Transconductance buffer circuit.

OTAs.   Assuming for simplicity that all OTAs are identical and that V_{out} is loaded by Y_L, we calculate

$$V_{out} = \frac{g_{m1}}{y_p} \frac{g_{m2}}{y_p} \frac{g_{m3}}{y_p} \cdots \frac{g_{mn}}{Y_L + y_p} (V_{in} - V_{out})$$

i.e.,

$$\frac{V_{out}}{V_{in}} = \frac{1}{1 + (y_p/g_m)^n (1 + Y_L/y_p)} \simeq 1 \qquad \left| \frac{y_p}{g_m} \right|^n \ll 1 \qquad (4\text{-}85a)$$

Similarly, the (Thévenin) output impedance $Z_{out}(s)$ is calculated as

$$Z_{out}(s) = \frac{1/y_p}{1 + (g_m/y_p)^n} \simeq \frac{1}{y_p} \left(\frac{y_p}{g_m} \right)^n \qquad \left| \frac{y_p}{g_m} \right|^n \ll 1 \qquad (4\text{-}85b)$$

Because it is easy to design the transconductance such that $g_m \gg |y_p|$, clearly $|V_{out}/V_{in}| \simeq 1$ and $|Z_{out}| \ll |1/y_p|$ for sufficiently large values of n.   Usually, $n = 2$ or 3 will be satisfactory.

4.5 SUMMARY

After having established the need for *gain* in the design of active filters, we have in this chapter discussed the two most important fundamental gain blocks, the *operational amplifier* and the *transconductance amplifier*.

Historically, the overwhelming majority of active filters have been developed around the ubiquitous op amp, and even today most discrete active *RC* filters are designed with op amps providing the needed gain.   We have, therefore, devoted much of this chapter to discussing those properties of op amps that are germane to active filter work and to developing the most frequently used op amp building blocks.   We emphasize again that the designer is well-advised to use realistic op amp models when evaluating a prospective filter.   In Chapters 5 and 6 we shall use the fundamental op amp building blocks to develop and discuss the active filter design schemes that have been found useful and practical.

Questions of economics and reliability make it appear unavoidable that in future an increasing amount of analog filter design work will be aimed at implementations in fully integrated form.   Chapter 7 will be devoted to this topic; we shall see that very important aspects of the design are simple circuitry, systematic and dense layout, and, especially, the possibility of automatic electronic tuning. All these requirements are satisfied more easily with transconductance amplifiers

than with op amps. Also, growing familiarity with OTAs and their wider availability will lead, we believe, to increased development of even discrete analog filter
circuits using OTAs to provide the necessary gain. We have, therefore, discussed
the relevant properties of transconductances and a number of useful g_m building
blocks in order to give the reader the needed background for such developments.
As with op amp designs, the g_m subcircuits will be used in later chapters to assemble
complete practical analog filter circuits.

Instead of presenting the reader simply with a catalog of existing op amp or
transconductance circuits, we have attempted to illustrate the underlying principles
and methods by which these circuits are derived. We hope that thereby we could
answer the so often heard student questions: How does one get this circuit? Where
does this circuit come from? Also, by studying and digesting the design methods,
we hope, the student will be better equipped to develop his or her own designs.

PROBLEMS

4.1 Assuming an ideal op amp, find the transfer function of the circuit in Fig. P4.1 and
determine the elements such that the circuit realizes (a) a lowpass, (b) a highpass,
and (c) a bandpass filter.

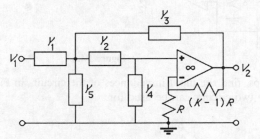

Figure P4.1

4.2 Assuming an ideal op amp, find the transmission matrix of the circuit in Fig. P4.2.
What can this circuit be used for?

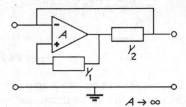

$A \to \infty$ **Figure P4.2**

4.3 Use the circuit of Fig. P4.2 with $Y_1 = Y_2 = G$ for the twoport negative impedance
converter (NIC) in Fig. P4.3. Show that

$$\frac{V_o}{V_i} = \frac{Y_2 - Y_1}{(Y_2 - Y_1) + (Y_4 - Y_3)}$$

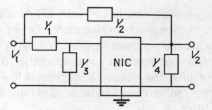

Figure P4.3

One can show that this circuit can realize entirely arbitrary transfer functions if the elements $Y_i(s)$ are RC admittances. However, sensitivities are quite high (remember difference effects!).

4.4 Find and compare the 3-dB bandwidths of the inverting and the noninverting unity-gain amplifiers in terms of the op amp parameter ω_t.

4.5 Calculate the exact poles of the Miller integrator transfer function Eq. (4-13a).

4.6 The twoport N in Fig. P4.6 is a second-order bandpass filter; the op amp is ideal. Discuss what type of filter functions the circuit may realize for the appropriate choice of resistors.

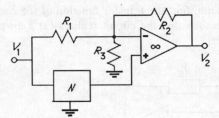

Figure P4.6

4.7 Assuming ideal op amps, find the input impedances of the circuits in Fig. P4.7a and b. Discuss what these two circuits can be used for.

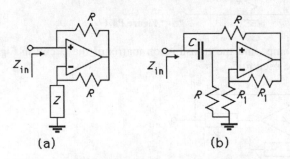

(a) (b) Figure P4.7

4.8 To demonstrate how the parameters of a real op amp can affect the transfer function of a circuit, show that the gain of the Miller integrator becomes

$$\frac{V_2}{V_1} = -\frac{1}{sCR} \frac{1 - sCR_o/A(s)}{1 + \frac{1}{A(s)}\left[1 + \left(\frac{1}{sC} + R_o\right)\left(\frac{1}{R} + \frac{1}{R_i}\right)\right]}$$

when the op amp model of Fig. 4-5 is used. Specifically, assume that $R = 1\ k\Omega$,

$C = 0.1 \mu F$, $R_i = 1.5 M\Omega$, $R_o = 50 \Omega$, and

$$A(s) = \frac{120,000}{(1 + s/30)(1 + s/7 \cdot 10^6)}$$

4.9 Determine the performance of the circuit in Fig. P4.9
   (a) Assuming ideal op amps
   (b) Assuming op amps with $A(s) = \omega_t/s$

Discuss your results. This is one example of the so-called *active R filters* which use op amps as integrators to avoid the need for external capacitors (see also Example 4.5 and Chapter 5).

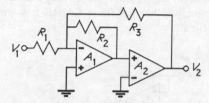

Figure P4.9

4.10 Determine the performance of an uncompensated Miller integrator designed with $R = 1200 \Omega$, $C = 5 nF$, and a $\mu A741$ op amp with dc gain of 120,000, 3-dB frequency equal to 12 Hz, and a second pole at $f_{t2} = 3.8$ MHz.
   At $f = 10$ kHz, find Q_I and the excess phase error $\Delta\phi(\omega)$.

4.11 Consider Fig. 4-1. Use sensitivity methods to determine the pole shifts of the normalized ideal transfer function

$$H(s) = \frac{3s}{s^2 + 0.1s + 1}$$

when the passive RC network has

$$\frac{N_{32}(s)}{D(s)} = \frac{s^2 + 0.1s + 1}{s^2 + 3s + 2}$$

and an op amp with $\omega_t = 2\pi - 10^6$ rad/s is used. s is normalized with respect to 10 kHz.

4.12 Evaluate as a function of frequency the output impedance of a noninverting amplifier (Fig. 4.17b) when the op amp has finite input and output resistors and the gain is modeled by Eq. (4-8).

4.13 Find the output impedance of the Sallen and Key bandpass of Fig. P4.1 when the op amp has finite gain and a finite output resistance R_o. Assume $R_i = \infty$.

4.14 An op amp has a slew rate of 0.5 V/μs and is connected to a power supply of ± 7.5 V. The device is used in a single-amplifier bandpass filter designed to have a midband gain of 6.9 at 9 kHz. What is the largest input signal amplitude that can be applied, if no distortion should occur? Repeat for a midband frequency of 90 kHz.

4.15 Calculate the output noise spectrum for a noninverting amplifier of gain 2. The noise models are as in Fig. 4-8 with $i_{np} = i_{nn} = 0.1$ pA/$\sqrt{Hz}$ and $e_n = 10$ nV/$\sqrt{Hz}$.

4.16 Assuming $A(s) \approx \omega_t/s$ with $\omega_t = 2\pi$ Mrad/s, design the lowpass filter of Fig. P4.9 such that $f_p = 100$ kHz and $Q_p = 3$; the dc gain K is arbitrary. Use the noise models of

Fig. 4-8 with numerical values as in Problem 4.15 to calculate the total rms output noise of the filter.   See references 76 and 77 for a discussion of the details.

4.17 **(a)** Calculate the output offset voltage in a noninverting amplifier of gain $K = 4$; the op amp offset voltage is known to be $V_{OS} = 12$ mV.

(b) The same op amp has an input bias current $I_{Bn} = 250$ nA and an offset current of $I_{OS} = \pm 80$ nA.   Calculate the expected output offset voltage caused by bias currents.

(c) Place a resistor R^+ into the noninverting input lead and determine its value such that the contribution due to bias currents is canceled.   Can this resistor be used to cancel all contributions to output offset voltage?

4.18 Figure P4.18 shows a first-order allpass filter; the op amp has a common mode rejection ratio of 75 dB.   Determine whether the finite CMRR has any significant effect on the performance of the circuit.

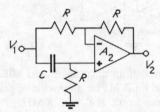

Figure P4.18

4.19 Determine the operation of the g_m-C circuits in Fig. P4.19a, b, and c.   Note that g_m may be positive or negative depending on input polarity!   This feature may be useful in Fig. P4.19c.

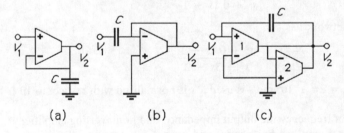

(a)   (b)   (c)

Figure P4.19   (a) Lowpass, (b) highpass, and (c) equalizer-allpass filters.

4.20 Determine the transfer function of the circuit in Fig. P4.20.   What is the effect of the finite input and output impedances of the transconductances?

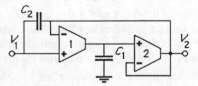

Figure P4.20   Second-order g_m-C filter.

4.21 Find the function of the circuit in Fig. P4.21.

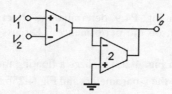

Figure P4.21

4.22 Determine the operation of the circuit in Fig. P4.22.

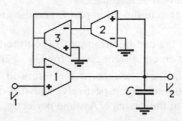

Figure P4.22

4.23 In an effort to save capacitor area, redesign the circuit of Fig. 4-14 such that the smallest capacitor (C_1) is 1 pF. For this case, investigate the effects of the parasitic input and output capacitors.

4.24 Prove that for the design discussed in Example 4-5, the bandwidth of the actively compensated "amplifier" is improved by a factor $\sqrt{1 + K}$ over that of the simple inverting amplifier of Fig. 4-17a.

4.25 Analyze the summing and differencing amplifier in Fig. P4.25 and show that the weighting factors cannot be tuned independently if the noninverting input is used to obtain summing.

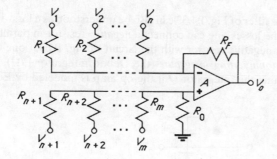

Figure P4.25

4.26 Analyze the passive compensation method of Fig. 4-21a and compare its performance to that of Fig. 4-21b.

4.27 Assume that, due to a change in op amps or changed operating conditions, ω_t decreases by 35%. For this case, compare the performance (i.e., Q_I and $\Delta\phi$) of the integrators in Figs. 4-20, 4-21b, and 4-22. For Fig. 4-22, assume a dual op amp, i.e., A_1 and A_2 on the same chip and substantially matched (but with a 2% ω_t matching error) and tracking.

4.28 Find the integrator quality factor of the Miller-inverter cascade of Fig. 4-23a.

4.29 For the active R filter in Fig. P4.9, determine the effect of the op amps' second pole, modeled by Eq. (4-11), on the performance. Discuss your results in view of Eqs. (4-63) and (4-64).

4.30 Show that the circuit in Fig. 4-26b realizes a floating inductor.

4.31 Equation (4-66a) shows the y-parameters and Fig. 4-25b the model of the *ideal* gyrator, i.e., $y_{11} = y_{22} = 0$.
   (a) How do the y matrix and the model change if in practice the transconductances have finite input and output impedances?
   (b) What is the effect of the conditions in part (a) on the grounded simulated inductor of Fig. 4-26a?
   (c) What is the effect of the conditions in part (a) on the floating simulated inductor of Fig. 4-26b? Find an equivalent circuit for Fig. 4-26b.

4.32 Discuss the effect on the performance of the filters in Fig. P4.32a and b, if the "inductors" of Fig. 4-26a and b, respectively, with the imperfections discussed in Problem 4.31, are used in the design. Assume perfect g_m matching.

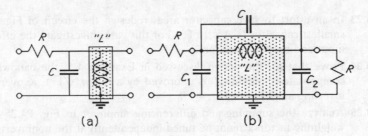

(a) (b)

Figure P4.32

4.33 The passive RC lowpass filter of Fig. P4.33 can also be understood as a lossy integrator. Show that to remove the losses one can connect a *negative* resistor in parallel with C. If one implements this negative resistor with the circuit of Fig. P4.7a, one obtains an ideal *single-amplifier noninverting integrator* (the Deboo integrator [91]). Analyze this circuit and determine the integrator Q if the op amp is modeled by Eq. (4-8).

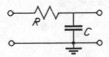

4.34 **(a)** Show that the circuit in Fig. P4.18 is a first-order allpass filter.
   (b) Use this circuit in the scheme illustrated in Fig. 4-28 to build an op amp inductance simulation circuit.
   Hint: Derive the input voltage V from another input V_1 via a differential op amp such that $V_2/V_1 = 1 - 1/s\tau$ as required by Eq. (4-67). To see what must be done, remember that the circuit is *linear;* thus, construct the voltage V as a linear combination of V_1 and V_2, i.e., $V = \alpha V_1 + \beta V_2$, and determine α and β. The resulting circuit is that of Fig. 4-32b.

4.35 Derive Eq. (4-72).

4.36 Derive the results corresponding to Eqs. (4-74) through (4-77) for the type I GIC.

4.37 Consider the circuit in Fig. P4.37 that consists of five identical OTAs and two capacitors. Assuming ideal OTAs, find its transfer function
    (a) By direct analysis and
    (b) More simply, by identifying its building blocks and comparing the result with Fig. P4.32a. Notice that this circuit is a g_m-C simulation of a simple RLC ladder! Also observe that the first OTA is not required if V_1 is an ideal voltage source.
    (c) Give expressions for the relevant filter parameters. Discuss any problems you perceive with this circuit.
    (d) Modify the circuit such that its midband gain is $H_M = 3$. Can you do this still with all identical OTAs?

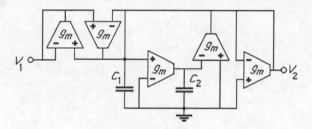

Figure P4.37

4.38 Use the method outlined in Problem 4.37 to design a first-order maximally flat highpass filter with a cutoff frequency of 24 kHz. Choose element values that are realistic for integration.

4.39 Eliminate the first (redundant) OTA in Fig. P4.37 and analyze the performance deviations of the remaining bandpass when the OTAs have finite output admittances $y_o = sC_o + G_o$. Assume perfect g_m matching and input admittances $y_i = 0$.

4.40 The noninverting phase-lead integrator will be used extensively in Chapters 5 and 6 in the design of active filters. It is, therefore, very important that these integrators be stable, i.e., that they not break into oscillations. As discussed in the text, there is no apparent problem when stability when the op amps are modeled as $A(s) \simeq \omega_t/s$. Nevertheless, these circuits are known to oscillate in certain circumstances depending on the type of op amp used in the design. The difficulties can be recognized when the more accurate op amp model of Eq. (4-11) with a second pole ω_2 is used. The second pole of the op amp is usually, but not always, a factor 2 or more larger than ω_t.
    (a) Determine the conditions to be satisfied by ω_2 so that the phase-lead integrator is stable.
    (b) If this condition is not satisfied, the circuit must be modified. A suitable configuration is shown in Fig. P4.40 [123]. Determine α and β as a function of op amp parameters such that
        (1) The important phase-lead condition $\Delta\phi \simeq +\omega/\omega_t$ is not violated.
        (2) The circuit remains stable.
        Assume that the two op amps are identical.

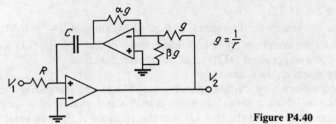

Figure P4.40

Second-Order Active Filter Sections

FIVE

5.1 INTRODUCTION

Active filters which realize the biquadratic transfer function

$$H_{(s)} = \frac{a_2 s^2 + a_1 s + a_0}{s^2 + b_1 s + b_0} = \frac{a_2 s^2 + a_1 s + a_0}{s^2 + s\dfrac{\omega_0}{Q_p} + \omega_0^2} \tag{5-1}$$

introduced in Chapter 1, Eqs. (1-80) and (1-81), continue to play an important role. These so-called *biquads* are often used in their own right in communications and measurement equipment, where, for example, a simple lowpass filter [Eq. (1-85)] may be used to reduce high-frequency noise in a system or to limit the signal bandwidth in preparation for sampled-data signal processing. Similarly, for measurements over a relatively narrow signal bandwidth, a second-order bandpass filter [Eq. 1-87) may be sufficient to eliminate dc and high-frequency noise components; or a notch filter [Eq. (1-88)] may be employed to block a troublesome signal at one frequency, such as 60-Hz power supply interference.

More important, however, biquads form a fundamental building block for the construction of filters of higher order in the form of cascade or multiple-loop feedback topologies because of the resulting reduced sensitivities to component variations and tolerances (Section 3.2.6). We also saw in Section 3.2.6 [Eqs. (3-71) and (3-81)] that the sensitivities of the total filter are reduced by choosing low-sensitivity second-order building blocks.

From the above discussion it follows that we need to investigate how to design the *best possible* biquad, where the meaning of "best possible" must be defined, of course, in the context of the application at hand. In our selection of suitable filters, we shall always insist on *low sensitivities* to component variations, but other criteria are also important for a design to be economical and useful in practice. To be *manufacturable*, the *component values* must be practical and the *component spread* should not be excessive. Since analog filters usually require some adjusting, the biquad must be *easy to tune*, preferably such that tuning the important filter parameters, ω_0 and Q_p, is noninteractive. Because the biquads must be able to drive loads, including other biquads in cascade or multiple feedback connections, we require *low output impedance* and, if possible, a *high input impedance $Z_{in}(s)$*. $|Z_{in}(j\omega)|$ in active filter sections may vary widely with frequency, so that it can pose a significant load for the driving biquad. Recall from Chapter 4 that the biquad's output impedance is mainly determined by the op amp output impedance, which, although small, is not zero; therefore, a frequency-dependent load may significantly alter the behavior of the filter.

For meeting the design requirements, the designer or user finds in the literature a great proliferation of different circuits, all of which are claimed by their respective inventors to be advantageous in one way or another. To provide guidance in the resulting confusion, several studies have proposed classifications of biquads according to some useful criteria in an attempt to discover common design principles and to identify the more practical circuits [6, 97, 98, 100]. We shall base our discussion on the development proposed in reference 100 because its systematic development serves the purpose of "bringing order into chaos" and at the same time shows clearly how the circuits are derived and how to optimize them for superior (low-sensitivity) performance. As mentioned, our main common design objective will be low sensitivities; the remaining criteria will be addressed on a case-by-case basis.

5.2 SINGLE-AMPLIFIER BIQUAD CONFIGURATIONS

We have shown in Fig. 4-1 a fairly general feedback configuration consisting of an *RC* circuit and an amplifier of gain *K*. This circuit and the transfer function given in Eq. (4-2) could be used as starting points for our discussion. However, one finds soon that the literature contains a number of circuits which cannot be brought into the form of Fig. 4-1, so that some modification seems advantageous. In Fig. 5-1 we show a more general feedback configuration which includes an op amp of gain $A(s)$ and an *RC* network. We have identified the nodes (or terminals) a, b, c, and d where the input voltage, the feedback connections, and the op amp input terminals, respectively, are attached. For greater clarity, we have placed arrows at those terminals that are inputs for the *RC* network and have also labeled the ground connection g. With all voltages referenced to ground, the *RC* transfer functions will be labeled

$$T_{kl}(s) = \frac{V_k}{V_l} = \frac{N_{kl}(s)}{D_0(s)} \qquad k = c, d; l = a, b \qquad (5\text{-}2)$$

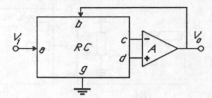

Figure 5-1 General active *RC* feedback configuration.

where $N_{kl}(s)$ and $D_0(s)$ are polynomials in s with coefficients determined by the values of the *RC* elements.   Simple analysis yields

$$\frac{1}{A(s)} V_0 = [T_{da}(s) - T_{ca}(s)]V_i + [T_{db}(s) - T_{cb}(s)]V_0$$

so that the transfer function of this network is given by

$$H(s) = \frac{V_o}{V_i} = \frac{T_{da}(s) - T_{ca}(s)}{T_{cb}(s) - T_{db}(s) + 1/A(s)} \qquad (5\text{-}3)$$

Equation (5-3) shows quite clearly that we require a second-order *RC* network if $H(s)$ is to be a second-order function.[1]   Also we observe that the *transmission zeros* of the *RC* active network *are determined by the forward path* through the network whereas the *poles are set by the feedback path*.   In the following we shall discuss in more detail how zeros and poles can be formed in the circuit and how they may be related.

5.2.1 Generation of Transfer Function Zeros

Clearly, the natural frequencies (the poles) of $H(s)$ are found from

$$T_{cb}(s) - T_{db}(s) + \frac{1}{A(s)} = 0 \qquad (5\text{-}4)$$

and we observe that this equation does not depend on node a; i.e., it is independent of the connection of V_i.   This result should not be unexpected for the reader, because we remember from basic circuits or systems courses that the natural frequencies are calculated for $V_i = 0$, i.e., from the zero-input response.   Having reminded ourselves that the *poles of a network are obtained for* $V_i = 0$, and realizing that the zeros of $H(s)$ depend on where the input signal is applied [the numerator of Eq. (5-3) depends on node a!], we can now state conversely that:

> The zeros of a transfer function are created *without* disturbing the poles, by feeding the input signal into any node or nodes that were previously connected to ground.

What this means in practice is illustrated in Fig. 5-2: if we have found a network with the desired poles (as discussed in Section 5.2.2), we can create transmission

---

[1]Third-order transfer functions are sometimes used; in that case a pole-zero cancellation is required to restore the final function to one of order 2.

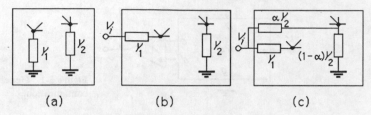

Figure 5-2 (a) Part of a pole-determining zero-input circuit showing two grounded elements; (b) Y_1 completely lifted off ground and connected to the input signal; (c) in addition, Y_2 lifted partially off ground and connected to V_i.

zeros by lifting any grounded element completely or partially off ground and connecting the resulting new node(s) to the input.

Example 5-1

Show that the circuit of Fig. 5-3a can have complex poles, and determine the location of the transfer function zeros obtainable by lifting the three resistors R_2, R_3, and R completely or partially off ground. Assume that the op amp is ideal.

Solution. From Fig. 5-3a we calculate by elementary analysis

$$T_{cb}(s) = \frac{V_c}{V_b} = \frac{V^-}{V_o} = \frac{N_{cb}}{D_1} = \frac{s^2 C^2 + 2sCG_1 + G_1G_2}{s^2C^2 + sC[2(G_1 + G_3) + G_2] + G_2(G_1 + G_3)}$$

which, with $1/A(s) = 0$ from Eq. (5-4), shows that the natural frequencies are the roots of

$$N_{cb}(s) = s^2 C^2 + 2sCG_1 + G_1G_2 = C^2\left(s^2 + s\frac{\omega_0}{Q} + \omega_0^2\right) = 0$$

Evidently, complex poles result if

$$Q = \frac{\omega_0 C}{2G_1} = \frac{1}{2}\sqrt{\frac{G_2}{G_1}} > \frac{1}{\sqrt{2}}$$

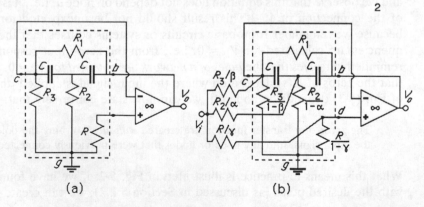

Figure 5-3 Example for zero generation: (a) "dead" circuit for desired poles; (b) method for applying input signal.

which can be obtained for $R_1 > 2R_2$. In Fig. 5-3b we have lifted the fractions α, β, and γ, respectively, of the resistors R_2, R_3, and R off ground. The resulting transfer function can be shown to be

$$H(s) = \frac{\gamma s^2 + s\dfrac{\gamma[2(G_1 + G_3) + G_2] - \alpha G_2 - 2\beta G_3}{C} + \dfrac{\gamma G_2(G_1 + G_3) - \beta G_2 G_3}{C^2}}{s^2 + s\omega_0/Q + \omega_0^2}$$

(5-5)

From Eq. (5-5) we can obtain a variety of transfer functions. For example, a bandpass with numerator $-\alpha s G_2/C$ is realized for $\beta = \gamma = G_3 = 0$; a highpass filter is found for $\beta = 1$, $\gamma = \alpha$, and $G_3 = G_1\alpha/(1 - \alpha)$ to give a numerator polynomial αs^2. We leave it the problems (Problem 5.2) to show that this structure is very versatile in that also allpass, highpass-notch, and lowpass-notch—but not lowpass—functions can be implemented by an appropriate choice of α, β, and γ. Note that as predicted, the poles are not affected by this process. For the following discussion of pole generation we will, therefore, concentrate on the "dead" network, i.e., the network with no input applied.

5.2.2 Generation of Transfer Function Poles

Returning to Fig. 5-1 and Eqs. (5-2) and (5-3), it is evident that establishing the system poles of $H(s)$ entails synthesizing a *threeport RC* network, which in general is quite a complicated task. In order to make the situation more tractable, we break the RC network into two, as shown in Fig. 5-4a. The reader can verify by simple analysis that the system poles are now the roots of the expression

$$T_{cb}(s) - T_{fe}(s) + \frac{1}{A(s)}$$

$$= \frac{1}{D_1(s)D_2(s)}\left[N_{cb}(s)D_2(s) - N_{fe}(s)D_1(s) + \frac{D_1(s)D_2(s)}{A(s)}\right] = 0 \quad (5\text{-}6)$$

where the definitions of the second-order RC transfer functions $T_{ij}(s)$ are analogous to Eq. (5-2). Observe that the $T_{ij}(s)$ are now *twoport* transfer functions, which

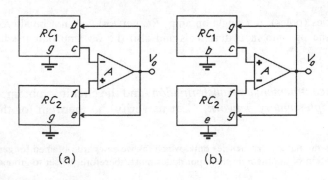

(a)   (b)

Figure 5-4 The system of Fig. 5-1 with the RC network split into two parts; signal input connections are not shown. (a) Original network; (b) complementary network.

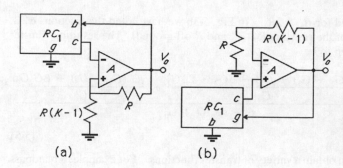

Figure 5-5 (a) Enhanced negative feedback (ENF) structure; (b) enhanced positive feedback (EPF) structure. Input connections are not shown.

implies that we have a much easier design task when compared to the previous case in Fig. 5-1. However, the simplification was obtained at a cost: the transfer function is now of *fourth* order (!) and the poles are determined by a fourth-order polynomial, Eq. (5-6), because we must assume that in general $D_1(s)$ and $D_2(s)$ are *different* even if the two RC networks are designed to have *nominally identical* poles; mismatches between elements and component tolerances or drifts will always result in $D_1 \neq D_2$. We choose, therefore, as a further simplification, to make one of the RC networks, say, RC_2, independent of frequency, as illustrated in Fig. 5-5a. Now, the system poles are determined by

$$T_{cb}(s) - \frac{K-1}{K} + \frac{1}{A(s)} = \frac{1}{D_1(s)}\left[N_{cb}(s) - \frac{K-1}{K}D_1(s) + \frac{D_1(s)}{A(s)}\right] = 0 \quad (5\text{-}7)$$

Since the main RC feedback path goes to the inverting op amp terminal but some positive feedback is applied that, as we shall see, causes Q enhancement, this structure has been called an *enhanced negative feedback* (ENF) circuit [11, 100].

At this point the student will correctly suspect that because of the symmetry of the situation we ought to get an *enhanced positive feedback* (EPF) network if in Fig. 5-4a we replace the RC network RC_1 by a resistive circuit and retain RC_2 as a frequency-dependent RC network. Rather than deriving the EPF circuit by repeating the previous steps, let us instead make use of an observation [101] which states:

> The natural frequencies (poles) in a single op amp–RC network will not be altered if one interchanges the op amp input terminals and also the op amp output with ground.[2]

This process is called the *complementary transformation*, and the circuit so obtained is referred to as the *complementary network*. Let us verify the theorem for the

---

[2]This theorem concerns the natural frequencies only, which, as we saw, are obtained for zero input ($V_i = 0$); for the application of the theorem the input nodes must, therefore, be set to ground.

circuit at hand, Fig. 5-4a.   Figure 5-4b shows the complementary network, whose natural frequencies are determined by the expression

$$T_{fg}(s) - T_{cg}(s) + \frac{1}{A(s)} = 0 \tag{5-8}$$

Observe now that

$$T_{fg} = \frac{V_{fe}}{V_{ge}} = \frac{V_{fg} + V_{ge}}{V_{ge}} = 1 - \frac{V_{fg}}{V_{eg}} = 1 - T_{fe} \tag{5-9a}$$

and

$$T_{cg} = \frac{V_{cb}}{V_{gb}} = \frac{V_{cg} + V_{gb}}{V_{gb}} = 1 - \frac{V_{cg}}{V_{bg}} = 1 - T_{cb} \tag{5-9b}$$

Thus, Eq. (5-8) becomes

$$T_{fg}(s) - T_{cg}(s) + \frac{1}{A(s)} = 1 - T_{fe} - (1 - T_{cb}) + \frac{1}{A(s)} = T_{cb} - T_{fe} + \frac{1}{A(s)}$$

which is identical to Eq. (5-6)!

Note that Eq. (5-9) says that *input and ground terminals in a network are complements* [100]; i.e., if the transfer function from input to output is $H(s)$, then interchanging input and ground connections gives the transfer function

$$H_c(s) = 1 - H(s) \tag{5-10}$$

The student is encouraged to keep this "trick" in mind because it proves useful for a variety of designs [101]; for example, if we have a bandpass circuit realizing

$$H_{BP}(s) = \frac{as}{s^2 + as + b} \tag{5-11a}$$

then interchanging input and ground results in a notch filter:

$$H_N(s) = 1 - H_{BP}(s) = \frac{s^2 + b}{s^2 + as + b} \tag{5-11b}$$

Similarly, if in the numerator of the bandpass filter of Eq. (5-11a) we replace a by $2a$, Eq. (5-10) gives an allpass circuit

$$H_{AP}(s) = 1 - 2H_{BP}(s) = \frac{s^2 - as + b}{s^2 + as + b} \tag{5-11c}$$

Returning now to the complementary transformation and the EPF circuit, we apply the transformation to the ENF circuit of Fig. 5-5a and obtain the EPF circuit of Fig. 5-5b.   The student can verify by direct analysis that the poles are determined by

$$\frac{1}{K} - T_{cg}(s) + \frac{1}{A(s)} = 0 \tag{5-12}$$

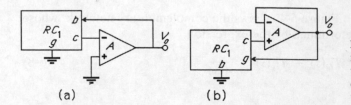

Figure 5-6 (a) Infinite-gain negative-feedback (NF) and (b) unity-gain positive-feedback (PF) circuits as special cases of the EPF and ENF configurations, respectively.

(a)   (b)

Use of Eq. (5-9b), i.e., $T_{cg} = 1 - T_{cb}$, shows that the roots of Eq. (5-12) are identical to those of Eq. (5-7). Therefore, the *dependence of the pole positions of the ENF and the EPF networks on circuit elements is identical*. Note the significance of this statement: ENF and EPF circuits are complementary; therefore, for identical passive RC networks, ENF, EPF, and all the numerous special cases that appear in the literature have identical pole sensitivities! Because pole sensitivity is the critical factor that determines the passband behavior, many of the arguments over the relative merits of the various circuits are, therefore, of little consequence. Observe further that the often encountered filters built with positive-feedback (PF) *unity-gain* or negative-feedback (NF) *infinite-gain* op amp configurations (for $|A| = \infty$) are nothing but special cases of the EPF and ENF circuits, respectively. Figure 5-6 shows the two structures which are obtained from Fig. 5-5 by setting $K = 1$. From Eqs. (5-7) and (5-12), their pole positions are given by

$$T_{cb}(s) + \frac{1}{A(s)} = 1 - T_{cg}(s) + \frac{1}{A(s)} = 0 \qquad (5\text{-}13)$$

which shows that the dependence of the poles on the op amp is the same in both configurations; thus, there is no apparent fundamental sensitivity advantage in designing the filter with a unity-gain instead of an infinite-gain op amp.

We are now ready to explore in more detail how transfer function poles are generated in the single-amplifier ENF and EPF networks of Fig. 5-5. More specifically, we want to determine which passive RC subnetworks are suitable and what restrictions they must satisfy in order to lead to practical second-order active RC sections.

Let us consider first the ENF circuit, Fig. 5-5a, whose denominator was given by Eq. (5-7). Because second-order filters are of concern here, it follows that the RC networks must be of second order,[3] i.e., in general,

$$T_{cb}(s) = \frac{N_{cb}}{D_1(s)} = \frac{a_2 s^2 + a_1 s + a_0}{s^2 + s\omega_1/q_p + \omega_1^2} \qquad (5\text{-}14)$$

where $q_p < 0.5$ because the poles of passive RC networks are simple and on the negative real axis [9, 10]. By Eq. (5-7), the desired complex poles of the active

---

[3]Or of third order with a pole-zero cancellation.

RC section are the roots of

$$D(s) = N_{cb} - \left(\frac{K-1}{K} - \frac{1}{A}\right)D_1(s)$$

$$= \frac{s^3}{\omega_t} + \left(a_2 - k_0 + \frac{\omega_1}{\omega_t}\frac{1}{q_p}\right)s^2 + \left(a_1 - k_0\frac{\omega_1}{q_p} + \frac{\omega_1^2}{\omega_t}\right)s + a_0 - k_0\omega_1^2$$

(5-15)

where we have for convenience introduced the abbreviation k_0 for the reciprocal of "the dc gain of the amplifier connected to RC_1" in Fig. 5-5a, i.e.,

$$k_0 = \frac{K-1}{K} - \frac{1}{A_0}$$

(5-16)

and we have used the op amp model introduced in Eq. (4-8):

$$A(s) \simeq \frac{\omega_t}{s + \sigma} = \frac{1}{s/\omega_t + 1/A_0}$$

Evidently, for amplifiers with infinite gain-bandwidth product, $\omega_t \to \infty$, $D(s)$ is of second order:

$$D(s) = d_2\left(s^2 + s\frac{\omega_0}{Q} + \omega_0^2\right)$$

$$= (a_2 - k_0)s^2 + \left(a_1 - k_0\frac{\omega_1}{q_p}\right)s + (a_0 - k_0\omega_1^2)$$

(5-17)

and we can determine the desirable properties to be satisfied by the *RC* network if we remember from Chapter 3 that the second-order transfer function is $2Q$ times more sensitive to changes in ω_0 than in Q. Therefore, small ω_0 sensitivities are more important for good section performance than small Q sensitivities. Because, evidently,

$$\omega_0^2 = \frac{a_0 - k_0\omega_1^2}{a_2 - k_0}$$

(5-18)

ω_0 can be made independent of the op amp's dc gain A_0, a parameter of high variability, in one of two ways: First, we can select $T_{cb}(s)$ as a bandpass function, i.e., $a_0 = a_2 = 0$, so that

$$T_{cb}(s) = \frac{(\omega_1/q_z)s}{s^2 + s\omega_1/q_p + \omega_1^2}$$

(5-19)

In this case

$$\omega_0 = \omega_1$$

(5-20a)

and

$$Q = \frac{k_0 q_z}{k_0 - 1 + k_0 q} = \frac{q_z}{q - (1 - k_0)/k_0}$$

(5-20b)

where we have labeled the ratio $q_z/q_p > 1$ as

$$\frac{q_z}{q_p} = 1 + q \tag{5-21}$$

However, observe that for *real* op amps the coefficients of s^3 and s^0 of the system polynomial $D(s)$ in Eq. (5-15) have opposite signs if $a_0 = 0$; i.e., $D(s)$ has a root in the right half-plane. Therefore, a bandpass function $T_{cb}(s)$ is not a possibility, because the active ENF filter would not be stable. Consequently, to make ω_0 independent of A_0, we choose the second possibility, that $a_0 = a_2\omega_1^2$; i.e., $T_{cb}(s)$ must be of the form

$$T_{cb}(s) = a_2 \frac{s^2 + s\omega_1/q_z + \omega_1^2}{s^2 + s\omega_1/q_p + \omega_1^2} \tag{5-22}$$

Then we have

$$\omega_0 = \omega_1 \tag{5-23a}$$

and

$$Q = \frac{q_z}{1 - \dfrac{k_0}{a_2 - k_0}\, q} \tag{5-23b}$$

with q given in Eq. (5-21). Observe that $Q > 0$ with $a_2 > k_0$ requires $qk_0/(a_2 - k_0) < 1$ or, with Eq. (5-16),

$$\frac{1}{K} > 1 - \frac{1}{A_o} - a_2 \frac{q_p}{q_z} \approx 1 - a_2 \frac{q_p}{q_z} \tag{5-24}$$

We note that the pole frequency ω_0 is identical to the pole frequency ω_1 of the passive RC network and is, therefore, determined only by passive components. Because ω_1 must be of the form $\omega_1 = 1/\sqrt{R_1R_2C_1C_2}$, it follows that, in addition to $S_{A_0}^{\omega_0} = 0$ nominally, the passive sensitivities have their theoretical minimum value 0.5. All good active filter sections should be designed to have these properties. At this point the student should refer back to Example 3-1 and the related comments!

The second-order RC circuit for the EPF network in Fig. 5-5b also must be of the form of Eq. (5-14), i.e.,

$$T_{cg}(s) = \frac{N_{cq}}{D_1(s)} = \frac{a_2 s^2 + a_1 s + a_0}{s^2 + s\omega_1/q_p + \omega_1^2} \tag{5-25}$$

From Eq. (5-12) the system polynomial is given by

$$D(s) = \frac{s^3}{\omega_t} + s^2\left(k_1 - a_2 + \frac{\omega_1}{\omega_t}\frac{1}{q_p}\right) + s\left(k_1\frac{\omega_1}{q_p} - a_1 + \frac{\omega_1^2}{\omega_t}\right) + k_1\omega_1^2 - a_0 \tag{5-26}$$

where we have defined

$$k_1 = \frac{1}{K} + \frac{1}{A_0} \tag{5-27}$$

For $\omega_t \to \infty$, the pole frequency ω_0 is obtained from

$$\omega_0^2 = \frac{a_0 - k_1\omega_1^2}{a_2 - k_1} \tag{5-28}$$

which leads to the same conclusions as for the ENF configuration: ω_0 can be made independent of A_0 by choosing $T_{cg}(s)$ as a *bandpass* function; i.e.; with $a_0 = a_2 = 0$,

$$T_{cg}(s) = \frac{(\omega_1/q_z)s}{s^2 + s\omega_1/q_p + \omega_1^2} \tag{5-29}$$

so that

$$\omega_0 = \omega_1 \tag{5-30a}$$

and

$$Q = q_z \frac{1}{q - (1 - k_1)/k_1} \tag{5-30b}$$

Note that for stability we require $Q > 0$, i.e., $q > (1 - k_1)/k_1$, or, with Eq. (5-27),

$$\frac{1}{K} + \frac{1}{A_0} > \frac{q_p}{q_z} \tag{5-31}$$

Alternatively, we can choose again $a_0 = a_2\omega_1^2$; i.e., $T_{cg}(s)$ must be of the form

$$T_{cg}(s) = a_2 \frac{s^2 + s\omega_1/q_z + \omega_1^2}{s^2 + s\omega_1/q_p + \omega_1^2} \tag{5-32}$$

in which case

$$\omega_0 = \omega_1 \tag{5-33a}$$

and

$$Q = q_z \frac{1}{1 - \frac{k_1}{a_2 - k_1}q} \tag{5-33b}$$

with q given in Eq. (5-21). In the latter case, $Q > 0$ requires, with Eq. (5-27), that the inequality

$$\frac{1}{K} + \frac{1}{A_0} < a_2 \frac{q_z}{q_p} \tag{5-34}$$

TABLE 5-1 SUMMARY OF SINGLE-AMPLIFIER BIQUAD DESIGNS RESULTING IN $S_A^{\omega_0} = 0$

Passive RC	Active circuit			
	ENF (Fig. 5-5a) $k = (K-1)/K$ $K < \dfrac{1}{1 - a_2 q_p/q_z}$	EPF (Fig. 5-5b) $k = 1/K$ $K < q_z/q_p$	NF (Fig. 5-6a) $k = 0,\ K = 1$	PF (Fig. 5-6b) $k = 1,\ K = 1$
$T_{cb} = a_2\,\dfrac{s^2 + s\dfrac{\omega_1}{q_z} + \omega_1^2}{s^2 + s\dfrac{\omega_1}{q_p} + \omega_1^2}$	$Q = \dfrac{q_z}{1 - \dfrac{k}{a_2 - k^q}}$	—	$Q = q_z$	—
$T_{cg} = \dfrac{(\omega_1/q_z)\,s}{s^2 + s\dfrac{\omega_1}{q_p} + \omega_1^2}$	—	$Q = \dfrac{q_z}{q - \dfrac{1-k}{k}}$	—	$Q = \dfrac{q_z}{q}$

Note: $\omega_0 = \omega_1$ for all cases; $q = q_z/q_p - 1$; $A_0 = \infty$ is assumed.

be satisfied.   This condition poses an unobvious problem: during power turn-on, A_0 increases from zero in some manner to a large value so that Eq. (5-34) will temporarily not be satisfied and the circuit may break into oscillations.   We shall, therefore, not consider this case any further; i.e., only bandpass-type RC networks will be permitted in EPF circuits.

Having observed that our designs lead to ω_0 being independent of the dc amplifier gain A_0 and of the positive feedback term K, we note from Eqs. (5-23b) and (5-30b) that Q is a function of *both* parameters and that the finite amplifier gains $1/k_0$ or $1/k_1$, respectively, serve to *enhance* Q above the value set by the passive RC circuit.   Note especially that in the ENF circuit we may choose $K = 1$ (the NF circuit of Fig. 5-6a) provided that the RC network has *complex* zeros ($q_z > 0.5$).   In that case $k_o = -1/A_0 \approx 0$ and $Q = q_z$.   Similarly, if we set $K = 1$ in the EPF circuit (the PF circuit of Fig. 5-6b with $k_1 = 1 + 1/A_0 \approx 1$), Q equals q_z/q.   Table 5-1 summarizes these results for the single-amplifier second-order filters designed such that the pole frequency is independent of the dc op amp gain.

In Section 5.2.3, we shall look in greater detail at the Q sensitivities and at ways to optimize the performance.   In Section 5.2.4 we shall look briefly at the effect of *finite* ω_t on the position of the filter poles.

5.2.3 Sensitivity Considerations and Optimization

We saw in Section 5.2.2 that the pole frequency ω_0 of (E)NF and (E)PF circuits satisfies

$$S_{A_0}^{\omega_0} = 0, \quad S_K^{\omega_0} = 0, \quad \text{and} \quad |S_x^{\omega_0}| = \frac{1}{2} \tag{5-35}$$

if the passive RC network is chosen to realize Eq. (5-22) or (5-29), respectively, and if the op amp gain A can be assumed constant, i.e., $A = A_0$.   In Eq. (5-35), x stands for any of the elements in the passive RC network.   To determine the corresponding Q sensitivities, let us initially consider the EPF configuration, Eq. (5-30b).   It is easy to calculate the passive sensitivities

$$S_{k_1}^Q = -\frac{1}{k_1}\frac{Q}{q_z} = -\frac{1}{k_1}\frac{1}{1+q}\frac{Q}{q_p} > -\frac{1}{k_1}\frac{1}{1+q}2Q \tag{5-36a}$$

$$S_x^Q = S_x^{q_z} - q\frac{Q}{q_z}S_x^q = S_x^{q_z} - \frac{q}{1+q}\frac{Q}{q_p}S_x^q \tag{5-36b}$$

where we have used Eq. (3-19) and the fact that $q_p < 0.5$.   Similarly, the gain-sensitivity product, defined in Eqs. (3-34) to (3-35), becomes

$$\Gamma_{A_0}^Q = A_0 S_{A_0}^Q = A_0 S_{k_1}^Q S_{A_0}^{k_1} = \frac{1}{k_1^2}\frac{Q}{q_z} = \frac{1}{k_1^2}\frac{1}{1+q}\frac{Q}{q_p} > \frac{1}{k_1^2}\frac{2Q}{1+q} \tag{5-36c}$$

The ratio $q_z/q_p = 1 + q$ realized by the passive RC network is a free design parameter that can be chosen to minimize the sensitivities.   To do this, we have to remember that k_1 and q are related by Eq. (5-30b), i.e.,

$$k_1 = \frac{1}{1 + q - q_z/Q} = \frac{1}{1 + q}\frac{1}{1 - q_p/Q} \tag{5-37}$$

With this equation, we obtain from Eqs. (5-36a) and (5-36c)

$$S_{k_1}^Q = 1 - (1 + q)\frac{Q}{q_z} = -\left(\frac{Q}{q_p} - 1\right) < -(2Q - 1) \tag{5-36d}$$

$$\Gamma_{A_0}^Q = (1 + q)\left(\frac{Q}{q_p} - 2 + \frac{q_p}{Q}\right) \approx (1 + q)\frac{Q}{q_p} = \frac{Q}{q_z} > (1 + q)2Q \tag{5-36e}$$

The approximation assumes $Q \gg q_p$. Note that both these sensitivities can become large! To be able to say anything definite about the remaining sensitivities in Eq. (5-36b), we need to look at specific cases because of the unknown passive sensitivities of q_z and q. We may only note that $\Gamma_{A_0}^Q$ increases with increasing values of q and that, from Eq. (5-37) with Eq. (5-27) and $|A| \gg 1$, q must satisfy the constraint

$$q \geq \frac{q_z}{Q} \tag{5-38}$$

because $K > 1$. The following example will illustrate the design and the trade-offs.

Example 5-2

Consider the "dead" pole-producing EPF circuit in Fig. 5-7a. Apply an input such that the circuit realizes a highpass with a gain of value 1 at the pole frequency. Determine the elements to realize $Q = 5$ and $\omega_0 = 6.28$ krad/s. Choose any available design parameters such that sensitivities are reduced and the element spread is reasonable. Assume for your calculations that $|A| \gg 1$ in the frequency range of interest.

Solution. To determine the poles that are realized by the circuit, we need the RC transfer function $T_{cg}(s)$; elementary analysis of Fig. 5-7 yields

$$T_{cg}(s) = \frac{V_c}{V_g} = \frac{sG_2/C_1}{s^2 + s\left(\frac{G_1 + G_2}{C_1} + \frac{G_1}{C_2}\right) + \frac{G_1G_2}{C_1C_2}} = \frac{s\omega_1/q_z}{s^2 + s\dfrac{\omega_1}{q_p} + \omega_1^2}$$

For the following discussion, we introduce the notation

$$C_1 = cC_2 \quad \text{and} \quad G_2 = gG_1$$

Evidently, we then have

$$\omega_1^2 = \frac{g}{c}\left(\frac{G_1}{C_2}\right)^2$$

$$q_z = \sqrt{\frac{c}{g}}$$

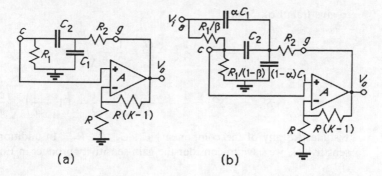

Figure 5-7   (a) Pole-producing circuit for EPF circuit; (b) creating transmission zeros.

and

$$\frac{q_z}{q_p} = 1 + q = 1 + \frac{1 + c}{g}$$

Q and ω_0 of the active circuit are determined via Eq. (5-12), $T_{cg}(s) - 1/K = 0$; i.e.,

$$s^2 + s \frac{1}{c}\frac{G_1}{C_2}[1 + c - g(K - 1)] + \frac{g}{c}\left(\frac{G_1}{C_2}\right)^2 = s^2 + s \frac{\omega_0}{Q} + \omega_0^2 = 0$$

Thus, as expected, we find $\omega_0 = \omega_1$ with

$$S_A^{\omega_0} = S_K^{\omega_0} = 0 \quad \text{and} \quad |S_x^{\omega_0}| = 0.5$$

Further, as in Table 5-1,

$$Q = \frac{\sqrt{cg}}{1 + c - g(1 - K)} = \frac{q_z}{q + 1 - K} \tag{5-39}$$

We have a total of three parameters, c, g, and K, related by Eq. (5-39), that must be chosen to get the prescribed value of Q. To make use of the available freedom, we calculate

$$S_{C_1}^q = -S_{C_2}^q = \frac{c}{gq}$$

$$S_{G_2}^q = -S_{G_1}^q = -\frac{1 + c}{gq}$$

and

$$|S_x^{q_z}| = \frac{1}{2}$$

to give from Eq. (5-36b)

$$S_{C_1}^Q = -S_{C_2}^Q = \frac{1}{2} - \frac{c}{g}\frac{Q}{q_z} = \frac{1}{2} - \sqrt{\frac{c}{g}}\, Q \tag{5-40a}$$

$$S_{G_1}^Q = -S_{G_2}^Q = \frac{1}{2} - \frac{1+c}{g}\frac{Q}{q_z} = \frac{1}{2} - \left(1 + \frac{1}{c}\right)\sqrt{\frac{c}{g}}\, Q \tag{5-40b}$$

x stands for any of the components C_1, C_2, R_1, R_2. In addition to these passive sensitivities, we have to consider the gain-sensitivity product in Eq. (5-36e),

$$\Gamma_{A_0}^Q = \left(1 + \frac{1+c}{g}\right)\left(\frac{Q}{q_p} - 2 + \frac{q_p}{Q}\right)$$

$$= \frac{1+c+g}{g}\left[\frac{Q}{\sqrt{cg}}(1+c+g) - 2 + \frac{\sqrt{cg}}{Q}\frac{1}{1+c+g}\right] \tag{5-40c}$$

$$\simeq \left(1 + \frac{1+c}{g}\right)^2 \sqrt{\frac{g}{c}}\, Q$$

The approximation is valid as long as $Q \gg q_p$. The two degrees of feeedom, c and g, can now be chosen to minimize sensitivities. Let us simplify matters by assuming c to be fixed. It follows that the passive sensitivities decrease with increasing g but that $\Gamma_{A_0}^Q$ goes through a minimum that can be found by setting the derivative of Eq. (5-40c) with respect to g to zero. Simple analysis for $Q \gg q_p$ finds this minimum to be

$$\Gamma_{A_0}^Q \simeq \frac{16}{3\sqrt{3}}\, Q \sqrt{\frac{1+c}{c}} \tag{5-41a}$$

at[4]

$$g_{min} \simeq 3(1+c) \tag{5-42}$$

For this value of g, the passive sensitivities are

$$S_{C_1}^Q = -S_{C_2}^Q = \frac{1}{2} - \frac{Q}{\sqrt{3}}\sqrt{\frac{c}{1+c}} \tag{5-41b}$$

$$S_{G_1}^Q = -S_{G_2}^Q = \frac{1}{2} - \frac{Q}{\sqrt{3}}\sqrt{\frac{1+c}{c}} \tag{5-41c}$$

[4]The exact value is

$$g_{min} = 3(1+c) + \frac{1}{2Q^2}\left[1 + \sqrt{1 + 12(1+c)Q^2}\right]$$

The choice of c is seen to affect the element spread and a trade-off between active and passive sensitivities. Let us simply choose $c = 1$ to yield, with $Q = 5$,

$$\Gamma^Q_{A_0} = 21.8 \qquad |S^Q_{C_i}| = 1.5 \qquad |S^Q_{G_i}| = 3.6$$

and, from Eq. (5-39),

$$K = 1 + \frac{1 + c}{g} - \sqrt{\frac{c}{g}\frac{1}{Q}} = 1.2517$$

Choosing $C_1 = C_2 = 5$ nF, we find further

$$R_1 = \frac{\sqrt{g}}{\omega_0 C_2} = 78 \text{ k}\Omega \quad \text{and} \quad R_2 = \frac{R_1}{6} = 13 \text{ k}\Omega$$

Finally, in order to create transmission zeros, we have to lift one or more of the grounded elements $[R_1, C_1, (K - 1)R]$ partially or completely off ground and connect them to the input, as shown in Fig. 5-2, such that the prescribed highpass filter is realized. The resulting transfer function will be of the form of Eq. (5-3). A little thought will convince the reader that feeding the input signal into a part of R will result in a fully biquadratic filter function, which is not useful for our example. Consequently, we lift off only the fractions αC_1 and βG_1 as shown in Fig. 5-7b. Analyzing the resulting RC network yields

$$T_{ca}(s) = \frac{V_c}{V_i} = \frac{\alpha s^2 + \beta \left[sG_1 \left(\frac{1}{C_1} + \frac{1}{C_2} \right) + \frac{G_1 G_2}{C_1 C_2} \right]}{s^2 + s \left(\frac{G_1 + G_2}{C_1} + \frac{G_1}{C_2} \right) + \frac{G_1 G_2}{C_1 C_2}}$$

Because we wish to realize a highpass function, we set $\beta = 0$ to obtain, with Eq. (5-3),

$$H(s) = \frac{V_0}{V_i} = K \frac{\alpha s^2}{s^2 + s \left(\frac{G_1 + G_2}{C_1} + \frac{G_1}{C_2} - K\frac{G_2}{C_1} \right) + \frac{G_1 G_2}{C_1 C_2}}$$

At the pole frequency we have required

$$|H(j\omega_1)| = \alpha KQ = 1$$

i.e.,

$$\alpha = \frac{1}{KQ} = 0.1598$$

and $\alpha C_1 = 800$ pF. Note that we may save one capacitor by setting $\alpha = 1$; in that case the pole-frequency gain equals $KQ = 6.259$.

It is also of interest to observe the effect of setting $K = 1$ in the EPF circuit. Using a unity-gain op amp configuration and the PF circuit in Fig. 5-6b gives, from

Table 5-1, $Q = q_z/q$. With the notation of Example 5-2, this results in

$$Q = \sqrt{\frac{g}{c} \frac{c}{1 + c}}$$

and, with Eq. (5-40c), in

$$\Gamma_{A_0}^Q \simeq \frac{g}{1 + c} \left(1 + \frac{1 + c}{g}\right)^2 = \frac{1 + c}{c} Q^2 \left(1 + \frac{c}{1 + c} \frac{1}{Q^2}\right) \simeq \frac{1 + c}{c} Q^2 \geq Q^2$$

The resistor spread becomes

$$g = \frac{(1 + c)^2}{c} Q^2 > 4Q^2$$

Finally, the passive sensitivities can be estimated from Eqs. (5-40a) and (5-40b) to be of the order of 0.5. We note, therefore, that at least for high-Q circuits the PF configuration is not appropriate, because it results in unacceptably high values of the gain-sensitivity product in addition to an inconveniently large component spread.

We turn now to the ENF circuit to investigate possible optimization of Q sensitivities. From Eq. (5-23b) the student will find it easy to verify the following sensitivity expressions:

$$S_{k_0}^Q = \frac{a_2}{a_2 - k_0} \left(\frac{Q}{q_z} - 1\right) \qquad \Gamma_{A_0}^Q = \frac{1}{k_0} S_{k_0}^Q = \frac{1}{k_0} \frac{a_2}{a_2 - k_0} \left(\frac{Q}{q_z} - 1\right)$$

$$S_x^Q = S_x^{q_z} + \frac{Q}{q_z} \frac{k_0 q}{a_2 - k_0} S_x^q \tag{5-43}$$

Before we can interpret these equations, we have to remember that k_0 and q are related by Eq. (5-23b),

$$k_0 = \frac{a_2}{1 + q/(1 - q_z/Q)} \tag{5-44}$$

where again $q > 1$ is a free design parameter. With Eq. (5-44), the sensitivities in Eq. (5-43) can be written as follows:

$$S_k^Q = \left(1 - \frac{q_p}{Q}\right)\left(\frac{Q/q_p - 1}{q} - 1\right) \tag{5-45a}$$

$$a_2 \Gamma_{A_0}^Q = \frac{Q}{q_p} \left(1 - \frac{q_p}{Q}\right)^2 \frac{1 + q}{q} \tag{5-45b}$$

$$S_x^Q = S_x^{q_z} + \left(\frac{Q}{q_z} - 1\right) S_x^q \tag{5-45c}$$

In general, q_z, $S_x^{q_p}$, and S_x^q are functions of q which must be known before an optimal value of q can be determined. Example 5-3, below, illustrates the procedure.

Frequently, the passive sensitivities decrease monotonically to zero with increasing values of q but the active sensitivities have a minimum which can be found by setting the derivative of Eq. (5-45b) to zero (compare Example 5-2). For the normally realistic assumption $Q \gg q_p$, we obtain, from $(d/dq)(a_2 \Gamma_{A_0}^Q) = 0$,

$$\frac{dq_p}{dq} \simeq -\frac{q_p}{q} \frac{1}{1+q} \tag{5-46}$$

If this equation has a positive solution for q, it will be the location of the minimum of $\Gamma_{A_0}^Q$. The value of q determines the amount of positive feedback applied to the ENF circuit: assuming $A_0 \gg 1$, we find from Eqs. (5-16) and (5-44) that

$$K = \frac{1 + 1/q - q_z/(qQ)}{1 + \dfrac{1}{q}(1 - a_2)\left(1 - \dfrac{q_z}{Q}\right)} \tag{5-47}$$

We note that because $K \geq 1$, Eq. (5-47) places an upper limit on the possible values of q. For instance, many (but not all) biquadratic passive RC circuits used for ENF realizations implement $a_2 = 1$; for that case, Eq. (5-47) with $K \geq 1$ results in

$$Q \geq q_z = q_p(1 + q)$$

i.e.,

$$q \leq \frac{Q}{q_p} - 1 \tag{5-48}$$

Example 5-3

Design the ENF circuit in Fig. 5-8 such that it realizes a bandpass function with $f_0 = 7.5$ kHz, $Q = 12.3$, and midband gain $H_m = 25$. Determine any free parameters such that the circuit operates at a point of low sensitivity.

Note: The reader can easily verify that the circuit in Fig. 5-8 is complementary to the one in Fig. 5-7; thus it is instructive to compare the following development and the equations with those in Example 5-2!

Solution. Rather than finding the transfer function by direct analysis, let us make use of Eq. (5-3) and the results obtained in Example 5-1. Obviously, $T_{da}(s) = 0$ and $T_{db}(s) = (K - 1)/K$; $T_{cb}(s)$ was given in Example 5-1:

$$T_{cb}(s) = \frac{V_c}{V_b} = \frac{N_{cb}}{D_1} = \frac{s^2 + 2s\dfrac{G_1}{C} + \dfrac{G_1 G_2}{C^2}}{s^2 + s\dfrac{2G_1 + G_2}{C} + \dfrac{G_1 G_2}{C^2}}$$

Further, by simple analysis,

$$T_{ca}(s) = \frac{V_c}{V_i} = \frac{N_{ca}}{D_1} = \frac{s\alpha G_2/C}{s^2 + s\dfrac{2G_1 + G_2}{C} + \dfrac{G_1 G_2}{C^2}}$$

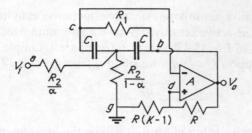

Figure 5-8 ENF bandpass circuit.

Therefore,

$$H(s) = \frac{V_0}{V_i} = \frac{-N_{ca}(s)}{N_{cb}(s) - \left(\frac{K-1}{K} - \frac{1}{A}\right)D_1(s)}$$

$$= -\frac{1}{1-k} \cdot \frac{s\alpha G_2/C}{s^2 + s\frac{G_1}{C}\left(2 - \frac{k_0}{1-k_0}\frac{G_2}{G_1}\right) + \frac{G_1 G_2}{C^2}}$$

$$(5\text{-}49)$$

where k was defined in Eq. (5-16). If we let $G_2 = gG_1$ and assume for now $|A| \gg 1$, we have

$$\frac{1}{1-k_0} = K \qquad \frac{k_0}{1-k_0} = K - 1 \qquad q_z = \frac{\sqrt{g}}{2} \qquad q_p = \frac{\sqrt{g}}{2+g} \qquad q = \frac{g}{2}$$

and, by Eq. (5-23) with $a_2 = 1$,

$$\omega_0 = \frac{\sqrt{g}}{R_1 C} \qquad Q = \frac{\sqrt{g}}{2 - g(K-1)}$$

The design parameter q, defined in Eq. (5-21), is restricted by Eq. (5-48):

$$q = \frac{g}{2} < \frac{Q}{\sqrt{g}}(2+g) - 1$$

i.e.,

$$g < 4Q^2 = 605.16$$

From Eqs. (5-45b) and (5-45c), the sensitivity expressions are

$$\Gamma_{A_0}^Q = \frac{2+g}{g}\left[\frac{Q(2+g)}{\sqrt{g}} - 2 + \frac{\sqrt{g}}{Q(2+g)}\right] \qquad |S_x^Q| = \frac{2Q}{\sqrt{g}} - \frac{1}{2} \quad (5\text{-}50)$$

The student should verify with the help of Eq. (5-46) that the gain-sensitivity product has a minimum of value $\simeq 51$ at $g_{min} \simeq 6$ [compare Eqs. (5-41a) and (5-42); note that here $c = 1$]. Figure 5-9 contains the corresponding plots of Eq. (5-50), from which it is evident that a choice of g to the left of this minimum would be quite inappropriate because both active and passive sensitivities increase rapidly. We see that $\Gamma_{A_0}^Q$ rises only slowly for $g > 6$, whereas the value of $|S_x^Q|$ still falls quite fast. Thus it is common practice to choose g somewhat larger than the value resulting in the minimum of Γ_x^Q. For our example, let us pick $g = 10$; this choice leads to

$$\Gamma_{A_0}^Q \simeq 53.6 \qquad |S_x^Q| \simeq 7.3$$

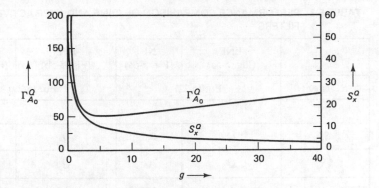

Figure 5-9 Sensitivity plots.

and, with $C = 5$ nF and $a_2 = 1$ from Eq. (5-47),

$$K = 1.1743 \qquad R_1 = \frac{\sqrt{g}}{\omega_0 C} = 13.42 \text{ k}\Omega \qquad R_2 = \frac{R_1}{g} = 1.342 \text{ k}\Omega$$

Finally, from Eq. (5-49),

$$|H_m| = \alpha \frac{gK}{2 - (K - 1)g} = \alpha\sqrt{g}KQ = 45.676\alpha = 25$$

i.e., $\alpha = 0.5473$.

Again, it is interesting to compare the ENF design with an NF design that applies no positive feedback (Fig. 5-6a). For $K = 1$ we obtain from Eq. (5-47)

$$g = g_{\max} = 4Q^2$$

and from Eq. (5-50)

$$|S_x^Q| \simeq \frac{1}{2} \qquad \Gamma_{A_0}^Q \simeq 2Q^2$$

Quite apart from the large element spread, such high values of gain-sensitivity product are usually unacceptable, especially in view of the fact that the deviations caused by the finite op amp bandwidth are also proportional to $\Gamma_{A_0}^Q$ as the next section will demonstrate. Table 5-2 gives a comparison of the relevant data for the different designs. For reference purposes we have also indicated the ω_0 and Q deviations caused by the op amp's finite bandwidth. These results will be derived in the following section.

5.2.4 The Effects of Finite Op Amp Bandwidth

In our previous discussion of transfer function poles, we assumed that the op amp gain $A(s)$ is independent of frequency; i.e., we set $A(s) = A_0 = \text{const}$ so that the system poles are determined from the second-order polynomial [Eq. (5-17)]. On

TABLE 5-2 PERFORMANCE COMPARISON OF (E)NF AND (E)PF ACTIVE FILTERS

	ENF (Fig. 5-5a)	NF (Fig. 5-6a)	EPF (Fig. 5-5b)	PF (Fig. 5-6b)
RC circuit	Eq. (5-22)		Eq. (5-29)	
$S_x^{\omega_0}$	0.5			
S_x^Q	Q	0.5	Q	0.5
$\Gamma_{A_0}^Q$	Q	$2Q^2$	Q	$2Q^2$
Element spread	independent of Q	$4Q^2$	independent of Q	$4Q^2$
$\dfrac{\Delta\omega_0}{\omega_0}$	$-\dfrac{1}{2Q}\Gamma_{A_0}^Q\dfrac{1}{\|A(j\omega_0)\|}$			
$\dfrac{\Delta Q}{Q_0}$	$\dfrac{1}{2Q}\Gamma_{A_0}^Q\dfrac{1}{\|A(j\omega_0)\|}$			

Note: x is any component in the passive *RC* circuit; orders of magnitude only are indicated.

the other hand, we pointed out in Chapter 4 that the frequency dependence of $A(s)$ is likely to cause noticeable deviations of the transfer function characteristics. The designer should, therefore, investigate what effect, if any, the frequency-dependent op amp gain has on the pole positions of the biquad. In this section we shall discuss in detail for the ENF circuit how the pole deviations can be determined. For EPF circuits or for any other configuration, the analysis proceeds along the same lines.

For finite ω_t, the system equation is given by Eq. (5-15) with

$$k_0 = \frac{K-1}{K} \tag{5-51}$$

and $a_0 = a_2\omega_1^2$; i.e., the poles are determined by the third-order polynomial

$$
\begin{aligned}
D(s) &= \frac{s^3}{\omega_t} + \left[(a_2 - k_0) + \frac{\omega_1}{\omega_t}\frac{1}{q_p}\right]s^2 \\
&\quad + \left[\frac{1}{q_z}\left(a_2 - k_0\frac{q_z}{q_p}\right) + \frac{\omega_1}{\omega_t}\right]\omega_1 s + \omega_1^2(a_2 - k_0) \\
&= \frac{s^3}{\omega_t} + (a_2 - k_0)\left[\left(1 + \frac{\omega_1/(q_p\omega_t)}{a_2 - k_0}\right)s^2 \right.\\
&\quad \left. + \frac{\omega_1}{q_z}\left(\frac{a_2 - k_0\dfrac{q_z}{q_p} + q_z\dfrac{\omega_1}{\omega_t}}{a_2 - k_0}\right)s + \omega_1^2\right]
\end{aligned} \tag{5-52}
$$

In writing Eq. (5-52), we assumed that $A_0 = \infty$ or, equivalently, that the op amp's 3-dB frequency σ is zero. We observe that due to the frequency dependence of $A(s)$ a parasitic pole was created, and that, as before, the nominal pole frequency and pole quality factor ω_0 and Q are given by Eq. (5-23) for $\omega_t \to \infty$. If we use this notation, Eq. (5-52) can be brought into a more meaningful form,

$$D(s) = \frac{s^3}{\omega_t} + (a_2 - k_0)\left[\left(1 + \frac{\omega_0/(q_p\omega_t)}{a_2 - k_0}\right)s^2 + \left(\frac{\omega_0}{Q} + \frac{\omega_0^2/\omega_t}{a_2 - k_0}\right)s + \omega_0^2\right] \quad (5\text{-}53)$$

which can be factored into its three roots: a distant negative real root at, say, $-\gamma\omega_t$ and a pair of dominant complex roots which determine the *actual* pole frequency and pole quality factor, ω_a and Q_a, respectively. Thus, we can formally factor Eq. (5-53):

$$D(s) = \left(\frac{s}{\omega_t} + \gamma\right)\left(s^2 + s\frac{\omega_a}{Q_a} + \omega_a^2\right)$$

$$= \frac{s^3}{\omega_t} + s^2\left(\gamma + \frac{\omega_a/\omega_t}{Q_a}\right) + s\left(\gamma\frac{\omega_a}{Q_a} + \frac{\omega_a^2}{\omega_t}\right) + \gamma\omega_a^2 \quad (5\text{-}54)$$

so that a comparison of Eqs. (5-53) and (5-54) permits the following equations to be written for the unknown quantities γ, ω_a, and Q_a:

$$\gamma = (a_2 - k_0)\left(\frac{\omega_0}{\omega_a}\right)^2 \quad (5\text{-}55a)$$

$$\gamma\frac{\omega_a}{Q_a} + \frac{\omega_a^2}{\omega_t} = (a_2 - k_0)\frac{\omega_0}{Q} + \frac{\omega_0^2}{\omega_t} \quad (5\text{-}55b)$$

$$\gamma + \frac{\omega_a}{\omega_t Q_a} = (a_2 - k_0) + \frac{\omega_0}{\omega_t q_p} \quad (5\text{-}55c)$$

Inserting γ from Eq. (5-55a) into Eq. (5-55b) results in

$$(a_2 - k_0)\omega_0^2\left(\frac{1}{\omega_a Q_a} - \frac{2}{\omega_0 Q}\right) = \frac{1}{\omega_t}(\omega_0^2 - \omega_a^2)$$

After setting $\omega_a = \omega_0 + \Delta\omega_0$ with the error $\Delta\omega_0$ assumed small, i.e., $\Delta\omega_0/\omega_0 \ll 1$, we obtain from this equation

$$\frac{\Delta\omega_0}{\omega_0}\left(2 - \frac{\omega_t}{\omega_0}\frac{a_2 - k_0}{Q_a}\right) \simeq \frac{\omega_t}{\omega_0}(a_2 - k_0)\left(\frac{1}{Q} - \frac{1}{Q_a}\right) \quad (5\text{-}56a)$$

Similarly, substituting γ from Eq. (5-55a) in Eq. (5-55c) leads to

$$(a_2 - k_0)\left[\left(\frac{\omega_0}{\omega_a}\right)^2 - 1\right] = \frac{\omega_0}{\omega_t}\left(\frac{1}{q_p} - \frac{\omega_a}{\omega_0}\frac{1}{Q_a}\right)$$

or

$$\frac{\Delta\omega_0}{\omega_0}\left[\frac{1}{Q_a} - 2(a_2 - k_0)\frac{\omega_t}{\omega_0}\right] \simeq \frac{1}{q_p} - \frac{1}{Q_a} \tag{5-56b}$$

For reasonable filter parameters, we can assume that $1/Q_a \ll 2(a_2 - k_0) \cdot \omega_t/\omega_0$ and $Q_a \gg q_p$ so that Eq. (5-56b) gives the following approximate expression for the pole frequency error:

$$\frac{\Delta\omega_0}{\omega_0} \simeq -\frac{1}{2(a_2 - k_0)}\frac{\omega_0}{\omega_t}\left(\frac{1}{q_p} - \frac{1}{Q_a}\right) \simeq -\frac{1}{2(a_2 - k_0)}\frac{1}{q_p}\frac{1}{|A(j\omega_0)|} \tag{5-57}$$

Further, making use of Eqs. (5-43) and (5-44), we can bring Eq. (5-57) into its final form:

$$\frac{\Delta\omega_0}{\omega_0} \simeq -\frac{1}{2Q}\Gamma_{A_0}^Q\frac{1}{|A(j\omega_0)|} \tag{5-58}$$

The Q error can be obtained from Eq. (5-56a) by substituting Eq. (5-58) for $\Delta\omega_0/\omega_0$ and setting $\Delta Q = Q_a - Q$ with $\Delta Q/Q \ll 1$:

$$\frac{\Delta Q}{Q} \simeq \frac{1}{2Q}\Gamma_{A_0}^Q\frac{1}{|A(j\omega_0)|}\left[1 - \frac{Q}{|A(j\omega_0)|(a_2 - k_0)}\right] \simeq \frac{1}{2Q}\Gamma_{A_0}^Q\frac{1}{|A(j\omega_0)|} \tag{5-59}$$

These two expressions show that the pole frequency and pole Q errors, as could be expected, are inversely proportional to the op amp gain at the frequency of interest and, moreover, that they both increase with increasing gain-sensitivity product! This insight serves as a further justification of our earlier efforts aimed at minimizing the gain-sensitivity product.

Example 5-4

The highpass and bandpass filters of Examples 5-2 and 5-3 are designed with op amps having $\omega_t/(2\pi) = 1.5$ MHz. Estimate the ω_0 and Q errors caused by the op amps' finite gain-bandwidth product. What would the deviations be for the corresponding PF and NF designs with $K = 1$?

Solution. In the highpass filter of Example 5-2, we had $\omega_0 = 6.28$ krad/s, $Q = 5$, and $\Gamma_{A_0}^Q = 21.8$; with these numbers we obtain, from Eqs. (5-58) and (5-59),

$$\frac{\Delta\omega_0}{\omega_0} \simeq -\frac{\Delta Q}{Q} \simeq -\frac{1}{2Q}\Gamma_{A_0}^Q\frac{1}{|A(j\omega_0)|} = -2.18\frac{6.28 \cdot 10^3}{2\pi \cdot 1.5 \cdot 10^6} = -1.44 \cdot 10^{-3} = -0.144\%$$

If a unity-gain amplifier is used ($K = 1$), we have $\Gamma_{A_0}^Q = 2Q^2 = 50$ and

$$\frac{\Delta\omega_0}{\omega_0} \simeq -\frac{\Delta Q}{Q} \simeq -0.33\%$$

a value 2.3 times larger.

In the bandpass circuit of Example 5-3 with $f_0 = 7.5$ kHz and $Q = 12.3$, we find analogously

$$\frac{\Delta\omega_0}{\omega_0} \simeq -\frac{\Delta Q}{Q} \simeq -\frac{1}{2Q}\Gamma_{A_0}^Q \frac{1}{|A(j\omega_0)|} = -\frac{53.6}{24.6}\frac{7.5}{1.5 \cdot 10^3} = -1.1\%$$

and for $K = 1$, with $\Gamma_{A_0}^Q = 302$,

$$\frac{\Delta\omega_0}{\omega_0} \simeq -\frac{\Delta Q}{Q} \simeq -6.1\%$$

a value 5.6 times larger. Note that this error in center frequency of -457 Hz is 75% of the total 3-dB bandwidth of the bandpass filter! An error of this magnitude would be entirely unacceptable in practice. Using a small amount of positive feedback ($K = 1.1743$ instead of $K = 1$) reduced the center frequency error to $\simeq 13\%$ of the bandwidth.

5.2.5 Passive RC Twoports

The information obtained so far in Section 5.2 about the design of good low-sensitivity second-order active filters is summarized in Tables 5-1 and 5-2. Note that these results were derived without any concern about the specifics of the passive *RC* section to be used in the design. We only noted that the *RC* network should realize a transfer function of the form of Eq. (5-14) or (5-25),

$$T(s) = \frac{N(s)}{D_1(s)} = \frac{a_2 s^2 + a_1 s + a_0}{s^2 + s\omega_1/q_p + \omega_1^2}$$

where either $a_0 = a_2\omega_1^2$ or $a_2 = a_0 = 0$. In the following we shall briefly discuss suitable *RC* sections and the type of active filter functions that can be obtained from them by lifting previously grounded components off ground as illustrated in Fig. 5-2.

We shall start by introducing a *loaded bridged T* (LBT) twoport that is "complementary to itself"—i.e., the circuit and its complement[5] are of the same topology. We shall show that a special case of the LBT circuit forms an *RC bridged T* suitable for ENF designs and that a second special case, the *RC ladder*, is complementary to the *RC* bridged T and can be used for EPF circuits. Next, we shall demonstrate that the two *active* networks are also related by the complementary transformation introduced in Section 5.2.2 so that the resulting EPF and ENF biquads have essentially identical behavior as far as their pole sensitivities are concerned (refer to Examples 5-2 and 5-3!). Differences arise only in the type of transfer functions that can be realized when creating input nodes by lifting circuit elements off ground. Finally, before leaving the discussion of *RC* twoports, we shall, for the sake of completeness, look at another often encountered passive *RC*

---

[5]Obtained by interchanging input and ground terminals; see Section 5.2.2.

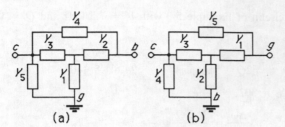

Figure 5-10 Loaded bridged T circuit (a) and its complement (b).

circuit, the *bridged twin T*. However, these latter circuits are slightly more complicated in their operation and afford the designer little additional generality.

The loaded bridged T circuit is shown in Fig. 5-10a, and its complement, in Fig. 5-10b. Simple analysis shows that the transfer functions realized are

$$T_{cb}(s) = \frac{N_{cb}(s)}{D(s)} = \frac{Y_2 Y_3 + Y_4(Y_1 + Y_2 + Y_3)}{(Y_1 + Y_2 + Y_3)(Y_4 + Y_5) + Y_3(Y_1 + Y_2)} \quad (5\text{-}60a)$$

$$T_{cg}(s) = \frac{N_{cg}(s)}{D(s)} = \frac{Y_1 Y_3 + Y_5(Y_1 + Y_2 + Y_3)}{(Y_1 + Y_2 + Y_3)(Y_4 + Y_5) + Y_3(Y_1 + Y_2)} \quad (5\text{-}60b)$$

As expected from Fig. 5-10, the two functions have the same poles, and their numerators are obtained by interchanging Y_1 with Y_2 and Y_4 with Y_5.

Figure 5-11 shows the generation of the transfer functions $T_{ca}(s)$ needed according to Eq. (5-3) to set the transmission zeros of the single-amplifier biquad. As discussed earlier, the additional input terminal a is created by lifting fractions of grounded admittances off ground. The two transfer functions $T_{ca}(s)$ are calculated as

$$T_{ca}(s) = \frac{V_c}{V_a}\bigg|_{V_b = 0} = \frac{\beta_1 Y_1 Y_3 + \beta_5 Y_5(Y_1 + Y_2 + Y_3)}{(Y_1 + Y_2 + Y_3)(Y_4 + Y_5) + Y_3(Y_1 + Y_2)} \quad (5\text{-}61a)$$

$$T_{ca}(s) = \frac{V_c}{V_a}\bigg|_{V_g = 0} = \frac{\beta_2 Y_2 Y_3 + \beta_4 Y_4(Y_1 + Y_2 + Y_3)}{(Y_1 + Y_2 + Y_3)(Y_4 + Y_5) + Y_3(Y_1 + Y_2)} \quad (5\text{-}61b)$$

for the circuits in Fig. 5-11a and b, respectively.

Let us now consider the specific form that the circuit of Fig. 5-10a must take on to realize the transfer function [Eq. (5-22)] suitable for ENF designs; since the high- and low-order coefficients of $N_{cb}(s)$ and $D(s)$, respectively, must be equal, it can be reasoned that Y_5 in Eq. (5-60) must be zero, because it affects only the coefficients of $D(s)$. With $Y_5 = 0$ we have two options for making $T_{cb}(s)$ a biquadratic transfer function:

Case A. $Y_1 = G_1$, $Y_2 = sC_2$, $Y_3 = sC_3$, and $Y_4 = G_4$, so that

$$T_{cb}(s) = \frac{s^2 C_2 C_3 + s(C_2 + C_3)G_4 + G_1 G_4}{s^2 C_2 C_3 + s[C_3(G_1 + G_4) + C_2 G_4] + G_1 G_4} \quad (5\text{-}62a)$$

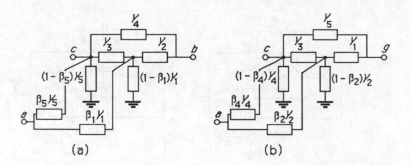

Figure 5-11 (a) The circuit of Fig. 5-10a with fractions β_1 and β_5, respectively, of the admittances, Y_1 and Y_5 lifted off ground. (b) The analogous situation for the circuit of Fig. 5-10b. *Note:* $\beta_i \leq 1$.

$$T_{ca}(s) = \frac{s\beta_1 G_1 C_3}{s^2 C_2 C_3 + s[C_3(G_1 + G_4) + C_2 G_4] + G_1 G_4} \tag{5-63a}$$

A comparison with Eq. (5-22) shows that

$$a_2 = 1 \qquad \omega_1^2 = \frac{G_1 G_4}{C_2 C_3} \qquad q_z = \frac{\sqrt{G_1/G_4}}{\sqrt{C_2/C_3} + \sqrt{C_3/C_2}}$$

$$q_p = \frac{\sqrt{G_1/G_4}}{\sqrt{C_2/C_3} + \sqrt{C_3/C_2}\,(1 + G_1/G_4)} \tag{5-64a}$$

The parameter q, defined in Eq. (5-21) is, therefore,

$$q = \frac{q_z}{q_p} - 1 = \frac{G_1/G_4}{1 + C_2/C_3} \tag{5-65a}$$

Figure 5-12a shows the circuit including possible feed-in components. The biquadratic transfer function realized is

$$H(s) = \frac{V_0}{V_i} = \alpha\,\frac{s^2 + s\left[G_4\left(\dfrac{1}{C_2} + \dfrac{1}{C_3}\right) + \dfrac{G_1}{C_2}\left(1 - \beta_1\dfrac{K}{\alpha}\right)\right] + \omega_0^2}{s^2 + s\omega_0/Q + \omega_0^2} \tag{5-66a}$$

with

$$\omega_0 = \omega_1 \qquad Q = \frac{q_z}{1 - (K - 1)q} \tag{5-67}$$

given in Eq. (5-64a) and via Eqs. (5-23b) and (5-16), respectively. The circuit can realize bandpass, allpass, and notch functions, depending on the choice of α and β_1.

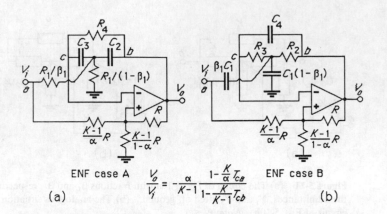

Figure 5-12 ENF circuits using bridged T RC twoports for cases A and B. The feed-in components and the realized transfer function are indicated.

Case B. $Y_1 = sC_1$, $Y_2 = G_2$, $Y_3 = G_3$, and $Y_4 = sC_4$, so that

$$T_{cb}(s) = \frac{s^2 C_1 C_4 + sC_4(G_2 + G_3) + G_2 G_3}{s^2 C_1 C_4 + s[C_4(G_2 + G_3) + C_1 G_3] + G_2 G_3} \tag{5-62b}$$

$$T_{ca}(s) = \frac{\beta_1 s C_1 G_3}{s^2 C_1 C_4 + s[C_4(G_2 + G_3) + C_1 G_3] + G_2 G_3} \tag{5-63b}$$

In this case we obtain

$$a_2 = 1 \qquad \omega_1^2 = \frac{G_2 G_3}{C_1 C_4} \qquad q_z = \frac{\sqrt{C_1/C_4}}{\sqrt{G_2/G_3} + \sqrt{G_3/G_2}}$$

$$q_p = \frac{\sqrt{C_1/C_4}}{\sqrt{G_2/G_3} + \sqrt{G_3/G_2}\,(1 + C_1/C_4)} \tag{5-64b}$$

with

$$q = \frac{q_z}{q_p} - 1 = \frac{C_1/C_4}{1 + G_2/G_3} \tag{5-65b}$$

The resulting circuit with possible feed-in components is shown in Fig. 5-12b; it realizes

$$H(s) = \frac{V_0}{V_i} = \alpha \frac{s^2 + s\left[\dfrac{G_2 + G_3}{C_1} + \dfrac{G_3}{C_4}\left(1 - \beta_1 \dfrac{K}{\alpha}\right)\right] + \omega_0^2}{s^2 + s\omega_0/Q + \omega_0^2} \tag{5-66b}$$

with ω_1 and Q given in Eq. (5-67).

Notice that the two cases are completely equivalent and either case may be taken when an RC generating network is needed to implement a biquadratic transfer function.   However, observe that case B may need a third capacitor in many situations.   An example for case A design was presented in Example 5-3.

Turning now to the second special case of *RC bridged T* network, we intend to realize a bandpass transfer function suitable for an active EPF design.   To keep with the earlier development of complementary networks, we choose Fig. 5-10b, which implements the function in Eq. (5-60b).   A bandpass function can be realized by setting $Y_5 = 0$ so that

$$T_{cg}(s) = \frac{N_{cg}(s)}{D(s)} = \frac{Y_1 Y_3}{(Y_1 + Y_2 + Y_3)Y_4 + Y_3(Y_1 + Y_2)} \qquad (5\text{-}68)$$

and, from Fig. 5-11b and Eq. (5-61b),

$$T_{ca}(s) = \frac{V_c}{V_a}\bigg|_{V_g=0} = \frac{\beta_2 Y_2 Y_3 + \beta_4 Y_4(Y_1 + Y_2 + Y_3)}{(Y_1 + Y_2 + Y_3)Y_4 + Y_3(Y_1 + Y_2)} \qquad (5\text{-}69)$$

It is apparent that there are again two possibilities for creating a bandpass-type function $T_{cg}(s)$:

Case A. $Y_1 = G_1$, $Y_2 = sC_2$, $Y_3 = sC_3$, and $Y_4 = G_4$, so that

$$T_{cg}(s) = \frac{sC_3 G_1}{s^2 C_2 C_3 + s[(C_2 + C_3)G_4 + C_3 G_1] + G_1 G_4} \qquad (5\text{-}70a)$$

$$T_{ca}(s) = \frac{V_c}{V_a}\bigg|_{V_g=0} = \frac{s^2 \beta_2 C_2 C_3 + \beta_4 G_4[s(C_2 + C_3) + G_1]}{s^2 C_2 C_3 + s[(C_2 + C_3)G_4 + C_3 G_1] + G_1 G_4} \qquad (5\text{-}71a)$$

The bandpass parameters defined in Eq. (5-29) are

$$\omega_1^2 = \frac{G_1 G_4}{C_2 C_3} \qquad q_z = \frac{\sqrt{G_4/G_1}}{\sqrt{C_3/C_2}}$$

$$q_p = \frac{\sqrt{G_4/G_1}}{\sqrt{C_3/C_2}} \frac{1}{1 + (G_4/G_1)(1 + C_2/C_3)} \qquad (5\text{-}72a)$$

The design parameter q, defined in Eq. (5-21), is

$$q = \frac{q_z}{q_p} - 1 = \frac{G_4}{G_1}\left(1 + \frac{C_2}{C_3}\right) \qquad (5\text{-}73a)$$

The circuit is shown in Fig. 5-13a.   Again we have indicated the possible feed-in components which lead to the general transfer function

$$H(s) = \frac{V_0}{V_i}$$

$$= -K \frac{s^2(\eta - \beta_2) + s\left[G_4\left(\dfrac{1}{C_2} + \dfrac{1}{C_3}\right)(\eta - \beta_4) + \eta\dfrac{G_1}{C_2}\right] + \omega_0^2(\eta - \beta_4)}{s^2 + s\omega_0/Q + \omega_0^2}$$

$$\qquad (5\text{-}74a)$$

where

$$\eta = \alpha \frac{K - 1}{K} \qquad \omega_0 = \omega_1 \qquad Q = \frac{q_z}{q + 1 - K} \tag{5-75}$$

by Eqs. (5-72a), (5-30b), and (5-27), respectively.

Case B. $Y_1 = sC_1$, $Y_2 = G_2$, $Y_3 = G_3$, and $Y_4 = sC_4$, so that

$$T_{cg}(s) = \frac{sC_1G_3}{s^2C_1C_4 + s[(G_2 + G_3)C_4 + C_1G_3] + G_2G_3} \tag{5-70b}$$

$$T_{ca}(s) = \frac{V_c}{V_a}\bigg|_{V_g=0} = \frac{\beta_4[s^2C_1C_4 + sC_4(G_2 + G_3)] + \beta_2G_2G_3}{s^2C_1C_4 + s[(G_2 + G_3)C_4 + C_1G_3] + G_2G_3} \tag{5-71b}$$

so that we obtain

$$\omega_1^2 = \frac{G_2G_3}{C_1C_4} \qquad q_z = \sqrt{\frac{G_2/G_3}{C_1/C_4}}$$

$$q_p = \frac{\sqrt{G_2/G_3}}{\sqrt{C_1/C_4}} \frac{1}{1 + (C_4/C_1)(1 + G_2/G_3)} \tag{5-72b}$$

with

$$q = \frac{q_z}{q_p} - 1 = \frac{C_4}{C_1}\left(1 + \frac{G_2}{G_3}\right) \tag{5-73b}$$

Figure 5-13b shows the circuit which realizes

$$H(s) = \frac{V_0}{V_i} = -K \frac{s^2(\eta - \beta_4) + s\left[\dfrac{G_2 + G_3}{C_1}(\eta - \beta_4) + \eta\dfrac{G_3}{C_4}\right] + \omega_0^2(\eta - \beta_2)}{s^2 + s\omega_0/Q + \omega_0^2} \tag{5-74b}$$

where the parameters are given in Eq. (5-75) with Eqs. (5-72b) and (5-73b).

Again, we note the symmetry in the equations describing cases A and B. Both functions $H(s)$ in Eqs. (5-74a) and (5-74b) can implement an arbitrary biquadratic transfer function; observe, however, that the numerators differ for the two cases and result in different types of transfer functions. We leave to the problems (Problem 5.10) for the student to determine which of the four circuits results in the most convenient realization for a given function. The design of a highpass filter, using case A, was illustrated in Example 5-2; Example 5-5, below, shows how a lowpass-filter EPF biquad can be obtained.

Finally, we want to point out that the circuits in Fig. 5-13a and b are complementary to those in Fig. 5-12a and b. Recall from Section 5.2.2 that the complementary transformation is performed after the inputs are grounded!

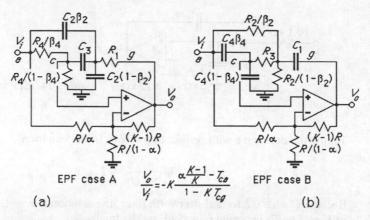

Figure 5-13 EPF circuits using RC ladder twoports for cases A and B. The feed-in components and the realized transfer function are indicated.

Example 5-5

Design a second-order Butterworth lowpass filter using the EPF topology to realize $f_0 = 2.9$ kHz; the realized dc gain is not important. Assume $|A(j\omega)| \to \infty$. Minimize the gain-sensitivity product.

Solution. From Eq. (5-74b), we find that $\eta = \beta_4 = 0$ is an appropriate choice for the feed-in elements. Note that case B is a better choice; case A requires an additional condition to be satisfied by the elements! The resulting transfer function is

$$H(s) = \frac{K\beta_2\omega_0^2}{s^2 + s\omega_0/Q + \omega_0^2}$$

with $\omega_0 = \omega_1$ given in Eq. (5-72b) and

$$Q = \frac{\sqrt{\dfrac{G_2/G_3}{C_1/C_4}}}{1 - K + \dfrac{C_1}{C_4}\left(1 + \dfrac{G_2}{G_3}\right)} = \frac{q_z}{q + 1 - K}$$

by Eq. (5-75). Compare also Table 5-1; note that $k = 1/K$. A convenient dc gain is K, which gives $\beta_2 = 1$.

Let us for simplicity assume $C_1 = C_4 = C$ and set $G_2 = gG_3$; then

$$\omega_1 = \frac{\sqrt{g}}{R_3 C} \quad \text{and} \quad Q = \frac{\sqrt{g}}{2 - K + g}$$

At this point we make use of Eq. (3-35), which says that the gain-sensitivity products pertaining to the open-loop and the closed-loop gains are the same; therefore, we calculate [compare Eq. (5-39c)]

$$\Gamma_{A_0}^Q = \Gamma_K^Q = KS_K^Q = K^2 \frac{Q}{\sqrt{g}} = \left(2 + g - \frac{\sqrt{g}}{Q}\right)^2 \frac{Q}{\sqrt{g}}$$

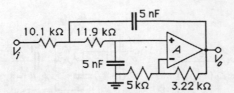

Figure 5-14 EPF lowpass filter.

Taking the derivative with respect to g results in the equation

$$3g - \frac{\sqrt{g}}{Q} - 2 = 0$$

Because $Q = 1/\sqrt{2}$ for a Butterworth filter, the solution is $g = 1.1784$, which results in $K = 1.6432$; choosing $C = 5$ nF yields, finally,

$$R_3 = 11.92 \text{ k}\Omega \qquad R_2 = \frac{R_3}{g} = 10.11 \text{ k}\Omega$$

The final circuit[6] is shown in Fig. 5-14.

The final RC twoport we shall discuss is the *general loaded bridged twin T* network shown in Fig. 5-15. The circuit can be understood as a parallel connection of two loaded bridged T structures in Fig. 5-10 where parallel admittances are, of course, combined into one. Evidently, the circuit is "complementary to itself," just as those in Fig. 5-10, as the student may demonstrate by interchanging terminals b and g. Because nothing will change in all the analysis except an appropriate permutation of subscripts, we shall not treat the general complementary structure. Routine analysis of the circuit in Fig. 5-15 results in

$$T_{cb}(s) = \frac{Y_1 Y_3 y_b + Y_4 Y_6 y_a + Y_7 y_a y_b}{(Y_3 + Y_6 + Y_7 + Y_8)y_a y_b - Y_3^2 y_b - Y_6^2 y_a} \tag{5-76}$$

where we have labeled

$$y_a = Y_1 + Y_2 + Y_3 \quad \text{and} \quad y_b = Y_4 + Y_5 + Y_6 \tag{5-77}$$

At this point we have to decide of which elements the admittances Y_i may consist so that $T_{cb}(s)$ is a second-order function of the form of Eq. (5-22). A little thought will convince the reader that to avoid the *products of three admittances* evident in Eq. (5-76) and the therefore likely s^3 terms requires that $y_a = ky_b$, where k is a

---

[6]This circuit belongs to the family of *Sallen and Key* filters [102]. The classical reference [102] lists of large number of second-order filter sections which use finite-gain amplifiers (note that the op amp with the two feedback resistors realizes an amplifier of gain $K = 1.64$). The circuits in Examples 5-2 and 5-3 also belong to this family, whose members, in general, are special cases of ENF and EPF configurations.

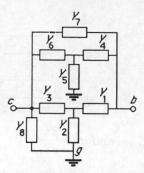

Figure 5-15 The bridged loaded twin T topology.

constant. This choice means that the admittances "seen" from the two *internal* nodes in Fig. 5-15 have the same time constant. Further, more careful analysis shows that $T_{cb}(s)$ of the form of Eq. (5-22) requires that $k = 1$ and $Y_8 = 0$. Using these conditions brings Eq. (5-76) into the form

$$T_{cb}(s) = \frac{Y_1Y_3 + Y_4Y_6 + Y_7y_a}{(Y_1 + Y_2)Y_3 + (Y_4 + Y_5)Y_6 + Y_7y_a} \tag{5-78}$$

which shows that an appropriate component choice is

$$Y_1 = sC_1 \qquad Y_2 = G_2 \qquad Y_3 = sC_3 \qquad Y_4 = G_4$$
$$Y_5 = sC_5 \qquad Y_6 = G_6 \qquad Y_7 = G_7$$

with

$$C_5 = C_1 + C_3 \quad \text{and} \quad G_2 = G_4 + G_6$$

to satisfy Eq. (5-77). Setting further

$$C_1 = C_3 = C$$
$$G_4 = G_6 = G$$
$$\text{and } G_7 = gG$$

results in

$$T_{cb}(s) = \frac{s^2 + 2sgG/C + \omega_1^2}{s^2 + 2(2 + g)\dfrac{G}{C}s + \omega_1^2} \tag{5-79}$$

where

$$\omega_1 = \frac{G}{C}\sqrt{1 + 2g} \tag{5-80a}$$

Also note that

$$q_z = \frac{\sqrt{1 + 2g}}{2g}$$

$$q_p = \frac{1}{2} \frac{\sqrt{1 + 2g}}{2 + g} \tag{5-80b}$$

$$\text{and } q = \frac{2}{g}$$

and observe that $g = 0$ results in a passive RC twin T notch filter (see Problem 1.1). With the equalities in Eq. (5-80b), the quality factor of the active ENF circuit is given by Eq. (5-67) as

$$Q = \frac{q_z}{1 - (K - 1)q} = \frac{\sqrt{1 + 2g}}{2[g - 2(K - 1)]} \tag{5-81}$$

from which we calculate

$$\Gamma_K^Q = K S_K^Q = 4K^2 \frac{Q}{\sqrt{1 + 2g}} \quad \text{with} \quad K = 1 + \frac{g}{2} - \frac{\sqrt{1 + 2g}}{4Q} \tag{5-82}$$

Also note that $g + 2 > 2K$ is required to keep $Q > 0$. To locate the minimum value of Γ_K^Q, we set its derivative with respect to g equal to zero to obtain

$$g_m = \frac{1}{6Q} \sqrt{1 + \frac{1}{36Q^2}} + \frac{1}{36Q^2} \approx \frac{1}{6Q}\left(1 + \frac{1}{6Q}\right) \tag{5-83}$$

However, we need to observe the restriction $K \geq 1$, which, with Eq. (5-82), results in

$$g_{min} = \frac{1}{2Q} \sqrt{1 + \frac{1}{4Q^2}} + \frac{1}{4Q^2} \approx \frac{1}{2Q}\left(1 + \frac{1}{2Q}\right) > g_m \tag{5-84}$$

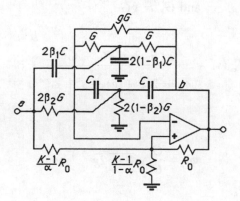

Figure 5-16 ENF circuit using an RC bridged twin T.

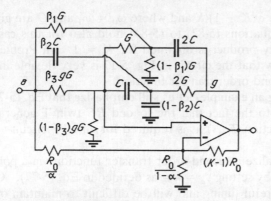

Figure 5-17 EPF circuit using an RC loaded twin T.

We conclude, therefore, that for a bridged twin-T generating network the negative feedback circuit has its minimum gain-sensitivity product when no positive feedback is applied; i.e., $K = 1$ and no Q enhancement. Also note that

$$S_g^Q = -\frac{1}{2}\frac{g}{1 + 2g} \tag{5-85}$$

is small at small values of g but that the element spread becomes large (of the order of $2Q$).

The active ENF circuit, including possible feed-in elements, is shown in Fig. 5-16.   It realizes the transfer function

$$H(s) = \alpha \frac{s^2 + 2\frac{G}{C}\left[2 + g - \frac{K}{\alpha}(\beta_1 + \beta_2)\right]s + \omega_0^2}{s^2 + s\omega_0/Q + \omega_0^2} \tag{5-86}$$

with $\omega_0 = \omega_1$ and Q given in Eqs. (5-80a) and (5-81), respectively.   Setting, for example, $\beta_1 = 0$, $\beta_2 = 1$, and $\alpha = K/(2 + g)$ results in a convenient realization of a notch filter.   Or choosing $\beta_1 = 0$, $\beta_2 = 1$, and $\alpha = K/(4 + 2g - 2K)$ results in an allpass circuit.

Highpass, lowpass, and bandpass functions cannot be implemented with the ENF circuit of Fig. 5-16.   For these latter cases, it is necessary to use the complementary topology, the EPF configuration of Fig. 5-17.   By direct analysis or by using the relationships between complementary networks derived earlier, Eqs. (5-10) and (5-11), the student may show that this EPF circuit with the indicated input connections realizes the transfer function

$$H(s) = -K \frac{(\eta - \beta_2)s^2 + 2s\frac{G}{C}[\eta(2 + g) - g\beta_3] + \omega_0^2\left(\eta - \frac{\beta_1 + g\beta_3}{1 + 2g}\right)}{s^2 + s\omega_0/Q + \omega_0^2}$$

$$\tag{5-87}$$

where we set again $\eta = \alpha(K - 1)/K$ and where $\omega_0 = \omega_1$ and Q are given in Eqs. (5-80a) and (5-81). Equations (5-82) to (5-84) hold also for this case; i.e., the minimum gain-sensitivity product is obtained for $K = 1$. In Problem 5.14 the student is asked to show that the circuit in Fig. 5-17 is very flexible in that it can realize an arbitrary second-order transfer function.

Before considering an example, let us still emphasize that Eq. (5-79) together with Eq. (5-10) points to the fact that the loaded RC twin T generating circuit realizes a bandpass function $T_{cg}(s)$ as is required for an EPF circuit according to Eq. (5-29).

Twin T circuits realize a second-order transfer function via a pole-zero cancellation made possible by setting $y_a = ky_b$ as defined in Eq. (5-77). Clearly, this relationship requires careful tuning and will be difficult to maintain over varying operating conditions. Furthermore, active and passive sensitivities are of the same order as those of the simpler ENF and EPF circuits based on RC bridged T or ladder networks [11]; thus there are no sensitivity advantages to make up for the difficulties caused by the pole-zero cancellation, nor are these filters more general in their realization capabilities than the earlier, simpler circuits. In some situations, however, component values and their adjustments may turn out to be more convenient.

Example 5-6

Realize a symmetrical notch filter with the following parameters: $f_0 = 2.3$ kHz, $Q = 12$, dc gain is arbitrary. Assume $g = 0.1$ and calculate the gain-sensitivity product obtained. For which value of g would the gain-sensitivity product be a minimum?

Solution. The circuit in Fig. 5-16 is suitable for our purpose. If we set $\beta_1 = 0$, $\beta_2 = 1$, and $\alpha = K/(2 + g)$, the transfer function is, from Eq. (5-86),

$$H(s) = \alpha \frac{s^2 + \omega_0^2}{s^2 + s\omega_0/Q + \omega_0^2}$$

with ω_0 and Q given in Eqs. (5-80a) and (5-81), respectively. With $g = 0.1$ we find, from Eq. (5-82), $K = 1.02717$ and $\Gamma_K^Q = 46.2$. Further, $\alpha = 0.48913$. Choosing $C = 5$ nF finally results in $R = 15.16$ kΩ.

The value of g_m to yield the minimum of Γ_K^Q was given in Eq. (5-83):

$$g_m \simeq \frac{1}{6Q}\left(1 + \frac{1}{6Q}\right) = 0.014$$

which, as predicted, results in $K \simeq 0.986 < 1$. Thus, the value of g to be taken is

$$g_{min} \simeq \frac{1}{2Q}\left(1 + \frac{1}{2Q}\right) = 0.043$$

according to Eq. (5-84), resulting in min $\{\Gamma_K^Q\} = 46.06$ and $K = 1$. For this optimal situation, the final circuit is shown in Fig. 5-18.

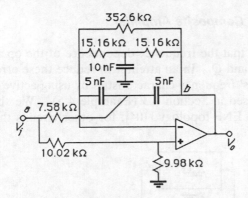

Figure 5-18 ENF notch filter for Example 5-6.

5.3 MULTIAMPLIFIER CIRCUITS

We saw in Section 5.2 that a second- or third-order *RC* generating network together with a single operational amplifier can be used to realize an arbitrary second-order transfer function. Evidently, therefore, there is no need to use more than one op amp per pole pair unless one can achieve in the resulting circuits a better performance than with the single-amplifier biquads (SABs) discussed so far. Nevertheless, the literature contains numerous multiamplifier circuits developed with the motivation to improve in some way the behavior of SABs, either by obtaining lower sensitivities or by making the circuits more versatile.

Clearly, to make the expense and increased power consumption of additional op amps worthwhile, the sensitivities of multiamplifier circuits should compare favorably with the numbers for SABs given in Table 5-2. In particular, given the importance of small ω_0 errors as discussed in Chapter 3, $S_x^{\omega_0}$ should be at its theoretical minimum and deviations caused by the op amp's frequency dependence should be small. A gain-sensitivity product of the order of Q is state-of-the-art design,[7] but one possible area for improvement is the fairly large passive Q sensitivity. However, it should not be obtained at the price of increased $S_x^{\omega_0}$ or $\Gamma_{A_0}^Q$. Indeed, the literature contains a large number of multiamplifier circuits, but analysis and a comparison show that many of them prove to behave no better, and some actually worse, than well-designed single-amplifier biquads [103]. A few of the better filters, most notably the biquad based on Antoniou's general impedance converter (GIC) [see Section 4.3.3] will be discussed in this section.

The second motivation for using additional op amps is the desire to realize several transfer functions with a given circuit topology and make them available *simultaneously* at different output nodes. The resulting circuits will be more versatile and could lead to convenient "universal" active-filter "building blocks." With this versatility in mind, we shall present some designs of three-amplifier biquads based on the state-variable approach.

---

[7] At least without paying other penalties.

5.3.1 Biquads Using a Composite Amplifier

We saw in Section 5.2.3 that the frequency dependence of the op amps gives rise to deviations in both ω_0 and Q. In an attempt to reduce these errors, we can try to improve the amplifiers' frequency characteristics by using active compensation, an idea that was addressed in Section 4.3 (Example 4.5). We shall discuss the method in detail for the ENF topology [104]; for other circuits the procedure is analogous.

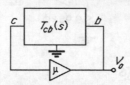

Figure 5-19 ENF feedback configuration.

To recapitulate, an ENF structure was designed by forming a feedback loop of a second-order RC circuit and an amplifier of closed-loop gain μ as shown in Fig. 5-19. For the single-amplifier ENF configuration, Fig. 5-5a, μ is given by

$$\mu(s) = \frac{1}{(K-1)/K - 1/A(s)} = \frac{1}{k - s/\omega_t} \tag{5-88}$$

where we have abbreviated $(K-1)/K$ by k and used the integrator approximation [Eq. (4-8)] for the op amp. The system poles are then determined by the expression

$$\mu T_{cb}(s) = \mu \frac{N_{cb}(s)}{D(s)} = 1 \tag{5-89}$$

i.e.,

$$N_{cb}(s) - \frac{1}{\mu} D(s) = 0$$

as in Eq. (5-7). If, further, $T_{cb}(s)$ is given by Eq. (5-22), the system poles are the roots of Eq. (5-52) and the pole deviations caused by the nonideal op amp, i.e., by finite ω_t, are obtained from Eqs. (5-58) and (5-59). We saw that the errors are proportional to (ω_0/ω_t) where ω_0 is the pole frequency. If we use instead of the single closed-loop amplifier of gain μ a *composite* amplifier of gain μ_c consisting of *two* op amps in a resistive feedback loop as discussed in Section 4.3, the gain will in general be of the form[8]

$$\mu_c(s) = \frac{1 + n_1 s/\omega_t}{k + d_1 s/\omega_t + d_2(s/\omega_t)^2} \tag{5-90}$$

The coefficients k, n_1, d_1, and d_2 are constants which depend on the resistive feedback network in the composite amplifier. Substituting μ_c for μ in Eq. (5-89)

---

[8]The frequency dependence of the gain will be *lowpass* in nature; it will not be a fully biquadratic function, because the output must be taken from an op amp whose gain goes to zero at high frequencies.

and using Eq. (5-22) with $a_2 = 1$ results in the fourth-order polynomial

$$
-d_2 \frac{s^4}{\omega_t^2} + \frac{s^3}{\omega_t}\left(n_1 - d_1 - \frac{\omega_0}{\omega_t}\frac{d_2}{q_p}\right)
$$

$$
+ (1 - k)\left\{s^2\left[1 + \frac{\omega_0}{\omega_t}\frac{\dfrac{n_1}{q_z} - \dfrac{d_1}{q_p}}{1 - k} - \left(\frac{\omega_0}{\omega_t}\right)^2\frac{d_2}{1 - k}\right]\right.
\tag{5-91}
$$

$$
\left. + s\omega_0\left(\frac{1}{Q_0} + \frac{\omega_0}{\omega_t}\frac{n_1 - d_1}{1 - k}\right) + \omega_0^2\right\} = 0
$$

In deriving Eq. (5-91), we made use of the fact that nominally $\omega_1 = \omega_0$ and that the nominal quality factor is

$$
Q_0 = \frac{1 - k}{1/q_z - k/q_p}
$$

[see Eq. (5-23)].   Observe that Eq. (5-53) is a special case of Eq. (5-91), obtained for $a_2 = 1$, $n_1 = d_2 = 0$, and $d_1 = -1$.   [Compare Eqs. (5-88) and (5-90).] Evidently, the circuit has now two parasitic poles, so that the polynomial of Eq. (5-91) can be factored into

$$
\left(\gamma_2 \frac{s^2}{\omega_t^2} + \gamma_1 \frac{s}{\omega_t} + \gamma_0\right)\left(s^2 + s\frac{\omega_a}{Q_a} + \omega_a^2\right) = 0
\tag{5-92}
$$

ω_a and Q_a determine the dominant pole pair, and the γ_i, $i = 0, 1, 2$, identify the parasitic poles.   Multiplying the two terms together to yield

$$
\gamma_2 \frac{s^4}{\omega_t^2} + \frac{s^3}{\omega_t}\left(\gamma_1 + \frac{\gamma_2}{Q_a}\frac{\omega_a}{\omega_t}\right) + s^2\left[\gamma_0 + \frac{\gamma_1}{Q_a}\frac{\omega_a}{\omega_t} + \gamma_2\left(\frac{\omega_a}{\omega_t}\right)^2\right]
$$

$$
+ s\omega_a\left(\gamma_1\frac{\omega_a}{\omega_t} + \frac{\gamma_0}{Q_a}\right) + \gamma_0\omega_a^2 = 0
$$

allows us to compare coefficients with Eq. (5-91):

$$
\gamma_0 = (1 - k)\left(\frac{\omega_0}{\omega_a}\right)^2
\tag{5-93a}
$$

$$
\omega_a\left(\gamma_1\frac{\omega_a}{\omega_t} + \frac{\gamma_0}{Q_a}\right) = (1 - k)\omega_0\left(\frac{1}{Q_0} + \frac{\omega_0}{\omega_t}\frac{n_1 - d_1}{1 - k}\right)
\tag{5-93b}
$$

$$
\gamma_0 + \frac{\gamma_1}{Q_a}\frac{\omega_a}{\omega_t} + \gamma_2\left(\frac{\omega_a}{\omega_t}\right)^2 = (1 - k)\left[1 + \frac{\omega_0}{\omega_t}\frac{\dfrac{n_1}{q_z} - \dfrac{d_1}{q_p}}{1 - k} - \left(\frac{\omega_0}{\omega_t}\right)^2\frac{d_2}{1 - k}\right]
\tag{5-93c}
$$

$$\gamma_1 + \frac{\gamma_2}{Q_a} \frac{\omega_a}{\omega_t} = n_1 - d_1 - \frac{\omega_0 d_2}{\omega_t q_p} \tag{5-93d}$$

$$\gamma_2 = -d_2 \tag{5-93e}$$

Note the similarity in this treatment with that of Section 5.2.4. We next substitute γ_0, γ_1, and γ_2 from Eq. (5-93a), (5-93d), and (5-93e), in Eq. (5-93c), and set $\omega_a = \omega_0 + \Delta\omega_0$; after a few lines of elementary algebra, this results in the expression

$$\frac{\Delta\omega_0}{\omega_0}\left[\frac{n_1 - d_1}{Q_a} - 2(1 - k)\frac{\omega_t}{\omega_0}\right] \simeq \frac{n_1}{q_z} - \frac{d_1}{q_p} - \frac{n_1 - d_1}{Q_a} + d_2\frac{\omega_0}{\omega_t}\frac{1}{q_p Q_a}$$

Because the term $2(1 - k)\omega_t/\omega_0$ in the square brackets is dominant, we obtain for the frequency error the approximate expression

$$\frac{\Delta\omega_0}{\omega_0} \simeq -\frac{1}{2(1 - k)}\left\{\frac{\omega_0}{\omega_t}\left[\frac{n_1}{q_z} - \frac{d_1}{q_p} - \frac{n_1 - d_1}{Q_a}\right] + d_2\left(\frac{\omega_0}{\omega_t}\right)^2\frac{1}{q_p Q_a}\right\} \tag{5-94a}$$

In a similar manner, inserting Eqs. (5-93a) and (5-93d) into Eq. (5-93b) and setting $Q_a = Q_0 + \Delta Q$ yields

$$\frac{\Delta Q}{Q_0} \simeq -\frac{Q_0}{1 - k}\left\{\frac{d_2}{q_p}\left(\frac{\omega_0}{\omega_t}\right)^2 + \frac{\Delta\omega_0}{\omega_t}\left[2(n_1 - d_1) - \frac{\omega_t}{\omega_0}\frac{1 - k}{Q_a}\right]\right\}$$

$$\simeq -\frac{d_2}{1 - k}\frac{Q_0}{q_p}\left(\frac{\omega_0}{\omega_t}\right)^2 - \frac{\Delta\omega_0}{\omega_0}\frac{Q_0}{Q_a} \tag{5-94b}$$

From Eq. (5-94) it is clear that the ω_0 and Q errors are again proportional to (ω_0/ω_t) unless we choose

$$\frac{n_1}{q_z} - \frac{d_1}{q_p} - \frac{n_1 - d_1}{Q_a} = 0 \tag{5-95}$$

i.e.,

$$\frac{n_1}{d_1} = \frac{Q_a/q_p - 1}{Q_a/q_z - 1}$$

in Eq. (5-94a). In that case we find finally

$$\frac{\Delta\omega_0}{\omega_0} \simeq -\frac{1}{2} d_2 K\left(\frac{\omega_0}{\omega_t}\right)^2\frac{1}{q_p Q_a} \tag{5-96a}$$

and

$$\frac{\Delta Q}{Q_0} \simeq -d_2 K\left(\frac{\omega_0}{\omega_t}\right)^2\frac{Q_a}{q_p} = 2Q_a^2\frac{\Delta\omega_0}{\omega_0} \tag{5-96b}$$

Equation (5-95) means that for improved filter performance the structure of the composite amplifier is *not arbitrary* but that it should be chosen such that its

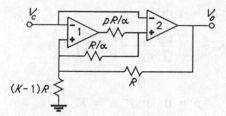

Figure 5-20 Composite amplifier [86, 104].

characteristics match the requirements of the filter [104]. Thus, the composite amplifier must realize a gain function of the form of Eq. (5-90) with coefficients that satisfy Eq. (5-95). Also note from Eq. (5-91) that stability requires $d_2 < 0$. With this condition and with $d_1 > 0$ it is easy to show that $\gamma_i > 0$, $i = 0, 1, 2$, in Eq. (5-92), so that the filter is indeed stable (Problem 5.18).

Note from Eqs. (5-88) and (5-90) that the closed-loop dc gains of the single amplifier and the composite amplifier are the same: $\mu(0) = \mu_c(0) = 1/k$. Because we did nothing but replace amplifiers in the ENF topology, all the equations including the optimization procedures derived earlier are still valid; the difference is only that that the *active* variabilities $\Delta\omega_0/\omega_0$ and $\Delta Q_0/Q_0$ are now proportional to $1/|A(j\omega_0)|^2$, whereas for the single-amplifier biquads they were proportional to $1/|A(j\omega_0)|$. Since the active sensitivities are now very small, we can afford to choose a design point farther to the right of the minimum of $\Gamma_{A_0}^Q$, say, $g = 20$ in Fig. 5-9, in order to reduce the passive sensitivities. The price paid for this choice is, of course, an increased component spread.

The literature contains a number of useful designs of composite amplifiers consisting of two or more op amps [85, 86]; one that is suitable for our purposes is shown in Fig. 5-20. Assuming two identical op amps, it realizes the gain

$$\mu_c(s) = \frac{1 + ms/\omega_t}{k + k(\alpha + p)s/\omega_t - m(s/\omega_t)^2} \tag{5-97}$$

where

$$m = 1 + p + \alpha k$$

The design parameters α and p of the composite amplifier must be determined according to Eq. (5-95):

$$\frac{n_1}{d_1} = \frac{Q_a/q_p - 1}{Q_a/q_z - 1} = \frac{m}{k(\alpha + p)} = 1 + \frac{1}{K - 1}\frac{K + p}{\alpha + p}$$

i.e., once the second-order RC section is chosen and K is determined as discussed earlier, the two resistor ratios α and p must satisfy the equation

$$\frac{K + p}{\alpha + p} = (K - 1)\left(\frac{Q_a/q_p - 1}{Q_a/q_z - 1} - 1\right) \tag{5-98}$$

Note from Eq. (5-47) with $a_2 = 1$ that

$$(K - 1)\left(\frac{Q_a/q_p - 1}{Q_a/q_z - 1} - 1\right) = 1$$

Therefore we obtain

$$p = 0 \quad \text{and} \quad \alpha = K \tag{5-99}$$

An example will illustrate the procedure and demonstrate the improvement in sensitivity attainable by using a composite amplifier.

Example 5-7

Design the circuit of Example 5-3 using an appropriate composite amplifier. Evaluate the passive sensitivities and the ω_0 and Q deviations caused by the op amps' finite gain-bandwidth product. Compare your results with those of the original design. Assume $\omega_t = 2\pi \cdot 1.5$ Mrad/s.

Solution. For the bandpass filter of Fig. 5-8 with $f_0 = 7.5$ kHz and $Q = 12.3$, we chose the resistor ratio $g = R_1/R_2 = 10$ to get $q_z = 1.5811$, $q_p = 0.26352$, and $K = 1.1743$. The passive sensitivities were $|S_x^Q| \approx 7.3$ and $|\Gamma_{A_0}^Q| \approx 53.6$. For these numbers, we obtain, from Eqs. (5-58) and (5-59), $\Delta\omega_0/\omega_0 \approx -1.1\%$ and $\Delta Q/Q \approx +1.1\%$.

Taking the composite amplifier of Fig. 5-20, we need to satisfy Eq. (5-99). We obtain, therefore, $p = 0$ and $\alpha = K = 1.1743$. The passive sensitivities have not changed; however, the ω_0 and Q deviations become now, from Eq. (5-96) with $d_2 = -m = -(1 + \alpha k) = -K = -1.1743$,

$$\frac{\Delta\omega_0}{\omega_0} \approx +\frac{1}{2} K^2 \left(\frac{\omega_0}{\omega_t}\right)^2 \frac{1}{q_p Q_a} \approx 0.00053\%$$

and

$$\frac{\Delta Q_a}{Q_a} \approx +K^2 \left(\frac{\omega_0}{\omega_t}\right)^2 \frac{Q_a}{q_p} \approx 0.16\%$$

Because these deviations, in particular the important ω_0 errors, are very small, we may try to reduce the passive sensitivities further by choosing a larger value for g (see Fig. 5-9!). For example, assume that the largest acceptable resistor ratio is $g = 60$; then, from Eq. (5-50), $|S_x^Q| \approx 2.68$, a reduction of 63%. The new value of K is, from Eq. (5-47), $K = 1.0228$, and we choose $p = 0$ and $\alpha = K$. Finally, with $d_2 = -m = -K$, we find from Eq. (5-96)

$$\frac{\Delta\omega_0}{\omega_0} \approx 0.00085\% \quad \text{and} \quad \frac{\Delta Q}{Q} \approx 0.26\%$$

At this stage, it could be claimed that the 2.28% change from the value $K = 1$ may not be worthwhile.[9] To investigate this factor, we set $K = 1$ and, therefore, the

---

[9]Note, however, that the second pole, ω_2, of the op amp, defined in Eq. (4-11), can be shown to cause instability unless $K > 2\omega_t/\omega_2$. Also, it has been shown [124] that the linear signal range is very limited and nonlinear oscillations are likely to occur if the op amps are overdriven.

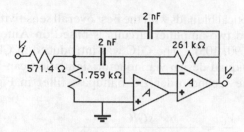

Figure 5-21 The NF circuit to realize Example 5-7.

resistor ratio $R_1/R_2 = g = 4Q^2 = 605.16$. For this case we obtain $|S_x^Q| \simeq 0.5$ and

$$\frac{\Delta\omega_0}{\omega_0} \simeq 0.0025\% \quad \text{and} \quad \frac{\Delta Q}{Q} \simeq 0.76\%$$

As expected, the passive Q sensitivities are reduced to 0.5 (see Table 5-2) but the ω_0 and Q deviations are increased by approximately a factor of 3, although they are still very small. Also note the large resistor ratio. The final circuit for this last design is shown in Fig. 5-21.

The reader should note that these extremely small ω_0 variabilities depend critically on satisfying Eq. (5-95) or, for the specific composite amplifier in Fig. 5-20, Equation (5-98). If the resistor ratios are not exact, the term

$$\frac{n_1}{q_z} - \frac{d_1}{q_p} - \frac{n_1 - d_1}{Q_a}$$

in Eq. (5-94a) will dominate and both variabilities will again be proportional to ω_0/ω_t. The multiplying factor, though, will normally be very small.

This problem does not arise, however, if a composite amplifier in the NF configuration with no Q enhancement is used, such as is shown in Fig. 5-21: in that case $q_z = Q_0$ and $p = 0$, $\alpha = K = 1$, so that the "resistor ratios" which determine the composite amplifier gain are exact. The resulting two-amplifier active filter section performs extremely well: all passive ω_0 and Q sensitivities are equal to 0.5 in magnitude and the ω_0 and Q deviations caused by finite ω_t are given by Eq. (5-96) with $K = -d_2 = 1$.

It should be apparent to the student from the earlier discussion of the complementary transformation that the analysis and the results for the (E)PF circuit are completely analogous.

5.3.2 Biquads Based on the General Impedance Converter

Because the performance of single-op amp active biquads is often unsatisfactory, we used in the previous section a second op amp to build a composite amplifier whose characteristics were matched to the requirements of the filter. We found that such a design could lead to a significant reduction of both active and passive sensitivities. An alternative strategy that is often employed searches for different topologies which promise better circuit performance than the single-amplifier biquads discussed in Section 5.2.

One of the more practical biquads with the best overall sensitivity performance among the many proposed two-amplifier circuits is based on Antoniou's general impedance converter [89, 90, 105]. The GIC was introduced in Chapter 4, Figs. 4-32 and 4-33; its use in biquad design via inductor simulation can best be understood by starting from the second-order *LC* bandpass filter in Fig. 5-22, which realizes the function

$$\frac{V_0}{V_i} = \frac{sG/C}{s^2 + sG/C + 1/(LC)}$$

(5-100)

A two-amplifier active *RC* filter is then easy to derive: we need only to replace the inductor by either the type I or the type II GIC shown in Fig. 4-32. Observe, though, that the filter output must be taken from the inductor node *n*, which is not an op amp output terminal. The solution lies in taking a type I GIC, where the output voltage of op amp A_1 is proportional to the inductor voltage (see Fig. 4-33):

$$V_4 = V_1\left(1 + \frac{Y_5}{Y_4}\right)$$

The resulting GIC bandpass biquad with

$$V_o = V_n\left(1 + \frac{G_5}{G_4}\right)$$

is shown in Fig. 5-23. With *L* replaced by $C_2G_4/G_1G_3G_5$ according to Eq. (4-71), it realizes the bandpass transfer function

$$\frac{V_o}{V_i} = \frac{s\dfrac{G}{C}\left(1 + \dfrac{G_5}{G_4}\right)}{s^2 + s\dfrac{G}{C} + \dfrac{G_1G_3G_5}{CC_2G_4}}$$

(5-101)

ω_0 and Q_0 are given by

$$\omega_0^2 = \frac{G_1G_3G_5}{CC_2G_4}$$

(5-102a)

and

$$Q_0 = \frac{1}{G}\sqrt{\frac{C}{C_2}}\sqrt{\frac{G_1G_3G_5}{G_4}}$$

(5-102b)

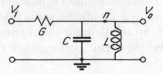

Figure 5-22 Second-order passive *LC* filter.

respectively, with

$$|S_x^{\omega_0}| = \frac{1}{2}, \quad |S_x^{Q_0}| = \frac{1}{2} \quad \text{and} \quad S_R^{Q_0} = 1 \tag{5-103}$$

x stands for the relevant passive components. Note that all passive sensitivities are very low, with $|S_x^{\omega_0}|$ at the theoretical minimum. Thus, the GIC biquad promises to be an excellent circuit, provided that the active sensitivities are also low. To investigate op amp effects, we analyze the simulated inductor with the op amp model $A_i(j\omega) \simeq \omega_{ti}/j\omega$, $i = 1, 2$, to obtain

$$Y_L(j\omega) \simeq \frac{1}{j\omega L_0} \frac{1 + \dfrac{H}{H - 1}\left(\dfrac{j\omega}{\omega_{t2}} - \dfrac{\omega^2 C_2}{\omega_{t1} G_3}\right)}{1 + H[j\omega/\omega_{t1} + G_3/(\omega_{t2} C_2)]} \tag{5-104}$$

In Eq. (5-104),

$$L_0 = \frac{C_2 G_4}{G_1 G_3 G_5} \tag{5-105}$$

and H is the midband gain of the filter; i.e., by Eq. (5-101),

$$H = 1 + \frac{G_5}{G_4} \tag{5-106}$$

Second-order small terms were neglected; i.e., we assumed $\omega^2 \ll \omega_{t1}\omega_{t2}$. Following the approach adopted in Section 4.3.3, i.e., approximating $1/(1 + x)$ by $1 - x$ for $x \ll 1$, Eq. (5-104) is rewritten as

$$\begin{aligned}
Y_L(j\omega) &\simeq \frac{1}{j\omega L_0}\left(1 - \frac{H}{H - 1}\frac{\omega^2 C_2}{\omega_{t1} G_3} + \frac{H}{H - 1}\frac{j\omega}{\omega_{t2}}\right)\left(1 - H\frac{G_3}{\omega_{t2} C_2} - H\frac{j\omega}{\omega_{t1}}\right) \\
&\simeq \frac{1}{j\omega L_0}\left\{1 - \frac{H}{H - 1}\frac{\omega}{\omega_{t2}}\left[(H - 1)\frac{G_3}{\omega C_2} + \frac{\omega_{t2}}{\omega_{t1}}\frac{\omega C_2}{G_3}\right]\right\} \\
&\quad - \frac{1}{\omega_{t2} L_0}\frac{H}{H - 1}\left[(H - 1)\frac{\omega_{t2}}{\omega_{t1}} - 1\right] \\
&= \frac{1}{j\omega L} + G_L
\end{aligned} \tag{5-107}$$

Terms proportional to $\omega^2/(\omega_{t1}\omega_{t2})$ were again neglected. Note that the inductor becomes *lossy*, repre: ented here by a parallel conductance G_L, and that the inductor value acquires an error. G_L can be reduced to zero by choosing

$$H - 1 = \frac{G_5}{G_4} = \frac{R_4}{R_5} = \frac{\omega_{t1}}{\omega_{t2}} \simeq 1 \tag{5-108a}$$

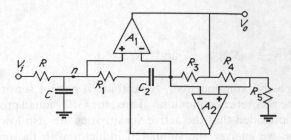

Figure 5-23 GIC bandpass biquad.

[see Eq. (4-78)], and the inductor error is minimized by selecting

$$C_2 = \frac{1}{\omega_c R_3} \tag{5-108b}$$

where ω_c is some critical frequency value; in our case, $\omega_c = \omega_0$. With these choices, the realized inductor is

$$L_0 = \frac{R_1}{\omega_0} \tag{5-108c}$$

The design is completed by choosing (for convenience)

$$R_1 = R_3 = R_4 = R_5 = R_0$$

$$C = C_2 = \frac{L_0}{R_0^2} = \frac{1}{\omega_0 R_0} \tag{5-108d}$$

$$G = \frac{1}{R} = \frac{\omega_0 C}{Q} = \frac{1}{R_0 Q}$$

R_0 sets the impedance level. The gain-bandwidth products of the two op amps will in general not be equal, so that, by Eq. (5-108a), some trimming of R_4 or R_5 may be required to maximize the inductor's quality factor.

Note that the above design process fixes the midband gain at $H = 2$. If H must be adjustable, for example, to be able to optimize the signal level in cascade realizations (see Chapter 6), the errors can become quite significant [106]: assuming that Eq. (5-108b) is satisfied and that $\omega_{t1} = \omega_{t2} = \omega_t$, we obtain from Eq. (107)

$$Y_L(j\omega) = \frac{1}{j\omega L} + G_L \simeq \frac{1}{j\omega L_0}\left(1 - \frac{\omega}{\omega_t}\frac{H^2}{H-1}\right) - \frac{\omega_0}{\omega_t}\frac{1}{\omega_0 L_0} H \frac{H-2}{H-1} \tag{5-109}$$

$$= \frac{1}{j\omega L_0}\left[1 - \frac{\omega}{\omega_t}\left(\sqrt{\frac{R_4}{R_5}} + \sqrt{\frac{R_5}{R_4}}\right)^2\right] - \frac{\omega_0}{\omega_t}\frac{1}{R_0}\left(\frac{R_4}{R_5} - \frac{R_5}{R_4}\right)$$

Evidently, the errors will be minimized for $R_4 = R_5$, i.e., $H = 2$. Gains of value $H < 2$ can be realized for the optimal design with $R_4 = R_5$ by splitting the input resistor $R = 1/G$ (Problem 5.19). If a midband gain of value larger than 2, i.e.,

$R_4 > R_5$, is required, Eq. (5-109) indicates a remaining inductor error of value, at $\omega = \omega_0$,

$$\Delta L \simeq L_0 \frac{\omega_0}{\omega_t} \frac{H^2}{H - 1} \geq 4L_0 \frac{\omega_0}{\omega_t} \tag{5-110a}$$

and a *negative* loss resistor

$$R_L \simeq -\omega_t L_0 \frac{H - 1}{H(H - 2)} = -\frac{\omega_t}{\omega_0} \frac{H - 1}{H(H - 2)} R_0 \tag{5-110b}$$

which appears functionally *in parallel* with the Q-determining resistor $R = 1/G$ in Fig. 5-23 (from node n to ground). Consequently, R_L can cause severe Q enhancement because Eq. (5-102b) will be modified as

$$Q_a = \frac{1}{G - |G_L|} \sqrt{\frac{C}{C_2}} \sqrt{\frac{G_1 G_3 G_5}{G_4}} = \frac{Q_0}{1 - |G_L|/G} \tag{5-111}$$

Fortunately, there exists a simple compensation method [106]: we need only connect a *positive* resistor R_c of value $R_c = |R_L|$ in parallel with C in Fig. 5-23.

The pole frequency and pole Q errors of this bandpass can be calculated quite easily by noting that with Eq. (5-109)

$$\omega_a = \frac{1}{\sqrt{LC}} = \frac{\sqrt{1 - \Delta L/L_0}}{CR_0} \simeq \omega_0 \left(1 - \frac{\Delta L/L_0}{2}\right)$$

$$= \omega_0 \left(1 - \frac{1}{2} \frac{\omega_0}{\omega_t} \frac{H^2}{H - 1}\right) \tag{5-112a}$$

Similarly, we find from

$$\frac{\omega_a}{Q_a} = \frac{G + G_L}{C} = \frac{1 + G_L/G}{Q_0 C R_0} \tag{5-112b}$$

that

$$Q_a = Q_0 \frac{1 - \dfrac{1}{2} \dfrac{\omega_0}{\omega_t} \dfrac{H^2}{H - 1}}{1 - Q_0 \dfrac{\omega_0}{\omega_t} H \dfrac{H - 2}{H - 1}} \tag{5-112c}$$

For the optimal design ($R_4 = R_5$, $H = 2$), the two values are

$$\omega_a \simeq \omega_0 \left(1 - 2\frac{\omega_0}{\omega_t}\right) \qquad Q_a \simeq Q_0 \left(1 - 2\frac{\omega_0}{\omega_t}\right)$$

i.e.,

$$\frac{\Delta \omega}{\omega_0} \simeq \frac{\Delta Q}{Q_0} \simeq -\frac{2}{|A(j\omega_0)|} \tag{5-112d}$$

which make this circuit comparable to the ENF single-amplifier biquad (see Table 5-2) but with lower passive Q sensitivities. Although the proposed compensation method is easy to apply, the student should not forget that we are trying to compensate the effect of an *active* parameter by a *passive* component; i.e., the method can work only for known (*measured*) values of ω_t and for fixed operating conditions. Also note that the cancellation of $-R_L$ by the compensation resistor R_c involves a *difference effect* and is, therefore, very sensitive.

Nevertheless, because of its many advantages, the GIC circuit of Fig. 5-23, for $H = 2$, has proven to be one of the best practical active biquads available. Note that the circuit:

- Has very low passive sensitivities
- Has small variabilities caused by the finite op amp bandwidth
- Can be built with identical capacitors and, except for the Q-determining resistor R, with all identical resistors
- Has a small component spread (equal to Q)
- Has gain, ω_0, and Q independently tunable
- Has infinite input impedance at the pole frequency (L_0 and C resonate at ω_0), an important property for cascade designs

In addition to a bandpass filter, the GIC circuit can be used to implement a variety of different transfer functions, as will be shown below. First, however, let us consider an example to illustrate the design.

Example 5-8

Design a second-order GIC bandpass filter to realize $f_0 = 12.5$ kHz, $Q = 16$, and a midband gain of 18 dB. Dual op amps with $f_t = 1.5$ MHz are available. Calculate the actual frequency and Q values obtained for your design.

Solution. For 18 dB gain, we find $H = 7.943$. Setting $R_1 = R_3 = R_5 = R_0$, we find

$$R_4 = 6.943R_0 \quad \text{and} \quad C = \frac{1}{2\pi \cdot f_0 R_0}$$

Further, from Eq. (5-110b),

$$R_L = -\frac{1500}{12.5}\frac{6.943}{7.942 \cdot 5.943}R_0 = -17.65R_0$$

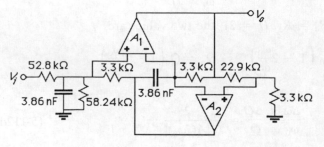

Figure 5-24 GIC bandpass for Example 5-8.

Note that $|R_L|$ and $R = QR_0$ have approximately the same size; the Q enhancement effect is, therefore, not negligible! If we choose $R_0 = 3.3$ kΩ, the final circuit with all elements is as shown in Fig. 5-24.

From Eq. (5-112) we obtain for the actually realized values, neglecting second-order effects,

$$f_a \simeq 12.5\left(1 - \frac{6.25}{1500}\frac{7.943^2}{6.943}\right) \text{ kHz} \simeq 12.027 \text{ kHz}$$

$$Q_a \simeq 16\left(1 - \frac{6.25}{1500}\frac{7.943^2}{6.943}\right) = 15.39$$

If no compensation resistor R_c had been applied, the Q error, by Eq. (5-112c), would have been

$$Q_a \simeq 16 \frac{1 - \dfrac{6.25}{1500}\dfrac{7.943^2}{6.943}}{1 - 16 \cdot \dfrac{12.5}{1500} \cdot 7.943 \cdot \dfrac{5.943}{6.943}} \simeq 165$$

i.e., the circuit would probably oscillate.

Observe that the f_0 error of 473 Hz is a significant fraction of the total bandwidth $f_0/Q = 781$ Hz! For most cases where $H > 2$ and high Q are required, some *predistortion* of the pole frequency may be necessary. In our case, smaller capacitors of value

$$C = \frac{1}{\omega_0(1 + \Delta\omega/\omega_0)R_0} = 3.72 \text{ nF}$$

would result in the *nominal* pole frequency $f_0 = 12.97$ kHz, which by virtue of the error $\Delta\omega/\omega_0 = -0.0379$, would be reduced to the *actual* pole frequency $f_0 \simeq 12.48$ kHz.

To investigate which other types of transfer functions can be realized by the GIC biquad, we use our well-known procedure of lifting grounded elements partially or completely off ground. The resulting circuit is shown in Fig. 5-25 [11, 105]; it realizes the function, with $\tau = R_0C$,

$$\frac{V_o}{V_i} = \frac{s^2[\alpha H - \gamma(H - 1)] + s\dfrac{1}{\tau Q}[\beta H - \gamma(H - 1)] + \dfrac{\gamma}{\tau^2}}{s^2 + s\dfrac{1}{\tau Q} + \dfrac{H - 1}{\tau^2}} \tag{5-113}$$

In Problem 5.21 the student is asked to determine which types of transfer functions can be realized by an appropriate choice of α, β, γ, and H. Clearly, a lowpass function cannot easily be obtained from this circuit. To realize a lowpass section with the GIC circuit, an alternative configuration is needed that the student is asked to derive in Problem 5.22; neither type I nor type II GICs will work.

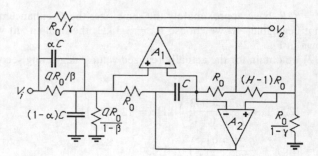

Figure 5-25   General GIC biquad.

5.3.3 Biquads Using Three Op Amps

The literature contains a number of second-order sections built with three or more op amps. Their main advantage is a somewhat greater versatility which permits different types of transfer functions to be realized simultaneously at the available op amp outputs. Also, tuning will often be noninteractive, because the additional op amps isolate different parts of the circuit from each other. These designs are derived from a *state-variable* realization first proposed by Kerwin, Huelsman, and Newcomb (KHN) [107] and later refined by a number of authors [108–110]. The circuits use *n* integrators for an *n*th-order transfer function and connect them with appropriate scaling coefficients and summers in a manner that is well known from the implementation of analog computers [47, 107]. For our second-order sections, this means we require two integrators and a summer, resulting in three–op amp filters. Before studying the actual circuits, a better understanding of the operation of these designs can be obtained by following their derivation via some fundamental block diagrams. The student will find it helpful to refer to Chapter 4, Sections 4.3.1 and 4.3.2, for the following discussion.

Consider the configuration in Fig. 5-26a, which shows a lossy and a lossless integrator connected in a loop. Remembering from our work in Chapter 4 that *inverting* integrators are easier to realize, we have connected two inverting integrators and have used a negative feedback factor, $-c_2$, to obtain negative loop gain for a stable system. The lossy integrator shown in the dashed box of this *two-integrator-loop filter* realizes $-a_1/(s + a_1 c_1)$, with all voltages referenced to ground. In practice, the two summers will, of course, be merged as shown in Fig. 5-26b and the three available transfer functions can be derived to equal

$$\frac{V_h}{V_i} = \frac{ks^2}{s^2 + a_1 c_1 s + a_1 a_2 c_2} \tag{5-114a}$$

$$\frac{V_b}{V_i} = \frac{-ka_1 s}{s^2 + a_1 c_1 s + a_1 a_2 c_2} \tag{5-114b}$$

$$\frac{V_l}{V_i} = \frac{ka_1 a_2}{s^2 + a_1 c_1 s + a_1 a_2 c_2} \tag{5-114c}$$

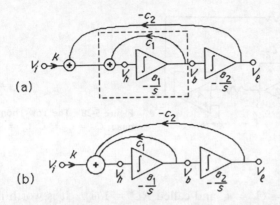

(a)

(b)

Figure 5-26 (a) Two-integrator-loop filter; (b) with merged summers.

which we recognize as highpass, bandpass, and lowpass functions, respectively. The summer and the first integrator in Fig. 5-26b can, of course, be combined to save one op amp by using the inverting integrator input as a summing node (Fig. 4-17c).   The resulting configuration, shown symbolically in Fig. 5-27a, points out a new difficulty: because $c_2 < 0$, the summer must use both inverting and noninverting inputs, which, as was shown in Chapter 4 (Problem 4.25), results in summing coefficients which are difficult to tune and adjust.   We make, therefore, one final modification by inserting an inverter into the outer feedback loop as shown in Fig. 5-27b.   This circuit realizes the transfer functions [Eqs. (5-114b) and (5-114c); the highpass output has been lost in the summer-integrator combination of the first block.   The actual circuit implementation, the Tow-Thomas (TT) biquad [108, 109], is shown in Fig. 5-28.

Disregarding for now the compensation capacitor C_c and assuming ideal op amps, simple analysis yields the bandpass and lowpass functions

$$\frac{V_b}{V_i} = - \frac{k\omega_0 s}{s^2 + s\omega_0/Q + \omega_0^2} \tag{5-115a}$$

$$\frac{V_l}{V_i} = - \frac{k\omega_0^2}{s^2 + s\omega_0/Q + \omega_0^2} \tag{5-115b}$$

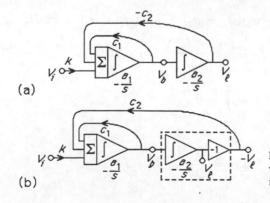

(a)

(b)

Figure 5-27 (a) Two-integrator loop with differential-input summer.   (b) Final modification, using one inverter.

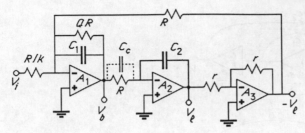

Figure 5-28 The Tow-Thomas (TT) bi-quad; $C_1 = C_2 = C$; $C_c = 4/(\omega_r R)$.

where we have set $C_1 = C_2 = C$ and called $RC = 1/\omega_0$. It is worth noting that one of the attractive features of the TT state-variable design is the low passive sensitivity: the ω_0 sensitivities are at their theoretical minimum,

$$S_x^{\omega_0} = -\frac{1}{2}$$

and

$$|S_x^Q| \leq 1$$

where x denotes the passive components. Note also that the element spread equals Q.

The deviations $\Delta\omega_0$ and ΔQ caused by the op amps' finite gain-bandwidth product can be determined most easily by remembering from Chapter 4, Eq. (4-52) and Problem (4.28), that finite ω_t results in integrators with finite quality factors, or, equivalently, finite phase errors. Using the notation

$$-\frac{a_i}{s} \rightarrow -\frac{a_i}{sf_i + \sigma_i}$$

in Eq. (5-114), we find that the poles of the two-integrator loop are determined by the expression

$$(sf_1 + \sigma_1)(sf_2 + \sigma_2) + a_1c_1(sf_2 + \sigma_2) + c_2a_1a_2$$

$$= f_1f_2s^2 + sa_1c_1\left(f_2 + \frac{f_2\sigma_1 + f\sigma_2}{a_1c_1}\right) + c_2a_1a_2\left(1 + \frac{c_1a_1\sigma_2 + \sigma_1\sigma_2}{c_2a_1a_2}\right) \quad (5\text{-}116)$$

$$= f_1f_2\left(s^2 + s\frac{\omega_a}{Q_a} + \omega_a^2\right) = 0$$

With

$$a_1c_1 = \frac{\omega_0}{Q_0} \quad \text{and} \quad c_2a_1a_2 = \omega_0^2$$

the actual pole frequency and pole Q terms in Eq. (5-116) can be determined from

$$\omega_a^2 = \frac{\omega_0^2}{f_1 f_2}\left(1 + \frac{1}{Q_0}\frac{\sigma_2}{\omega_0} + \frac{\sigma_1\sigma_2}{\omega_0^2}\right) \tag{5-117a}$$

$$Q_a = \frac{\omega_a}{\omega_0}\frac{Q_0}{f_2 + Q_0\dfrac{f_2\sigma_1 + f_1\sigma_2}{\omega_0}} \tag{5-117b}$$

The terms f_1 and σ_1 for the Miller integrator were derived in Eq. (4-52).

Taking care of the modifications appropriate for Fig. 5-28—the total input conductance is $G(1 + k + 1/Q)$ and not just G as in the Miller integrator—we obtain

$$f_1 = 1 + \frac{\omega_0}{\omega_{t1}}\left(1 + k + \frac{1}{Q}\right) \quad\text{and}\quad \sigma_1 = -\frac{\omega^2}{\omega_{t1}}$$

f_2 and σ_2 for the Miller-inverter cascade (the dashed box in Fig. 5-27b) are, from Problem 4.28,

$$f_2 = 1 + \frac{\omega_0}{\omega_{t2}} \quad\text{and}\quad \sigma_2 = -\frac{\omega^2}{\omega_{t2}} - 2\frac{\omega^2}{\omega_{t3}}$$

Thus, we find from Eq. (5-117) at $\omega \simeq \omega_0$, for $Q \gg 1$ and assuming matched op amps with $\omega_0^2 \ll \omega_t^2$,

$$\frac{\omega_a - \omega_0}{\omega_0} = \frac{\Delta\omega_0}{\omega_0} \simeq -\frac{2 + k}{2}\frac{\omega_0}{\omega_t} = -\frac{2 + k}{2}\frac{1}{|A(j\omega_0)|} \tag{5-118a}$$

and

$$\frac{Q_a}{Q_0} \simeq \frac{1}{1 - 4Q_0\omega_0/\omega_t} \tag{5-118b}$$

Evidently, although the frequency deviations are very small and comparable with our previous good circuits, the standard three-amplifier biquad suffers from severe Q enhancement. Recall that this problem was pointed out already in connection with Eq. (4-64)!

The above analysis has made it apparent that the cause of the Q enhancement can be found in the finite integrator quality factors introduced in Eq. (4-53): both the Miller integrator and the Miller-inverter cascade have *negative* Q factors,

$$Q_M(\omega_0) \simeq \frac{\omega_0}{\sigma_1} \simeq -\frac{\omega_t}{\omega_0} \quad\text{and}\quad Q_{M\text{-}inv}(\omega_0) \simeq \frac{\omega_0}{\sigma_2} \simeq -\frac{1}{3}\frac{\omega_t}{\omega_0}$$

respectively, whose effects *add* to cause the Q enhancement expressed in Eq. (5-118b). The student may recall from Fig. 4-23b that a simple wiring change increases the quality factor of the Miller-inverter cascade by a factor of 3; thus, connecting the noninverting input terminal of op amp A_3 in Fig. 5-28 to the inverting input terminal of op amp A_2 instead of grounding it reduces the Q enhancement problem by a factor of 2.

If this improvement proves insufficient, passive compensation can be used. Recall from Fig. 4-21a and Problem 4.26 that a phase-lead capacitor of value $1/(\omega_t R)$ restores the integrator Q of the Miller integrator nominally to infinity. Because in the TT circuit we need to correct *four* phase lags, one from the Miller integrator and three from the Miller-inverter cascade, the required compensation capacitor C_c in Fig. 5-28 equals

$$C_c = \frac{4}{\omega_t R} \qquad (5\text{-}119a)$$

Alternatively, we may use the compensation method of Fig. 4-21b and use a resistor of value

$$R_c = \frac{4}{\omega_t C} \qquad (5\text{-}119b)$$

in *series* with C_2 in Fig. 5-28.

It was discussed before that passive compensation is often unsatisfactory in practice because it attempts to match the behavior of electrically dissimilar parameters, in this case a compensation capacitance or resistance with the gain-bandwidth product of op amps. Far more reliable performance could be expected from *active* compensation, such as the one proposed in Fig. 4-24a, where two op amps were used to realize a noninverting integrator with *positive Q* factor. The student will appreciate that this configuration has an additional advantage in that the Q enhancement problem will be largely eliminated because the two phase errors in the two-integrator loop of the TT biquad just cancel. Recall that the circuit in Fig. 4-24a realizes a function of the form [see Eq. (4-61)]

$$\frac{1/\omega_0}{sf_2 + \sigma_2}$$

with

$$f_2 = 1 + \frac{\omega_0}{\omega_{t2}} \quad \text{and} \quad \sigma_2 = +\left(\frac{2\omega^2}{\omega_{t3}} - \frac{\omega^2}{\omega_{t2}}\right)$$

Inserting these expressions into Eq. (5-117) and assuming $Q_0 \gg 1$ results in

$$\frac{\omega_a - \omega_0}{\omega_0} = \frac{\Delta\omega_0}{\omega_0} \approx -\frac{1}{2}\left[(1 + k)\frac{\omega_0}{\omega_{t1}} + \frac{\omega_0}{\omega_{t2}}\right] \qquad (5\text{-}120a)$$

and

$$\frac{Q_a}{Q_0} \approx \frac{1 + \Delta\omega_0/\omega_0}{1 + \frac{\omega_0}{\omega_{t2}} + Q_0\left[\frac{2\omega_0}{\omega_{t3}} - \frac{\omega_0}{\omega_{t1}} - \frac{\omega_0}{\omega_{t2}} - \frac{\omega_0^2}{\omega_{t1}\omega_{t2}} + \frac{\omega_0^2(1 + k)}{\omega_{t1}}\left(\frac{2}{\omega_{t3}} - \frac{1}{\omega_{t2}}\right)\right]} \qquad (5\text{-}120b)$$

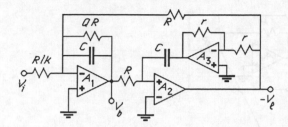

Figure 5-29 The Åckerberg-Mossberg (AM) circuit.

For *matched* op amps, Eq. (5-120b) simplifies to

$$\frac{Q_a}{Q_0} \simeq \frac{1 - \frac{1}{2}(2 + k)\omega_0/\omega_t}{1 + \omega_0/\omega_{t2} + Q_0 k\omega_0^2/\omega_t^2} \tag{5-120c}$$

The circuit, first proposed by Åckerberg and Mossberg [110], is shown in Fig. 5-29. We said earlier that one of the advantages of multi–op amp designs is the greater versatility in realizing general transfer functions. To demonstrate this fact for the two-integrator-loop filters, we observe that a general biquadratic function can be obtained from the generic configuration in Fig. 5-27b by adding the outputs V_b and $-V_l$ to the input V_i as shown in Fig. 5-30. Using Eqs. (5-114a) through (5-114c), the realized function is found to equal

$$\frac{V_o}{V_i} = -\frac{\alpha s^2 + a_1 s(\alpha c_1 - \beta k) + a_1 a_2(\alpha c_2 - \gamma k)}{s^2 + a_1 c_1 s + a_1 a_2 c_2} \tag{5-121}$$

so that by an appropriate choice of the summing coefficients α, β, and γ the transfer function zeros can be placed anywhere in the s-plane.

The reader will appreciate that for the cost of an additional op amp summer this approach can be used for *any* two-integrator-loop design. This summer can be saved by an alternative approach which injects the input signal into additional nodes in the filter section. Of course, this should be done in such a way that the transfer function poles are not disturbed; i.e., as discussed earlier, additional inputs

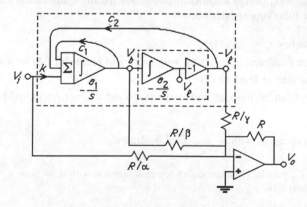

Figure 5-30 General biquad design obtained by summing the two-integrator loop outputs.

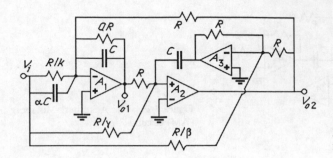

Figure 5-31 General biquad based on the AM circuit using feed-forward.

can be obtained by *lifting elements off ground*, or additional components can be connected to *virtual ground nodes* so that the system polynomial remains unchanged. Figure 5-31 shows the resulting circuit for the Åckerberg-Mossberg biquad. The biquad realizes the function

$$\frac{V_{o1}}{V_i} = -\frac{\alpha s^2 + s\omega_0(k - \beta) + \gamma\omega_0^2}{s^2 + s\omega_0/Q + \omega_0^2} \tag{5-122}$$

Observe from Eq. (5-122) that at the output V_{o1} we can obtain all types of second-order transfer functions: highpass ($k = \beta = \gamma = 0$), bandpass ($\alpha = \beta = \gamma = 0$), lowpass[10] ($\alpha = k = \beta = 0$), notch [$k = \beta = 0$, $\alpha = \gamma$, $\alpha > \gamma$ (highpass notch); $\alpha < \gamma$ (lowpass notch)], and allpass ($\alpha = \gamma = k = 1$, $\beta = 1 + 1/Q$). It is easy to verify that the active and passive zero sensitivities are low. At the same time, the choice of components to set the transmission zeros does not affect the poles, so that the pole sensitivities are as small as those of the AM circuit derived above. Also note that the element spread is low and that all filter coefficients are independently tunable by trimming resistors: the effect of varying k, α, β, and γ is obvious from Eq. (5-122); ω_0 is tuned by trimming the input resistor of the second integrator; and Q is determined by the resistor QR.

Example 5-9

Using the circuit of Fig. 5-31, design a fourth-order geometrically symmetrical bandpass filter to satisfy the following specifications:

Bandcenter frequency $f_0 = 22$ kHz

0.45 dB maximum passband attenuation with passband bandwidth $\Delta f = 4$ kHz

Stopband attenuation: at least 13 dB

A transmission zero in both upper and lower stopband to improve out-of-band rejection

Available are matched op amps with $\omega_t = 2\pi \cdot 2.3$ Mrad/s.

---

[10]Of course, the output V_{o2} provides a more direct lowpass function with $\alpha = \beta = \gamma = 0$, realizing Eq. (5-115b) with a minus sign.

Solution. The specifications suggest an elliptic filter transfer function. Since the total filter must be of fourth order and geometrically symmetrical, a second-order elliptic lowpass function is to be taken as the prototype. Table III-3a in Appendix III gives

$$H_{LP}(p) = 0.213409 \frac{p^2 + 7.46394}{p^2 + 1.28475p + 1.67671}$$

which is transformed into the desired bandpass function by the lowpass-to-bandpass transformation

$$p = 5.5 \frac{s^2 + 1}{s}$$

s is the bandpass frequency normalized with respect to $\omega_n = 2\pi \cdot 22$ krad/s. Consequently, the desired bandpass transfer function, in factored form, equals

$$H_{BP}(s) = 0.462 \frac{s^2 + 1.6352}{s^2 + 0.12869s + 1.2268} \, 0.462 \frac{s^2 + 0.61155}{s^2 + 0.10490s + 0.81513}$$

$$= H_1(s)H_2(s)$$

We need to design, therefore, two second-order notch sections: a lowpass notch $H_1(s)$ and a highpass notch $H_2(s)$. For both sections, we set $k = \beta = 0$ and select $C = 3.3$ nF. With this choice we find for $H_1(s)$:

$$\alpha C = 1.524 \text{ nF} \qquad R = \frac{1}{\omega_1 C} = 1.98 \text{ k}\Omega \qquad Q_1 R = 17.04 \text{ k}\Omega \qquad \frac{R}{\gamma_1} = 1.485 \text{ k}\Omega$$

and for $H_2(s)$

$$\alpha C = 1.524 \text{ nF} \qquad R = \frac{1}{\omega_2 C} = 2.43 \text{ k}\Omega \qquad Q_2 R = 20.90 \text{ k}\Omega \qquad \frac{R}{\gamma_2} = 3.236 \text{ k}\Omega$$

All passive sensitivities are less than or equal to 1 and the expected ω_0 and Q errors caused by finite ω_t are of the order of 22 kHz/2.3 MHz $\simeq 1\%$.

Before leaving the two-integrator-loop filters, let us recall the circuit in Fig. 4-18, which was introduced as a method for active compensation of a finite-gain amplifier. We pointed out at that time that the resulting circuit, repeated here as Fig. 5-32 for convenience, could be regarded as a "capacitorless"[11] active RC filter, a so-called active R filter. The student should also understand now that with the op amps modeled as integrators, that is with

$$A(s) \simeq \frac{\omega_t}{s}$$

---

[11]The capacitors are, of course, the internal compensation capacitors of the op amps.

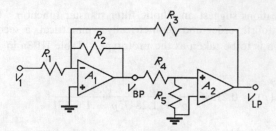

Figure 5-32 Active R filter.

the structure in Fig. 5-32 is nothing but a two-integrator-loop filter where A_1 forms the lossy and A_2 the lossless integrator.[12] Although variations of the configuration with as few as two resistors have been reported [111, 112], we use here five resistors to permit greater freedom [87] in setting the parameters of the lowpass and bandpass filter functions

$$\frac{V_{LP}}{V_1} = -\frac{ad\omega_t^2}{s^2 + \dfrac{b}{\sqrt{cd}}\sqrt{cd}\,\omega_t s + cd\omega_t^2} \tag{5-123a}$$

$$\frac{V_{BP}}{V_1} = -\frac{\dfrac{a}{b}\sqrt{cd}\,\dfrac{b}{\sqrt{cd}}\,s\omega_t}{s^2 + \dfrac{b}{\sqrt{cd}}\sqrt{cd}\,\omega_t s + cd\omega_t^2} \tag{5-123b}$$

In Eqs. (5-123a) and (5-123b), we used again the abbreviations

$$a = \frac{G_1}{G} \qquad b = \frac{G_2}{G} \qquad c = \frac{G_3}{G} \qquad d = \frac{G_4}{G_4 + G_5} \qquad G = G_1 + G_2 + G_3$$

Quality factor and pole frequency, respectively, are given by

$$Q = \frac{\sqrt{cd}}{b} \quad \text{and} \quad \omega_0 = \omega_t \sqrt{cd} \tag{5-124}$$

and we recognize that all passive sensitivities are very small (≤ 0.5 in magnitude) and that $S_{\omega_t}^{\omega_0} = 1$. Active R filters were introduced primarily to achieve more easily integrable circuits with reduced capacitor values[13] and further to obtain improved high-frequency performance by including the more realistic op amp integrator model directly into the design. Note that the typical pole-frequency and pole-Q errors that are normally contributed by the op amp's finite gain-bandwidth product ω_t are now caused by the second pole that occurs at much higher fre-

---

[12]This was already pointed out in Section 4.3.

[13]The internal compensation capacitors are multiplied by the Miller effect to yield a much larger effective size than their actual value. A few picofarads is sufficient for filter operation as low as the audio range.

quencies: using the model of Eq. (4-11), i.e., $A(s) = s(1 - s/\omega_2)/\omega_t$, the student may show that the poles are the roots of the polynomial

$$s^2\left[1 - b\frac{\omega_t}{\omega_2} + cd\left(\frac{\omega_t}{\omega_2}\right)^2\right] + s\omega_t\left(b - 2cd\frac{\omega_t}{\omega_2}\right) + cd\omega_t^2 = 0$$

Thus, using Eq. (5-124), pole frequency and pole Q are found to be

$$\omega_a \simeq \omega_0\sqrt{1 + \frac{1}{Q}\frac{\omega_0}{\omega_2} - \left(\frac{\omega_0}{\omega_2}\right)^2} \simeq \omega_0 \qquad (5\text{-}125a)$$

and

$$Q_a \simeq \frac{Q}{\sqrt{1 - 2Q\dfrac{\omega_0}{\omega_2}}} \qquad (5\text{-}125b)$$

Note that as in previous observations we find the frequency error to be quite small, whereas the quality factor can be severely enhanced.

Example 5-10

Determine the frequency and Q errors in an active R filter designed for $f_0 = 100$ kHz and $Q = 15$. The available op amps have $\omega_t = 2\pi \cdot 1$ Mrad/s and $\omega_2 = 2\pi \cdot 3.5$ Mrad/s. Discuss any potential stability problems in the circuit. Determine the maximum permitted excess integrator phase (refer to Section 4.2.2.1) if the Q errors must be controlled to less than 10%.

Solution. For the given parameters, Eqs. (5-125a) and (5-125b) result in $\Delta f_0 \simeq 54$ Hz, a negligible 0.054% error, while $Q_a \simeq 40$, an error of $\simeq 165\%$!

To express this problem differently, recall that $\Delta\phi \simeq \omega_0/\omega_2$ was defined in Section 4.2.2.1 as the excess phase (below $-90°$) of the integrator at the operating frequency ω_0. We find, therefore, from Eq. (5-125b), that a phase error larger than

$$\Delta\phi \simeq \frac{1}{2Q}\frac{180°}{\pi}$$

will cause the circuit to oscillate ($Q < 0$). For the numbers in this example, this means that $\Delta\phi$ must be less than 1.9°.

If the Q error must remain less than 10%, we find from Eq. (5-125b)

$$\frac{Q_a - Q}{Q} \simeq \frac{1}{\sqrt{1 - 2Q\,\Delta\phi}} - 1 \leq 0.1$$

i.e.,

$$\Delta\phi \leq \frac{1}{2Q}\left(1 - \frac{1}{1.21}\right)\frac{180°}{\pi} \simeq 0.33°$$

The numbers obtained in the above example are typical for any high-frequency active filter design and are *not* indicative of active R filters only. The main information that the student should retain from this example is that *filter operation at high frequencies is extremely sensitive to phase errors in the feedback loops.* Because these errors are determined not only by the active devices used in the design but also by such factors as wiring and layout parasitics which are very difficult to control and model, it should be apparent that these problems cannot be solved in the design phase. Clearly, at frequencies of several hundred kilohertz or even megahertz it is not possible to predetermine parasitic phase shifts by design to an accuracy of a fraction of one degree; rather, postfabrication tuning becomes unavoidable. Our discussion in Chapter 7 will treat suitable automatic tuning methods in detail. The procedures will at the same time address the more obvious problem of frequency tuning: note from Eq. (5-124) that ω_0 is proportional to ω_t, a parameter that varies widely from op amp to op amp and depends furthermore on operating conditions, such as bias and temperature. Obviously, then, because ω_t is not a parameter that should dominantly affect the filter's operation, active R filters cannot be considered practical as discrete filter circuits. They can, however, be used as a model or concept for *integrated-circuit* realizations of analog filters [113], where on-chip automatic tuning methods can be used to solve the tolerance and drift problems.

Before we conclude our discussion of second-order sections which use three op amps we would like to note that there are other three-op amp biquads [117, 121, 122] which have sensitivity and stability properties superior to those of the HKN and TT circuits when real op amps are used. Several such circuits are compared in reference 47 with respect to Q and ω_0 enhancements and will not be pursued here.

5.4 CIRCUITS USING TRANSCONDUCTANCES

We discussed in Section 4.4 that using operational transconductance amplifiers (OTAs) results in convenient realizations of active biquads because OTAs lend themselves readily to the construction of fundamental active building blocks. To demonstrate the ease of low-sensitivity OTA biquad design, let us investigate one particular circuit which is based on the two-integrator-loop configuration of Fig. 5-26b. It is redrawn along with a modified version in Fig. 5-33. Simple analysis shows that for both feedback structures the system poles are given by the polynomial

$$s^2 + sa_1c_1 + a_1a_2c_2 = 0$$

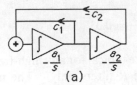
(a)

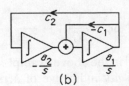
(b)

Figure 5-33 Fundamental two-integrator loops for transconductance biquads.

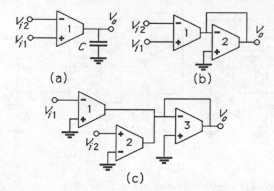

Figure 5-34 Fundamental transconductance building blocks: (a) integrating, (b) scaling, (c) summing.

To realize the filter, we have to investigate only how to implement the different building blocks identified in Fig. 5-33: inverting and noninverting integrators, summers, and scaling factors. While we could assemble these blocks with OTAs, capacitors, and resistors (in analogy to our previous designs that relied on op amps, capacitors, and resistors), let us choose here to use *only* OTAs and capacitors, and furthermore only *grounded* capacitors in anticipation of our discussion of fully integrated realizations in Chapter 7: for IC implementation, transconductance-capacitor filters are easier to design, and grounded capacitors are more convenient and less affected by parasitic errors than floating capacitors. In addition, we shall see that tuning procedures are simplified and problems caused by mismatch or lack of tracking of dissimilar components tend to be reduced.

Figure 5-34 shows simple methods for the realization of the needed building blocks [65]. An inverting or noninverting integrator realizing

$$V_o = \frac{g_{m1}}{sC}(V_{i1} - V_{i2}) \tag{5-126}$$

is shown in Fig. 5-34a; Fig. 5-34b demonstrates scaling of either polarity:

$$V_o = \frac{g_{m1}}{g_{m2}}(V_{i1} - V_{i2}) \tag{5-127}$$

and a convenient summing (or differencing) circuit with different scale factors realizing

$$V_o = -\frac{g_{m1}}{g_{m3}}V_{i1} + \frac{g_{m2}}{g_{m3}}V_{i2} \tag{5-128}$$

is shown in Fig. 5-34c. Connecting these blocks with appropriate signs as indicated in Fig. 5-33 results in the pole-determining loops shown in Fig. 5-35 [114]. Elementary analysis shows that the system polynomials are

$$s^2 + s\frac{\omega_0}{Q} + \omega_0^2 = s^2 + s\frac{1}{C_1}\frac{g_{m1}g_{m3}}{g_{m5}} + \frac{g_{m1}g_{m2}g_{m4}}{C_1C_2g_{m5}} \tag{5-129a}$$

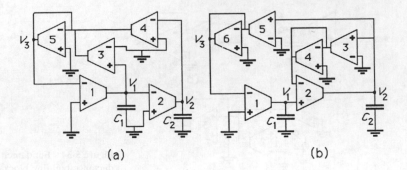

(a)   (b)

Figure 5-35   OTA implementation of Fig. 5-33a and b.

for Fig. 5-35a and

$$s^2 + s\frac{\omega_0}{Q} + \omega_0^2 = s^2 + s\frac{1}{C_2}\frac{g_{m2}g_{m3}}{g_{m4}} + \frac{g_{m1}g_{m2}g_{m5}}{C_1 C_2 g_{m6}} \qquad (5\text{-}129\text{b})$$

for Fig. 5-35b. The two capacitors may be chosen equal for convenience in discrete designs; however, it is probably preferable, especially for IC implementations, to choose all transconductances identical.[14] In that case, with $g_{mi} = g_m$ for all i, we find from Eq. (5-129a)

$$\omega_0 = \frac{g_m}{\sqrt{C_1 C_2}} \quad \text{and} \quad Q = \sqrt{\frac{C_1}{C_2}}$$

Note that all sensitivities are low, i.e., equal to $\frac{1}{2}$ in magnitude, but that the capacitor ratio equals Q^2. Obviously, for a circuit with identical g_m values, the g_{m3}–g_{m4} and the g_{m5}–g_{m6} combinations in Fig. 5-35b can be replaced by short circuits for a saving of four OTAs, but the more general circuits shown permit convenient voltage scaling and greater flexibility in setting transmission zeros, as will be demonstrated in the following.

To obtain the desired transmission zeros *without* disturbing the system poles, we use the by now well-known technique of connecting input *voltages* to previously *grounded* terminals; note though, that we may *also* inject input *currents* into any finite-impedance node, such as the nodes labeled V_1, V_2, and V_3 in Fig. 5-35.[15] Evidently, with seven grounded plus three finite-impedance nodes there are many possibilities for obtaining different transmission zeros. The student is strongly

---

[14]In discrete designs, all OTAs will also be nominally identical, and different g_{mi} values must be set by different bias currents [65]; thus, even in this case, advantages may result from a design with all identical g_m values.

[15]Input currents of either polarity are obtained, of course, with a transconductance via $I = g_m(V_{i1} - V_{i2})$.

encouraged to explore the alternatives. In Fig. 5-36, we show a subset of the possibilities for Fig. 5-35a which retains our goal of grounded capacitors. Considering V_a, V_b, and V_c as inputs, the output voltages of the circuit can be derived to equal

$$V_1 = \frac{sC_2g_{m1}(g_{m5}V_a - g_{m4}V_c) + g_{m1}g_{m2}g_{m4}V_b}{D(s)} \tag{5-130a}$$

$$V_2 = \frac{(sC_1g_{m2}g_{m5} + g_{m1}g_{m2}g_{m3})V_b + g_{m1}g_{m2}(g_{m4}V_c - g_{m5}V_a)}{D(s)} \tag{5-130b}$$

$$V_3 = \frac{s^2C_1C_2g_{m4}V_c + s(C_2g_{m1}g_{m3}V_a - C_1g_{m2}g_{m4}V_b) + g_{m1}g_{m2}g_{m4}V_a}{D(s)} \tag{5-130c}$$

where

$$D(s) = C_1C_2g_{m5}\left(s^2 + s\frac{1}{C_1}\frac{g_{m1}g_{m3}}{g_{m5}} + \frac{g_{m1}g_{m2}g_{m4}}{C_1C_2g_{m5}}\right) \tag{5-130d}$$

Choosing, for example, $V_a = V_b = 0$ and $V_c = V_i$ results in

$$\frac{V_1}{V_i} = H_{BP}(s) = -\frac{sC_2g_{m1}g_{m4}}{D(s)}$$

$$\frac{V_2}{V_i} = H_{LP}(s) = \frac{g_{m1}g_{m2}g_{m4}}{D(s)}$$

$$\frac{V_3}{V_i} = H_{HP}(s) = \frac{s^2C_1C_2g_{m4}}{D(s)}$$

with $D(s)$ given in Eq. (5-130d). A general biquadratic transfer function can be obtained by setting $V_a = V_b = V_c = V_i$, so that from Eq. (5-130c)

$$\frac{V_3}{V_i} = \frac{s^2C_1C_2g_{m4} + s(C_2g_{m1}g_{m3} - C_1g_{m2}g_{m4}) + g_{m1}g_{m2}g_{m4}}{D(s)} \tag{5-131}$$

Clearly, $C_2g_{m1}g_{m3} = C_1g_{m2}g_{m4}$ results in a band-rejection function, and setting $C_1g_{m2}g_{m4} = 2C_2g_{m1}g_{m3}$ and $g_{m4} = g_{m5}$ yields an allpass filter. The indicated equality

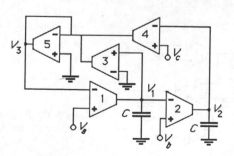

Figure 5-36 General OTA biquad based on Fig. 5-35a.

conditions may be difficult to maintain in discrete circuits without tuning; they are, however, easily implemented in integrated form.

Using different inputs or the topology of Fig. 5-35b results in different ways for realizing second-order transfer characteristics which may have advantages or disadvantages in number of components, element spread, ease of tuning, or sensitivity to parasitics.

Note that the voltage transfer functions given in the above equations are *ideal*, and recall from Sections 4.2.3 and 4.4 that parasitic input and, especially, output impedances cause errors in filter performance that should be investigated from case to case (Problem 5.33) [125]. For example, the student may show that the parasitic input and output capacitors of the OTAs result in one parasitic pole in the configuration of Fig. 5-35a and in two parasitic poles in Fig. 5-35b, in addition to the effects of the output conductances. At high operating frequencies for which OTA designs are intended, the deviations of these poles may well not be negligible. For filters in the audio-frequency range, however, errors caused by parasitic components and also by the finite bandwidth of OTAs are usually negligible.

A final very simple lowpass-bandpass configuration that uses OTAs and avoids many of the parasitic problems is shown in Fig. 5-37. As indicated, it is derived from a passive RLC resonance circuit by making use of the grounded resistor of Fig. 4-13, here $R = 1/g_{m3}$, and the grounded inductor simulation of Fig. 4-36a, here $L = C/(g_{m1}g_{m2})$. The input current I is obtained from $I = g_{m3}V_i$. Thus, setting $g_{m1} = g_{m2} = g_m$, we obtain in complete analogy to the passive circuit in Fig. 5-37, given by

$$\frac{V_{BP}}{I} = \frac{1}{sC + 1/(sL) + G} = \frac{sL}{s^2LC + sLG + 1}$$

the following active realization:

$$\frac{V_{BP}}{g_{m3}V_i} = \frac{1}{sC + g_m^2/(sC) + g_{m3}}$$

i.e.,

$$\frac{V_{BP}}{V_i} = \frac{sg_{m3}C}{s^2C^2 + sg_{m3}C + g_m^2} \tag{5-132a}$$

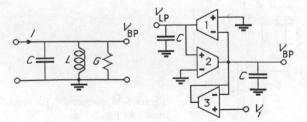

Figure 5-37 Passive RLC resonance circuit and the active OTA-C simulation.

and

$$\frac{V_{LP}}{V_i} = -\frac{g_m g_{m3}}{s^2 C^2 + s g_{m3} C + g_m^2} \qquad (5\text{-}132b)$$

Note from Fig. 5-37 that all parasitic OTA capacitors are in parallel with the main circuit capacitors; thus, they do not increase the order of the transfer function and can be absorbed by predistortion. Also, the output conductances of OTAs 2 and 3 are in parallel with the "conductance" g_{m3} and can be absorbed by predistortion. Thus, the only "separate" parasitic is g_o of OTA 1; the student may show that its effect is to lower Q and increase the pole frequency. Evidently, the circuit realizes nominally

$$\omega_0 = \frac{g_m}{C} \quad \text{and} \quad Q = \frac{g_m}{g_{m3}} \qquad (5\text{-}133)$$

and has the additional advantage that, at $\omega = \omega_0$, where the maximum output voltages occur, both voltage gains are equal to unity:

$$\left| \frac{V_{BP}}{V_i}(j\omega_0) \right| = 1 \qquad \left| \frac{V_{LP}}{V_i}(j\omega_0) \right| = \frac{g_m}{\omega_0 C} = 1$$

The maximum internal voltage seen by the OTAs is, therefore, given by V_i, which can be controlled easily without scaling; this latter fact is convenient in practice because all available OTA designs suffer from a limited linear signal range.

5.5 SUMMARY

In this chapter we have presented the design and performance of single-amplifier and multiple-amplifier biquads. A few additional good circuits are presented in the problems at the end of this chapter, and many more active second-order sections with reasonable performance or with special properties can be found in the literature. We have chosen to concentrate on the few circuits that are useful for pedagogical reasons and some that are generally accepted as the "best" second-order active filters, and we have discussed their design and evaluation in considerable detail in order to give the student the tools with which to evaluate or optimize other circuits that may seem appropriate for the intended applications. The student should remember that it is in general preferable to use circuits whose pole frequency is determined by the minimum number (*four*) of passive components and is independent of the parameters of the active devices,[16] which normally have large

[16]This condition is, of course, violated by OTA-based designs, where ω_0 is proportional to an expression of the form $\sqrt{g_{m1}g_{m2}}$. Circuits which use OTAs, or more generally transconductances, are especially useful for fully integrated filters (see Chapter 7), where on-chip tuning schemes are available to control g_m. Apart from yielding simpler, more intuitive design approaches, the main advantage of OTA-based circuits is the wider useful frequency range. Until OTAs with 'designable' g_m become available, the engineer should remember that g_m is in general a quantity with wide tolerances. It is set by applying a bias current which in turn must be carefully controlled if the frequency parameters are to be stable.

variabilities. The passive sensitivities of the filters' critical pole frequencies are then at their theoretical minimum value, 0.5. All circuits discussed in this book satisfy this property.

The frequency and Q errors caused by the op amps' finite gain-bandwidth product for all good filters are of the form $k\omega_0/\omega_t$; i.e., they are proportional to $1/|A(j\omega_0)|$ with k of the order of unity. Note, though, the behavior of two-integrator-loop filters, where the quality factor may suffer severe deviations due to small phase errors in the loop. Generally, different good circuits *differ* only in the multiplying factor k. An exception is the design based on a composite amplifier where the errors are very small, ideally proportional to $1/|A(j\omega_0)|^2$. It should be apparent to the student that the composite-amplifier idea can also be applied to multiamplifier circuits. For example, a quad–op amp chip could be used to design a GIC-type filter with extremely low active *and* passive sensitivities *and* convenient noninteractive tuning, albeit using four op amps.

The only real advantages of multiamplifier circuits are (1) the simultaneous availability of different transfer functions at different outputs and (2) the ease of tuning and noninteractive adjusting of filter parameters, because the design equations are seen to be decoupled by the insertion of additional op amps. As far as practical performance in terms of sensitivity is concerned, multi-op amp filters were seen to perform not significantly better than well-designed single-amplifier circuits. The increased expense of additional op amps and power consumption should, therefore, be carefully weighed against the advantages gained. If low sensitivity— in particular to the active device tolerances—is of major concern, it may well be preferable to build a *single* amplifier filter with the appropriate *composite* amplifier.

A detailed comparison of the circuits described in this book as well as of many other configurations can be found in reference 47. The comparison is based on statistical multiparameter sensitivity measures (Chapter 3) and essentially supports the rating arrived at in this book with simpler, more intuitive means.

PROBLEMS

5.1 Active filters can, of course, become unstable. To demonstrate this fact, discuss the stability of the circuit in Fig. P5.1 as a function of the amplifier gain K. Draw a locus of the poles in the s-plane as a function of K.

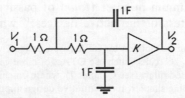

Figure P5.1

5.2 Consider the circuit in Fig. 5-3 described by Eq. (5-5). Determine which transfer functions can be realized with this circuit by an appropriate choice of the available parameters α, β, and γ.

5.3 Use the ideas expressed in Eqs. (5-10) and (5-11) to build a second-order allpass filter out of the GIC bandpass in Fig. 5-23.   Compare your result with Fig. 5-25.

5.4 Consider the circuit in Fig. P5.4.

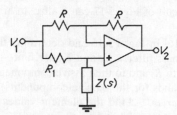

Figure P5.4

(a) What are the conditions to be satisfied by the impedance $Z(s)$ so that $T(s) = V_2/V_1$ is an allpass function?

(b) Configure the circuit to realize a first-order allpass function (see Problem 4.18).

(c) Use the circuit to design a third-order allpass to have a maximally flat delay of $\tau = 55$ μs.

5.5 The PF circuit of Fig. 5-6b uses a unity-gain amplifier (voltage follower) whose phase lag for small values of ω was shown in Eq. (4-49) to be equal to $\Delta\phi \approx -\omega/\omega_t$.   Find a passive compensation network to correct this nominal phase error and thereby extend the useful frequency range of the PF circuit.

    Hint: Use a resistor in the feedback path and a capacitor to ground.

5.6 Use the ENF circuit, case A (Fig. 5-12), to realize a bandpass function with $f_0 = 4.5$ kHz, $Q = 3.8$, and midband gain $H_B = 1.4$ dB.   Use any free parameters to minimize sensitivities.   What are the ω_0 and Q errors if op amps with $\omega_t = 2\pi \cdot 1$ Mrad/s are available?   What are the ω_0 and Q errors if the resistors and capacitors, respectively, may be assumed to match and track but their nominal values have, respectively, 1% and 5% errors?

5.7 Use the ENF circuit, case A, to realize a notch filter with $f_0 = 1.2$ kHz, $Q = 6$, and a dc gain $H_0 = 1$.

5.8 Verify that the circuits in Fig. 5-12 are complementary to those in Fig. 5-13, and thus that they are equivalent in terms of their pole sensitivity behavior.

5.9 For the circuits in Fig. 5-13, determine the coefficients α and β_i, $i = 2, 4$, necessary to realize an LP, an HP, a BP, a notch, or an allpass function.

5.10 Consider the four second-order filter sections in Figs. 5-12 and 5-13.

(a) Determine which of the circuits is preferable for realizing an HP, an LP, a BP, a notch, or an allpass section.   Note the results of Problem 5.8!   See reference 115 for a detailed discussion.

(b) What effect, if any, has the finite frequency-dependent op amp gain on the numerator coefficients?

5.11 Use the most appropriate one of the circuits in Fig. 5-12 to realize a general biquadratic function with $\omega_p = \omega_z = \omega_0 = 2\pi \cdot 2000$ rad/s, $Q = 3$, and (a) $Q_z = kQ$ and (b) $Q_z = Q/k$.   For $k = 3$, sketch the magnitudes of the two functions versus frequency.

    Such circuits are called *gain equalizers*; they are used alone or together with delay equalizers to correct or compensate for distortions introduced in transmission systems.

5.12 Use the EPF circuit of Fig. 5-13a for building a second-order allpass to realize a maximally flat delay of 1 ms. The pole frequency is $f_0 = 2.2$ kHz.

5.13 Realize a notch filter with $Q = 8$ and $f_z = 800$ Hz with the ENF circuit of Fig. 5-16. Choose any available free parameters to minimize the gain-sensitivity product.

5.14 Show that the EPF circuit of Fig. 5-17 can realize an arbitrary second-order transfer function.

5.15 Analyze the circuit [116] in Fig. P5.15 and determine the admittances such that the circuit realizes a bandpass filter. K_1 and K_2 are finite-gain amplifiers. Find the ω_0 and the Q sensitivities to K_i and to the passive components. Determine the elements such that $S_x^Q = 0$ (x stands for the passive components). For this design the circuit has been called *Q-invariant*. Find the element values for $f_0 = 1$ kHz, $Q = 33.3$. What, if any, is the limit on midband gain?

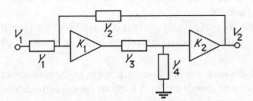

Figure P5.15

5.16 Use the ENF circuit of Fig. 5-12 to realize the two-section delay equalizer of Example 1-8.

5.17 A batch of active bandpass filters has to be designed to satisfy $Q = 15$ and $f_0 = 9.8$ kHz. The available op amps have nominally $\omega_t/2\pi = 1.2$ MHz, but the expected variations due to varying operating conditions and tolerances from unit to unit are -500 kHz $\le \Delta f_t \le +800$ kHz. If multiple op amps are used, they may be assumed to be matched and to track. Design the circuit by the following methods, calculate the passive Q sensitivities, and estimate the expected pole-frequency and pole-Q errors caused by the mentioned ω_t tolerances. Prepare a table with your results and comment on the different realizations.

(a) A single-amplifier ENF circuit with free parameters chosen to minimize the active sensitivity

(b) An NF circuit (Fig. 5-6a), i.e., the design of part (a) with an infinite-gain amplifier, $K = 1$

(c) The design of part (a), but with the op amp replaced by the composite amplifier of Fig. 5-20 with $p = 0$

(d) The design of part (c), but assuming that the resistor ratio α has a 1% error

(e) The design of part (b), but with the op amp replaced by the composite amplifier of Fig. 5-20 with $p = 0$ and $\alpha = K = 1$ as in Fig. 5-21

(f) The GIC bandpass circuit of Fig. 5-23 with $H = 2$

(g) The Åckerberg-Mossberg circuit of Fig. 5-29

(h) The active R filter of Fig. 5-32

5.18 With the comments made after Eq. (5-96), prove that the ENF filter using the composite amplifier is stable.

5.19 Consider the GIC bandpass filter of Fig. 5-23. Show that gains of value less than 2 can be obtained by splitting the input resistor R.

5.20 Determine the effect of finite capacitor quality factors on the performance of the GIC bandpass filter. Determine the minimum value of the capacitor's quality factor if the filter Q is not to change by more than 5% from a nominal design value of 50.

Hint: Finite Q_c means lossy capacitors; model these losses as a parallel conductance for each of the two capacitors.

5.21 Determine which types of transfer functions (BP, LP, or HP, for example) can be realized with the GIC circuit of Fig. 5-25.

5.22 Develop a GIC circuit to realize a lowpass function (see reference 11, Section 9.8).

Hint: Form a voltage divider from a resistor R and the grounded impedance of Fig. 4-33 [Eq. (4-71)]. When selecting the elements, remember that the output must be taken from an op amp output.

5.23 Use the method of Fig. 5-30 and the Tow-Thomas biquad to realize a lowpass notch with dc gain $H_l = 2.6$, $\omega_z = 7.2$ krad/s, $\omega_0 = 4$ krad/s, $Q = 4.6$, and high-frequency gain $H_h = 1$.

5.24 Realize a bandpass filter with $f_0 = 3.4$ kHz, $Q = 14$, and gain $H = 2.6$ using the Tow-Thomas biquad. Give all element values.

5.25 Repeat Problem 5.24 for the Åckerberg-Mossberg circuit. Find the Q and ω_0 deviations if the op amps have a gain-bandwidth product of $\omega_t/2\pi = 1.5$ MHz.

5.26 Use Fig. 5-31 to build a circuit with the specifications given in Problem 5.23.

5.27 Use Fig. 5-31 to build a circuit with the specifications given in Problem 5.11.

5.28 Use Fig. 5-31 to build a circuit with the specifications given in Problem 5.12.

5.29 Design an active R bandpass filter (Fig. 5-32) with center frequency $f_0 = 100$ kHz, a bandwidth of 3 kHz between the 3-dB points, and midband gain $H_B = 150$. Use μA741 op amps with $f_t = 1.3$ MHz; you may assume that they are *nominally* equal.

Make a complete sensitivity analysis with respect to all the "elements" $x = R_i$, $i = 1, \ldots, 5$, and ω_{tj}, $j = 1, 2,$; i.e., find S_x^P, where P stands for ω_0, Q, and H_B. Interpret your results in view of the fact that S_x^P is a small-change measure and ω_{tj} may vary over wide ranges. Discuss the effect of the op amp's second pole, assumed to be located at $\omega_2 \simeq 2\pi \cdot 3.2$ Mrad/s.

5.30 Figure 5-36 was derived from Fig. 5-35a by injecting the input signals V_a, V_b, and V_c as shown. Use the analogous approach to derive a general second-order transfer function from Fig. 5-35b. Apply inputs at all possible points in the circuit (of course, without changing the poles) and then decide which inputs and which OTAs must be kept to retain a general biquadratic function.

5.31 Use Fig. 5-36 to design an allpass function with maximally flat delay of 40 μs; the pole frequency equals 82 kHz.

5.32 Use Fig. 5-36 to design a magnitude equalizer with a "bump" of (a) $+2$ dB and (b) -3 dB. Refer to Problem 5.11 for definitions.

5.33 Design an OTA notch filter by use of Fig. 5-36 for $f_z = f_p = 50$ kHz and $Q_p = 25$. Choose identical OTAs as far as possible. Investigate the effect of finite OTA output conductances $1/g_o = 1$ MΩ.

5.34 The so-called Tarmy-Ghausi (TG) circuit, shown in Fig. P5.34, was introduced in reference 117. It can be shown to have very low sensitivities to both the active and the passive elements; its disadvantage is the need for op amps with differential output. Although available, they are not as popular as single-ended output op amps. Determine the transfer functions V_k/V_i, $k = 2, 3, 4$, and the function V_o/V_i obtained by summing the outputs V_2, V_3, and V_4 as shown in the dashed box. Calculate the pole frequency and the pole Q as a function of the circuit parameters. Determine the sensitivities as a function of the circuit parameters. What types of functions can one obtain at the output V_o?

See reference 47 or 117 for a detailed discussion of the TG circuit.

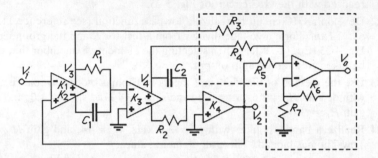

Figure P5.34

5.35 Figure P5.35 shows a modified version of the Tarmy-Ghausi (TG) circuit in Fig. P5.34 which avoids the disadvantage of differential op amp outputs. It was recognized by Moschytz [118, 119] that the amplifier stages in the TG circuit are first-order allpass functions which can also be realized with single-ended output op amps (see Fig. P5.4). Placing two such allpass stages into a loop yields the circuit in Fig. P5.35.

Determine the transfer functions V_k/V_i, $k = 1, 2, 3$; calculate the pole frequency and pole Q as a function of the elements; and determine the relevant sensitivity expressions.

Although the sensitivity measures of this circuit are not quite as low as those of the original TG circuit, it is much easier to realize.

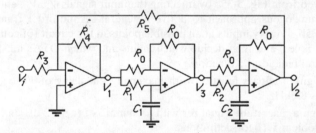

Figure P5.35

5.36 Use Moschytz' circuit in Fig. P5.35 together with the summing technique of Fig. P5.34 to realize the function

$$H(s) = \frac{s^2 + 3}{s^2 + 0.1s + 1}$$

The frequency parameter s is normalized with respect to $\omega_0 = 2\pi \cdot 9$ krad/s.

THE DESIGN
OF HIGH-ORDER FILTERS

SIX

6.1 INTRODUCTION

After having discussed in Chapter 5 the realization of second-order filters in great detail, we shall treat in this chapter the design methods that lead to practical filter implementations of order higher than 2. Specifically, we wish to investigate how the transfer function

$$H(s) = \frac{V_o}{V_i} = \frac{N(s)}{D(s)} = \frac{a_m s^m + a_{m-1} s^{m-1} + \ldots + a_1 s + a_0}{s^n + b_{n-1} s^{n-1} + \ldots + b_1 s + b_0} \qquad (6\text{-}1)$$

with $n \geq m$ and $n > 2$ can be realized in an efficient way with low sensitivities to component tolerances. For the most part, we shall in the following assume that both n and m are *even*, so that both numerator and denominator polynomials, $N(s)$ and $D(s)$, can be factored into the product of second-order pole-zero pairs [compare Eq. (3-67)]. Without loss of generality we may assume *even* polynomials in our discussion of *active* filters because an odd function can always be factored into the product of an even function and a *first-order* function. The latter can easily be realized by a *passive RC* network (Problem 6.1) and can be appended to the higher-order active filter as an additional section.

Our discussion of sensitivity behavior of high-order filter realizations in Chapter 3 has shown that, in general, it is not advisable to realize the transfer function $H(s)$ in the so-called *direct form*. By *direct form* we mean a network that uses only one or perhaps two op amps embedded in a high-order passive *RC* network such that Eq. (6-1) is realized. Although it is possible in principle to realize Eq. (6-1) in direct form, the resulting circuits are normally so sensitive to component tolerances that reliable performance in practice cannot be expected. A further disadvantage of these synthesis methods is their use of a very large number of passive components to realize a function of given order. Because of these problems, we shall in this book not discuss any direct realization methods. However, for historical reasons and to show that a direct realization is indeed possible, we give the reader in Problem 6.2 an opportunity to develop a suitable synthesis method and to demonstrate that high sensitivities must normally be expected; the procedure leads to the realization of the transfer function coefficients via *difference effects*, which, according to our sensitivity discussion in Chapter 3, should be avoided if at all possible.

Having ruled out the *direct realization*, we are left with the following methods which were identified in Chapter 3 as resulting in acceptable low-sensitivity designs:

1. The *cascade approach*, where the high-order function $H(s)$ is factored into subnetworks of second order. The resulting *biquads* are realized by the methods discussed in Chapter 5 and connected in cascade such that their product implements the prescribed function $H(s)$. The cascade method is used widely in industry because it is well understood, very easy to implement, and efficient in its use of active devices (as few as one op amp per pole pair). It uses a modular approach and results in filters that for the most part show satisfactory performance in practice. One of the main advantages of cascade filters is that they are very *easy to tune* because each biquad is responsible for the realization of only *one* pole pair (and zero pair): the realizations of the individual critical frequencies of the filter are *decoupled* from each other. The disadvantage of this decoupling is that for filters of high order, say, $n > 8$, with stringent requirements and tight tolerances, cascade designs are often found to be still too sensitive to component variations in the passband. In these cases, the following approaches lead to more reliable circuits.

2. The *multiple-loop feedback* or *coupled-biquad approach*, which also splits the high-order transfer function into second-order subnetworks. However, they are then interconnected in some type of feedback configuration. This feedback introduces *coupling* between the biquads and is carefully chosen to reduce the transfer function sensitivities (see Section 3.2.6.3). The multiple-loop feedback approach retains the modularity of cascade designs but at the same time yields high-order filter realizations with noticeably better passband sensitivities (see Fig. 3-13). Of the numerous topologies that have been proposed in the literature [e.g., 36], we shall discuss only the FLF (follow-the-leader feedback) method of Fig. 3-12a and the LF (leapfrog) method of Fig. 3-12b, along with a few special cases. Both methods are particularly well suited for all-pole functions but can be extended to realizations of arbitrary high-order transfer functions $H(s)$. Although based on coupling of biquads in a feedback configuration, the LF procedure is derived by simulating an *LC* ladder and will, therefore, be treated as part of the following method.

3. The *ladder simulation approach*, which attempts to find active realization methods that inherit in one way or another the recognized excellent low-sensitivity properties of passive doubly terminated *LC* ladder filters (see Section 3.2.6.2). Broadly speaking, the ladder simulation approaches fall into two groups. One is based on *element substitution*, where the inductors are simulated, e.g., via gyrators or general immittance converters (see Section 4.3.3), and the resulting active "components" are inserted into the "*LC*" filter topology. The second group may be labeled *operational simulation* of the *LC* ladder, where the active circuit is configured to realize the *internal operation*, i.e., the equations, of the *LC* prototype without any regard to the circuit's topology. Active filters simulating the behavior of *LC* ladders have been found to have the lowest sensitivities and, consequently, to be the most appropriate for filters with stringent requirements. They have the additional advantage that they can draw on the wealth of knowledge gained in the area of analysis, synthesis, and fabrication of lossless filters. For example, the many passive filter tables [13, 18, 21] can be used directly in the design of active ladder simulations. A disadvantage of this design method is that a passive *LC* prototype must, of course, exist[1] before an active simulation can be attempted. Also, we shall find that usually a relatively large number of active devices are required.

In this chapter, we shall discuss these methods in sufficient detail to permit the reader to design practical manufacturable active filters. We shall start with the cascade approach because of its simplicity and wide use. After cascade designs, we shall present the FLF multiple-loop feedback topology, which has been found to yield useful filters with a reasonable amount of design effort. Both cascade and multiple-loop feedback techniques are modular, with active biquads used as the fundamental building blocks [36]. As a further practical advantage we mentioned that both these types of realizations are very efficient in their use of active elements (one op amp per pole pair), whereas ladder simulations, to be discussed last, often require many more active devices to realize a transfer function of given order, resulting in increased power consumption and more noise sources.

6.2 CASCADE REALIZATIONS

As discussed in Section 3.2.6.1, the high-order transfer function [Eq. (6-1)] is factored into the product of second-order functions

$$T_i(s) = k_i \frac{\alpha_{2i}s^2 + \alpha_{1i}s + \alpha_{0i}}{s^2 + s\omega_{0i}/Q_i + \omega_{0i}^2} = k_i t_i(s) \qquad (6-2)$$

---

[1]The realizability conditions for passive *LC* filters are more restrictive than those for active *RC* filters.

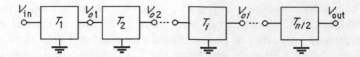

Figure 6-1 Cascade realization of nth-order transfer function.

such that

$$H(s) = \prod_{i=1}^{n/2} T_i(s) = \prod_{i=1}^{n/2} k_i \frac{\alpha_{2i}s^2 + \alpha_{1i}s + \alpha_{0i}}{s^2 + s\omega_{0i}/Q_i + \omega_{0i}^2} = \prod_{i=1}^{n/2} k_i t_i(s) \qquad (6\text{-}3)$$

Recall that n is assumed to be even. The transfer functions of the individual biquads are labeled $T_i(s)$ in this chapter, and k_i is a suitably defined *gain constant*, e.g., such that the leading coefficient in the numerator of the *gain-scaled* transfer function $t_i(s)$ is unity or such that $|t_i(j\omega_{0i})| = 1$. The coefficients α_{2i}, α_{1i}, and α_{0i} determine the type of second-order function $T_i(s)$ which can be realized by an appropriate choice of biquad from the circuits discussed in Chapter 5. If we then may assume that the output impedances of the biquads are sufficiently small (compared to the input impedances), all second-order blocks can be connected in cascade (Fig. 6-1) without causing mutual interactions due to loading, and the product of the biquadratic functions is realized as required by Eq. (6-3).

Although this simple process leads in a straightforward way to a possible cascade design, it leaves several questions unanswered:

1. Which zero should be assigned to which pole when the biquadratic functions $T_i(s)$ are formed? Since we have $n/2$ pole pairs and $n/2$ zero pairs (counting zeros at 0 and at ∞), we can select from $(n/2)!$ possible *pole-zero pairings*.
2. In which order should the biquads be cascaded? Does the *cascading sequence* make a difference? For $n/2$ biquads, we have $(n/2)!$ possible sequences.
3. How should the gain constants k_i in Eq. (6-2) be chosen to determine the signal level for each biquad? In other words, what is the optimum *gain distribution*?

Because the total transfer function is simply the product of the biquad sections, the selections in steps 1, 2, and 3 are quite arbitrary as far as $H(s)$ is concerned. However, they do affect to some extent the sensitivity performance.[2] More important, they determine significantly the dynamic range, i.e., the distance between the maximum possible undistorted signal and the noise floor, because the maximum and minimum signal levels throughout the cascade filter will be seen to depend on pole-zero pairings, cascading sequence, and gain distribution.

We saw in Chapter 4 that the maximum undistorted signal voltage that a filter

---

[2]Although sensitivities are a function of pole-zero pairing, the effect usually is not very strong and will not be discussed in this book. For a detailed treatment, see references 47 and 97. Also, the selection of pole-zero pairing for best sensitivity often conflicts with the choice necessary for best dynamic range.

can process is limited, depending on the operating frequency, either by the power supply or by the slew rate of the op amps. Let us label this maximum signal level V_{op} and assume that it is measured at the *output* of the biquadratic section.[3] We have to make sure, then, that the signal level at any section output, $|V_{oi}(j\omega)|$, satisfies

$$\max |V_{oi}(j\omega)| < V_{op} \qquad 0 \le \omega < \infty, i = 1, \ldots, n/2 \qquad (6\text{-}4a)$$

Note that this condition must indeed be satisfied for *all* frequencies and not only in the passband, because large signals even outside the passband must not be allowed to overload and saturate the op amps.[4]

The lower limit of the useful signal range is set by the noise floor. If *in the passband* of a cascade filter the signal at an internal stage becomes very small, it must be amplified back up to the prescribed output level. Since from any point in the cascade of filter stages, say, at the output of stage i, signal and noise are amplified by the same amount, namely,

$$H_i^+(s) = \prod_{j=i+1}^{n/2} T_j(s) \qquad (6\text{-}5)$$

we may conclude that the signal-to-noise ratio will suffer if in the cascade filter the signal suffers inband attenuation, i.e., if it is permitted to become small. The function $H_i^+(s)$, defined in Eq. (6-5), is the *noise gain* from the output of section i to the filter output. Thus, the second condition to be satisfied by the output voltage of any biquad is

$$\min |V_{oi}(j\omega)| \to \max \qquad \omega_L \le \omega \le \omega_U, i = 1, \ldots, n/2 \qquad (6\text{-}4b)$$

In this case we are, of course, only concerned with signal frequencies between the lower and upper passband corners, ω_L and ω_U, respectively, because in the stopband, signal-to-noise ratio is of no interest.[5]

---

[3]This assumption will always be correct in single-amplifier biquads, where section output and op amp output are the same. In multiamplifier biquads, *each* op amp output has to be evaluated and the maximum op amp output voltage in the biquad section must be determined. To avoid overdriving any op amp sooner than any other one inside a biquad, it is intuitively reasonable that any available design freedom in the biquads should be chosen such that all op amps "see" the same signal level. If this is not possible, i.e., if the undistorted output voltage V_{oi} of section i is only a fraction $1/q_i$ of the maximum internal voltage, with $q_i > 1$, V_{oi} in all the following equations must be replaced by $q_i V_{oi}$. For simplicity, we shall assume in our discussion that the sections can be designed such that $q_i = 1$.

[4]When the op amps are being overdriven, the circuit still acts as a filter which removes the higher harmonics that are generated due to nonlinear operation. The problem that arises when saturating the op amps is, therefore, not so much harmonic distortion of the signal but intermodulation distortion and deviations of the transfer function shape; refer to Fig. 4-7c!

[5]The noise spectrum of a filter section usually has the same shape as the transfer function magnitude, which means that the highest noise peaks occur at the pole frequencies with the highest Q values. Since these are mostly found just beyond the specified corners of the passband, they would not be included in the measurement defined in Eq. (6-4b). Therefore, to avoid decreased dynamic range caused by possibly large noise peaks at the passband corners, it is advisable to extend ω_L and ω_U beyond the specified passband into the transition band to cover the pole frequencies with the highest Q values.

Pole-zero pairing, section ordering, and gain assignments will now be chosen such that the conditions in Eq. (6-4) are satisfied.

6.2.1 Pole-Zero Pairing

We show in Fig. 6-2 the magnitude $|t_i(j\omega)|$ of a gain-scaled transfer function $t_i(s)$ with finite poles and zeros and finite pole and zero quality factors. Also indicated are the relevant values of the maximum, M_i, and the minimum, m_i. Notice that, as is often the case, M_i lies outside and m_i at the edge of the passband, $\omega_L \leq \omega \leq \omega_U$. The actual magnitude minimum at ω_m lies in the stopband and is of no concern. From Fig. 6-2 it is apparent that the values of M_i and m_i will change if the relative positions and quality factors of the pole and the zero of $t_i(s)$ are altered (see also Fig. 1-18). Assuming a *flat input spectrum* at the input of each biquad section, i.e., $|V_{in}(j\omega)|$ = const, the pole-zero pairing should now be chosen such that M_i is minimized and m_i is maximized, or, in other words, $|t_i(j\omega)|$ *should be as flat as possible in the frequency range of interest.* Expressed differently, we would like to have the ratio M_i/m_i as close to unity as possible, which means that for each biquad the "*measure of flatness*"

$$d_i = \log \frac{M_i}{m_i} \qquad i = 1, 2, \ldots, n/2 \qquad (6\text{-}6)$$

should be minimized. The optimal pole-zero assignment for the total nth-order cascade filter is then the one which minimizes the maximum value of d_i:

$$d_{\max} = \max \{d_i\} \to \min \qquad i = 1, 2, \ldots, n/2 \qquad (6\text{-}7)$$

In the following, we present an algorithm which finds this optimal pairing [125–127, 11, 97].

Assume that the mth-order numerator polynomial $N(s)$ in Eq. (6-1) has r real zeros (including those at 0 and ∞) and $(m - r)/2$ complex zero pairs for a total of

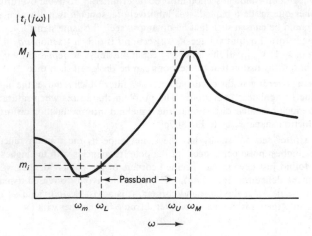

Figure 6-2 Magnitude of a gain-scaled biquadratic transfer function.

m roots z_j.   Note that r is even because we assume m to be even.   The denominator $D(s)$ of Eq. (6-1) has $n/2$ complex root pairs which we label p_k.

1. Construct a table containing all possible pole-zero pairings and the corresponding values of d_{kj}, defined as in Eq. (6-6), for pairing pole number k with zero number j.   The numbers d_{kj} in the table are calculated, either approximately with the help of the equations in Fig. 1-18, or exactly via a computer routine.   The result is shown in Table 6-1 for a tenth-order function $H(s)$ with only complex poles, three complex zeros z_1, z_2 and z_3, and four real zeros z_4 through z_7.   Remember that each second-order section has *two* zeros (including those at 0 and ∞); i.e., two real zeros must be assigned to any complex pole pair.   Thus, in the table, pole p_k can be paired with (z_4-z_5), labeled z_{45}, or (z_6-z_7), labeled z_{67}.   In addition, we have to consider the permutations z_{46}, z_{57}, and z_{47}, z_{56}.

    Each row shows the measures d_{kj} for all possible combinations of zeros with the particular pole: the first five columns contain the pairings if (z_4, z_5) and (z_6, z_7) are kept together; similarly, columns 1 to 3 and 6 to 7 contain the situation for the grouping z_{46} and z_{57}, and columns 1 to 3 and 8 to 9 show the results for z_{47} and z_{56}.

2. Select that pairing which results in the minimum value of $d_{\max}$ according to Eq. (6-7).   This selection process is much simplified by drawing a diagram of the values d_{kj} as a function of z_j as shown in Fig. 6-3.

    The abscissa is marked with the zero numbers (uniformly spaced for convenience); the computed values d_{kj}, identified by the pole number k, are entered on vertical lines above the zero marking j.

3. Obtain the pole-zero assignment by drawing horizontal *assignment lines*:

    Remembering that each pole has to be associated with exactly one complex zero or one *pair* of real zeros and that each zero can, of course, be assigned only once, the *first* assignment line (line 1) is drawn as low as possible

TABLE 6-1   TABLE OF FLATNESS MEASURES d_{kj} FOR ALL POSSIBLE POLE-ZERO PAIRINGS OF A TENTH-ORDER TRANSFER FUNCTION WITH COMPLEX POLES, THREE COMPLEX ZEROS, AND FOUR REAL ZEROS

d_{kj} is the measure in Eq. (6-6) for pairing pole number k with zero number j.

	z_1	z_2	z_3	z_{45}	z_{67}	z_{46}	z_{57}	z_{47}	z_{56}
p_1	d_{11}	d_{12}	d_{13}	$d_{1,45}$	$d_{1,67}$	$d_{1,46}$	$d_{1,57}$	$d_{1,47}$	$d_{1,56}$
p_2	d_{21}	d_{22}	d_{23}	$d_{2,45}$	$d_{2,67}$	$d_{2,46}$	$d_{2,57}$	$d_{2,47}$	$d_{2,56}$
p_3	d_{31}	d_{32}	d_{33}	$d_{3,45}$	$d_{3,67}$	$d_{3,46}$	$d_{3,57}$	$d_{3,47}$	$d_{3,56}$
p_4	d_{41}	d_{42}	d_{43}	$d_{4,45}$	$d_{4,67}$	$d_{4,46}$	$d_{4,57}$	$d_{4,47}$	$d_{4,56}$
p_5	d_{51}	d_{52}	d_{53}	$d_{5,45}$	$d_{5,67}$	$d_{5,46}$	$d_{5,57}$	$d_{5,47}$	$d_{5,56}$

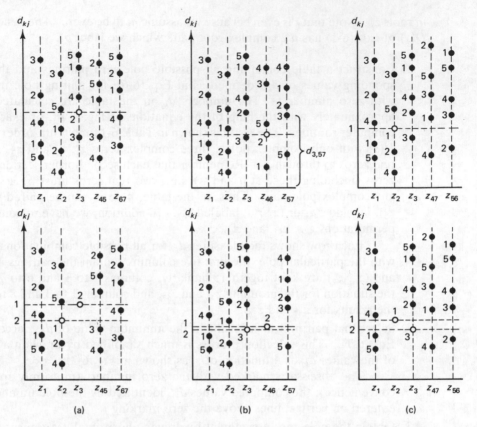

Figure 6-3 Diagrams showing the values d_{kj} for the different possible pole-zero assignments. The numbers 1 through 5 refer to the pole numbers. The diagrams in column (a) correspond to the pole-zero assignments where the real zeros (z_4-z_5) and (z_6-z_7) are grouped together; similarly, column (b) is valid for the groupings z_{46}-z_{57} and column (c) for z_{47}-z_{56}. d_{kj} is the height of the "pole point" above the abscissa on the vertical dashed "zero line." The horizontal dashed lines, numbered 1 through 4, are the pole-zero "assignment lines."

such that on and below this line *each pole* exists in a *different* zero column *at least once*. If there are more than two *distinct* real zeros, a d_{kj} diagram must be drawn for each possible pairing of real zeros. The first assignment line identifies the first pole-zero pairing, i.e., the minimum of d_{max} as per Eq. (6-7). If any pole number appears more than once below the first assignment line, we exclude the first pole-zero pair from further considerations and draw the *second* assignment line (line 2) in the same manner; the process continues with the *third* assignment line (line 3), and so forth, until the pairing is complete, i.e., until the remaining pole numbers appear only once.

4. In some circumstances, such as the availability of preferred types of biquads or ease of tuning, certain poles and zeros must be paired regardless of the

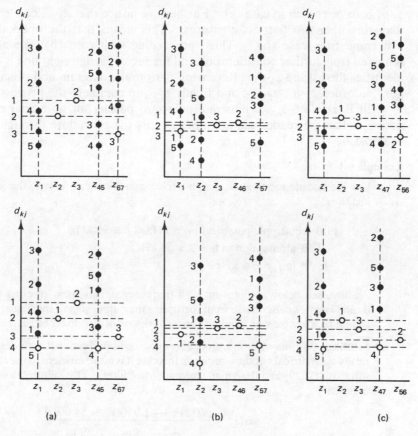

Figure 6-3 *(Continued)*

optimum pole-zero assignment procedure.   In these cases, we simply preassign the appropriate poles and zeros and leave them out of the remaining pairing process.   If, on the other hand, a particular pole k and zero j should *not* be paired, we simply assign an arbitrarily high value d_{kj} to this pair so that it will not fall below any assignment line.[6]

The pole-zero pairing process is illustrated by Fig. 6-3, where we have assigned arbitrary values to the different d_{kj} measures in Table 6-1.   Because we assumed four distinct real zeros, we have three possible ways of pairing them in groups of two (see Table 6-1) and consequently there are three different d_{kj} diagrams as given in the first row of Fig. 6-3 and identified as columns (a), (b), and (c).   The first assignment lines for the three cases are also shown in the figure.   They indicate that z_3-p_2 is the best choice for case (a), that z_{46}-p_2 is best for case (b), and that

[6]For example, z_1 and p_3 in Fig. 6-3 have such a large value d_{31} that their pairing will not take place.

z_2-p_1 is to be chosen in case (c). Further, we notice that $d_{2,46}$, case (b), is smaller than the other two flatness measures. This means that the optimal assignment will come from case (b).[7] Thus, proceeding with case (b), the z_{46} column is removed from further consideration and the second assignment line is drawn which identifies the pairing z_3-p_3. Removing z_3 from the diagram and proceeding yields next, in order, z_1-p_1, z_{57}-p_5, and finally z_2-p_4 to complete the process.

In Example 6-1, we illustrate pole-zero pairing for an actual cascade filter design so that the reader can attempt to verify the procedure through a numerical example.

Example 6-1

Use the cascade topology to design a telecommunications filter to the following specifications:

> 1 dB–equiripple passband in 68 kHz $\leq f \leq$ 96 kHz
> \>60 dB attenuation in $0 \leq f \leq$ 56 kHz
> \>48 dB in $f \geq$ 200 kHz

As discussed below, to prevent high-frequency signals from entering the filter, and dc and low-frequency noise from corrupting the output signal, the first and last sections should be a lowpass or bandpass and a highpass or bandpass block, respectively.

Solution. We may use the procedures discussed in Section 1.6 to design a geometrically symmetrical bandpass filter; in this case it is more efficient to use the appropriate software [11, 29] and design an unsymmetrical filter. This step results in the transfer function

$$H(s) = \frac{Ks^2 (s^2 + 41.00^2)(s^2 + 54.092^2)}{\displaystyle\prod_{i=1}^{4}\left(s^2 + \frac{\omega_i}{Q_i}s + \omega_i^2\right)} \tag{6-8}$$

where

$$\omega_1 = 68.093 \qquad Q_1 = 32.047 \qquad \omega_2 = 73.094 \qquad Q_2 = 10.404$$

$$\omega_3 = 83.794 \qquad Q_3 = 7.500 \qquad \omega_4 = 95.528 \qquad Q_4 = 14.604$$

All frequency parameters are normalized with respect to $\omega_0 = 2\pi$ krad/s; i.e., the above numbers give the pole and zero frequencies in kHz.

Bearing in mind the first and last section requirements and at the same time trying to minimize the number of different filter blocks, we choose to build two notch and two bandpass filters. At this point we have to decide on the best pole-zero assignment, i.e., we have to construct a table of the form of Table 6-1. To this end we split the denominator into four second-order terms and assign in turn the four possible numerator terms

$$s \qquad s \qquad (s^2 + 41.00^2) \qquad (s^2 + 54.092^2)$$

---

[7]Although cases (a) and (c) need not be considered after the first assignment line, we have completed the corresponding diagrams for purposes of illustration and practice.

TABLE 6-2 d_{kj} VALUES FOR EXAMPLE 6-1

	$z_1 \ (\pm \ j41)$	$z_2 \ (\pm \ j54.09)$	$z_{34} \ (0, \infty)$	$z_{43} \ (0, \infty)$
p_1 (68.093/32.047)	1.16	1.11	1.35	1.35
p_2 (73.049/10.404)	0.61	0.57	0.77	0.77
p_3 (83.794/7.5000)	0.66	0.74	0.52	0.52
p_4 (95.528/14.604)	1.26	1.43	1.01	1.01

to the four denominator factors.   As an example, we may choose

$$H(s) = \frac{k_1 83.704s}{s^2 + (83.704/7.500)s + 83.704^2} \frac{k_2(s^2 + 41.00^2)}{s^2 + (73.049/10.404)s + 73.049^2}$$

$$\frac{k_3(s^2 + 54.092^2)}{s^2 + (68.093/32.047)s + 68.093^2} \frac{k_4 95.528s}{s^2 + (95.528/14.604)s + 95.528^2} \tag{6-9}$$

$$= T_1(s)T_2(s)T_3(s)T_4(s) = \prod_{i=1}^{4} k_i t_i(s)$$

The parameters k_i are suitably defined gain constants.   A simple computer program permits us to calculate the maxima and the minima (at the band edges) of the four second-order sections and their permutations; a total of sixteen biquad functions must be evaluated.   Note, however, that all functions are of the type sketched in Fig. 1-18, so that the equations in Fig. 1-18 can be used for finding the maxima.   Either way, the process results in Table 6-2.

With the numbers we can now draw the assignment line graph in Fig. 6-4 to help us select a best pole-zero pairing.   The four assignment lines are shown in the figure; they identify the optimum pairings (z_1, p_2), (z_2, p_1), (z_{34}, p_4), and (z_{43}, p_3). Equation (6-9) shows the result.   Gain assignment and section ordering will be treated in Examples 6-2 and 6-3 below.

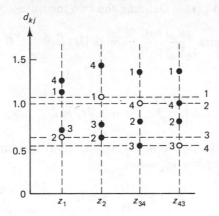

Figure 6-4 Assignment line graph for Example 6-1.

The example demonstrates that even in fairly simple low-order cases the problem of pole-zero assignment can be quite computation-intensive. If the appropriate facilities are not available, the engineer can often arrive at a good suboptimal solution by simply assigning each zero or zero pair to the closest pole [11, 97, 126]. In our example, this choice would have resulted in (z_1, p_1), (z_2, p_2), (z_{34}, p_4), and (z_{43}, p_3) with almost the same dynamic range!

6.2.2 Section Ordering

After the pole-zero assignment has been solved, the designer has to determine the optimal ordering out of the $(n/2)!$ possibilities in which the biquads can be connected to form the cascade network. For example, for an eighth-order network with four sections, there exist 24 possible ways to cascade the biquads:

$$
\begin{array}{cccccc}
T_1T_2T_3T_4 & T_1T_2T_4T_3 & T_1T_3T_2T_4 & T_1T_3T_4T_2 & T_1T_4T_2T_3 & T_1T_4T_3T_2 \\
T_2T_1T_3T_4 & T_2T_1T_4T_3 & T_2T_3T_1T_4 & T_2T_3T_4T_1 & T_2T_4T_1T_3 & T_2T_4T_3T_1 \\
T_3T_1T_2T_4 & T_3T_1T_4T_2 & T_3T_2T_1T_4 & T_3T_2T_4T_1 & T_3T_4T_1T_2 & T_3T_4T_2T_1 \\
T_4T_1T_2T_3 & T_4T_1T_3T_2 & T_4T_2T_1T_3 & T_4T_2T_3T_1 & T_4T_3T_1T_2 & T_4T_3T_2T_1
\end{array}
$$

As mentioned earlier, the best sequence is chosen as the one which maximizes the dynamic range. The procedure is completely analogous to our earlier discussion where we kept the transfer functions of the individual biquads as flat as possible: We now endeavor to design the cascade connection such that the transfer functions from input to the output (recall, however, note 3 in Section 6.2) of all intermediate biquads is as flat as possible so that the maximum signal voltages do not overdrive the op amps and so that, over the passband, the smallest signal stays well above the noise floor. Therefore, the relationships of Eq. (6-4) must be satisfied,

$$\max |V_{oi}(j\omega)| < V_{\text{out}} \qquad 0 \le \omega \le \infty \qquad (6\text{-}10\text{a})$$

$$\min |V_{oi}(j\omega)| \to \max \qquad \omega_L \le \omega \le \omega_U \qquad (6\text{-}10\text{b})$$

where $V_{oi}(s)$ is now the output voltage of the cascade of the first i sections when driven by an input signal $V_{\text{in}}(s)$. Defining the two measures

$$M_i = \frac{\max |V_{oi}(j\omega)|}{|V_{\text{in}}(j\omega)|} = \max \left| \frac{V_{oi}(j\omega)}{V_{\text{in}}(j\omega)} \right| = \max |H_i(j\omega)| \qquad 0 \le \omega \le \infty \quad (6\text{-}11\text{a})$$

and

$$m_i = \frac{\min |V_{oi}(j\omega)|}{|V_{\text{in}}(j\omega)|} = \min \left| \frac{V_{oi}(j\omega)}{V_{\text{in}}(j\omega)} \right| = \min |H_i(j\omega)| \qquad \omega_L \le \omega \le \omega_U \quad (6\text{-}11\text{b})$$

we require again that the flatness criterion of Eq. (6-6) be minimized, now, however, *by choice of the cascading sequence:*

$$d_i = \log \frac{M_i}{m_i} \to \min \qquad (6\text{-}12)$$

$H_i(s)$ is, of course, the transfer function, also referred to as the signal gain, from the input to the output of the first i sections, i.e.,

$$H_i(s) = \prod_{j=1}^{i} T_j(s) \qquad (6\text{-}13)$$

The optimal sequence is the one which minimizes the maximum number d_i:

$$d_{max} = \max [d_i] \rightarrow \min \qquad i = 1, \ldots, n/2 - 1 \qquad (6\text{-}14)$$

Note that we do not have to consider $d_{n/2}$, because, with all sections connected in the cascade filter, $d_{n/2}$ is nothing but a measure of the prescribed passband ripple. With the problem identified, the optimum cascading sequence can be found in principle by calculating d_i for all $(n/2)!$ sequences and selecting the one that satisfies Eq. (6-14). Since this direct approach involves a considerable amount of computation, more efficient methods have been developed which use either linear programming techniques, such as the "branch and bound" method [125–127], or "backtrack programming" [128]. The necessary computer algorithms are described in the literature. Fundamentally, the methods proceed as follows:

Starting with a cascading sequence that, based on experience or intuition, appears to be a good choice,[8] we calculate the measures d_i and $d_{max.0}$ according to Eqs. (6-12) and (6-14). Any other sequence of biquads is then discarded as soon as a value $d_i > d_{max.0}$ is encountered. If the initial choice was good, this implies that a number of d_i values for different cascading choices will never have to be computed. If a later sequence is found whose d_{max} is less than $d_{max.0}$, it becomes the new optimum and the process continues until all sequences have been tried.

The possible choices are frequently further limited by other considerations. For example, it is often desirable to have as the first section in the cascade a lowpass or a bandpass biquad so that high-frequency signal components are kept from the amplifiers in the filter in order to minimize slew-rate problems. Similarly, the designer may wish to employ a highpass or a bandpass biquad as the last section in order to eliminate low-frequency noise, dc offset, or power supply ripple from the filter output. In such situations, the optimum sequencing is performed only on the remaining sections.

The following example will illustrate some of the steps discussed.

Example 6-2

Continue Example 6-1 by finding the optimal cascading sequence for the four second-order sections.

Solution. The computations in this case are relatively simple because it is required that the final circuit starts and terminates in a bandpass section. Thus, using the notation in Eq. (6-9), we have the four possibilities

$$T_1 T_2 T_3 T_4 \qquad T_1 T_3 T_2 T_4 \qquad T_4 T_2 T_3 T_1 \qquad T_4 T_3 T_2 T_1$$

---

[8]A first guess that is often very close to the optimum is the one which sequences the sections in the order of increasing values of Q_i, i.e., $Q_1 < Q_2 \ . \ \ . \ \ < Q_{n/2}$, so that the section with the flattest transfer function magnitude comes first, the next flattest one follows, and so on.

which means we have to compute only the maxima M_i and minima m_i, i.e., the values d_i per Eq. (6-12), of the six functions

$$T_1 T_2 \qquad T_1 T_3 \qquad T_1 T_2 T_3 \qquad T_4 T_2 \qquad T_4 T_3 \qquad T_4 T_2 T_3$$

because the corresponding values of $T_1 T_2 T_3$ and $T_1 T_3 T_2$ are identical, as are those of $T_4 T_2 T_3$ and $T_4 T_3 T_2$, and the value d_4 obtained from all four sections in any case is just a measure of the prescribed 1 dB ripple. The possible paths through the filter are shown in Fig. 6-5: the figures above the lines are the section numbers; the figures below the nodes (the section outputs) are the numbers d_i of Eq. (6-12) as computed below. Our task is now to find that path through the filter with the smallest maximum d_i per Eq. (6-14).

As a starting sequence, let us take $T_1 T_2 T_3 T_4$ as in Eq. (6-9); i.e., we order the sections with increasing values of Q. The computations lead to:

$$T_1 \rightarrow d_1 = 0.52 \qquad T_1 T_2 \rightarrow d_{12} = 0.50$$

$$T_1 T_3 \rightarrow d_{13} = 0.74 \qquad T_1 T_2 T_3 \rightarrow d_{123} = 1.07$$

with

$$\min \{d_{1i,\max}\} = 1.07$$

$$T_4 \rightarrow d_4 = 1.01 \qquad T_4 T_2 \rightarrow d_{42} = 0.75$$

$$T_4 T_3 \rightarrow d_{43} = 0.60 \qquad T_4 T_2 T_3 \rightarrow d_{423} = 0.53$$

It follows that the optimal ordering is identified by d_4 as $T_4 T_3 T_2 T_1$, i.e., the order is reversed from that given in Eq. (6-9). But observe that the order $T_1 T_2 T_3 T_4$ in Eq. (6-9) results in almost the same dynamic range: $d_{123} = 1.07$ as opposed to $d_4 = 1.01$.

6.2.3 Gain Assignment

As the final step in the realization of a cascade filter, we have to assign the gain constants of the biquads. Generally, the selection is again based on dynamic range concerns with the goal of keeping the signals below amplifier saturation limits and above the system noise. Assuming as before that the output voltage of the biquads

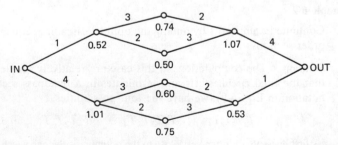

Figure 6-5 Possible paths through the eighth-order filter with the constraints that the first and last sections must be bandpass types.

reaches the critical magnitudes,[9] we require that none of the voltages $V_{oi}(s)$, $i = 1, \ldots, n/2$, in Fig. 6-1 overdrive any op amp sooner than any other one. Because the circuit is linear and all voltages rise proportionally to V_{in}, it is clear that the maximum undistorted input signal can be processed if we choose the gain constants such that all internal output voltages V_{oi} are equal in magnitude to the presumably prescribed magnitude of the output voltage:[10]

$$\max |V_{oi}(j\omega)| = \max |V_{o.n/2}(j\omega)| = \max |V_{\text{out}}(j\omega)| \qquad i = 1, \ldots, n/2 - 1$$

For the analysis, it is convenient to use the notation of Eqs. (6-3) and (6-13), i.e.,

$$H(s) = \prod_{i=1}^{n/2} T_i(s) = \prod_{i=1}^{n/2} k_i t_i(s) = \prod_{i=1}^{n/2} k_i \prod_{i=1}^{n/2} t_i(s) \tag{6-15}$$

and

$$H_i(s) = \prod_{j=1}^{i} T_j(s) = \prod_{j=1}^{i} k_j t_j(s) = \prod_{j=1}^{i} k_j \prod_{j=1}^{i} t_j(s) \tag{6-16}$$

and to label the total prescribed gain constant

$$K = \prod_{i=1}^{n/2} k_i \tag{6-17}$$

such that

$$\max \left| \prod_{i=1}^{n/2} t_i(j\omega) \right| = M_{n/2} \tag{6-18a}$$

is some given value. Further, let us denote the maxima of the intermediate gain-scaled transfer functions by M_i, i.e.,

$$\max \left| \prod_{k=1}^{i} t_k(j\omega) \right| = M_i \qquad i = 1, \ldots, n/2 - 1 \tag{6-18b}$$

Then we obtain $k_1 M_1 = K M_{n/2}$, i.e.,

$$k_1 = K \frac{M_{n/2}}{M_1} \tag{6-19a}$$

to make $\max|V_{oi}(j\omega)| = \max |V_{\text{out}}(j\omega)|$. Similarly, $k_1 k_2 M_2 = K M_{n/2}$; i.e., with Eq. (6-19a),

$$k_2 = \frac{M_1}{M_2}$$

---

[9]As said before, this assumption is reasonable in most cases, but recall note 3 in Section 6.2.

[10]The section numbering, $i = 1, \ldots, n/2 - 1$, is that of the section ordering obtained in the previous step!

results in $\max|V_{o2}(j\omega)| = \max |V_{out}(j\omega)|$. Proceeding in the same manner yields

$$k_j = \frac{M_{j-1}}{M_j} \qquad j = 2, \ldots, n/2 \qquad\qquad (6\text{-}19b)$$

where $k_{n/2}$ can also be obtained from the previous values via

$$k_{n/2} = K \prod_{j=1}^{n/2-1} k_j^{-1} \qquad\qquad (6\text{-}19c)$$

Choosing the gain constants as in Eq. (6-19) guarantees that all op amps see the same maximum voltage to assure that the largest possible signal can be processed without distortion.

Example 6-3

Complete the design of Examples 6-1 and 6-2 by assigning gain constants k_i, $i = 1$, $\ldots$, 4, such that the maximum signal level to be handled by the op amps in each section is equalized. The output voltage should be maximized. Assume that the maximum expected input signal level is 60 mV and that the op amp output voltage, in the frequency range where the filter must operate, is limited to 1.4 V. Realize all sections by Åckerberg-Mossberg circuits (Fig. 5-31).

Solution. With pole-zero pairing and section ordering solved in Examples 6-1 and 6-2, we address the gain assignment as discussed in Eqs. (6-18) and (6-19). The total transfer function with as yet unknown gain constants and the definitions of the gain-scaled functions $t_i(s)$ were given in Eq. (6-9). But note that per Example 6-2 the section ordering is *reversed* from that in Eq. (6-9): $T_4T_3T_2T_1$!

The proposed second-order section, appropriately modified for this example, is repeated for convenience in Fig. 6-6; it realizes, with $\omega_0 = 1/(RC)$,

$$T_{o1}(s) = \frac{V_{o1}}{V_i} = -\frac{\alpha s^2 + s\omega_0 k + \gamma\omega_0^2}{s^2 + s\omega_0/Q + \omega_0^2}$$

and

$$T_{o2}(s) = \frac{V_{o2}}{V_i} = -\frac{s\omega_0(\gamma - \alpha) + \omega_0^2(\gamma/Q - k)}{s^2 + s\omega_0/Q + \omega_0^2}$$

respectively. Naturally, for the three-amplifier Åckerberg-Mossberg section we have to be concerned also with the level of the output voltage V_{o2} so that op amp A_2 stays in the linear range of operation. For the *bandpass* sections we set $\alpha = \gamma = 0$ to obtain

$$T_{o1}(s) = \frac{V_{o1}}{V_i} = -\frac{k\omega_0 s}{s^2 + s\omega_0/Q + \omega_0^2} \qquad\qquad (6\text{-}20a)$$

and

$$T_{o2}(s) = \frac{V_{o2}}{V_i} = +\frac{k\omega_0^2}{s^2 + s\omega_0/Q + \omega_0^2}$$

with

$$|T_{o1}|_{max} \approx |T_{o2}|_{max} \approx kQ$$

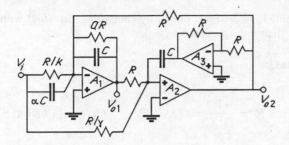

Figure 6-6 Åckerberg-Mossberg section for Example 6-3.

for high Q. Therefore, all op amps see the same voltage maxima. For the *band-rejection* sections we set $k = 0$ to yield

$$T_{o1}(s) = \frac{V_{o1}}{V_i} = -\frac{\alpha(s^2 + \omega_0^2\gamma/\alpha)}{s^2 + s\omega_0/Q + \omega_0^2} \qquad (6\text{-}20b)$$

and

$$T_{o2}(s) = \frac{V_{o2}}{V_i} = -\frac{s\omega_0(\gamma - \alpha) + \omega_0^2\gamma/Q}{s^2 + s\omega_0/Q + \omega_0^2}$$

with, for high Q and at $\omega = \omega_0$,

$$|T_{o1}|_{\max} \simeq Q|\gamma - \alpha| \qquad |T_{o2}|_{\max} \simeq \sqrt{Q^2(\gamma - \alpha)^2 + \gamma^2} > |T_{o1}|_{\max}$$

Therefore, in the band-rejection sections, the output of op amps A_2 and A_3 is larger and must be considered when maximizing dynamic range. However, for both band-rejection sections in Eq. (6-9) we have

$$Q^2(\gamma - \alpha)^2 \gg \gamma^2$$

as the student can readily verify. Thus, in our example we may look only at the output of op amp A_1.

Using an approximate evaluation or, better, a computer routine, we find

$$|t_4|_{\max} = M_1 \simeq 14.6 \qquad |t_4t_3|_{\max} = M_2 \simeq 20.2$$

$$|t_4t_3t_2|_{\max} = M_3 \simeq 62.5 \qquad |t_4t_3t_2t_1|_{\max} = M_4 \simeq 139$$

Also, we have

$$\prod_{j=1}^{4} k_j M_4 V_i = K M_4 V_i = V_o \qquad \text{i.e.,} \qquad K = \frac{V_o}{V_i}\frac{1}{M_4} = \frac{1.4}{0.06}\frac{1}{139} \simeq 0.168$$

and

$$k_4 = \frac{K M_4}{M_1} = 1.60 \qquad k_3 = \frac{M_1}{M_2} = 0.72$$

$$k_2 = \frac{M_2}{M_3} = 0.32 \qquad k_1 = \frac{M_3}{M_4} = \frac{K}{k_1 k_2 k_3} = 0.45$$

With these numbers and $C = 5$ nF we obtain by comparing Eq. (6-9) with Eq. (6-20) for the biquads in their correct cascading order:

Section 4 (bandpass):

$$\alpha = k_4 = 1.60 \qquad R = \frac{1}{\omega_{04}C} = 333.2 \ \Omega \qquad Q_4R = 4.688 \ \text{k}\Omega$$

Section 3 (bandreject):

$$\alpha = k_3 = 0.72 \qquad \gamma = \left(\frac{\omega_{z3}}{\omega_{03}}\right)^2 \alpha = 1.01$$

$$R = \frac{1}{\omega_{03}C} = 467.5 \ \Omega \qquad Q_3R = 14.98 \ \text{k}\Omega$$

Section 2 (bandreject):

$$\alpha = k_2 = 0.32 \qquad \gamma = \left(\frac{\omega_{z2}}{\omega_{02}}\right)^2 \alpha = 1.915$$

$$R = \frac{1}{\omega_{02}C} = 435.7 \ \Omega \qquad Q_2R = 4.534 \ \text{k}\Omega$$

Section 1 (bandpass):

$$\alpha = k_1 = 0.45 \qquad R = \frac{1}{\omega_{01}C} = 380.3 \ \Omega \qquad Q_1R = 2.852 \ \text{k}\Omega$$

The simulated output voltages of the four sections are shown in Fig. 6-7.

6.2.4 An Exact Approach to Dynamic Range Optimization

The dynamic range optimization procedure discussed in Sections 6.2.1 through 6.2.3 is based on the assumption that pole-zero pairing, section ordering, and gain assignment are independent of each other and can, therefore, be separated. Further, the method neglects to take into account any noise generated *in the biquads themselves* and the effect, if any, that pole-zero pairing and gain assignment may have on this noise. This approach was taken, of course, because it is conceptually easier to understand and because it is considerably easier computationally. The assumptions are, however, fundamentally at least questionable.

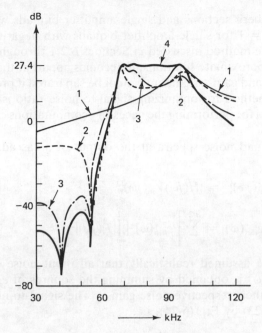

Figure 6-7 Simulated output after 1, 2, 3, and 4 sections, respectively.

The problem was solved partially in reference 128 by defining a different measure to be optimized which includes both the signal gain [Eq. (6-13)] from input to the output of section i *and* the noise gain [Eq. (6-5)] from section i to the output. The method of reference [128], however, still neglects the effect of pole-zero pairing and gain assignment on section noise. Recall from Chapter 4 (Problem 4.16) that the shape of the output noise spectrum looks approximately like the transfer function magnitude [77, 129]. More specifically, if we model the noise contributed by a second-order section $T_j(s)$ as an equivalent input noise source $e_{j,\text{in}}(\omega)$ followed by a noise-free section, then that section's output noise equals

$$e_{j,\text{out}}(\omega) = e_{j,\text{in}}(\omega)|T_j(j\omega)| \qquad (6\text{-}21)$$

In the vicinity of the filter passband, the equivalent input noise $e_{j,\text{in}}(\omega)$, in general, is approximately *flat*, and its magnitude depends on the topology, the type of transfer function realized, the section parameters, such as gain and quality factor, and the impedance level. For example, for second-order bandpass sections one can show [129] that

$$e_{j,\text{in}}(\omega) \simeq k \, \frac{Q_j^n}{H_{Bj}^m} \qquad (6\text{-}22)$$

where H_{Bj} and Q_j are midband gain and quality factor, respectively, and where k is a constant that depends on the impedance level and the type of active devices. The value of n is either 1 or 2, and m equals 0 or 1 depending on the kind of section used. For example, one finds $n = 2$ and $m = 0$ for GIC biquads, $n = m$

= 1 for Äckerberg-Mossberg sections and single-amplifier biquads with positive feedback, and $n = 2$, $m = 1$ for single-amplifier biquads with negative feedback [129]. Thus, although the method discussed in Sections 6.2.1 through 6.2.3 gives in most situations good approximate solutions, it becomes apparent that pole-zero pairing, section ordering, and gain assignment cannot be separated if precise results are needed. An exact method for optimizing signal-to-noise ratio is outlined in the following; an algorithm for performing the necessary computations can be found in reference 130.

The squared signal and noise spectra at the output of a cascade filter are, respectively,

$$|V_{out}(j\omega)|^2 = |H(j\omega)V_{in}(j\omega)|^2 \tag{6-23}$$

$$e_{out}^2(\omega) = \sum_{j=1}^{n/2} \left[e_{j,in}^2(\omega) \prod_{i=j}^{n/2} |T_i(j\omega)|^2 \right] \tag{6-24}$$

In writing Eq. (6-24), we assumed realistically that all input noise sources are uncorrelated so that $e_{out}^2(\omega)$ is obtained by summing the squares of $e_{j,in}(\omega)$, multiplied by the squares of their respective noise gains. The signal-to-noise ratio is found by dividing Eq. (6-23) by Eq. (6-24), i.e.,

$$SN(\omega) = \frac{|V_{in}(j\omega)|^2}{e_{1,in}^2(\omega) + \sum_{j=2}^{n/2} \left[e_{j,in}^2(\omega) \prod_{i=1}^{j-1} \frac{1}{|T_i(j\omega)|^2} \right]} \tag{6-25}$$

where we have made use of $H(s) = \prod_{i=1}^{n/2} T_i(s)$ per Eq. (6-3). Further recall that we defined the section transfer functions in Eq. (6-2) as $T_i(s) = k_i t_i(s)$ where the gain constants k_i have to satisfy Eq. (6-19) in order to avoid nonlinear performance in any of the op amps.[11] Using these relationships along with Eqs. (6-17) and (6-18) in Eq. (6-25), the student can show (Problem 6-12) that the signal-to-noise ratio becomes

$$SN(\omega) = \frac{|V_{in}(j\omega)|^2}{e_{1,in}^2(\omega) + \sum_{j=2}^{n/2} \left[e_{j,in}^2(\omega) \prod_{i=1}^{j-1} \frac{1}{k_i^2} \prod_{i=1}^{j-1} \frac{1}{|t_i(j\omega)|^2} \right]}$$

$$= \frac{|V_{in}(j\omega)|^2 K^2 M_{n/2}^2}{e_{1,in}^2(\omega) K^2 M_{n/2}^2 + \sum_{j=2}^{n/2} \left[e_{j,in}^2(\omega) M_{j-2}^2 \prod_{i=1}^{j-1} \frac{1}{|t_i(j\omega)|^2} \right]} \tag{6-26}$$

Finally, note that the numerator of Eq. (6-26) contains only prescribed known quantities. Consequently, maximizing the signal-to-noise ratio requires that the

---

[11]Note that Eq. (6-19) must be valid even in the present, more general procedure because the maximum amplifier voltages must still be all equal regardless of pole-zero pairing and section ordering.

maximum of the denominator of Eq. (6-26) be minimized over the frequency band of interest:

$$\max \left\{ e_{1,\text{in}}^2(\omega) K^2 M_{n/2}^2 + \sum_{j=2}^{n/2} \left[e_{j,\text{in}}^2(\omega) M_{j-2}^2 \prod_{i=1}^{j-1} \frac{1}{|t_i(j\omega)|^2} \right] \right\} \qquad (6\text{-}27)$$

Equation (6-27) minimizes the peak noise; if broadband rms noise is of interest, an additional integration over frequency has to be performed [77, 130].

Observe from Eqs. (6-26) and (6-27) that section 1 provides the largest contribution to the output noise, section 2 is next, and so on. Also note, that the expressions permit the designer to include arbitrary *nonwhite* input and noise spectra. Further details and a relatively efficient algorithm for computing Eq. (6-27) can be found in the literature [130].

For applications with low-level signals where noise is the primary concern, we need to minimize the total output noise defined in Eq. (6-24):

$$e_{\text{out}}^2(\omega) = |H(j\omega)|^2 \left\{ e_{1,\text{in}}^2(\omega) + \sum_{j=2}^{n/2} \left[e_{j,\text{in}}^2(\omega) \prod_{i=1}^{j-1} \frac{1}{|T_i(j\omega)|^2} \right] \right\}$$

$$= |H(j\omega)|^2 \left[e_{1,\text{in}}^2(\omega) + \frac{e_{2,\text{in}}^2(\omega)}{|T_1(j\omega)|^2} + \frac{e_{3,\text{in}}^2(\omega)}{|T_1(j\omega) T_2(j\omega)|^2} \right. \qquad (6\text{-}28)$$

$$\left. + \ldots + \frac{e_{n/2,\text{in}}^2(\omega)}{|T_1(j\omega) \ldots T_{n/2-1}(j\omega)|^2} \right]$$

It is evident from this expression that, for prescribed $H(s)$, the total output noise will be reduced if, over the passband, $|T_1(j\omega)|$ is made as large as possible, then $|T_2(j\omega)|$ as large as possible, and so forth; or, simply, the maximum amount of gain should be assigned to the leading section, then as much as possible to section 2, and so forth.

6.3 MULTIPLE-LOOP FEEDBACK REALIZATIONS

As mentioned earlier, multiple-loop feedback topologies are also based on biquad building blocks which are then embedded into a resistive feedback configuration. The resulting coupling between sections is selected such that transfer function sensitivities are reduced. Several studies have shown that the different available configurations have approximately equal sensitivity behavior [53, 36, 47]; we shall, therefore, concentrate our discussion only on the follow-the-leader feedback (FLF) and, as part of the ladder simulation techniques, leapfrog (LF) topologies, which have the advantage of being relatively easy to derive without any sacrifice in performance. Our derivation will reflect the fact that both configurations[12] are particularly convenient for geometrically symmetrical bandpass functions (see Sec-

---

[12]As indeed all other multiple-loop feedback circuits.

tion 1.6) and that the LF topology is obtained from a direct simulation of an LC lowpass ladder.

6.3.1 The FLF Topology

The follow-the-leader feedback topology was introduced in Section 3.2.6 and Fig. 3-12a. The actual implementation of the summer and the feedback factors is shown in Fig. 6-8. We assumed that there are n noninteracting sections so that the order of the realized transfer function $H(s)$ is $2n$. Assuming further that the two summer op amps are ideal, simple analysis yields

$$-V_0 = \frac{R_{F0}}{R_{in}} V_{in} + \sum_{i=1}^{n} \frac{R_{F0}}{R_{Fi}} V_i = \alpha V_{in} + \sum_{i=1}^{n} F_i V_i \qquad (6\text{-}29)$$

where we define α and the feedback factors F_i as

$$\alpha = \frac{R_{F0}}{R_{in}} \quad \text{and} \quad F_i = \frac{R_{F0}}{R_{Fi}} \qquad (6\text{-}30)$$

respectively. Similarly, we find for the output summer

$$V_{out} = - \sum_{i=0}^{n} K_i V_i = - \sum_{i=0}^{n} \frac{R_A}{R_{oi}} V_i \qquad (6\text{-}31)$$

where the definition of the resistor ratios K_i is apparent. Any of the parameters F_i and K_i may, of course, become zero by replacing the corresponding resistor, R_{Fi}

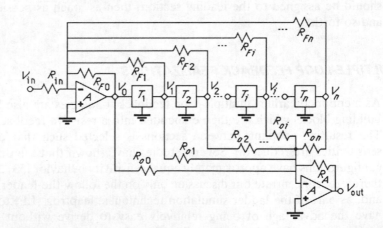

Figure 6-8 FLF circuit built from second-order sections $T_i(s)$ and a feedback network consisting of an op amp summer with resistors R_{Fi}. Also shown is an output summer with resistors R_{oi} to facilitate the realization of arbitrary transmission zeros.

or R_{oi}, respectively, by an open circuit. Finally, the internal voltages V_i can be computed from

$$V_i = V_0 \prod_{j=1}^{i} T_j(s) \qquad i = 1, \ldots, n \qquad (6\text{-}32)$$

so that with Eq. (6-29)

$$H_0(s) = \frac{V_0}{V_{in}} = - \frac{\alpha}{1 + \sum_{k=1}^{n} \left[F_k \prod_{j=1}^{k} T_j(s) \right]} \qquad (6\text{-}33)$$

which in turn with Eq. (6-32) yields

$$H_i(s) = \frac{V_i}{V_{in}} = - \frac{\alpha \prod_{j=1}^{i} T_j(s)}{1 + \sum_{k=1}^{n} \left[F_k \prod_{j=1}^{k} T_j(s) \right]} \qquad i = 1, \ldots, n \qquad (6\text{-}34)$$

Note that

$$H_i(s) = H_n(s) \prod_{j=i+1}^{n} \frac{1}{T_j(s)} \qquad (6\text{-}35)$$

where

$$H_n(s) = \frac{V_n}{V_{in}} = - \frac{N_n(s)}{D(s)} = - \frac{\alpha \prod_{j=1}^{n} T_j(s)}{1 + \sum_{k=1}^{n} \left[F_k \prod_{j=1}^{k} T_j(s) \right]} \qquad (6\text{-}36)$$

is the transfer function of the FLF network *without* the output summer, i.e., with $R_{oi} = \infty$ for all i.

From Eq. (6-36) it is seen that the transmission zeros of $H_n(s)$ are set by the zeros of $T_j(s)$, i.e., by the *feed-forward* path, whereas the poles of $H_n(s)$ are determined by the *feedback* network and involve both the poles and zeros of the biquads $T_j(s)$ and the feedback factors F_k. Designing an FLF network to realize a transfer function $H_n(s)$ with arbitrary zeros requires, therefore, second-order sections with finite transmission zeros. It is generally quite difficult and will not be discussed in this book. Instead, we refer the interested reader to the literature [137, 47], where this case is treated and where a suitable algorithm can be found. The situation becomes much simpler if we observe from Eq. (6-34) that the numerator of $H_i(s)$ is proportional to s^i if $T_i(s)$ realizes the second-order bandpass filter:

$$T_i(s) = A_i \frac{s/Q_i}{s^2 + s/Q_i + 1} = A_i t_i(s) \qquad (6\text{-}37)$$

In Eq. (6-37), Q_i and A_i are the section's pole quality factor and midband gain, respectively, and s is the normalized bandpass frequency parameter. In that case, a transfer function $H(s)$ with a general numerator polynomial can be obtained by adding the intermediate output voltages V_i by means of an additional summer as is shown in Fig. 6-8 and Eq. (6-31). Inserting Eqs. (6-33) and (6-34) into Eq. (6-31) yields

$$H(s) = \frac{V_{\text{out}}}{V_{\text{in}}} = \frac{N(s)}{D(s)} = \alpha \frac{K_0 + \sum_{k=1}^{n} \left[K_k \prod_{j=1}^{k} T_j(s) \right]}{1 + \sum_{k=1}^{n} \left[F_k \prod_{j=1}^{k} T_j(s) \right]} \tag{6-38}$$

The structure of these equations becomes clearer by considering a specific case. Let $n = 4$; then Eq. (6-38) reads

$$H(s) = \alpha \frac{K_0 + K_1 T_1 + K_2 T_1 T_2 + K_3 T_1 T_2 T_3 + K_4 T_1 T_2 T_3 T_4}{1 + F_1 T_1 + F_2 T_1 T_2 + F_3 T_1 T_2 T_3 + F_4 T_1 T_2 T_3 T_4} \tag{6-39}$$

Equation (6-38) is a ratio of two polynomials whose roots can be set by an appropriate choice of the functions $T_i(s)$, the parameters K_i for the transmission zeros, and the feedback factors F_i for the poles. The computations are easiest if the function to be realized is an arithmetically symmetrical bandpass so that we can use the lowpass-to-bandpass transformation, Eq. (1-104),

$$p = Q \frac{s^2 + 1}{s}$$

where $Q = \omega_0/B$ is the "quality factor" of the high-order bandpass with bandwidth B and p is the normalized lowpass frequency. In that case, the bandpass functions in Eq. (6-37) with *all identical* pole frequencies are transformed into the first-order prototype lowpass functions

$$T_{i\text{LP}}(p) = \frac{A_i Q/Q_i}{p + Q/Q_i} = A_i \frac{q_i}{p + q_i} \tag{6-40}$$

The parameter q_i equals Q/Q_i and A_i is now the dc gain of the lowpass section. Applying the lowpass-to-bandpass transformation also to the prescribed bandpass function [Eq. (6-38)] of order $2n$ converts it into a lowpass function $H_{\text{LP}}(p)$ of order n. Substituting Eq. (6-40) in the denominator expression shows that the poles are determined by

$$D(p) = \prod_{j=1}^{n} (p + q_j) + \sum_{k=1}^{n-1} \left[f_k \prod_{i=k+1}^{n} (p + q_i) \right] + f_n \tag{6-41}$$

where we have introduced the abbreviation

$$f_i = F_i \prod_{j=1}^{i} A_j q_j \tag{6-42a}$$

It is evident from Eqs. (6-38) and (6-39) that the transmission zeros are determined by a polynomial analogous to Eq. (6-41),

$$N(p) = \alpha K_0 \prod_{j=1}^{n} (p + q_j) + \sum_{j=1}^{n-1} \left[k_j \prod_{i=j+1}^{n} (p + q_i) \right] + k_n \qquad (6\text{-}43)$$

where

$$k_i = \alpha K_i \prod_{j=1}^{i} A_j q_j \qquad (6\text{-}42b)$$

Again, to be specific, the eighth-order example of Eq. (6-39) becomes a fourth-order lowpass function

$$H_{\text{LP}}(p) = \frac{V_o}{V_i} = \frac{a_4 p^4 + a_3 p^3 + a_2 p^2 + a_1 p + a_0}{b_4 p^4 + b_3 p^3 + b_2 p^2 + b_1 p + b_0} \qquad (6\text{-}44)$$

whose coefficients have to be equated with those of the polynomials in Eqs. (6-41) and (6-43) for $n = 4$. For the denominator, we obtain:

$$b_4 = 1$$

$$b_3 = b_4(q_1 + q_2 + q_3 + q_4) + f_1$$

$$b_2 = b_4[q_1(q_2 + q_3 + q_4) + q_2(q_3 + q_4) + q_3 q_4]$$
$$\quad + f_1(q_2 + q_3 + q_4) + f_2$$

$$b_1 = b_4[(q_1 + q_2)q_3 q_4 + (q_3 + q_4)q_1 q_2]$$
$$\quad + f_1[q_2 q_3 + (q_2 + q_3)q_4] + f_2(q_3 + q_4) + f_3$$

$$b_0 = b_4(q_1 q_2 q_3 q_4) + f_1 q_2 q_3 q_4 + f_2 q_3 q_4 + f_3 q_4 + f_4$$

These are four equations in eight unknowns, f_i and q_i, $i = 1, \ldots, 4$, which can be written more conveniently in matrix form:

$$
\begin{pmatrix}
1 & 0 & 0 & 0 \\
q_2 + q_3 + q_4 & 1 & 0 & 0 \\
q_2 q_3 + (q_2 + q_3)q_4 & q_3 + q_4 & 1 & 0 \\
q_2 q_3 q_4 & q_3 q_4 & q_4 & 1
\end{pmatrix}
\begin{pmatrix}
f_1 \\ f_2 \\ f_3 \\ f_4
\end{pmatrix} \qquad (6\text{-}45)
$$

$$
=
\begin{pmatrix}
b_3 - b_4(q_1 + q_2 + q_3 + q_4) \\
b_2 - b_4[q_1(q_2 + q_3 + q_4) + q_2(q_3 + q_4) + q_3 q_4] \\
b_1 - b_4[(q_1 + q_2)q_3 q_4 + (q_3 + q_4)q_1 q_2] \\
b_0 - b_4(q_1 q_2 q_3 q_4)
\end{pmatrix}
$$

The transmission zeros are found via an identical process: the unknown parameters k_i are computed from an equation of the form of Eq. (6-45) with f_i replaced by k_i and b_i replaced by a_i, $i = 1, \ldots, 4$. Also, $K_0 = a_4/\alpha$.

The matrix expression of Eq. (6-45) is written in a form to indicate that the unknown parameters f_i can be solved from a set of *linear* equations whose coefficients are functions of the given values b_i and of the numbers q_i which for given Q are determined by the quality factors Q_i of the second-order sections $T_i(s)$. This fact indicates that the Q_i are *free* parameters that may be chosen to satisfy any criteria which the designer feels might lead to a better-working circuit. It was pointed out repeatedly in Chapter 3 that free design parameters can beneficially be chosen to reduce a circuit's sensitivity to element variations. In our present case, this leads to a multiparameter (the n Q_i values of the n second-order sections) optimization problem that has been solved to minimize the statistical sensitivity measures defined in Section 3.3 [53, 131–134]. The curve for the FLF design in Fig. 3-13 is a result of such an optimization; it shows that in the passband the sensitivities of an FLF circuit can be significantly better than those of a cascade realization. Unfortunately, however, it requires the solution of the mentioned quite difficult optimization problem and the availability of the corresponding algorithm or software [132].

To simplify the procedure, other designers have investigated the performance of "FLF" structures for specific values of the quality factors Q_i: the design becomes particularly simple if all the Q_i factors are chosen to be equal, a choice which has the additional practical advantage of resulting in *all identical* second-order building blocks. For this reason, the resulting approach has been referred to as the *primary resonator block* (PRB) technique [135]. The passband sensitivity performance of PRB circuits is much better than that of cascade designs; indeed it is almost, but not quite, as good as that of fully optimized FLF structures [47, 133]. The relevant equations are derived in the following.

With $q_i = q$ for all i we find from Eq. (6-45)

$$
\begin{pmatrix}
1 & 0 & 0 & 0 \\
3q & 1 & 0 & 0 \\
3q^2 & 2q & 1 & 0 \\
q^3 & q^2 & q & 1
\end{pmatrix}
\begin{pmatrix}
f_1 \\ f_2 \\ f_3 \\ f_4
\end{pmatrix}
=
\begin{pmatrix}
b_3 - 4qb_4 \\
b_2 - 6q^2b_4 \\
b_1 - 4q^3b_4 \\
b_0 - q^4b_4
\end{pmatrix}
\tag{6-46}
$$

which shows that

$$
F_1 A_1 q = f_1 = b_3 - 4qb_4
$$

$$
F_2 A_1 A_2 q^2 = f_2 = b_2 - 6q^2 b_4 - 3qf_1
$$

$$
F_3 A_1 A_2 A_3 q^3 = f_3 = b_1 - 4q^3 b_4 - 3q^2 f_1 - 2qf_2
\tag{6-47}
$$

$$
F_4 A_1 A_2 A_3 A_4 q^4 = f_4 = b_0 - q^4 b_4 - q^3 f_1 - q^2 f_2 - qf_3
$$

The system in Eq. (6-47) represents four equations for the five unknowns, q, f_i, $i = 1, \ldots 4$, (in general, n equations for $n + 1$ unknowns), which indicates that one parameter, q, can still be used for optimization purposes. This single degree of freedom is often eliminated by choosing

$$q = \frac{b_{n-1}}{nb_n} \qquad (6\text{-}48a)$$

i.e., $q = b_3/(4b_4)$ in Eq. (6-47), which means $f_1 = 0$. The remaining feedback factors can then be computed recursively from Eq. (6-47). The systematic nature of the equations makes it apparent how to proceed for $n > 4$. As a matter of fact, it is not difficult to show [135, 11] that, with $f_1 = 0$, in general

$$f_2 = b_{n-2} - \frac{n(n-1)}{2!} q^2 b_n \qquad (6\text{-}48b)$$

$$f_i = b_{n-i} - \frac{q^i}{(n-i)!} \left[\frac{n!}{i!} b_n + \sum_{j=2}^{i-1} \frac{f_i}{q^j} \frac{(n-j)!}{(i-j)!} \right] \qquad i = 3, \ldots, n \quad (6\text{-}48c)$$

As mentioned earlier, equations of identical form, with f_i replaced by k_i and b_i by a_i with $K_0 = a_n/\alpha$, are used to determine the summing coefficients K_i of the output summer, which establishes the transmission zeros.[13] Thus, given a geometrically symmetrical bandpass function, Eqs. (6-48a) to (6-48c) can be used to calculate the parameter q and all feedback and summing coefficients required for a PRB design: all second-order bandpass sections, [Eq. (6-37)] are tuned to the same pole frequency $\omega = \omega_0$ and have the same pole quality factor $Q_p = Q/q$, where $Q = \omega_0/B$. ω_0, Q, and B, respectively, are the center frequency, the quality factor, and the bandwidth of the high-order bandpass filter.

Note that the design procedure computes only the *products* f_i and k_i; the actual values of the resistor ratios F_i and K_i, as well as the gain constants A_i, are not yet uniquely determined. As a matter of fact, the gain constants A_i are *free* parameters that, as the reader will correctly suspect, are selected to maximize the circuit's dynamic range in much the same way as for cascade designs. For practical FLF (PRB) designs, this step of scaling the signal levels is very important because an inadvertently poor choice of gain constants can result in *very* large internal signals and, consequently, very poor dynamic range. The simple diagram in Fig. 6-9 illustrates the potential problem: we have assumed a PRB design of an eighth-order bandpass whose midband gain as well as that of all second-order sections was set equal to unity. For simplicity, we have represented all transfer functions by straight-line approximations. Observe from Eq. (6-35) that the internal output voltages are obtained by dividing the total filter output consecutively by $T(s)$. Because we scaled all midband gains to unity, all voltages at $\omega = \omega_0$ reach the value $1 \cdot V_{in}$; however, toward the bandedge, the signals become increasingly larger because the second-order bandpass function $|T(j\omega)|$ decreases as $|\omega - \omega_0|$ increases.

---

[13]Although b_n is normally equal to unity, we have retained this coefficient in all the above equations so that the direct substitution of a_i for b_i can be performed.

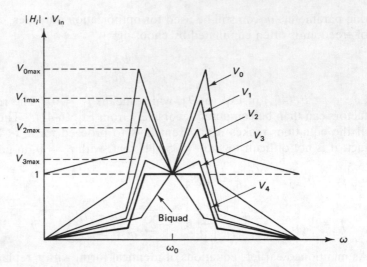

Figure 6-9 Signal levels in a PRB design where all gain constants are set equal to unity.

After our earlier study of the corresponding problem in cascade filters, the solution should then be apparent: We equalize again the maximum signal levels. Label the transfer function $H_n(s)$ as

$$H_n(s) = M_n h_n(s)$$

where M_n is a given gain constant defined such that

$$\max |h_n(j\omega)| = 1 \tag{6-49}$$

i.e.,

$$\max |H_n(j\omega)| = M_n \qquad 0 \le \omega \le \infty$$

Then, using Eq. (6-35), we obtain

$$H_i(s) = H_n(s) \prod_{j=i+1}^{n} \frac{1}{T_j(s)} = \left[h_n(s) \prod_{j=i+1}^{n} \frac{1}{t_j(s)} \right] \left(M_n \prod_{j=i+1}^{n} \frac{1}{A_j} \right) \tag{6-50}$$

and define, using Eq. (6-37),

$$M_i = \max \left| h_n(j\omega) \prod_{j=i+1}^{n} \frac{1}{t_j(j\omega)} \right|$$

$$= \max \left\{ |h_n(j\omega)| \prod_{j=i+1}^{n} \sqrt{1 + Q_j^2 \left(\omega - \frac{1}{\omega} \right)^2} \right\} \tag{6-51}$$

$$0 \le \omega \le \infty, i = 0, 1, \ldots, n - 1$$

M_0 takes into account that also the input summing amplifier must not be overdriven. The values M_i can be calculated approximately, or accurately by computer. With M_i known, we compute A_i such that all signal maxima throughout the FLF filter are identical.[14]

$$M_i M_n \prod_{j=i+1}^{n} \frac{1}{A_j} = M_n \qquad (6\text{-}52)$$

i.e.,

$$M_i = \prod_{j=i+1}^{n} A_j \qquad i = 0, 1, \ldots, n-1$$

This expression yields, in order,

$$A_n = M_{n-1} \qquad A_i = \frac{M_{i-1}}{M_i} \qquad i = n-1, n-2, \ldots, 1 \qquad (6\text{-}53)$$

A simple approximate solution of the dynamic range maximization problem can be obtained [136] by observing that the maxima M_i occur close to the bandedge of the high-order bandpass filter (see Fig. 6-7), where by definition

$$\left(\omega - \frac{1}{\omega}\right)^2 \simeq Q^{-2}$$

If we further assume that throughout the passband $|h_n(j\omega)| \simeq 1$, Eq. (6-51) shows that

$$M_i \simeq \prod_{j=i+1}^{n} \sqrt{1 + \left(\frac{Q_j}{Q}\right)^2} \qquad i = 0, 1, \ldots, n-1 \qquad (6\text{-}54)$$

so that from Eq. (6-53)

$$A_i \simeq \sqrt{1 + \left(\frac{Q_i}{Q}\right)^2} = \sqrt{1 + \frac{1}{q_i^2}} \qquad i = 1, \ldots, n \qquad (6\text{-}55)$$

The same equations hold for the PRB case where $Q_i = Q_p$ for all i so that $A_i = A \simeq \sqrt{1 + (Q_p/Q)^2}$.

Additional details about signal-level scaling and dynamic range optimization of multiple-loop feedback (MF) circuits can be found in the references [36, 47, 131–134]. In particular, reference 129 discusses the calculation of the noise level and presents a comparative study of dynamic range limitations in MF bandpass filters. It is shown there that, based on ease of design, sensitivity, and dynamic range performance, the FLF topology results in the most practical multiple-loop feedback designs.

---

[14]Note that the method assumes again that the critical signal levels are measured at the biquad outputs. As noted earlier, if an internal voltage becomes larger, appropriate scaling factors can be used to avoid any problems.

The following example illustrates the complete FLF (PRB) design process.

Example 6-4

Design a sixth-order Chebyshev bandpass filter with center frequency $f_0 = 5$ kHz and a 0.5 dB–ripple bandwidth $\Delta f = 300$ Hz. Realize the filter as a PRB configuration with GIC bandpass sections. Available are op amps with gain-bandwith product $\omega_t = 2\pi \cdot 1.2$ Mrad/s.

Solution. From Table III-2a in Appendix III, we find the transfer function of the prototype lowpass function to be

$$H_{\mathrm{LP}}(p) = \frac{-0.716}{p^3 + 1.253p^2 + 1.535p + 0.716} = \frac{-0.716}{(p + 0.626)(p^2 + 0.626p + 1.142)}$$

The prototype lowpass frequency p is converted to the bandpass frequency s by the transformation

$$p = Q\frac{s^2 + 1}{s} = \frac{5000}{300}\frac{s^2 + 1}{s} = 16.667\left(\frac{s^2 + 1}{s}\right)$$

where s is normalized with respect to $\omega_0 = 2\pi \cdot 5$ krad/s and the quality factor $Q = 16.667$.

For the PRB circuit, the transfer function equals, from Eq. (6-36) with $T_i = T$ and with $T_{\mathrm{LP}}(p)$ given in Eq. (6-40),

$$H_{\mathrm{LP}}(p) = \frac{-\alpha T_{\mathrm{LP}}^3(p)}{1 + F_1 T_{\mathrm{LP}}(p) + F_2 T_{\mathrm{LP}}^2(p) + F_3 T_{\mathrm{LP}}^3(p)}$$

$$= \frac{-\alpha A^3 q^3}{(p + q)^3 + f_1(p + q)^2 + f_2(p + q) + f_3}$$

The parameters q and f_i were defined in Eqs. (6-40) and (6-42); they are obtained from Eq. (6-48) as follows:

$$q = \frac{b_2}{3b_3} = \frac{1.253}{3} = 0.4177$$

$$f_1 = 0 \qquad f_2 = b_1 - 3q^2 = 1.0116 \qquad f_3 = b_0 - q^3 - qf_2 = 0.2206$$

In addition, we find from Eq. (6-55)

$$A = \sqrt{1 + \frac{1}{q^2}} = 2.595$$

so that, after applying the lowpass-to-bandpass transformation, the second-order bandpass sections become

$$T_{\mathrm{BP}}(s) = \frac{2.595(q/Q)s}{s^2 + (q/Q)s + 1} = \frac{2.595s/39.902}{s^2 + s/39.902 + 1}$$

T_{BP} is to be realized three times by GIC filter sections (Fig. 5-23) in the PRB topology

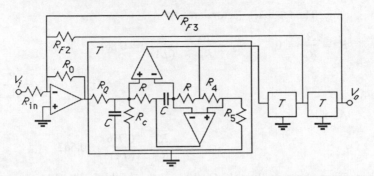

Figure 6-10 PRB topology with three identical GIC bandpass sections. The circuit for one section is shown explicitly, as are the input summer and the feedback resistors.

shown in Fig. 6-10. Each section realizes Eq. (5-101), repeated here for convenience as

$$\frac{V_{\text{out}}}{V_{\text{in}}} = \frac{s\dfrac{G_Q}{C}\left(1 + \dfrac{G_5}{G_4}\right)}{s^2 + s\dfrac{G_Q}{C} + \dfrac{G^2 G_5}{C^2 G_4}}$$

Evidently, the midband gain equals

$$A = 1 + \frac{R_4}{R_5} = 2.595$$

and ω_0 and Q_p are obtained from Eq. (5-102):

$$\omega_0 = \frac{1}{RC}\sqrt{A-1} \quad \text{and} \quad Q_p = \frac{R_Q}{R}\sqrt{A-1}$$

Let us choose $C = 100$ nF and $R_4 = 4$ kΩ; then

$$R = \frac{\sqrt{A-1}}{\omega_0 C} = 810\ \Omega \qquad R_Q = \frac{Q_p}{\sqrt{A-1}}\,R = 25.6\ \text{k}\Omega$$

$$R_5 = \frac{R_4}{A-1} = 2.51\ \text{k}\Omega$$

To cancel the nominal effect of the "negative loss resistor," Eq. (5-110b), we connect a compensation resistor of value

$$R_c = \frac{\omega_t}{\omega_0}\,\frac{A-1}{A(A-2)}\,R = \frac{1.2\cdot10^6}{5\cdot10^3}\,\frac{1.595}{2.59\cdot0.595} = 248\ \Omega$$

as shown in Fig. 6-10.

The feedback factors are

$$F_1 = \frac{R_0}{R_{F1}} = 0 \qquad F_2 = \frac{R_0}{R_{F2}} = \frac{f_2}{(qA)^2} = 0.861$$

$$F_3 = \frac{R_0}{R_{F3}} = \frac{f_3}{(qA)^3} = 0.173$$

Also,

$$\alpha = \frac{R_0}{R_{in}} = \frac{0.716}{(qA)^3} = 0.562$$

Thus, choosing $R_0 = 1 \text{ k}\Omega$, the resistors of the feedback network are

$$R_{F1} = \infty \qquad R_{F2} = 1.16 \text{ k}\Omega \qquad R_{F3} = 5.78 \text{ k}\Omega \qquad R_{in} = 1.78 \text{ k}\Omega$$

6.4 LC LADDER SIMULATIONS

We saw in Chapter 3 that lossless filters designed for maximum power transfer have the best possible passband sensitivities. Because such circuits are normally realized as *LC* ladders, a considerable amount of effort has been devoted in recent years to the development of active circuits which in one way or another simulate the performance of passive ladders and thereby inherit their good sensitivity performance. As mentioned in the introduction, ladder simulations can be classified into two groups: *operational simulation* and *element substitution*. Both methods start from an existing *LC* prototype ladder; operational simulation endeavors to represent the internal operation of the ladder by simulating the equations describing the circuit's performance, i.e., Kirchhoff's voltage and current laws and the *I-V* relationships of the ladder arms. Fundamentally, this procedure is based on simulating the *signal-flow graph* of the ladder where all voltages and all currents are considered signals which propagate through the circuit [138, 139, 11]. The signal-flow graph method will be developed in Section 6.4.1.

The element substitution procedure replaces all inductors by an active network whose input impedance over the appropriate frequency range is inductive. The circuits were illustrated in Figs. 4-16 and 4-26. Various approaches to this method will be presented in Section 6.4.2. We shall, however, discuss the technique only briefly because it is less general and methodical than the operational simulation of ladders. Also, the element substitution method does not lend itself very easily to signal-level scaling for dynamic range optimization. For further details, the reader is referred to the references [11, 47, 140, 144].

6.4.1 LC Ladder Simulation by Signal-Flow Graphs

For our purposes, the signal-flow graph (SFG) method can be understood most easily by considering a section of a ladder as shown in Fig. 6-11. The circuit is

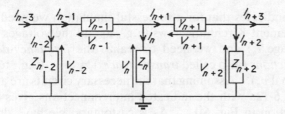

Figure 6-11 Section of a ladder network.

analyzed readily by writing Kirchhoff's laws and the *I-V* relationships for the ladder arms as follows:

$$\cdots$$

$$
\begin{aligned}
I_{n-2} &= I_{n-3} - I_{n-1} & V_{n-2} &= Z_{n-2}I_{n-2} = Z_{n-2}(I_{n-3} - I_{n-1}) \\
V_{n-1} &= V_{n-2} - V_n & I_{n-1} &= Y_{n-1}V_{n-1} = Y_{n-1}(V_{n-2} - V_n) \\
I_n &= I_{n-1} - I_{n+1} & V_n &= Z_nI_n = Z_n(I_{n-1} - I_{n+1}) \\
V_{n+1} &= V_n - V_{n+2} & I_{n+1} &= Y_{n+1}V_{n+1} = Y_{n+1}(V_n - V_{n+2}) \\
I_{n+2} &= I_{n+1} - I_{n+3} & V_{n+2} &= Z_{n+2}I_{n+2} = Z_{n+2}(I_{n+1} - I_{n+3})
\end{aligned}
\tag{6-56}
$$

$$\cdots$$

In the active simulation of this circuit, all currents and voltages are to be represented as *voltage* signals. In order to achieve this goal, we use a resistive scaling factor R as shown in one of these equations as an example,

$$
I_nR = I_{n-1}R - I_{n+1}R \qquad V_n = \frac{Z_n}{R}I_nR = \frac{Z_n}{R}(I_{n-1}R - I_{n+1}R)
$$

and introduce the notation

$$
I_kR = i_k \qquad V_k = v_k \qquad \frac{Z_k}{R} = z_k \qquad Y_kR = y_k
\tag{6-57}
$$

The lowercase symbols are used to represent the *scaled* quantities; note that z_k and y_k are now dimensionless voltage transfer functions and that both i_k and v_k are voltages. We have retained the symbol i_k in order to remind ourselves of the origin of that signal as a current in the original ladder. With Eq. (6-57), equation group (6-56) takes on the following form:

$$\cdots$$

$$
\begin{aligned}
i_{n-2} &= i_{n-3} - i_{n-1} & v_{n-2} &= z_{n-2}i_{n-2} = z_{n-2}(i_{n-3} - i_{n-1}) \\
v_{n-1} &= v_{n-2} - v_n & i_{n-1} &= y_{n-1}v_{n-1} = y_{n-1}(v_{n-2} - v_n) \\
i_n &= i_{n-1} - i_{n+1} & v_n &= z_ni_n = z_n(i_{n-1} - i_{n+1}) \\
v_{n+1} &= v_n - v_{n+2} & i_{n+1} &= y_{n+1}v_{n+1} = y_{n+1}(v_n - v_{n+2}) \\
i_{n+2} &= i_{n+1} - i_{n+3} & v_{n+2} &= z_{n+2}i_{n+2} = z_{n+2}(i_{n+1} - i_{n+3})
\end{aligned}
\tag{6-58}
$$

This group of equations indicates that for a successful simulation we need to build *voltage summers* to implement Kirchhoff's laws (e.g., to add the voltages i_{n-1} and $-i_{n+1}$ to form the voltage i_n) and we need to realize the *frequency-dependent multipliers* or *transfer functions* (also called *transmittances*) z_k or y_k (e.g., to convert the voltage i_n into the signal v_n). Assuming that the necessary circuits are available, the flow diagram in Fig. 6-12a with the indicated interconnections gives the realization of the ladder section in Fig. 6-11. As is customary, we have drawn the "current signals" and their summing nodes in the top line and the "voltage signals" with their summing nodes in the bottom line.

The implementation is slightly inconvenient because it requires taking the *difference* of two signals. Recall from Chapter 4 that *summing* of signals is preferable! The student should verify that the necessary correction is obtained quite easily by making all downward-pointing signal paths,[15] the blocks representing the ladder's shunt arms, *inverting*. This choice results in just the right signal inversions throughout the flow diagram as shown in Fig. 6-12b so that only *additions* are required, and it guarantees also that all internal loop gains are negative. The only price to be paid is that in some cases the overall transfer function suffers a sign inversion (a 180° phase shift), which in most cases is of no consequence.

The SFG diagram of Fig. 6-12b is also referred to as *leapfrog* (LF) topology; the reason for this name becomes apparent when the circuit is redrawn as shown in Fig. 6-13. At this occasion we also wish to emphasize that the correct realization of the transfer function poles implies that *all loop gains*, such as $z_n y_{n+1}$, *in the signal-flow graph must be realized correctly*. For example, assuming i_{n-3} to be the input, v_{n+2} the output, and $i_{n+3} = 0$, the function realized by the graph in Fig. 6-12b can be shown to equal

$$\frac{v_{n+2}}{i_{n-3}} = \frac{z_{n+2}y_{n+1}z_n y_{n-1}z_{n-2}}{(1 + D_1)\left(1 + \dfrac{D_2}{1 + D_1}\right)\left(1 + \dfrac{D_3}{1 + \dfrac{D_2}{1 + D_1}}\right)\left(1 + \dfrac{D_4}{1 + \dfrac{D_3}{1 + \dfrac{D_2}{1 + D_1}}}\right)}$$

$$(6\text{-}59)$$

where

$$D_1 = y_{n+1}z_{n+2} \qquad D_2 = z_n y_{n+1} \qquad D_3 = y_{n-1}z_n \qquad D_4 = z_{n-2}y_{n-1}$$

This expression shows quite clearly that the transfer function poles can be expected

---

[15]The upward-pointing paths, i.e., the ladder's series arms, could equally well be taken as inverting. Recall that inverting integrators can be built with one op amp whereas good noninverting integrators require two op amps for their realization. Thus, the designer should select the method which minimizes the number of noninverting integrators in order to save op amps.

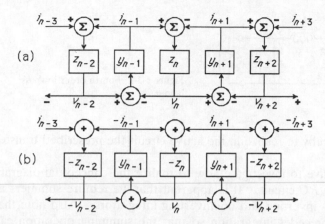

(a)

(b)

Figure 6-12 (a) Signal-flow graph block diagram representation of the ladder section in Fig. 6-11; (b) transformation resulting in only positive input summers.

to be accurate if all loop gains are realized correctly. It will serve as a guide in our later implementation of general signal-flow graph filters.

6.4.1.1 Realization of all-pole lowpass filters.

Let us illustrate this process on a simple fourth-order all-pole lowpass filter. The student will recall from Chapter 2 that this circuit is realized by an LC ladder with two series inductors and two shunt capacitors along with two terminating resistors as is shown in Fig. 6-14. As discussed, let us normalize all the elements by dividing them by a scaling resistor R and label

$$C_i R = c_i \qquad \frac{L_i}{R} = l_i \qquad \frac{R_i}{R} = r_i = \frac{1}{g_i} \qquad I_i R = i_i \qquad V_i = v_i \qquad (6\text{-}60)$$

With this notation, the required equations are

$$i_1 = \frac{Kv_{\text{in}} - v_2}{sl_1 + r_S} \qquad -v_2 = \frac{-1}{sc_2}(i_1 - i_3) \qquad -i_3 = \frac{1}{sl_3}(-v_2 + v_4)$$

$$v_4 = \frac{-1}{sc_4 + g_L}(-i_3) \qquad (6\text{-}61)$$

The corresponding signal-flow graph implementation is shown in Fig. 6-15. We have used the integrator symbolism introduced in earlier figures (see, for example, Fig. 5-33) because each transfer function block is recognized as an inverting or noninverting integrator. Note from Fig. 6-14 that V_4 and V_{out} are identical. We have introduced an arbitrary constant K at the input which multiplies all signals

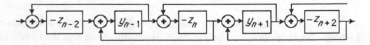

Figure 6-13   The diagram of Fig. 6-12b redrawn in the *leapfrog* (LF) configuration.

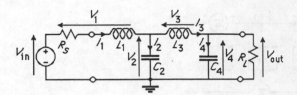

Figure 6-14 Fourth-order lowpass ladder.

by K and permits us thereby to realize in the active circuit the prescribed transfer function with a gain K.

At this point we have to investigate how to realize the conceptual diagram of Fig. 6-15 as an active RC circuit. It is apparent that we require summers as well as lossy and lossless inverting and noninverting integrators. Also note that summing in each case precedes integration so that the summing operation can always be performed at the integrator summing node, as was discussed in Chapter 4.

Observe that the ladder simulation consists of a number of two-integrator loops, containing an inverting and a noninverting integrator each. From the material discussed in Chapter 5 the student will remember that this situation poses a potential problem if the integrators have phase errors (or finite Q factors), because the loop quality factors can become severely enhanced (see Fig. 5-27 and the related discussion). At that occasion we solved the problem by combining the inverting phase-lag Miller integrator of Fig. 4-20 with the noninverting actively compensated phase-lead integrator of Fig. 4-24a.[16] Thereby, the total phase error around the loop is approximately reduced to zero. The two integrators are repeated here in Fig. 6-16. They realize

$$V_0 = \pm \frac{G_1 V_1 + G_2 V_2}{sC + G_3} \tag{6-62}$$

where the minus sign is valid for the inverting lossy Miller integrator and the plus sign must be used for the phase-lead integrator. Note from Fig. 6-15 that the internal ladder arms are realized by lossless integrators ($R_3 = \infty$) whereas the two end branches require lossy integrators (R_3 finite) in order to account for load resistors. Thus, it remains only to interconnect the appropriate versions of Fig. 6-16 in the manner prescribed in Fig. 6-15 to arrive at the final circuit shown in Fig. 6-17. At the internal output nodes, we have indicated the signals of Fig. 6-15 (e.g., i_1) which correspond to the voltages in the final realization.

All capacitors were chosen equal for convenience. One method for finding the values of the resistors from the *known* components of the LC ladder requires

---

[16]This method was used to improve the performance of the Tow-Thomas biquad; it led us to the Åckerberg-Mossberg biquad. See Figs. 5-28 and 5-29. Note also the results of Problem 4.40, which point out the possibly necessary modifications to the unity-gain inverter in Fig. 6-16b to avoid oscillations caused by the op amp's second pole [123].

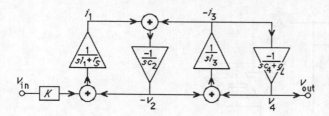

Figure 6-15 Simulation flow diagram of the ladder in Fig. 6-14.

comparing the equations realized by Fig. 6-17 with the corresponding equations, Eq. (6-61), that describe the original ladder:

$$i_1 = \frac{G_1 V_{in} + G_2(-v_2)}{sC + G_3} \longrightarrow \frac{Kv_{in} - v_2}{sl_1 + r_S} = \frac{Kv_{in} - v_2}{sL_1/R + R_S/R} \qquad (6\text{-}63a)$$

$$-v_2 = -\frac{G_4 i_1 + G_5(-i_3)}{sC} \longrightarrow -\frac{i_1 - i_3}{sc_2} = -\frac{i_1 - i_3}{sC_2 R} \qquad (6\text{-}63b)$$

$$-i_3 = \frac{G_6(-v_2) + G_7 v_4}{sC} \longrightarrow \frac{-v_2 + v_4}{sl_3} = \frac{-v_2 + v_4}{sL_3/R} \qquad (6\text{-}63c)$$

$$v_4 = -\frac{G_8(-i_3)}{sC + G_9} \longrightarrow -\frac{-i_3}{sc_4 + g_L} = -\frac{-i_3}{sC_4 R + R/R_L} \qquad (6\text{-}63d)$$

Considering the needed equality of the time constants in Eq. (6-63a), we find

$$CR_3 = \frac{L_1}{R_S} \quad \text{or} \quad R_3 = \frac{L_1}{C}\frac{1}{R_S} \qquad (6\text{-}64a)$$

From the dc gain factors of the signals V_{in} and $(-v_2)$ we obtain

$$\frac{R_3}{R_1} = K\frac{R}{R_S} \quad \text{and} \quad \frac{R_3}{R_2} = \frac{R}{R_S}$$

i.e.,

$$R_1 = R_3 \frac{R_S}{KR} = \frac{L_1}{C}\frac{1}{KR} \quad \text{and} \quad R_2 = R_3 \frac{R_S}{R} = \frac{L_1}{C}\frac{1}{R}$$

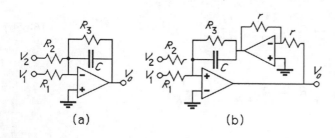

Figure 6-16 (a) Inverting lossy Miller integrator; (b) noninverting lossy phase-lead integrator.

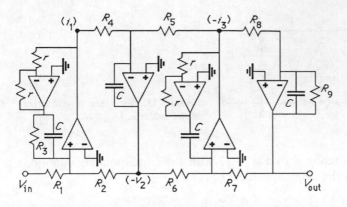

Figure 6-17 Active realization of the LC ladder of Fig. 6-14.

Note that the value of R_1 determines the realized gain K. Similarly, we find from Eqs. (6-63b) through (6-63d):

$$CR_4 = CR_5 = C_2R \qquad \text{i.e., } R_4 = R_5 = \frac{C_2}{C} R \qquad (6\text{-}64\text{b})$$

$$CR_6 = CR_7 = \frac{L_3}{R} \qquad \text{i.e., } R_6 = R_7 = \frac{1}{R} \frac{L_3}{C} \qquad (6\text{-}64\text{c})$$

$$CR_8 = C_4R \qquad \text{i.e., } R_8 = \frac{C_4}{C} R$$

and

$$\frac{R_8}{R_9} = \frac{R_L}{R} \qquad \text{i.e., } R_9 = R_8 \frac{R_L}{R} = \frac{C_4}{C} R_L \qquad (6\text{-}64\text{d})$$

The scaling resistor R and the capacitor C have arbitrary values and can be chosen to obtain convenient and practical components.

Note that these component values result in the correct realization of all loop gains as required per our earlier discussion:

$$\frac{l_1}{r_S} = CR_3 \qquad l_1c_2 = C^2R_2R_4 \qquad l_3c_2 = C^2R_5R_6$$

$$l_3c_4 = C^2R_7R_8 \qquad c_4r_L = CR_9 \qquad (6\text{-}65\text{a})$$

Also, R_1 can be obtained via the dc gain from V_{in} through the first integrator:

$$\frac{K}{r_S} = \frac{G_1}{G_3} \qquad (6\text{-}65\text{b})$$

i.e.,

$$R_1 = R_3 \frac{R_S}{KR} = \frac{L_1}{C} \frac{1}{KR}$$

as before.

6.4.1.2 Maximization of dynamic range.

The remaining task in our design of a signal-flow graph simulation of an *LC* ladder is that of voltage-level scaling for dynamic range maximization. It may be accomplished by noting that a scale factor can be inserted into each signal line *as long as the loop gains are not changed* [see Eq. (6-59)]. Employing signal-level scaling in this manner will not change the transfer function except for an overall gain factor. The procedure is illustrated in the flow diagram in Fig. 6-18, where we have employed the scale factors α, β, and γ to the diagram of Fig. 6-15. (Ignore the contour C for the time being.) Its utility for dynamic range optimization is explained as follows.

With the output voltage prescribed by the given transfer function, we see from Fig. 6-15 that

$$-i_3 = -v_4(sc_4 + g_L)$$

Assume that the maximum of the voltage $-i_3$ is α times as large as the maximum of v_4, i.e.,

$$\max |i_3| = \max\{|v_4|\sqrt{g_L^2 + (\omega c_4)^2}\} = \alpha \max |v_4|$$

Clearly, multiplying the integrator gain by α equates the two maxima:

$$\max |\hat{i}_3| = \max\left\{|v_4|\frac{\sqrt{g_L^2 + (\omega c_4)^2}}{\alpha}\right\} = \frac{\alpha \max |v_4|}{\alpha} = \max |v_4|$$

For finding the maxima, the signals are, of course, evaluated in $0 \le \omega \le \infty$. To keep the loop gain constant, we multiply v_4 by $1/\alpha$ before adding this signal to $-\hat{v}_2$ to form $-\hat{i}_3$. Next, assume that the maximum of $-\hat{v}_2$ is β times as large as that of $-\hat{i}_3$; proceeding as before, we multiply $-\hat{v}_2$ by β and correct the loop gain by multiplying $-\hat{i}_3$ by $1/\beta$ before adding it to $\hat{i}_1$. At this point, all signals to the right of $\hat{i}_1$ have equal maxima. If now the maximum of $\hat{i}_1$ is γ times as large as the remaining (already equalized) maxima, we scale $\hat{i}_1$ by γ and correct the loop gain by a scale factor $1/\gamma$. This process has multiplied the total transfer function by $\alpha\beta\gamma$, and we can correct the gain by rescaling the factor K appropriately as shown in Fig. 6-18. As always, the maximum allowable signal level is determined by the given op amps. The factors α, β, and γ are found in the usual way, either approximately, or exactly by computer. Note, in particular, that the integrator output

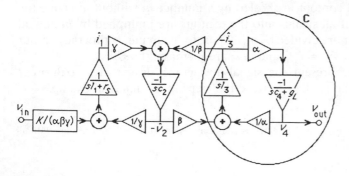

Figure 6-18 Illustrating dynamic range scaling for a signal-flow graph simulation of a lowpass filter.

voltages correspond directly to the currents in the ladder series arms and the voltages across the shunt arms; thus, any ladder analysis program [11, 29] can be used to compute the relevant maxima. To determine how the scale factors affect the element values in the active realization, we return to Eq. (6-63), rewritten for Fig. 6-18:

$$i_1 = \frac{G_1 V_{in} + G_2(-v_2)}{sC + G_3} \longrightarrow \frac{\dfrac{K}{\alpha\beta\gamma} v_{in} - \hat{v}_2 \dfrac{1}{\gamma}}{sL_1/R + R_S/R} \qquad (6\text{-}66a)$$

$$-v_2 = -\frac{G_4 i_1 + G_5(-i_3)}{sC} \longrightarrow -\frac{\hat{i}_1\gamma - \hat{i}_3 \dfrac{1}{\beta}}{sC_2 R} \qquad (6\text{-}66b)$$

$$-i_3 = \frac{G_6(-v_2) + G_7 v_4}{sC} \longrightarrow \frac{-\hat{v}_2\beta + \hat{v}_4 \dfrac{1}{\alpha}}{sL_3/R} \qquad (6\text{-}66c)$$

$$v_4 = -\frac{G_8(-i_3)}{sC + G_9} \longrightarrow -\frac{-\hat{i}_3\alpha}{sC_4 R + R/R_l} \qquad (6\text{-}66d)$$

Thus, we find by comparing the elements in the active simulation with the required coefficients in the signal-flow graph:

$$R_1 = \frac{L_1}{CR}\frac{\alpha\beta\gamma}{K} \qquad R_2 = \gamma\frac{L_1}{CR} \qquad R_3 = \frac{L_1}{C}\frac{1}{R_S} \qquad R_4 = \frac{1}{\gamma}\frac{C_2}{C}R$$

$$R_5 = \beta\frac{C_2}{C}R \qquad R_6 = \frac{1}{\beta}\frac{L_3}{CR} \qquad R_7 = \alpha\frac{L_3}{CR} \qquad (6\text{-}67)$$

$$R_8 = \frac{1}{\alpha}\frac{C_4}{C}R \qquad R_9 = \frac{C_4}{C}R_L$$

Although we have discussed the dynamic range maximization procedure specifically for an all-pole lowpass filter, the method is entirely general and is valid for *any* signal-flow graph realization. It is based on the fact that in a flow graph all signals inside a closed contour are scaled by a number m without affecting the remainder of the graph, if all signals into the contour are multiplied by m and all signals leaving the contour are divided by m. Thus, for example, within the contour C in Fig. 6-18, the signals $v_4 = v_{out}$ and $-\hat{i}_3$ are increased by β.[17]

In the following we present the signal-flow graph design of a sixth-order lowpass filter to permit the student to trace the steps in a numerical example.

---

[17]The conversion of Fig. 6-12a to Fig. 6-12b is a special case of this scaling procedure with all scale factors equal to -1.

Example 6-5

Realize the sixth-order Chebyshev lowpass filter of Fig. 6-19 (Problem 2.23) as an active SFG circuit. Design the filter such that the dc gain equals unity.

Solution. Let us first perform a source transformation in order to reduce the number of loops by 1; Fig. 6-20 shows the result. Next, using the procedure discussed in the text, convert the circuit into a signal-flow graph (Fig. 6-21); note that all elements are normalized by a suitably chosen resistor R_p and that the scaling factors of dynamic range optimization are already included. Using noninverting and inverting lossless and lossy integrators (Fig. 6-16), the signal-flow graph is readily converted into the final active RC realization as shown in Fig. 6-22.

Note that all these steps were performed without the need for any calculations. In order to obtain the component values, we write the equations for the scaled signal-flow graph (Fig. 6-21) and compare them with those describing the active circuit in Fig. 6-22:

$$V_1 = -\frac{G_0 V_{in} + G_2 V_2}{sC + G_{11}} \quad\longrightarrow\quad -\frac{i_{in}\dfrac{K}{\alpha\beta\gamma\delta\varepsilon} + (-i_2)\dfrac{1}{\varepsilon}}{sc_1 + g_s}$$

$$V_2 = \frac{G_1 V_1 + G_3 V_3}{sC} \quad\longrightarrow\quad \frac{-v_1\varepsilon + v_3\dfrac{1}{\delta}}{sl_2}$$

$$V_3 = -\frac{G_4 V_2 + G_6 V_4}{sC} \quad\longrightarrow\quad -\frac{-i_2\delta + i_4\dfrac{1}{\gamma}}{sc_3}$$

$$V_4 = \frac{G_5 V_3 + G_7 V_5}{sC} \quad\longrightarrow\quad \frac{v_3\gamma + (-v_5)\dfrac{1}{\beta}}{sl_4}$$

$$V_5 = -\frac{G_8 V_4 + G_{10} V_6}{sC} \quad\longrightarrow\quad -\frac{i_4\beta + (-i_6)\dfrac{1}{\alpha}}{sc_5}$$

$$V_6 = \frac{G_9 V_5}{sC + G_{12}} \quad\longrightarrow\quad \frac{-v_5\alpha}{sl_6 + r_L}$$

(6-68)

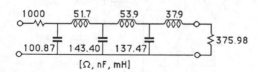

Figure 6-19 Sixth-order Chebyshev LC low-pass filter.

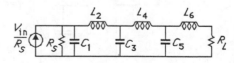

Figure 6-20 The circuit of Fig. 6-19 after a source transformation.

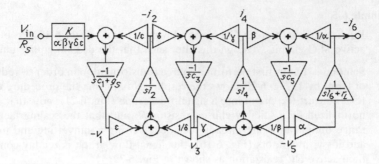

Figure 6-21 The signal-flow graph representing the circuit of Fig. 6-20.

Recall that $V_{in}/R_S = i_{in}$. At this point we have to determine the values of the scale factors; from a ladder analysis program [11] or from a special purpose program, we obtain, with $R_p = 1$ kΩ,

$$\alpha = \frac{\max |V_5|}{\max |I_6 R_p|} = \frac{0.702}{0.810} = 0.867 \qquad \beta = \frac{\max |I_4 R_p|}{\max |V_5|} = \frac{1.32}{0.702} = 1.88$$

$$\gamma = \frac{\max |V_3|}{\max |I_4 R_p|} = \frac{0.865}{1.32} = 0.655 \qquad \delta = \frac{\max |I_2 R_p|}{\max |V_3|} = \frac{1.39}{0.865} = 1.61$$

$$\varepsilon = \frac{\max |V_1|}{\max |I_2 R_p|} = \frac{0.722}{1.39} = 0.555$$

The multiplying constant K is obtained as follows.

From Fig. 6-19, the ladder's dc gain equals $R_L/(R_L + R_S) = 0.2733$; to bring the gain up to the desired value we need, therefore,

$$K = \frac{1}{0.2733} = 3.66$$

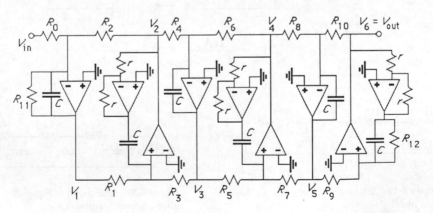

Figure 6-22 Active implementation of the LC ladder in Fig. 6-19.

To determine the component values of the active circuit, it is convenient to scale the impedance level of each integrator by a suitably chosen resistor R_{ai} (note that R_{ai} may be different for each integrator; this will allow us to choose equal capacitors throughout the active circuit as was assumed in Fig. 6-22).

Let us choose $C = 10$ nF. Then we obtain from Eq. (6-68) by comparing coefficients:

$$\frac{R_{a1}}{R_0} = \frac{K}{\alpha\beta\gamma\delta\varepsilon} = \frac{3.66}{0.954} = 3.84$$

$$\frac{R_{a1}}{R_2} = \frac{1}{\varepsilon} = 1.81 \qquad CR_{a1} = C_1 R_p \qquad \frac{R_{a1}}{R_{11}} = \frac{R_p}{R_s}$$

Thus,

$$R_{a1} = \frac{C_1}{C} R_p = 10.09 \text{ k}\Omega \qquad R_0' = \frac{R_{a1}}{3.84} = 2.63 \text{ k}\Omega$$

$$R_2 = \varepsilon R_{a1} = 5.60 \text{ k}\Omega \qquad R_{11} = \frac{R_s R_{a1}}{R_p} = 10.09 \text{ k}\Omega$$

Similarly, from the second integrator,

$$CR_{a2} = \frac{L_2}{R_p} \qquad \text{i.e.,} \qquad R_{a2} = \frac{L_2}{C} \frac{1}{R_p} = 5.17 \text{ k}\Omega$$

and

$$R_1 = \frac{R_{a2}}{\varepsilon} = 9.32 \text{ k}\Omega \qquad R_3 = \delta R_{a2} = 8.32 \text{ k}\Omega$$

Analogously, we obtain from integrators 3, 4, 5, and 6:

$$R_{a3} = \frac{C_3}{C} R_p = 14.34 \text{ k}\Omega \qquad R_4 = \frac{R_{a3}}{\delta} = 8.91 \text{ k}\Omega \qquad R_6 = \gamma R_{a3} = 9.39 \text{ k}\Omega$$

$$R_{a4} = \frac{L_4}{C} \frac{1}{R_p} = 5.36 \text{ k}\Omega \qquad R_5 = \frac{R_{a4}}{\gamma} = 8.18 \text{ k}\Omega \qquad R_7 = \beta R_{a4} = 10.08 \text{ k}\Omega$$

$$R_{a5} = \frac{C_5}{C} R_p = 13.75 \text{ k}\Omega \qquad R_8 = \frac{R_{a5}}{\beta} = 7.31 \text{ k}\Omega \qquad R_{10} = \alpha R_{a5} = 11.90 \text{ k}\Omega$$

$$R_{a6} = \frac{L_6}{C} \frac{1}{R_p} = 3.79 \text{ k}\Omega \qquad R_9 = \frac{R_{a6}}{\alpha} = 4.37 \text{ k}\Omega \qquad R_{12} = \beta \frac{R_p R_{a5}}{R_L} = 10.08 \text{ k}\Omega$$

6.4.1.3 Design of all-pole bandpass filters.

If an all-pole bandpass filter is required, we may consider Fig. 6-14 to be the lowpass prototype and Fig. 6-15 the corresponding signal-flow graph realization. The lowpass is obtained by applying the lowpass-to-bandpass transformation

$$p = Q \frac{s^2 + 1}{s}$$

to the prescribed bandpass function with center frequency ω_0 and bandwidth equal to $B = \omega_0/Q$. As always, s is normalized with respect to ω_0. Because the lowpass ladder arms are realized by integrators of the general form

$$T_i(p) = \frac{\pm a_i}{p + q_i} \tag{6-69}$$

we may apply the lowpass-to-bandpass transformation directly to the signal-flow diagram to obtain

$$T_i(s) = \frac{\pm a_i \dfrac{1}{Q} s}{s^2 + sq_i/Q + 1} \tag{6-70}$$

because, as in the FLF configuration, it leads to realizable bandpass functions. For example, in the fourth-order lowpass of Fig. 6-14 we need the four blocks

$$T_1(s) = \frac{s/(l_1Q)}{s^2 + sr_s/(l_1Q) + 1} \qquad T_2(s) = -\frac{s/(c_2Q)}{s^2 + 1}$$

$$T_3(s) = \frac{s/(l_3Q)}{s^2 + 1} \qquad\qquad T_4(s) = -\frac{s/(c_4Q)}{s^2 + sg_L/(c_4Q) + 1} \tag{6-71}$$

Observe that we obtain both inverting and noninverting second-order bandpass sections and that all sections except the two end blocks must be able to realize ideally *infinite Q factors* if the original ladder is lossless. Applying this procedure to the scaled ladder in Fig. 6-18 shows that the four second-order blocks must realize the functions

$$\hat{i}_1(s) = \left[\frac{K}{\alpha\beta\gamma} v_{\text{in}} + (-\hat{v}_2)\frac{1}{\gamma}\right]\frac{s/(l_1Q)}{s^2 + sr_s/(l_1Q) + 1}$$

$$-\hat{v}_2(s) = \left[\hat{i}_1\gamma + (-\hat{i}_3)\frac{1}{\beta}\right]\frac{-s/(c_2Q)}{s^2 + 1}$$

$$-\hat{i}_3(s) = \left[(-\hat{v}_2)\beta + \hat{v}_4\frac{1}{\alpha}\right]\frac{s/(l_3Q)}{s^2 + 1} \tag{6-72}$$

$$\hat{v}_4(s) = [(-\hat{i}_3)\alpha]\frac{-s/(c_4Q)}{s^2 + sg_L/(c_4Q) + 1}$$

Any appropriate biquad, such as those discussed in Chapter 5, can now be used to realize the functions in Eq. (6-72). Note only that we need in general two inputs and that the circuits must be able to realize an infinite Q factor. For example, the ENF circuit of Fig. 5-8 and a slightly modified version of the EPF circuit in Fig. 5-13a are suitable bandpass filter sections. In both sections, the input resistor must be split further in order to permit us to sum the required signals.

The resulting two circuits are shown in Fig. 6-23; the realized functions are

$$V_o = - \frac{Ks\frac{1}{C}(G_a V_1 + G_b V_2)}{s^2 + s\frac{G_1}{C}\left[2 - (K-1)\frac{G_2}{G_1}\right] + \frac{G_1 G_2}{C^2}} \qquad (6\text{-}73a)$$

for Fig. 6-23a, and, for Fig. 6-23b:

$$V_o = + \frac{Ks\frac{1}{C}(G_a V_1 + G_b V_2)}{s^2 + s\frac{G_3}{C}\left[2 + \frac{G_2}{G_3} - (K-1)\frac{G_1}{G_3}\right] + \frac{G_3(G_1 + G_2)}{C^2}} \qquad (6\text{-}73b)$$

Note that $G_a + G_b + G_c = G_2$. By comparing Eq. (6-73) with the prescribed functions in Eq. (6-72), all the elements of the final realization can be computed.

In Fig. 6-24, we show the final circuit structure of the eighth-order bandpass function realized with single-amplifier biquads. The four sections $T_i(s)$ are separated by dashed lines to facilitate identifying the blocks. As shown, the realization has the advantage of economical use of op amps; a disadvantage is the dependence of gain, quality factor, and pole frequency on the same resistors, as is clear from Eq. (6-73). Thus, adjusting and tuning the circuit will be complicated. In this respect, a better realization would use, for example, Åckerberg-Mossberg biquads at the cost of 3 times as many op amps. See Problem 6.26 for the modifications necessary to obtain a noninverting AM bandpass section [11].

6.4.1.4 Signal-flow graph (SFG) design based on general LC ladder filters.
We have discussed so far how to obtain an active realization of an all-pole LC lowpass filter by representing the ladder as a signal-flow graph and sim-

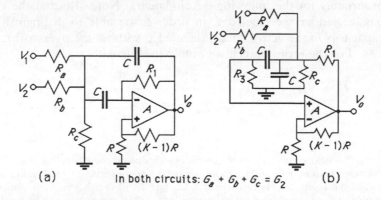

(a)   In both circuits: $G_a + G_b + G_c = G_2$ (b)

Figure 6-23 Single-amplifier biquads suitable for realizing the functions in Eq. (6-72).

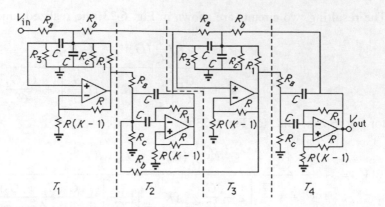

Figure 6-24 Eighth-order bandpass filter realized with single-amplifier biquads.

ulating the series inductors and the shunt capacitors through integrators. We have also seen that this procedure can be extended to all-pole bandpass filters by use of the lowpass-to-bandpass transformation which converts all integrators into band-pass sections, some of which must be able to realize infinite Q factors. In this section we shall extend the powerful SFG technique to the situation of general-parameter LC ladders with finite transmission zeros where the ladder arms may contain, in principle, arbitrary LC immittances. The realization of such lossless ladders was discussed in Chapter 2; a typical example is shown in Fig. 6-25. From the figure it is apparent that the most general series and shunt arms which the simulation must be able to handle are those shown in Fig. 6-26.[18] Recall from our discussion in the beginning of Section 6.4.1 that all voltages and currents will be represented by voltage signals and that their algebraic signs will be correct for future summing if we select to model all series branches by *noninverting transmit-tances* $y(s)$ and all shunt branches by *inverting transmittances* $z(s)$. We have, therefore, chosen the signs of the relevant voltages and currents in Fig. 6-26 ap-propriately for the following development. Note also that the capacitors in the passive ladder are labeled $\hat{C}$ in order to be able to distinguish them from the capacitors in the active circuit (labeled C without the overscore).

For the series branch we obtain the current

$$I_o = Y(s)(V_1 + V_2) = \cfrac{1}{R_k + sL_1 + \cfrac{1}{s\hat{C}_2 + \cfrac{1}{s\hat{C}_3 + \cfrac{1}{sL_4}}}} (V_1 + V_2) \quad (6\text{-}74a)$$

[18]A ladder with even more general branches could possibly be imagined but we shall limit our discussion to those in Fig. 6-26. These types of ladder arms arise in typical realizations from consecutive partial and complete pole removals (see Chapter 2). Nevertheless, the technique can be extended readily to more complicated ladder arms as long as their immittances can be represented as continued series parallel connections. At the end of our presentation it should become quite clear to the reader how to proceed in such a more general situation.

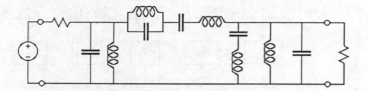

Figure 6-25 Typical resistively terminated *LC* ladder.

which is converted into a voltage through multiplication with a scaling resistor R_p. (*p* stands for *passive*; R_p is the resistor used to scale the *passive* circuit.) Also recall that in the *active* simulation the input signals into each branch may in general be multiplied by a constant to effect signal-level scaling for improved dynamic range (compare Fig. 6-18 and Fig. 6-27 below). Anticipating the need for these scaling factors, let us multiply V_1 and V_2 by two real numbers a and b. Thus, Eq. (6-74a) becomes

$$I_o R_p = Y(s)R_p(aV_1 + bV_2) \tag{6-74b}$$

$$= \cfrac{1}{\cfrac{R_k}{R_p} + \cfrac{sL_1}{R_p} + \cfrac{1}{s\hat{C}_2 R_p} + \cfrac{1}{s\hat{C}_3 R_p + \cfrac{1}{sL_4/R_p}}} (aV_1 + bV_2)$$

Using, as before, *lower case* symbols for the *normalized* variables, we obtain finally for the series branch

$$i_o = y(s)(av_1 + bv_2) = \cfrac{1}{r_k + sl_1 + \cfrac{1}{sc_2} + \cfrac{1}{sc_3 + 1/(sl_4)}} (av_1 + bv_2) \tag{6-74c}$$

In a completely analogous fashion, we find for the shunt branch in Fig. 6-26

$$V_o = -Z(s)(I_1 + I_2) = -\cfrac{1}{G_k + s\hat{C}_1 + \cfrac{1}{sL_2} + \cfrac{1}{sL_3 + \cfrac{1}{s\hat{C}_4}}} (I_1 + I_2) \tag{6-75a}$$

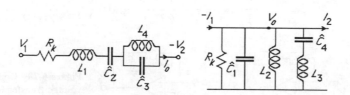

Figure 6-26 Typical ladder series and shunt arms.

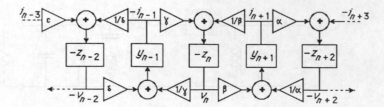

Figure 6-27 Signal-flow graph of a section of a general ladder. Scale factors for signal-level scaling are also indicated.

which by impedance-level scaling with R_p and signal-level scaling with a and b becomes

$$V_o = -\frac{Z(s)}{R_p}(al_1R_p + bl_2R_p)$$

$$= -\frac{1}{G_kR_p + s\hat{C}_1R_p + \dfrac{1}{sL_2/R_p} + \dfrac{1}{sL_3/R_p + \dfrac{1}{s\hat{C}_4R_p}}}(al_1R_p + bl_2R_p) \qquad (6\text{-}75b)$$

Finally, lower case notation gives

$$v_o = -z(s)(ai_1 + bi_2) = -\frac{1}{g_k + sc_1 + \dfrac{1}{sl_2} + \dfrac{1}{sl_3 + 1/(sc_4)}}(ai_1 + bi_2) \qquad (6\text{-}75c)$$

Our task is then to realize the transmittances $-z(s)$ and $y(s)$ as active RC circuits with the appropriate signal scale factors such that a signal-flow graph of the form shown in Fig. 6-27 is implemented. Recall that lower case v and i represent *voltages* and that all lower case r, g, sc, and sl symbols are dimensionless *voltage transfer functions (transmittances)* and *not* immittances.

The realization of the transmittances in Eqs. (6-74c) and (6-75c) is accomplished easily if we remember that an admittance in the feedback path of an op amp becomes inverted, and converted into a transfer function as is illustrated in Fig. 6-28. To clarify the direction of the transfer function $T(s)$ we have indicated its input by an arrow. Simple analysis shows that the circuits realize

$$\frac{V_o}{V_i} = -\frac{1}{Y(s)RT(s)} \quad \text{and} \quad \frac{V_o}{V_i} = \frac{1}{Y(s)RT(s)} \qquad (6\text{-}76)$$

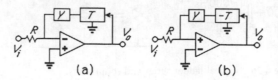

Figure 6-28 Conceptual realization of a voltage transfer function inversely proportional to $T(s)Y(s)$ with an op amp circuit.

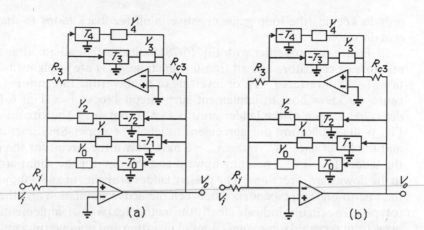

Figure 6-29 Conceptual realization of the transmittances $y(s)$ and $-z(s)$ of Eqs. (6-74b) and (6-75b).

for Fig. 6-28a and b, respectively. This concept can be extended [145] to more than one branch by making use of current summing at the op amp input node and "nesting." Figure 6-29 shows the circuits, again for the inverting and noninverting transfer functions. Assuming for now ideal op amps, the reader will find it easy to verify that the two circuits realize

$$\frac{V_o}{V_i} = \pm \frac{G_i}{Y_0 T_0 + Y_1 T_1 + Y_2 T_2 + \dfrac{G_3 G_{c3}}{Y_3 T_3 + Y_4 T_4}} \qquad (6\text{-}77)$$

where the plus sign is valid for Fig. 6-29a and the minus sign for Fig. 6-29b. Clearly, the form of Eq. (6-77) is the same as those of Eqs. (6-74c) and (6-75c); to make the circuits realize the prescribed ladder arms, we have only to choose

$$Y_0 = G_0 \qquad T_0 = 1 \qquad Y_1 = sC_1 \qquad T_1 = 1 \qquad Y_2 = G_2 \qquad T_2 = \frac{1}{s\tau_2} \qquad (6\text{-}78)$$

$$Y_3 = sC_3 \qquad T_3 = 1 \qquad Y_4 = G_4 \qquad T_4 = \frac{1}{s\tau_4}$$

Also, in order to obtain the required two inputs, let us label the first input voltage V_{i1} and the feed-in resistor R_{i1} and add a second input voltage, V_{i2}, through a resistor R_{i2}. The output voltages of the circuits in Fig. 6-29 become then

$$V_o = \pm \frac{G_{i1} V_{i1} + G_{i2} V_{i2}}{G_0 + sC_1 + \dfrac{G_2}{s\tau_2} + \dfrac{G_3 G_{c3}}{sC_3 + G_4/(s\tau_4)}} \qquad (6\text{-}79)$$

which is exactly of the desired form. As was mentioned earlier, this scheme can easily be extended to functions of higher order, if required; attention must be paid only to keeping the loop gains negative in all feedback loops so that the circuits remain stable.

Figure 6-29 together with Eq. (6-78) indicates that all admittances are either resistors or capacitors and all transfer functions $T_i(s)$ are straight-through connections $(+1)$, inverters (-1), or inverting or noninverting integrators $(\pm 1/s\tau)$. Because we know how to implement such circuit blocks (see Fig. 6-16), the final active realization of the ladder arms is easy to obtain. The circuits are shown in Fig. 6-30. Following our agreement to label the upper SFG lines as "currents" and the lower ones as "voltages," we have drawn the circuit for the series arm of the ladder $(v_{in} \rightarrow i_{out})$ pointing upward and the one for the shunt arm $(i_{in} \rightarrow v_{out})$ in the downward direction. The passive ladder branches are also shown, to indicate the one-to-one correspondence between the active simulation and the passive prototype. The circuits indicate clearly the methodical way of implementing the ladder arms from repeated interconnections of inverting and noninverting integrators. As mentioned before, extensions to transmittances of higher order are trivial. Analysis of the two circuits, assuming ideal op amps, results in

$$V_o = \pm \frac{G_{i1}V_{i1} + G_{i2}V_{i2}}{G_0 + sC_1 + \dfrac{G_2}{sC_2R_{c2}} + \dfrac{G_3G_{c3}}{sC_3 + G_4/(sC_4R_{c4})}} \tag{6-80}$$

as predicted by Eq. (6-79). Finally, let us multiply numerator and denominator of the right-hand side of Eq. (6-80) by a normalizing resistor R_a. (a stands for *active*; R_a is the resistor used to scale the *active* circuit.) The result is for the *series* arm

$$V_o = + \frac{R_aG_{i1}V_{i1} + R_aG_{i2}V_{i2}}{R_aG_0 + sC_1R_a + \dfrac{R_aG_2}{sC_2R_{c2}} + \dfrac{R_aG_3}{sC_3R_{c3} + \dfrac{G_4R_{c3}}{sC_4R_{c4}}}} \tag{6-81}$$

It is to be equated to Eq. (6-74b):

$$I_oR_p = + \frac{aV_1 + bV_2}{\dfrac{R_k}{R_p} + \dfrac{sL_1}{R_p} + \dfrac{1}{s\hat{C}_2R_p} + \dfrac{1}{s\hat{C}_3R_p + \dfrac{1}{sL_4/R_p}}}$$

Comparing coefficients and assuming all equal capacitors in the active circuit leads to the equations for the components:

$$R_{i1} = \frac{R_a}{a} \qquad R_{i2} = \frac{R_a}{b} \qquad R_0 = \frac{R_aR_p}{R_k} \qquad C = \frac{L_1}{R_aR_p}$$

$$R_{c2}R_2 = \frac{\hat{C}_2}{C}R_aR_p \qquad R_{c3}R_3 = \frac{\hat{C}_3}{C}R_aR_p \qquad R_{c4}R_4 = \frac{L_4\hat{C}_3}{C^2} \tag{6-82a}$$

R_a and R_p are chosen to obtain convenient element values.

Note that the last three equations each determine only the *product* of two resistors, leaving three degrees of freedom. They are normally used to maximize dynamic range by equalizing the signal levels at all op amp outputs. The method proceeds as follows. Recall that we used already SFG scaling to equalize the signal levels at the outputs of the transmittances $y(s)$ and $-z(s)$ in Fig. 6-27 (i.e., V_o in Fig. 6-30). The remaining task is, therefore, only to equalize the maxima of the output voltages V', V'', and V''' at the three integrating op amps in both Fig. 6-30a and Fig. 6-30b. This is accomplished easily by noting that in both circuits

$$|V'| = \frac{G_{c2}}{\omega C_2}|V_o| \qquad |V''| = \left|\frac{G_{c3}}{\omega C_3 - [G_4 G_{c4}/(\omega C_4)]}\right||V_o|$$

and

$$|V'''| = \frac{G_{c4}}{\omega C_4}|V''| = \left|\frac{G_{c3}G_{c4}}{\omega^2 C_3 C_4 - G_{c4}G_4}\right||V_o|$$

Now, assume that for $0 \le \omega \le \infty$

$$m' = \frac{\max |V'|}{\max |V_o|} \qquad m'' = \frac{\max |V''|}{\max |V_o|} \qquad m''' = \frac{\max |V'''|}{\max |V_o|} \tag{6-83}$$

where the constants m', m'', and m''' may be less than 1 or greater than 1. Clearly, the voltage maxima of V_o, V', and V'' are made equal by multiplying the resistors R_{c2} and R_{c3} by m' and m'', respectively, to yield:

$$|V'| \longrightarrow \frac{G_{c2}/m'}{\omega C_2}|V_o| \quad \text{and} \quad |V''| \longrightarrow \left|\frac{G_{c3}/m''}{\omega C_3 - [G_4 G_{c4}/(\omega C_4)]}\right||V_o| \tag{6-84a}$$

Now note that scaling R_{c3} also decreases $|V'''|$ by m''. Therefore, to equalize the voltage maxima of V''' and V_o, we need to multiply R_{c4} by (m'''/m''):

$$|V'''| \longrightarrow \left|\frac{(G_{c3}/m'')\,(G_{c4}m''/m''')}{\omega^2 C_3 C_4 - (G_{c4}m''/m''')\,(G_4 m'''/m'')}\right||V_o| \tag{6-84b}$$

and at the same time multiply the resistor R_4 by (m''/m''') in order not to shift the transmission zero at

$$\omega_0 = \sqrt{\frac{(G_{c4}m''/m''')\,(G_4 m'''/m'')}{C_3 C_4}} = \sqrt{\frac{G_{c4}G_4}{C_3 C_4}}$$

Finally, to keep the expression in Eq. (6-80) unaltered by this process, we need only scale also the resistors R_2 and R_3 appropriately:

$$V_o = \pm \frac{G_{i1}V_{i1} + G_{i2}V_{i2}}{G_0 + sC_1 + \dfrac{m'G_2}{sC_2(m'R_{c2})} + \dfrac{(m''G_3)(G_{c3}/m'')}{sC_3 + \dfrac{G_4 m'''/m''}{sC_4(R_{c4}m'''/m'')}}} \tag{6-85}$$

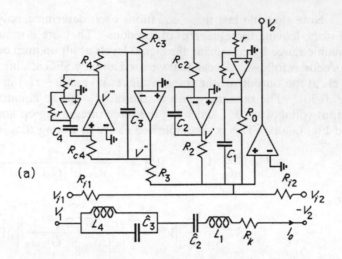

(a)

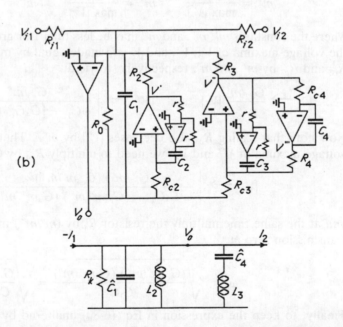

(b)

Figure 6-30 Active realization of the ladder branches: (a) the series arm of Fig. 6-26a [plus sign in Eq. (6-80)]; (b) the shunt arm of Fig. 6-26b [minus sign in Eq. (6-80)]. For ease of reference, the corresponding passive ladder branches are drawn below the active simulations.

Thus, the remaining degrees of freedom are just sufficient to permit us to equalize the output voltages of all op amps in the SFG circuit. The resistor values are still obtained from Eq. (6-82a) after relabeling the resistors appropriately:

$$m'R_{c2}\frac{R_2}{m'} = \frac{\hat{C}_2}{C}R_aR_p \qquad m''R_{c3}\frac{R_3}{m''} = \frac{\hat{C}_3}{C}R_aR_p$$

$$\frac{m'''}{m''}R_{c4}\frac{R_4}{m'''/m''} = \frac{L_4\hat{C}_3}{C^2} \qquad (6\text{-}82b)$$

A possible choice for the element values is:

$$R_{c2} = \frac{1}{m'}\sqrt{\frac{\hat{C}_2}{C}R_aR_p} \qquad R_2 = m'\sqrt{\frac{\hat{C}_2}{C}R_aR_p}$$

$$R_{c3} = \frac{1}{m''}\sqrt{\frac{\hat{C}_3}{C}R_aR_p} \qquad R_3 = m''\sqrt{\frac{\hat{C}_3}{C}R_aR_p} \qquad (6\text{-}82c)$$

and

$$R_{c4} = \frac{m''}{m'''}\sqrt{\frac{L_4\hat{C}_3}{C^2}} \qquad R_4 = \frac{m'''}{m''}\sqrt{\frac{L_4\hat{C}_3}{C^2}}$$

Proceeding in an entirely similar fashion, we obtain for the *shunt* arm of the ladder, *before* scaling for dynamic range maximization,

$$V_o = -\frac{R_aG_{i1}V_{i1} + R_aG_{i2}V_{i2}}{R_aG_0 + sC_1R_a + \dfrac{R_aG_2}{sC_2R_{c2}} + \dfrac{R_aG_3}{sC_3R_{c3} + \dfrac{G_4R_{c3}}{sC_4R_{c4}}}} \qquad (6\text{-}86)$$

which must be equated to Eq. (6-75b):

$$V_o = -\frac{aI_1R_p + bI_2R_p}{G_kR_p + s\hat{C}_1R_p + \dfrac{1}{sL_2/R_p} + \dfrac{1}{sL_3/R_p + 1/(s\hat{C}_4R_p)}}$$

Thus, using again identical capacitors in the active circuit, the equations for the elements are, using Eq. (6-84),

$$R_{i1} = \frac{R_a}{a} \qquad R_{i2} = \frac{R_a}{b} \qquad R_0 = R_k\frac{R_a}{R_p}$$

$$C = \hat{C}_1\frac{R_p}{R_a} \qquad m'R_{c2}\frac{R_2}{m'} = \frac{L_2}{C}\frac{R_a}{R_p} \qquad (6\text{-}82d)$$

$$m''R_{c3}\frac{R_3}{m''} = \frac{L_3}{C}\frac{R_a}{R_p} \qquad \frac{m'''}{m''}R_{c4}\frac{R_4}{m'''/m''} = \frac{L_3\hat{C}_4}{C^2}$$

From Eq. (6-82d), the scaled resistor values are obtained in the same way as was indicated in Eq. (6-82c):

$$R_{c2} = \frac{1}{m'}\sqrt{\frac{L_2}{C}\frac{R_a}{R_p}} \qquad R_2 = m'\sqrt{\frac{L_2}{C}\frac{R_a}{R_p}}$$

$$R_{c3} = \frac{1}{m''}\sqrt{\frac{L_3}{C}\frac{R_a}{R_p}} \qquad R_3 = m''\sqrt{\frac{L_3}{C}\frac{R_a}{R_p}} \qquad (6\text{-}82e)$$

$$R_{c4} = \frac{m''}{m'''}\sqrt{\frac{L_3\hat{C}_4}{C^2}} \qquad R_4 = \frac{m'''}{m''}\sqrt{\frac{L_3\hat{C}_4}{C^2}}$$

The numbers m', m'', and m''' are defined in Eq. (6-83); they can be obtained readily from a ladder analysis program [e.g., 11], by noting from Eq. (6-80) and the correspondence between the active and passive circuits that

> V', V'', and V''', respectively, represent,
>
> in the *series arm* of Fig. 6-30a, the voltage on capacitor $\hat{C}_2$, the voltage on capacitor $\hat{C}_3$, and the current (times R_p) through inductor L_4;
>
> and in the *shunt arm* of Fig. 6-30b, the current (times R_p) through inductor L_2, the current (times R_p) through inductor L_3, and the voltage on capacitor $\hat{C}_4$

of the passive circuits in Fig. 6-26 or Fig. 6-30.

Observe again the one-to-one correspondence between the elements of the active circuit and those of the passive ladder. The absence of any component in the passive ladder leads to the absence of the corresponding branch in the active simulation. For example, if the passive shunt arm being simulated has no inductor L_2 (i.e., $L_2 = \infty$), the complete noninverting integrator with R_2, R_{c2}, and C_2 in Fig. 6-30 is deleted.

The Effect of Finite Op Amp Bandwidth.   Before our design task is complete, we must investigate the effect of the finite gain-bandwidth product ω_t of the op amps. Until now we have neglected op amp imperfections in order to be able to concentrate on the principle of signal-flow graph simulations without obscuring the circuit development with presumably small errors caused by nonideal op amps. Let us assume that all op amps are identical, modeled, as always, by

$$A(s) \simeq \frac{\omega_t}{s}$$

and look specifically at the shunt branch, Fig. 6-30b [Eq. (6-86)]. Our task is simplified if we recall from Chapter 4 the following relationships for the

Inverter: $\dfrac{V_{\text{out}}}{V_{\text{in}}} = -\dfrac{1}{1 + 2/A}$

Inverting integrator: $\dfrac{V_{\text{out}}}{V_{\text{in}}} = -\dfrac{1}{sCR}\dfrac{1}{1 + \left(1 + \dfrac{1}{sCR}\right)\dfrac{1}{A}}$

Noninverting integrator: $\dfrac{V_{\text{out}}}{V_{\text{in}}} = +\dfrac{1}{sCR}\dfrac{1}{\dfrac{1}{1 + 2/A} + \left(1 + \dfrac{1}{sCR}\right)\dfrac{1}{A}}$

Thus, for nonideal op amps we use these equations and obtain instead of the ideal denominator of Eq. (6-86), i.e.,

$$G_0 + sC_1 + \frac{G_2}{sC_2R_{c2}} + \frac{G_3}{sC_3R_{c3} + \dfrac{G_4R_{c3}}{sC_4R_{c4}}}$$

the expression

$$G_0\left(1 + \frac{1}{A}\right) + sC_1\left(1 + \frac{1}{A}\right) + G_2\left(\frac{V'}{V_o} + \frac{1}{A}\right)$$

$$+ G_3\left(\frac{V''}{V_o} + \frac{1}{A}\right) + (G_{i1} + G_{i2})\frac{1}{A} \quad (6\text{-}87)$$

where

$$\frac{V'}{V_o} = \frac{1}{s\tau_2}\frac{1}{\dfrac{1}{1 + 2/A} + \left(1 + \dfrac{1}{s\tau_2}\right)\dfrac{1}{A}} \simeq \frac{1}{s\tau_2}\frac{1}{1 + \dfrac{1}{\omega_t\tau_2} - \dfrac{2}{A} + \dfrac{1}{A}} \quad (6\text{-}88)$$

and

$$\frac{V''}{V_o} = \frac{1}{s\tau_3\left[\dfrac{1}{1 + 2/A} + \left(1 + \dfrac{1}{s\tau_3}\right)\dfrac{1}{A}\right] + \dfrac{G_4R_{c3}}{s\tau_4}\left[\dfrac{1}{1 + [1 + 1/(s\tau_4)]/A} + \dfrac{s\tau_4}{A}\right]}$$

$$\simeq \frac{1}{s\tau_3\left(1 + \dfrac{1}{\omega_t\tau_3} - \dfrac{2}{A} + \dfrac{1}{A}\right) + \dfrac{G_4R_{c3}}{s\tau_4}\left(1 - \dfrac{1}{\omega_t\tau_4} - \dfrac{1}{A} + \dfrac{s^2\tau_4}{\omega_t}\right)} \quad (6\text{-}89)$$

V'' and V' are identified in Fig. 6-30. In these equations we have labeled

$$C_2R_{c2} = \tau_2 \qquad C_3R_{c3} = \tau_3 \qquad C_4R_{c4} = \tau_4 \quad (6\text{-}90)$$

and assumed that $|A(j\omega)| \gg 1$, so that $1/(1 + 1/A) \simeq 1 - 1/A$; if we further assume that $\omega_t\tau_2 \gg 1$ and $\omega_t\tau_4 \gg 1$, we obtain

$$\frac{V'}{V_o} \simeq \frac{1}{s\tau_2}\frac{1}{1 - 1/A} \simeq \frac{1}{s\tau_2}\left(1 + \frac{1}{A}\right) \tag{6-91}$$

and

$$\frac{V''}{V_o} \simeq \frac{1}{\left(s\tau_3 + \dfrac{G_4R_{c3}}{s\tau_4}\right)\left(1 - \dfrac{1}{A}\right)} \simeq \frac{1}{s\tau_3 + \dfrac{G_4R_{c3}}{s\tau_4}}\left(1 + \frac{1}{A}\right) \tag{6-92}$$

Inserting Eqs. (6-91) and (6-92) into Eq. (6-87) results in

$$G_0\left(1 + \frac{1}{A}\right) + sC_1\left(1 + \frac{1}{A}\right) + \frac{G_2}{s\tau_2}\left(1 + \frac{s^2\tau_2}{\omega_t} + \frac{1}{A}\right)$$

$$+ \frac{G_3}{s\tau_3 + \dfrac{G_4R_{c3}}{s\tau_4}}\left[\left(1 + \frac{1}{A}\right) + \left(s\tau_3 + \frac{G_4R_{c3}}{s\tau_4}\right)\frac{1}{A}\right] + (G_{i1} + G_{i2})\frac{1}{A} \tag{6-93}$$

Because the ladder arm will always contain either a capacitor or an inductor or a resonance circuit, the last term in Eq. (6-93) will usually be negligible compared to the previous terms; for example, combining the second and last terms yields, with the same approximations as before,

$$sC_1\left(1 + \frac{1}{A}\right) + (G_{i1} + G_{i2})\frac{1}{A} = sC_1\left(1 + \frac{1}{A} + \frac{G_{i1} + G_{i2}}{sC_1}\frac{1}{A}\right)$$

$$= sC_1\left(1 + \frac{G_{i1} + G_{i2}}{\omega_tC_1} + \frac{1}{A}\right) \simeq sC_1\left(1 + \frac{1}{A}\right) \tag{6-94a}$$

Similarly, we may assume that

$$\left|\left(s\tau_3 + \frac{G_4R_{c3}}{s\tau_4}\right)\frac{1}{A}\right|_{s=j\omega} = \left|-\frac{\omega^2\tau_3}{\omega_t} + \frac{G_4R_{c3}}{\omega_t\tau_4}\right| \ll 1 \tag{6-94b}$$

in the frequency range of interest. Consequently, the main effect of nonideal op amps is to multiply each term in the denominator for Eq. (6-86) by $(1 + 1/A)$; i.e., the output voltage of the *shunt* transmittance, Fig. 6-30b, becomes

$$V_o \simeq -\frac{R_aG_{i1}V_{i1} + R_aG_{i2}V_{i2}}{R_aG_0 + sC_1R_a + \dfrac{R_aG_2}{sC_2R_{c2}} + \dfrac{R_aG_3}{sC_3R_{c3} + \dfrac{G_4R_{c3}}{sC_4R_4}}}\left(1 - \frac{1}{A}\right) \tag{6-95a}$$

The *series* transmittance, Fig. 6-30a, with imperfect op amps equals

$$V_o \simeq +\frac{R_aG_{i1}V_{i1} + R_aG_{i2}V_{i2}}{R_aD(s)} \tag{6-96}$$

where

$$D(s) = (G_0 + sC_1)\left(\frac{1}{1 + \dfrac{2}{A}} + \frac{1}{A}\right) + G_2\left[\frac{1}{A} + \frac{1}{s\tau_2\left(1 + \dfrac{1}{A} + \dfrac{1}{s\tau_2 A}\right)}\right]$$

$$+ G_3\left[\frac{1}{A} + \frac{1}{s\tau_3\left(1 + \dfrac{1}{A}\right) + \dfrac{1 + R_{c3}G_4}{A} + \dfrac{R_{c3}G_4}{s\tau_4}\left(\dfrac{1}{1 + 2/A} + \dfrac{1}{A} + \dfrac{1}{s\tau_4 A}\right)}\right]$$

$$+ \frac{G_{i1} + G_{i2}}{A}$$

Assuming as before that $|A(j\omega)| \gg 1$, this expression reduces to

$$D(s) \simeq (G_0 + sC_1)\left(1 - \frac{1}{A}\right) + \frac{G_2}{s\tau_2}\left(1 - \frac{1}{A} + \frac{s^2\tau_2}{\omega_t} - \frac{1}{\omega_t\tau_2}\right)$$

$$+ \frac{G_3}{s\tau_3\left(1 + \dfrac{1}{A} + \dfrac{1 + R_{c3}G_4}{\omega_t\tau_3}\right) + \dfrac{R_{c3}G_4}{s\tau_4}\left(1 + \dfrac{2}{A} - \dfrac{1}{A} - \dfrac{1}{\omega_t\tau_4}\right)}$$

$$+ \frac{G_{i1} + G_{i2} + G_3}{A}$$

Thus, with $\omega_t\tau_2 \gg 1$, $\omega_t\tau_4 \gg 1$, and $\omega_t\tau_3 \gg 1 + R_{c3}G_4$, we obtain for the *series* transmittance (Fig. 6-30a)

$$V_o \simeq + \frac{R_aG_{i1}V_{i1} + R_aG_{i2}V_{i2}}{R_aG_0 + sC_1R_a + \dfrac{R_aG_2}{sC_2R_{c2}} + \dfrac{R_aG_3}{sC_3R_{c3} + \dfrac{G_4R_{c3}}{sC_4R_{c4}}}}\left(1 + \frac{1}{A}\right) \quad (6\text{-}95b)$$

$(G_{i1} + G_{i2} + G_3)/A$ can be absorbed in the remaining terms, as was illustrated in Eq. (6-94).

With $\omega_t \gg \omega$, $1 \pm 1/A$ on the $j\omega$-axis approximates to

$$1 \pm \frac{j\omega}{\omega_t} \simeq e^{\pm j\Delta\phi(\omega)} \qquad \Delta\phi(\omega) = \frac{\omega}{\omega_t}$$

which means that the *shunt* branches have a *phase-lag* error and the *series* branches a *phase-lead* error of equal magnitude. Thus, within the approximations made, the first-order effects of the op amps cancel and all loop gains are correct as is required according to Eq. (6-59):

$$1 - [-z(j\omega)e^{+j\Delta\phi(\omega)}][y(j\omega)e^{-j\Delta\phi(\omega)}] = 1 + z(j\omega)y(j\omega) \qquad (6\text{-}97)$$

This result justifies our earlier development of SFG design based on ideal op amps!

The student should note the systematic way in which the signal-flow graph technique leads to an active simulation of a passive lossless ladder. Observe, also, that the method applies equally well to *lossy* ladders where zeros are shifted away from the $j\omega$-axis in order to obtain a desired phase response. Such ladders often simply have internal resistors which can easily be handled by our method.

Designers of lossless ladders have found it convenient and practical to tune their filters by adjusting the transmission zeros via tuning the corresponding LC resonance circuits. Note that the same procedure can be used in the active simulation: we may easily tune the "active tank circuits," the three-op amp structures between R_3 and R_{c3} in Fig. 6-30, until the filter behaves as desired.

A final comment about SFG ladder simulations pertains to the sensitivity of these circuits to component tolerances and losses. Because the SFG method simulates the behavior of a passive ladder in all details, the well-known low passband sensitivity (see Chapter 3) to circuit elements that represent ladder components is maintained. These "elements" are R_0, C_1, and all integrator time constants in Fig. 6-30. Any effects that op amp imperfections have on the *value* of circuit "elements" is, therefore, of relatively little concern to the designer as long as these effects are small.[19] We have seen, however, in Chapter 4 and in Eqs. (6-91) through (6-94) that the finite op amp gain-bandwidth product results in phase errors, which we have modeled as a finite error in quality factor [see Eqs. (4-53) and (4-62)] of the inverting and noninverting integrators. Note that in SFG filters these integrators simulate the behavior of capacitors and inductors, respectively, so that we can use Eq. (3-86),

$$d\alpha(\omega) \simeq \frac{1}{2}\left(\frac{1}{Q_L} + \frac{1}{Q_c}\right)\omega\tau(\omega)$$

to evaluate their effect on the filter. Observe further that due to this correspondence we have

$$Q_L = -Q_C \simeq \frac{\omega_t}{\omega}$$

which means that in the passband, $d\alpha \simeq 0$. This result is entirely consistent with Eq. (6-97), which indicated that the denominator polynomial, i.e., the poles of the filter transfer function, is independent of $\Delta\phi(\omega) \simeq \omega/\omega_t$. We may conclude, therefore, that an SFG simulation of an LC ladder filter will be very insensitive to all

---

[19]Note that most of the neglected terms in Eqs. (6-91) through (6-94) result in small element *value* errors whose effects, as just pointed out, are negligible.

circuit components if our adopted design guidelines (inverting phase-lag integrators for capacitors and noninverting phase-lead integrators for inductors) are followed.

Design Summary.   Let us summarize this procedure and show step by step with the help of Fig. 6-25 how an SFG simulation of an *LC* ladder is designed.

Step 1. Using the procedures discussed in Chapter 2 or available tables [13, 18, 21] or software [11, 29], design the *LC* ladder to meet the prescribed specifications.

Step 2. For $0 \leq \omega \leq \infty$, calculate the maxima of all inductor currents and capacitor voltages in the ladder. These values are used later in scaling to equalize the signal levels at the op amp outputs.

Step 3. If the ladder starts with a shunt arm, perform a source transformation. This step reduces the number of meshes in the ladder and saves components. Scale the passive ladder by R_p (e.g., $R_p = 1\ \Omega$) in preparation for the active simulation where all signals are voltages.

Step 3 transforms the circuit of Fig. 6-25 into the "three-branch ladder" in Fig. 6-31.

Step 4. Form the signal-flow graph with one transmittance for each ladder arm. Use *inverting* transmittances for the *shunt* arms and *noninverting* transmittances for the *series* arms. This choice (or its dual) guarantees that all loop gains are negative and that the phase errors caused by finite ω_t cancel out. No more attention need be paid to the algebraic signs of the signals in any of the further development. Include in the SFG the *global* scaling factors computed from the ratios of two consecutive maxima of series branch currents and shunt branch voltages as obtained in step 2. Also include the factor K, which determines the overall gain of the realization.

Figure 6-32a shows a block diagram representation of our ladder, and Fig. 6-32b gives the signal-flow graph obtained in step 4, including the necessary *global*

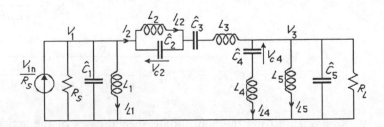

Figure 6-31   The ladder of Fig. 6-25 after source transformation.

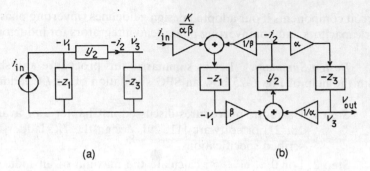

(a)   (b)

Figure 6-32 (a) Block diagram of the ladder and (b) scaled signal-flow graph.

scaling factors, defined as

$$\alpha = \frac{\max |i_2|}{\max |v_3|} = \frac{\max |I_2 R_p|}{\max |V_3|} \qquad \beta = \frac{\max |v_1|}{\max |i_2|} = \frac{\max |V_1|}{\max |I_2 R_p|}$$

Step 5. By inspection of the impedance-scaled ladder, write the expressions of the transmittances and set up the equations for the scaled SFG.

For our example circuit, this step leads to

$$i_{in} = \frac{V_{in} R_p}{R_S}$$

$$-v_1 = -V_1 = \frac{\dfrac{K}{\alpha\beta} i_{in} - \dfrac{1}{\beta} i_2}{G_S R_p + s\hat{C}_1 R_p + \dfrac{1}{sL_1/R_p}} \tag{6-98a}$$

$$i_2 = I_2 R_p = -\frac{\beta v_1 - (1/\alpha)v_3}{sL_3/R_p + \dfrac{1}{s\hat{C}_3 R_p} + \dfrac{1}{s\hat{C}_2 R_p + \dfrac{1}{sL_2/R_p}}} \tag{6-98b}$$

$$v_3 = V_3 = -\frac{-\alpha i_2}{G_L R_p + s\hat{C}_5 R_p + \dfrac{1}{sL_5/R_p} + \dfrac{1}{sL_4/R_p + 1/(s\hat{C}_4 R_p)}} \tag{6-98c}$$

Step 6. Construct the appropriate special cases of the active circuits of Fig. 6-30 to realize the required ladder arms.

Step 7. Using normalizing resistors R_{ai}, $i = 1, 2, \ldots$, set up by inspection the equations for the transfer functions of the active circuits. Note that R_{ai} may be different for each active ladder arm; this freedom permits the designer to obtain, for example, equal capacitors throughout the active circuit.

Step 8. Compare the equations for the active circuits with those for the passive branches in order to obtain expressions for the element values.

For the example under consideration, Steps 6, 7, and 8 lead to the following results:
The circuit for the $-z_1$ block, from Fig. 6-30b, is shown in Fig. 6-33a; it realizes

$$v_1 = -\frac{R_{a1}G_{i1}i_{\text{in}} + R_{a1}G_{i2}i_2}{R_{a1}G_0 + sC_1R_{a1} + \dfrac{R_{a1}G_2}{sC_2R_{c2}}} \tag{6-99a}$$

Comparing Eqs. (6-98a) and (6-99a) leads to

$$R_{i1} = \frac{\alpha\beta}{K} R_{a1} \qquad R_{i2} = \beta R_{a1} \qquad R_0 = R_S \frac{R_{a1}}{R_p} \qquad C_1 = \hat{C}_1 \frac{R_p}{R_{a1}} \tag{6-100a}$$

$$R_{c2} = R_2 = \sqrt{\frac{L_1}{C_2} \frac{R_{a1}}{R_p}}$$

Because R_{a1} is a free parameter, it may now be chosen to obtain suitable component values; a frequent choice is to select a convenient capacitor of value C and then make all capacitors in the active circuit the same: $C_i = C$, all i. The required value of R_{a1} to be inserted into Eq. (6-100a) is

$$R_{a1} = \frac{\hat{C}_1}{C} R_p \tag{6-101a}$$

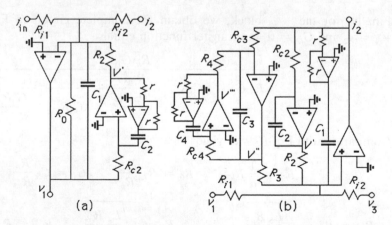

Figure 6-33 Circuits to realize (a) $-z_1$ and (b) y_2 of the summary example.

In a similar fashion, we obtain from Fig. 6-30a the circuit realization for the y_2 block; it is shown in Fig. 6-33b and has the transfer function

$$i_2 = +\cfrac{R_{a2}G_{i1}v_1 + R_{a2}G_{i2}v_3}{sC_1R_{a2} + \cfrac{R_{a2}G_2}{sC_2R_{c2}} + \cfrac{R_{a2}G_3}{sC_3R_{c3} + \cfrac{G_4R_{c3}}{sC_4R_{c4}}}} \qquad (6\text{-}99\text{b})$$

Note that for ease of reference we have retained the element names introduced in Fig. 6-30; an element of a particular name has meaning only in the circuit block under consideration, that is, R_2 in Fig. 6-33b is *not* the same as R_2 in Fig. 6-33a. Comparing Eq. (6-99b) to Eq. (6-98b) yields

$$R_{i1} = \frac{1}{\beta}R_{a2} \qquad R_{i2} = \alpha R_{a2} \qquad C_1 = \frac{L_3}{R_{a2}R_p} \qquad (6\text{-}100\text{b})$$

$$R_{c2} = R_2 = \sqrt{\frac{\hat{C}_3}{C_2}R_pR_{a2}} \qquad R_{c3} = R_3 = \sqrt{\frac{\hat{C}_2}{C_3}R_pR_{a2}} \qquad R_{c4} = R_4 = \sqrt{\frac{L_2\hat{C}_2}{C_3C_4}}$$

Because we chose all capacitors to be identical, we obtain from Eqs. (6-100a) and (6-100b)

$$C_1 = C = \hat{C}_1\frac{R_p}{R_{a1}} = \frac{L_3}{R_{a2}R_p}$$

i.e.,

$$R_{a2} = R_{a1}\frac{L_3}{\hat{C}_1R_p^2}$$

or with Eq. (6-101a)

$$R_{a2} = \frac{L_3}{CR_p} \qquad (6\text{-}101\text{b})$$

Finally, for the $-z_3$ block, we obtain the complete circuit in Fig. 6-30b with $V_{i1} \triangleq i_2$ and $G_{i2} = 0$; the transfer function equals

$$v_3 = -\cfrac{R_{a3}G_{i1}i_2}{R_{a3}G_0 + sC_1R_{a3} + \cfrac{R_{a3}G_2}{sC_2R_{c2}} + \cfrac{R_{a3}G_3}{sC_3R_{c3} + \cfrac{G_4R_{c3}}{sC_4R_{c4}}}} \qquad (6\text{-}99\text{c})$$

Comparing with Eq. (6-98c) yields

$$R_{i1} = \frac{1}{\alpha}R_{a3} \qquad R_0 = R_L\frac{R_{a3}}{R_p} \qquad C_1 = \hat{C}_5\frac{R_p}{R_{a3}} \qquad (6\text{-}100\text{c})$$

$$R_{c2} = R_2 = \sqrt{\frac{L_5}{C_2}\frac{R_{a3}}{R_p}} \qquad R_{c3} = R_3 = \sqrt{\frac{L_4}{C_3}\frac{R_{a3}}{R_p}} \qquad R_{c4} = R_4 = \sqrt{\frac{L_4\hat{C}_4}{C_3C_4}}$$

Again, making all capacitors identical results in

$$C_1 = C = \hat{C}_5 \frac{R_p}{R_{a3}} = \hat{C}_1 \frac{R_p}{R_{a1}}$$

i.e.,

$$R_{a3} = R_{a1} \frac{\hat{C}_5}{\hat{C}_1} = R_p \frac{\hat{C}_5}{C} \tag{6-101c}$$

Step 9. Optimize the dynamic range by equating the op amp output signal levels V', V'', and V''' in Fig. 6-30 to the previously equalized levels at the outputs V_o of the active circuit blocks which simulate the ladder arms. This is accomplished by replacing ($\Rightarrow$) the previously calculated resistor values R_i and R_{ci}, $i = 2, 3, 4$, by the new values as follows:

$$R_{c2} \Rightarrow \frac{R_{c2}}{m'} \qquad R_{c3} \Rightarrow \frac{R_{c3}}{m''} \qquad R_{c4} \Rightarrow \frac{m''}{m'''} R_{c4} \tag{6-102}$$

$$R_2 \Rightarrow m' R_2 \qquad R_3 \Rightarrow m'' R_3 \qquad R_4 \Rightarrow \frac{m'''}{m''} R_4$$

Here, per Eq. (6-83), the scaling factors are defined as

$$m' = \frac{\max |V'|}{\max |V_o|} \qquad m'' = \frac{\max |V''|}{\max |V_o|} \qquad m''' = \frac{\max |V'''|}{\max |V_o|}$$

i.e.,

$$m' = \frac{\max |V_{\hat{C}2}|}{\max |I_{\text{Sc}}R_p|} \qquad m'' = \frac{\max |V_{\hat{C}3}|}{\max |I_{\text{Sc}}R_p|} \qquad m''' = \frac{\max |I_{L4}R_p|}{\max |I_{\text{Sc}}R_p|} \tag{6-103a}$$

for the series arm, and

$$m' = \frac{\max |I_{L2}R_p|}{\max |V_{\text{Sh}}|} \qquad m'' = \frac{\max |I_{L3}R_p|}{\max |V_{\text{Sh}}|} \quad \text{and } m''' = \frac{\max |V_{\hat{C}4}|}{\max |V_{\text{Sh}}|} \tag{6-103b}$$

for the shunt arm. I_{Sc} is the total current through the series branch and V_{Sh} is the total voltage across the shunt branch of the LC ladder; the components $\hat{C}_2$, $\hat{C}_3$, $\hat{C}_4$, L_2, L_3, and L_4 are identified in Fig. 6-30. Note that the values of the parameters m', m'', and m''' are known from step 2 of the design process.

Step 10. Interconnect the circuit blocks obtained in the previous nine steps.

We shall terminate our discussion of the SFG simulation of LC ladders by illustrating the whole design process with a numerical example.

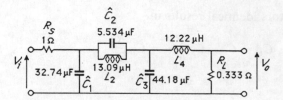

Figure 6-34 Fourth-order elliptic LC lowpass ladder.

Example 6-6

Design an SFG circuit to simulate the fourth-order elliptic lowpass ladder filter shown in Fig. 6-34 [139]. Use signal-level scaling to optimize the dynamic range. Choose all capacitors in the SFG circuit equal to 5 nF and select reasonable resistor values for a discrete realization. The dc gain should equal $H_0 = 1$. Table 6-3 lists the relevant maxima of the LC ladder currents and voltages needed for signal-level scaling.

Solution. The results of steps 1 and 2 are given in Fig. 6-34 and Table 6-3. Because our ladder starts with a shunt arm, we perform a source transformation, and scaling by R_p to obtain Fig. 6-35. Also note that we have combined the load resistor with the last ladder arm because the desired output voltage V_o is proportional to the current I_4. In the signal-flow graph simulation all signals are voltages; consequently, a separate transmittance and often several op amps and resistors may be saved by deriving the output voltage from that op amp which simulates the series current through the last ladder arm and using scaling to implement the desired voltage gain.

The result of step 4, the scaled signal-flow graph, is shown in Fig. 6-36; the parameters α, β, and γ, choosing $R_p = 1\ \Omega$, are determined from

$$\alpha = \frac{\max |v_3|}{\max |i_4 R_p|} = \frac{0.69}{0.8661} = 0.797 \qquad \beta = \frac{\max |i_2 R_p|}{\max |v_3|} = \frac{1.125}{0.69} = 1.630$$

$$\gamma = \frac{\max |v_1|}{\max |i_2 R_p|} = \frac{0.699}{1.125} = 0.621$$

TABLE 6-3 VOLTAGE AND CURRENT MAXIMA

Voltage or current	Maximum of voltage or current
Voltage across $\hat{C}_1$	0.6992 V
Current through L_2	1.550 A
Voltage across $\hat{C}_2$	1.262 V
Current through $(L_2 \| \hat{C}_2)$	1.125 A
Voltage across $\hat{C}_3$	0.6900 V
Current through L_4	0.8661 A

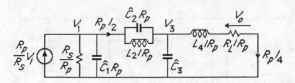

Figure 6-35

To determine K so that the dc gain H_0 equals unity, we note from the prescribed ladder that at dc

$$R_p I_4 = \frac{R_p}{R_s + R_L} V_i = \frac{1}{1.333} V_i = 0.75 V_i$$

To make the output voltage $V_o = R_p I_4$ equal to V_i in magnitude at dc, we multiply V_i by K such that

$$|V_i(j0)| = 0.75K \, |V_i(j0)|$$

i.e.,

$$K = 1.333$$

The initial scale factor equals, therefore,

$$\frac{K}{\alpha\beta\gamma} = \frac{1.333}{0.807} = 1.652$$

Next, we find from the scaled ladder

$$v_1 = \frac{1.652 i_{in} - \dfrac{1}{0.621} i_2}{s\hat{C}_1 R_p + G_s R_p} \qquad i_2 = \frac{0.621 v_1 - \dfrac{1}{1.630} v_3}{\dfrac{1}{s\hat{C}_2 R_p + R_p/(sL_2)}}$$

$$\qquad (6\text{-}104)$$

$$v_3 = \frac{1.630 i_2 - \dfrac{1}{0.797} i_4}{s\hat{C}_3 R_p} \qquad i_4 = \frac{0.797 v_3}{sL_4/R_p + R_L/R_p}$$

Comparing these equations with Fig. 6-30 in order to recognize the type of active branches needed in the simulation shows that the final circuit will take on the form shown in Fig. 6-37; this is step 6 of the design procedure.

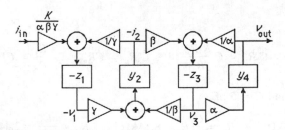

Figure 6-36

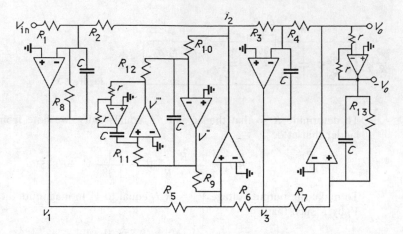

Figure 6-37

The values of the elements in the active circuit are obtained by following steps 7 and 8; the resulting equations for the active circuit are compared to those of the passive circuit, Eq. (6-104). We obtain for each SFG branch:

$$v_1 = \frac{G_1 R_{a1} i_{in} + G_2 R_{a1} i_2}{sCR_{a1} + G_8 R_{a1}} \qquad i_2 = \frac{G_5 R_{a2} v_1 + G_6 R_{a2} v_3}{G_9 R_{a2}}$$

$$\qquad\qquad\qquad\qquad\qquad sCR_{10} + \frac{G_{12} R_{10}}{sCR_{11}}$$

$$\qquad\qquad\qquad\qquad\qquad\qquad\qquad\qquad\qquad\qquad\qquad\qquad (6\text{-}105)$$

$$v_3 = \frac{G_3 R_{a3} i_2 + G_4 R_{a3} i_4}{sCR_{a3}} \qquad i_4 = v_o = \frac{G_7 R_{a4} v_3}{sCR_{a4} + G_{13} R_{a4}}$$

Comparing the coefficients in the numerators of Eqs. (6-104) and (6-105) results in

$$G_1 R_{a1} = 1.652 \qquad G_2 R_{a1} = \frac{1}{0.621} \qquad G_5 R_{a2} = 0.621 \qquad G_6 R_{a2} = \frac{1}{1.630}$$

$$G_3 R_{a3} = 1.630 \qquad G_4 R_{a3} = \frac{1}{0.797} \qquad G_7 R_{a4} = 0.797$$

Further,

$$\hat{C}_1 R_p = CR_{a1} \qquad \hat{C}_2 R_p = C\frac{R_9 R_{10}}{R_{a2}} \qquad \hat{C}_3 R_p = CR_{a3} \qquad \frac{L_4}{R_p} = CR_{a4}$$

Thus, with $R_p = 1\ \Omega$ and $C = 5$ nF,

$$R_{a1} = \frac{\hat{C}_1}{C} R_p = 6.55\ k\Omega \qquad R_{a3} = \frac{\hat{C}_3}{C} R_p = 8.84\ k\Omega \qquad R_{a4} = \frac{L_4}{CR_p} = 2.44\ k\Omega$$

R_{a2} is undetermined; let us choose $R_{a2} = 5\ k\Omega$. With these values, we find for the feed-in resistors into the transmittances:

$$R_1 = \frac{R_{a1}}{1.652} = 3.965\ k\Omega \qquad R_2 = 0.621R_{a1} = 4.068\ k\Omega \qquad R_3 = \frac{R_{a3}}{1.630} = 5.423\ k\Omega$$

$$R_4 = 0.797R_{a3} = 7.053\ k\Omega \qquad R_5 = \frac{R_{a2}}{0.621} = 8.052\ k\Omega$$

$$R_6 = 1.630R_{a3} = 8.15\ k\Omega \qquad R_7 = \frac{R_{a4}}{0.797} = 3.061\ k\Omega$$

The remaining equations are

$$R_9R_{10} = R_pR_{a2}\frac{\hat{C}_2}{C} = (2.352\ k\Omega)^2$$

$$R_8 = R_s\frac{R_{a1}}{R_p} = 6.554\ k\Omega \qquad R_{13} = \frac{R_{a4}R_p}{R_L} = 7.327\ k\Omega$$

$$R_{11}R_{12} = \frac{L_2}{C}\frac{R_9R_{10}}{R_{a2}R_p} = (1.702\ k\Omega)^2$$

Finally, we calculate

$$m'' = \frac{\max|v_{c2}|}{\max|i_2|} = \frac{1.262}{1.125} = 1.122 \quad \text{and} \quad m''' = \frac{\max|i_{L2}|}{\max|i_2|} = \frac{1.55}{1.125} = 1.38$$

to yield

$$R_9 \Rightarrow m''(2.352\ k\Omega) = 2.639\ k\Omega \qquad R_{10} \Rightarrow \frac{2.352\ k\Omega}{m''} = 2.096\ k\Omega$$

$$R_{11} \Rightarrow \frac{m''}{m'''}(1.702\ k\Omega) = 1.384\ k\Omega \qquad R_{12} \Rightarrow \frac{m'''}{m''}(1.702\ k\Omega) = 2.093\ k\Omega$$

6.4.2 LC Ladder Simulation by Element Substitution

6.4.2.1 Inductor replacement. An intuitively appealing technique for obtaining an active circuit from an *LC* filter starts from an existing *LC* ladder and replaces all inductors by active simulations, such as a capacitively loaded gyrator (Fig. 4-26) or the general impedance converter (GIC) shown in Fig. 4-33. Specifically, we saw in Chapter 4, Section 4.3.3, that a high-quality *grounded* inductor is obtained from a type II GIC. The circuit is repeated in Fig. 6-38; it realizes ideally

$$Z_{in}(j\omega) = j\omega CR_1R_5 = j\omega L_0 \qquad (6\text{-}106)$$

The errors in component value and quality factor were derived in Eqs. (4-75)

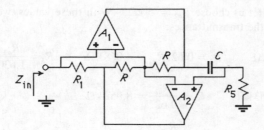

Figure 6-38 Type II GIC for inductance simulation.

through (4-77), where it was shown that the inductor errors are minimized by setting

$$R_5 = \frac{1}{\omega_c C} \tag{6-107}$$

ω_c is some critical frequency parameter, such as the passband corner. The resistor R_1 is then determined from the desired inductor as

$$R_1 = \omega_c L_0 \tag{6-108}$$

As an example, consider the realization of a fifth-order all-pole highpass LC ladder in Fig. 6-39a; the corresponding active simulation is shown in Fig. 6-39b. Note that the capacitors and resistors of the *RLC prototype* are unchanged and that the inductors are simply replaced by the circuit of Fig. 6-38.

Because the simulated inductor in Fig. 6-38 is *grounded*, the technique works best for highpass and a restricted class of bandpass circuits where the passive prototype has no floating inductors.[20] For lowpass and bandpass ladders with floating inductors, impedance transformations are first applied to eliminate these components and replace them by others that are easier to realize. Both transformations, introduced by Bruton [143] and by Gorski-Popiel [142], use *impedance scaling* by a factor $1/(ks)$. Bruton scales the entire ladder, whereas Gorski-Popiel transforms only parts of the circuit.

6.4.2.2 Bruton's transformation.

As will be seen shortly, *Bruton's transformation* is especially useful for *RLC* prototype circuits which have only *grounded* capacitors. In this way, the method can be understood as the dual of the previous procedure. As mentioned, each impedance of the passive circuit is scaled by $1/(ks)$; the resulting impedance transformations are indicated in Fig. 6-40, i.e., resistors become capacitors, inductors become resistors, and a capacitor is converted into a *frequency-dependent negative resistor (FDNR)* as in

$$Z_c(j\omega) = \frac{1}{j\omega C} \longrightarrow \hat{Z}_c(j\omega) = Z_c(j\omega) \cdot \frac{1}{j\omega k} = -\frac{1}{\omega^2 C k} \tag{6-109}$$

---

[20]Also, transformers should be avoided. Although floating inductors and transformers can be simulated in principle (see Chapter 4 and Section 6.4.2.3, below), the realizations are not very desirable in that they use too many op amps and result in reactances encumbered by parasitics and losses.

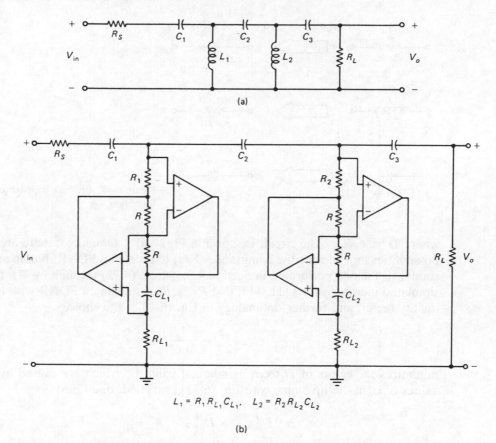

Figure 6-39 (a) LC highpass ladder; (b) active simulation.

Thus, $\hat{Z}_c(j\omega)$ is real and negative and a function of frequency. The generally used symbol for an FDNR or "supercapacitor" is also shown in Fig. 6-40. It should be clear that this transformation does not alter the transfer function of the circuit, because, as a dimensionless voltage ratio, it is independent of impedance scaling.

At this point the question arises how an FDNR element can be implemented. The answer is found again in Antoniou's GIC of Fig. 4-33, which realizes the *grounded* impedance, Eq. (4-71),

$$Z_{in} = \frac{Z_1 Z_3 Z_5}{Z_2 Z_4} = \frac{Y_2 Y_4}{Y_1 Y_3 Y_5}$$

Thus, setting $Y_1 = Y_5 = sC$ and $R_2 = R_3 = R$, we obtain

$$Z_{in}(j\omega) = -\frac{1}{\omega^2 C^2 R_4} \triangleq -\frac{1}{\omega^2 D} \qquad (6\text{-}110)$$

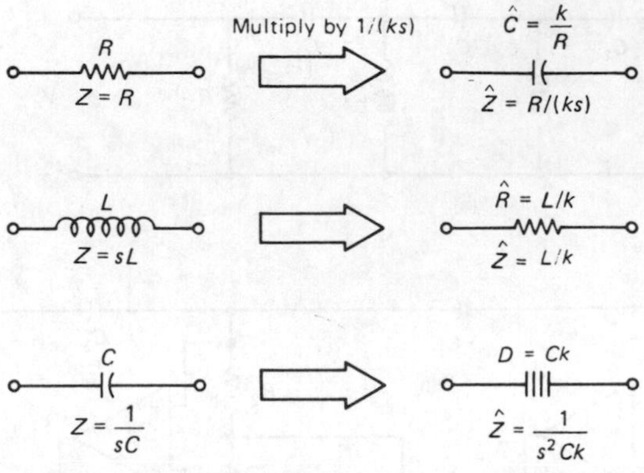

Multiply by $1/(ks)$ $\hat{C} = \dfrac{k}{R}$

$Z = R$ $\Longrightarrow$ $\hat{Z} = R/(ks)$

$Z = sL$ $\Longrightarrow$ $\hat{R} = L/k$ $\hat{Z} = L/k$

$Z = \dfrac{1}{sC}$ $\Longrightarrow$ $D = Ck$ $\hat{Z} = \dfrac{1}{s^2 Ck}$

Figure 6-40 Bruton's impedance transformation.

where $D = C^2 R_4$. The circuit is shown in Fig. 6-41. Of course, there are other possibilities for choosing the admittances $Y_i(s)$ to obtain an FDNR, but an analysis similar to the one performed in Section 4.3.3, Eqs. (4-72) through (4-77), for the simulated inductor shows [11, 140] that $R_2 = R_3$ results in an FDNR with infinite quality factor, and further, in analogy to Eq. (6-107), the choice

$$R_4 = \frac{1}{\omega_c C} \tag{6-111}$$

minimizes the errors of D from its nominal value D_0 which are caused by finite values of ω_t of the op amps; with Eq. (6-111) satisfied, one finds:

$$D\big|_{\omega = \omega_c} \simeq D_0\left(1 + 4\,\frac{\omega_c}{\omega_t}\right) \tag{6-112}$$

ω_c is again a suitably chosen critical frequency, such as the filter's passband corner or the resonant frequency of a tank circuit that is determined by the simulated inductor. Note that the FDNR element is grounded. This is the reason for our earlier observation that the method is particularly useful for prototype circuits with only grounded capacitors because the transformation converts each capacitor into an FDNR. Similar to the GIC-based realizations of floating inductors, floating FDNRs are possible, but they are uneconomical, using twice as many op amps, and encumbered by parasitic components and losses.

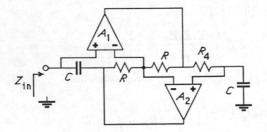

Figure 6-41 GIC realization of a *grounded* FDNR.

As an example for the circuit structure, consider the design of an active simulation of a fifth-order LC lowpass ladder: the LC prototype is shown in Fig. 6-42a, the transformed circuit in Fig. 6-42b, and the final FDNR configuration in Fig. 6-42c.

Note that the realization in Fig. 6-42c has two practical difficulties. First, because the *entire* ladder must be transformed, the active circuit no longer contains a source and load resistor. If these two components are prescribed and must be maintained as in the original passive circuit, the simple modification of buffering input and output as shown in Fig. 6-43 will solve the problem.

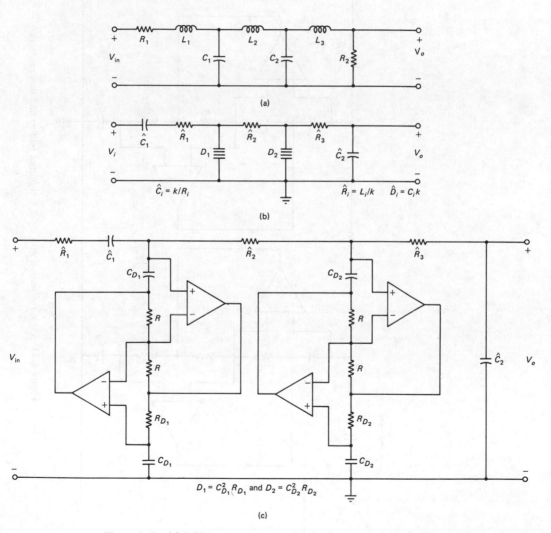

Figure 6-42   (a) LC lowpass prototype; (b) circuit after applying the Bruton transformation; (c) active simulation.

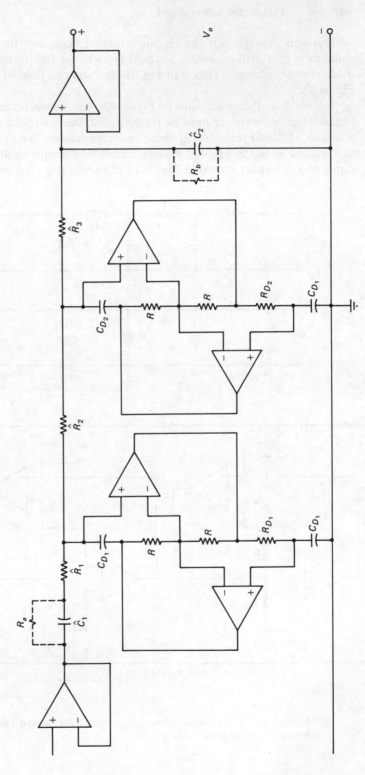

Figure 6-43 Final realization with input and output buffers and bias resistors.

Second, we note that the noninverting input terminals of the upper two op amps have no dc path to ground, which means that the required dc bias currents for the input transistors cannot be provided. To remedy this problem, we could connect a *large* bias resistor R_B from any of the nodes of the resistive subnetwork $\hat{R}_1$, $\hat{R}_2$, $\hat{R}_3$ to ground. Note, however, that this bias resistor represents in the original ladder a parasitic inductor of value $L = kR_B$ from that node to ground. The solution is, therefore, generally unsatisfactory because this parasitic inductor results in gain deviations at low frequencies and, in particular, it contributes a transmission zero at dc. A better approach consists of adding *two* resistors, R_a and R_b, across the terminating capacitors as shown in Fig. 6-43 [11]. In that case we obtain at dc

$$H(j0) = \frac{R_b}{R_b + R_a + \hat{R}_1 + \hat{R}_2 + \hat{R}_3}$$

which must equal, from the original ladder,

$$H(j0) = \frac{R_L}{R_L + R_S}$$

From these two equations we obtain

$$R_b = \frac{R_L}{R_S}(R_a + \hat{R}_1 + \hat{R}_2 + \hat{R}_3) = \frac{R_L}{R_S}\left(R_a + \sum_i \hat{R}_i\right) \qquad (6\text{-}113)$$

Further, because the two resistors R_a and R_b correspond to parasitic inductors, $L_a = kR_a$ and $L_b = kR_b$, respectively, which shunt the source and load resistors of the LC ladder, it is clear that their values will affect the filter transfer function at low frequencies. To minimize their effect, it can be shown that R_a and R_b should be chosen to satisfy

$$R_a, R_b \gg \hat{R}_1 + \hat{R}_2 + \hat{R}_3 = \sum_i \hat{R}_i \qquad (6\text{-}114)$$

i.e., by Eq. (6-113),

$$\frac{R_a}{R_b} = \frac{L_a}{L_b} \simeq \frac{R_S}{R_L}$$

In Eqs. (6-113) and (6-114), the sums go over all the series resistors in the simulated ladders. Example 6-7 illustrates the circuit design using Bruton's transformation.

Example 6-7

The transfer function of a lowpass filter with two prescribed transmission zeros was found in Problem 1.26 and realized in Problem 2.36(d). A possible realization is shown in Fig. 6-44. Implement this filter in active form using the Bruton transformation. Rescale the impedance level such that $R_S = R_L = 1$ kΩ. Note that the circuit is realized as a minimum-capacitor ladder in order to be convenient for Bruton's transformation.

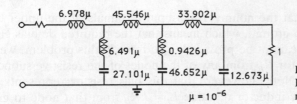

Figure 6-44 LC lowpass filter for Example 6-7. The impedance level is normalized so that $R_S = R_L = 1$.

Solution. First we rescale the impedance level to 1 kΩ; in Fig. 6-44, this results in inductor units of mH and capacitor units of nF. Next, let us choose the parameter k in Fig. 6-40 as $k = 1$ μs. Then, by comparing Figs. 6-44 and 6-45, we obtain:

$$R_1 = 6.98 \text{ k}\Omega \qquad R_2 = 6.49 \text{ k}\Omega \qquad R_3 = 45.55 \text{ k}\Omega$$

$$R_4 = 943 \ \Omega \qquad R_5 = 33.90 \text{ k}\Omega \qquad C_S = C_L = 1 \text{ nF}$$

Also, choose $R = 3.3$ kΩ. Further, from Fig. 6-40 and Eq. (6-110), we have

$$D_2 = 27.101 \cdot 10^{-15} \text{ s}^2/\Omega = C_2^2 R_{42} \quad \text{with} \quad R_{42} = \frac{1}{\omega_{c2} C_2}$$

by Eq. (6-111). The first shunt branch resonates at 12 kHz; therefore let us set $\omega_{c2} = 2\pi \cdot 12$ krad/s to yield

$$C_2 = 2.04 \text{ nF} \qquad R_{42} = 6.5 \text{ k}\Omega$$

Similarly, we find for the other two shunt branches

$$D_4 = 46.652 \cdot 10^{-15} \text{ s}^2/\Omega = C_4^2 R_{44} \qquad R_{44} = \frac{1}{\omega_{c4} C_4}, \ \omega_{c4} = 2\pi \cdot 24 \text{ krad/s}$$

$$D_6 = 12.673 \cdot 10^{-15} \text{ s}^2/\Omega = C_6^2 R_{46} \qquad R_{46} = \frac{1}{\omega_{c6} C_6}, \ \omega_{c6} = 2\pi \cdot 6.25 \text{ krad/s}$$

ω_{c4} and ω_{c6}, respectively, are chosen from the second transmission zero at 24 kHz and a frequency close to the bandedge, 6.25 kHz. Thus, we obtain

$$C_4 = 7.04 \text{ nF} \qquad R_{44} = 942 \ \Omega \qquad C_6 = 497.7 \text{ pF} \qquad R_{46} = 51.2 \text{ k}\Omega$$

Finally, with $R_1 + R_3 + R_5 = 86.43$ kΩ and choosing $R_a = 640$ kΩ, Eq. (6-113) results in $R_b = 726$ kΩ. The final active realization is shown in Fig. 6-45.

6.4.2.3 Gorski-Popiel's embedding technique.

As mentioned earlier, more general ladders which do not have either all inductors or all capacitors grounded lead to inefficient realizations with direct inductor replacement or with Bruton's transformation. For those cases, Gorski-Popiel proposed [142] an embedding technique which separates all inductor subnetworks from an RLC circuit by general impedance converters and thereby converts all inductors into resistors. The number of GICs used in this technique is equal only to the *number of connections* to the inductive subnetworks rather than requiring one or two GICs per inductor.

Before presenting an outline of the method, let us digress for a moment and discuss the optimum design of a general impedance converter when it is not ter-

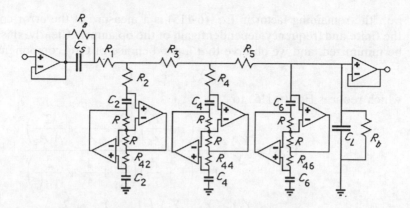

Figure 6-45 FDNR realization of the circuit in Fig. 6-44.

minated by a given fixed element (normally a resistor or a capacitor; see, for
example, Figs. 6-38 and 6-41) but by a general admittance as is the case in Gorski-
Popiel's method.

    Optimum Design of a GIC [11, 140].   We have presented Antoniou's GIC
and its analysis in Section 4.3.3.   The circuit of Fig. 4-33 is repeated in Fig. 6-46,
but now the admittance $Y_5(s)$ is to be understood as a load of the impedance
converter rather than as an element of the GIC.   Our objective is to minimize the
dependence of the GIC's operation on the op amps' gain-bandwidth product ω_t.
To this end we may use Eq. (4-72), repeated here as Eq. (6-115),

$$Y_{in}(s) = \frac{Y_1 Y_3 Y_5}{Y_2 Y_4} \frac{1 + \left(1 + \dfrac{Y_4}{Y_5}\right)\left[\dfrac{1}{A_2} + \dfrac{1}{A_1}\dfrac{Y_2}{Y_3} + \dfrac{1}{A_1 A_2}\left(1 + \dfrac{Y_2}{Y_3}\right)\right]}{1 + \dfrac{Y_3}{Y_2}\left(1 + \dfrac{Y_5}{Y_4}\right)\left[\dfrac{1}{A_2} + \dfrac{1}{A_1}\dfrac{Y_2}{Y_3} + \dfrac{1}{A_1 A_2}\left(1 + \dfrac{Y_2}{Y_3}\right)\right]} \tag{6-115}$$

and recall that ideally we wish to realize the admittance

$$Y_{in}(s) = \frac{Y_1 Y_3 Y_5}{Y_2 Y_4}$$

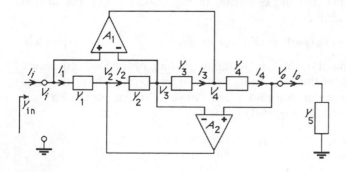

Figure 6-46 Antoniou's impedance
converter with a load $Y_5(s)$.

i.e., the remaining factor in Eq. (6-115) is a measure of the error contributed by the finite and frequency-dependent gain of the op amps. Clearly, this error should be minimized, and we observe that a good choice of GIC components is

$$Y_2 = Y_3 = G \tag{6-116}$$

which reduces Eq. (6-115) to

$$
Y_{in}(s) = \frac{Y_1 Y_5}{Y_4} \frac{1 + \left(1 + \dfrac{Y_4}{Y_5}\right)\left(\dfrac{1}{A_1} + \dfrac{1}{A_2} + \dfrac{2}{A_1 A_2}\right)}{1 + \left(1 + \dfrac{Y_5}{Y_4}\right)\left(\dfrac{1}{A_1} + \dfrac{1}{A_2} + \dfrac{2}{A_1 A_2}\right)}
$$

$$
\simeq \frac{Y_1 Y_5}{Y_4}\left(\frac{Y_4}{Y_5} - \frac{Y_5}{Y_4}\right)\left(\frac{1}{A_1} + \frac{1}{A_2} + \frac{2}{A_1 A_2}\right) \tag{6-117}
$$

With $Y_4 = sC$ and $Y_1 = G_1$ we have, ideally,

$$Y_{in}(s) = \frac{I_i}{V_i} = \frac{1}{sCR_1} \quad Y_5 = \frac{1}{sk}\frac{I_o}{V_o} \tag{6-118}$$

where we have labeled the conversion factor

$$CR_1 = k \tag{6-119}$$

For the approximation in Eq. (6-117), we assumed that $|A(j\omega)| \gg 1$ and used that $1/(1 + x) \simeq 1 - x$ for small x. Because the element Y_4 is a capacitor and $Y_5(s) = I_o/V_o$ is, in general, an arbitrary function of frequency, it is usually not possible to set $Y_4 = Y_5$, which, by Eq. (6-117), would be the necessary choice for reducing the GIC's op amp dependence to zero. However, we can choose to equate at least their magnitudes at a frequency ω_c:

$$|Y_5(j\omega_c)| = |Y_4(j\omega_c)| = \omega_c C \tag{6-120a}$$

where, as always, ω_c is a critical frequency selected, e.g., near the passband corner of the filter. The reader is encouraged to compare this analysis and the results with the treatment in Section 4.3.3 and Eqs. (6-107) and (6-111); as can be expected, the simulated inductor and the FDNR are special cases of the more general development in this section.

Note from Eq. (6-117) that the condition in Eq. (6-120a) with the further assumption $|A_i(j\omega)| \gg 1$ implies

$$|Y_{in}(j\omega_c)| \simeq |Y_1(j\omega_c)| = G_1 \tag{6-120b}$$

i.e., matching the end element of the GIC to the magnitude of the load impedance results in the first element of the GIC being matched approximately to the magnitude of the impedance seen at that end. The converse statement is also true, as can be seen readily by solving Eq. (6-117) for

$$-Y_5(s) = \frac{-I_o}{V_o}$$

which is the admittance seen *into* the output of the GIC of Fig. 6-46, as a function of

$$-Y_{in}(s) = \frac{-I_i}{V_i}$$

which is the admittance seen from its input:

$$-Y_5(s) = \frac{-Y_{in}Y_4}{Y_1} \frac{1 + \left(1 + \dfrac{Y_1}{-Y_{in}}\right)\left(\dfrac{1}{A_1} + \dfrac{1}{A_2} + \dfrac{2}{A_1A_2}\right)}{1 + \left(1 + \dfrac{-Y_{in}}{Y_1}\right)\left(\dfrac{1}{A_1} + \dfrac{1}{A_2} + \dfrac{2}{A_1A_2}\right)}$$

$$\simeq \frac{-Y_{in}Y_4}{Y_1}\left(\frac{Y_1}{-Y_{in}} - \frac{-Y_{in}}{Y_1}\right)\left(\frac{1}{A_1} + \frac{1}{A_2} + \frac{2}{A_1A_2}\right) \qquad (6\text{-}121)$$

The parallelism with Eq. (6-117) is apparent: choosing Y_1 and Y_{in} to satisfy Eq. (6-120b) results in Eq. (6-120a). We conclude, therefore, that the GIC performance may be optimized by setting $Y_2 = Y_3$ [(Eq. 6-116)] and satisfying the matching condition [Eq. (6-120)] at *either end*!

Figure 6-47a shows the resulting impedance converter for use in Gorski-Popiel's embedding technique; also shown in Fig. 6-47b is a symbolic representation introduced [11] to keep the later circuit diagrams reasonably simple. With function and optimal design of the impedance converter understood, we are now ready to consider the embedding technique.

The Gorski-Popiel Technique.   The method is a generalization of the inductance simulation of Fig. 6-38, where a resistor R_5 is used to terminate the impedance converter of Fig. 6-47a in order to obtain an inductor $L_0 = kR_5$ [see Eq. (6-106)]. Consider a purely resistive network R with impedance matrix $\mathbf{Z}_r$ such that the vectors of input currents, $\mathbf{I}$, and input voltages, $\mathbf{V}$, are related by

$$\mathbf{V} = \mathbf{Z}_r\mathbf{I} \qquad (6\text{-}122)$$

Note that all elements in $\mathbf{Z}_r$ are frequency-independent real numbers because R is resistive. Suppose, now, we connect to each of the terminals of R (apart from

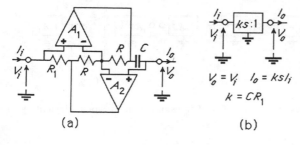

(a)   (b)

Figure 6-47 Antoniou's impedance converter for use in Gorski-Popiel's technique. *Note:* For optimal design, set $\omega_c C = |I_o/V_o|_{\omega=\omega_c}$ or $R_1 = |V_i/I_i|_{\omega=\omega_c}$. (a) Circuit; (b) symbolic representation.

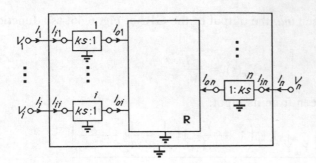

Figure 6-48 Simulation of inductance network after Gorski-Popiel [142].

ground) a GIC of the form of Fig. 6-47a as is shown in Fig. 6-48. The combined network is then described by the matrix equation

$$\mathbf{V} = \mathbf{Z}_r(ks\mathbf{I}_i) = (ks\mathbf{Z}_r)\mathbf{I}_i \qquad (6\text{-}123)$$

which implies that the matrix of the combined network

$$\mathbf{Z}_I = sk\mathbf{Z}_r \qquad (6\text{-}124)$$

is that of a purely inductive circuit. Evidently, this means for our simulation of *LC* filters that each connection to an inductor-only subnetwork in an *RLC* circuit can be "cut" and the cut "repaired" by a $ks:1$ impedance converter. If each inductor of value L is then replaced by a resistor of value L/k, the behavior of the total network will not be altered. The procedure is illustrated for an inductive "T," a floating inductor, and an inductive "Π" in Fig. 6-49a, b, and c, respectively. The realization of the floating inductor in Fig. 6-49b is, of course, completely analogous to that of Fig. 4-26b, where two *gyrators* and a grounded capacitor were used to implement a floating inductor. Finally, we present in Fig. 6-50 an illustration of how Gorski-Popiel's technique is applied to the active realization of an *LC* ladder filter.

The passive circuit is, slightly redrawn, the ladder of Fig. 6-25. We have indicated the five locations where the passive ladder must be "cut" in order to

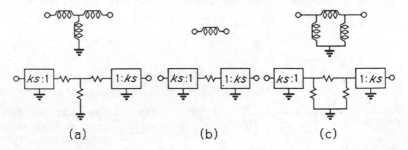

| (a) | (b) | (c) |

Figure 6-49 Illustrating the Gorski-Popiel transformation for some elementary inductance networks.

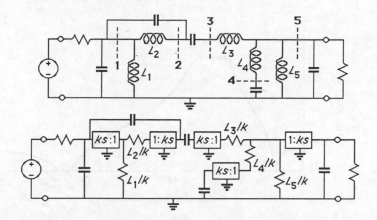

Figure 6-50 Illustrating the ladder-embedding technique on the filter circuit of Fig. 6-25.

separate the inductor subnetworks and how the cuts are to be "repaired" after the inductors are replaced by resistors of value L_i/k.

Recall that for optimal performance each GIC must be designed such that Eq. (6-120) is satisfied; i.e., the resistor R_{1i} and capacitor C_i of the ith GIC must be chosen such that

$$\frac{1}{R_{1i}} = |-Y_{\text{Cut},i}(j\omega_c)| \quad \text{or} \quad C_i = \frac{1}{\omega_c}|Y_{\text{Cut},i}(j\omega_c)| \qquad (6\text{-}125)$$

$Y_{\text{Cut},i}(s)$ is the admittance seen into the RLC circuit at the cut where the ith GIC is to be inserted. It and its value at ω_c can be found with the aid of a filter analysis program, such as in references 11 or 29. Note, on the other hand, that all conversion constants k_i used in embedding a particular inductive subnetwork must be the same, $k_i = k$. Thus, R_{1i} and C_i, as found in Eq. (6-125), are constrained by

$$R_{1i}C_i = k \qquad (6\text{-}126)$$

A design example, along with further scaling possibilities for obtaining convenient element values, can be found in reference 140. In the following, we present a numerical example to illustrate the technique.

Example 6-8

The design of a geometrically symmetrical bandpass filter to realize the attenuation specifications of Fig. 6-51 was discussed in Problem 2.21, and the realization was found to be that of Fig. 6-52a. Use Gorski-Popiel's technique to implement this circuit in active form.

Solution. We have indicated in Fig. 6-52a the three locations where an impedance converter must be inserted. As discussed in the text, this step permits us to replace the inductors L_i by resistors of value L_i/k, $i = 1, 2, 3, 4$. The resulting circuit is

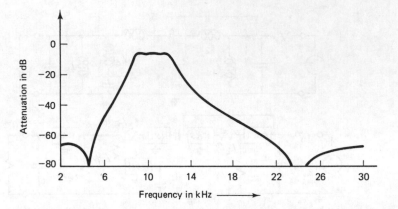

Figure 6-51 Bandpass specifications for Example 6-8.

shown in Fig. 6-52b. The values of the capacitors and of the source and load resistors are the same as those in the passive circuit. The resistors representing the inductors in the passive filter are obtained from $R_i = L_i/k$. Let us choose $k = 10$ μs; then we obtain

$$R_{L1} = R_{L4} = \frac{L_1}{k} = \frac{L_4}{k} = 84\,\Omega \qquad R_{L2} = \frac{L_2}{k} = 5.09\,\text{k}\Omega \qquad R_{L3} = \frac{L_3}{k} = 954\,\Omega$$

The ks:1 blocks are realized as shown in Fig. 6-47, where we must remember that for best performance Eq. (6-125) should be satisfied. As critical frequency ω_c we choose a point just beyond the upper passband corner: $\omega_c = 2\pi \cdot 12.5$ krad/s. At this frequency, we compute with the aid of a suitable analysis program

$$|Z_1(j\omega_c)| = 55.8\,\Omega \qquad |Z_4(j\omega_c)| = 45.6\,\Omega \qquad |Z_{23}(j\omega_c)| = 599\,\Omega$$

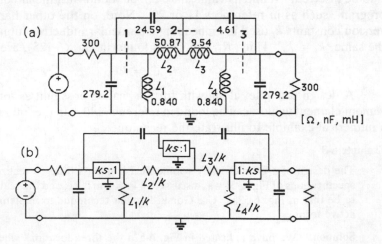

Figure 6-52 (a) Passive LC realization of the specifications in Fig. 6-51; (b) active implementation by Gorski-Popiel's technique.

Z_1 and Z_4 are the impedances "seen into" the ladder at the terminals of L_1 and L_4, respectively; Z_{23} is the impedance "seen into" the node joining L_2 and L_3; all are computed via $|V/I|$ at the appropriate locations in the ladder. Thus, we obtain for the left converter from Eqs. (6-125) and (6-126)

$$C = \frac{1}{\omega_c |Z_1(j\omega_c)|} = 228 \text{ nF} \qquad R_{11} = \frac{k}{C} = 43.86 \ \Omega$$

Similarly, we find for the center and the right converters, respectively,

$$C = \frac{1}{\omega_c |Z_{23}(j\omega_c)|} = 21.3 \text{ nF} \qquad R_{12} = \frac{k}{C} = 469.5 \ \Omega$$

$$C = \frac{1}{\omega_c |Z_4(j\omega_c)|} = 279 \text{ nF} \qquad R_{13} = \frac{k}{C} = 35.84 \ \Omega$$

6.5 SUMMARY

In this chapter, we have discussed the more practical techniques for the design of active filters of order higher than 2: *cascade* design, *multiple-loop feedback* approaches, and methods which *simulate the behavior of LC ladder filters*. We have pointed out that, although possible, direct realization methods are impractical because they result in high sensitivities to component values. In many applications, a cascade design where the high-order function is factored into the product of first- and second-order sections will lead to satisfactory results. The practical advantages of these circuits are modularity, ease of design, flexibility, very simple tuning procedures, and economical use of op amps, with as few as one op amp per pole pair. Also, we should point out that *the cascade design method is general* in that an arbitrary transfer function can be realized, with no restrictions placed on the permitted locations of poles and zeros.

For more challenging filter requirements where the simple cascade topologies may still be too sensitive to parameter changes, the designer can use multiple-loop feedback (MF) configurations or, for best performance and *if* a passive prototype ladder exists, ladder simulations. MF circuits retain the advantage of modularity; also, they can be built with a minimum number of op amps (one per pole pair) if single-amplifier biquads are used for the second-order building blocks. Generally, if optimal performance is desired, computer-aided optimization routines must be used to adjust the available free design parameters. However, excellent performance with very simple design procedures, no optimization, and high modularity can be obtained by use of the *primary-resonator-block* (PRB) method, where all biquad building blocks are identical.

It was pointed out in Chapter 3 that properly designed *LC* ladder filters give the lowest passband sensitivities. Consequently, from the point of view of minimum passband sensitivity to component tolerances, the best active filters are obtained by simulating *LC* ladders. If the prescribed transfer characteristic can at all be realized as a passive LC ladder, then the designer can make use of the wealth of available information about the design of such circuits. For that case, a number of different procedures have been developed that are helpful for "translating" the passive LC circuit into its active counterpart. Some of them simply take the passive circuit and—possibly after a transformation—replace the "offending" components, the inductors, by active networks; others imitate the mathematical behavior of the whole LC circuit by realizing the integrating action of inductors and capacitors via active RC integrators.

Both the component substitution and the flow-graph simulation approaches result in active circuits of high quality; a disadvantage is that usually many more op amps are used than in cascade or MF methods. This drawback is offset, however, by the fact that the sensitivities to component tolerances are "almost" as low as those of the originating ladder.

An important practical aspect of active filters is their *limited dynamic range*, restricted at the low end by noise and at the upper end by the finite linear signal swing of op amps. For each type of realization, we have, therefore, discussed the methods that lead to a maximized linear signal range. The procedures always proceeded to equalize the op amp output voltages by exploiting available free gain constants or impedance-scaling factors. We have stressed as a disadvantage of the element substitution method that no general dynamic range scaling method appears to be available.

All the methods discussed so far in this book have dealt with the design of filters in *discrete* form; that is, we have assumed that separate passive components and operational or transconductance amplifiers are assembled, for example, on a printed circuit board to make up the desired filter. Prompted by tremendous advances in digital integrated circuits, however, problems of compatibility with digital systems, cost containment, reliability, and reduced size provide powerful incentives for designers to search for ways that permit more and more of a system to be integrated on a silicon chip. In particular, we wish to combine *analog* and *digital* circuits on the same integrated circuit (IC). Of specific interest within the theme of this book is the question: Can one build an analog filter in fully integrated form? The next two chapters will deal with several aspects of this topic.

In Chapter 7, we shall discuss a number of methods that have been proposed for the design of *continuous-time* analog filters. At the time of this writing (1989), this matter is by no means completely solved but is subject to intensive research.

In Chapter 8, we present some necessary background for and the design of *sampled-data switched-capacitor* (SC) *filters*. As the name implies, in these circuits the signals are no longer continuous functions of time but instead are sampled; that is, as in digital filters, the signals are represented by their values at a finite number of distinct points in time. The filtering operation is still *analog*, however, because the signal values are not digitized as in digital filters; only the time axis is divided into discrete points. Chapter 8 will discuss the motivation for this approach and some of the mathematical and theoretical background and will present some proven techniques for the design of second- and higher-order SC filters.

PROBLEMS

6.1 Consider the passive RC section in Fig. P6.1. Show how the components should be chosen so that a general first-order function is realized that has a pole and a zero on the negative real axis.

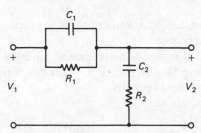

Figure P6.1

6.2 Show that the active network in Fig. P6.2 with RC admittances $Y_i(s)$ realizes

$$H(s) = \frac{N_m(s)}{D_n(s)} = \frac{Y_1 - Y_2}{Y_3 - Y_4}$$

If $N_m(s)$ and $D_n(s)$ are two arbitrary polynomials of degree m and n, respectively, the RC admittances can be found by dividing N_m and D_n by a common polynomial $Q_k(s)$ with degree $k \geq n - 1$ so that

$$H(s) = \frac{N_m(s)/Q_k(s)}{D_n(s)/Q_k(s)} = \frac{Y_1 - Y_2}{Y_3 - Y_4}$$

Evidently, this step has not changed $H(s)$. Now assume that $Q_k(s)$ has only simple negative real roots, i.e.,

$$Q_k(s) = \prod_{i=1}^{k} (s + \sigma_i)$$

and perform partial fraction expansions of N_m/Q_k and D_n/Q_k. After multiplying both expansions by s, the resulting expressions allow you to recognize the RC admittances. Use this technique to realize a fifth-order Butterworth filter with the circuit of Fig. P6.2. Note how the coefficients of the given transfer function are determined by difference effects!

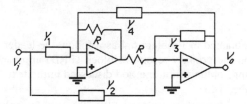

Figure P6.2

6.3 To be designed is a sixth-order Chebyshev bandpass filter with center frequency $f_0 = 5$ kHz and a 0.5 dB–ripple bandwidth of $\Delta f = 300$ Hz. Realize the filter as a cascade connection of ENF single-amplifier biquads (Fig. 5-12). Optimize the dynamic range.

6.4 A geometrically symmetrical bandpass filter is to be designed to meet the following specifications:

> Passband: 1 dB equiripple in 12 kHz $\leq f \leq$ 36 kHz
>
> Stopbands: at least 25 dB attenuation in $f \leq$ 4.8 kHz and $f \geq$ 72 kHz

Realize the circuit as a cascade of Åckerberg-Mossberg sections. Maximize the signal magnitude that can be processed.

6.5 Realize a fourth-order Chebyshev lowpass function having a passband ripple of 3 dB as a cascade of Åckerberg-Mossberg sections. The cutoff frequency is 5 kHz. The gain constant must be realized correctly.

6.6 Realize the transfer function

$$H(s) = \frac{V_{\text{out}}}{V_{\text{in}}} = \frac{H_M s(s^2 + 1)}{(s^2 + 0.2s + 0.87)(s^2 + 0.5s + 1)}$$

as a cascade connection of two appropriate single-amplifier biquads. Choose equal capacitors and select any free parameters such that the active sensitivities are minimized. The gain constant H_M should be maximized. The frequency is normalized with respect to $\omega_z = 2\pi \cdot 3.6$ krad/s.

6.7 The transfer function of a fifth-order elliptic filter used in telephone channel bank systems is shown below:

$$H(s) = \frac{100(s^2 + 29.2^2)(s^2 + 43.2^2)}{(s + 16.8)(s^2 + 19.4s + 400.34)(s^2 + 4.72s + 507.33)}$$

The frequency s is normalized with respect to 1 krad/s. Realize this function by the cascade method, using two single-amplifier biquads in addition to a first-order section.

6.8 Derive a cascade realization of the function

$$H(s) = \frac{s(s^2 + 0.25)(s^2 + 2.25)}{(s^2 + 0.09s + 0.83)(s^2 + 0.1s + 1.18)(s^2 + 0.2s + 1.01)}$$

with single-amplifier biquads. Find the optimal pole-zero pairing, section ordering, and gain assignment.

6.9 Repeat Problem 6.8, but use three-amplifier Åckerberg-Mossberg sections.

6.10 Repeat Problem 6.8 for an FLF (PRB) configuration; show only the values of ω_{0i}, Q_i, H_i, and F_i.

6.11 Verify Eq. (6-19b).

6.12 Verify Eq. (6-26).

6.13 Realize the function obtained in Problem 6.4 as a follow-the-leader feedback (FLF) configuration based on the primary resonator block (PRB) approach. Use the approximate dynamic range optimization method discussed in the text. Use the single-amplifier biquads of Fig. 5-13.

6.14 Realize a geometrically symmetrical bandpass filter obtained from the prototype lowpass function

$$H(p) = \frac{1.4491(p^2/3.2^2 + 1)}{p^3 + 2.2201p^2 + 2.4544p + 1.4491}$$

by the lowpass-to-bandpass transformation $p = Q(s^2 + 1)/s$. The bandpass filter's center frequency equals 8 kHz; its bandwidth is 2 kHz. Use the FLF topology (PRB special case), maximize the dynamic range, and realize the transmission zero by the feed-forward method (Fig. 6-6).

6.15 Use the FLF topology with single-amplifier biquads to realize a fourth-order Butterworth bandpass filter with center frequency $f_0 = 10$ kHz and a 3 dB bandwidth of 1 kHz. Maximize the dynamic range. Sensitivities do not have to be minimized. You do not have to design the second-order sections; it is sufficient to identify their design parameters (Q_i, ω_{0i}, gain).

6.16 Realize a sixth-order 0.5 dB–ripple Chebyshev bandpass filter, using GIC second-order blocks in an FLF configuration. The center frequency is $f_0 = 8$ kHz; the 0.5-dB bandwidth $\Delta f = 500$ Hz. Select the section gain constants to equalize the signal levels in the circuit.

6.17 For the design of Example 6-4, give an expression for $S_{\omega_0}^H$, where $H = V_o/V_i$ and ω_0 is the bandpass center frequency. Assume that the sections track, i.e., $\omega_{0i} = \omega_0$ for all i. Evaluate $S_{\omega_0}^{|H|}$ at $\omega = \omega_0$.

6.18 Realize a sixth-order FLF bandpass filter derived from a third-order Butterworth prototype lowpass function. The bandpass is to have a center frequency $f_0 = 7.2$ kHz and a 3-dB bandwidth $\Delta f = 700$ Hz. The minimum statistical gain sensitivity can be shown to occur when the three second-order sections have the following quality factors: $Q_1 = 43.100$, $Q_2 = 44.650$, and $Q_3 = 29.075$ [47]. Design the circuit, using GIC biquads. Optimize the dynamic range.

6.19 To be realized is a sixth-order Butterworth bandpass filter with passband in 4 kHz $\leq f \leq 4.9$ kHz, $A_p \leq 3$ dB. Use an FLF topology (PRB special case) and GIC sections. The available op amps have $\omega_t = 2\pi \cdot 1.2$ Mrad/s. Optimize the dynamic range by the approximate method discussed in the text and be sure to compensate for any ω_t-induced Q errors.

6.20 Implement the function obtained in Problem 6.4 by use of the signal-flow graph (SFG) technique. The realized midband gain must be equal to unity.

6.21 The minimum-inductance ladder realization of a fifth-order Chebyshev lowpass filter with a 1-dB bandwidth of 33.5 krad/s is shown in Fig. P6.21. Implement this circuit as an active network:
(a) By the signal-flow graph technique
(b) By use of the Bruton transformation
(c) By use of Gorski-Popiel's transformation

Compare the different realizations in terms of expected performance and use of op amps.

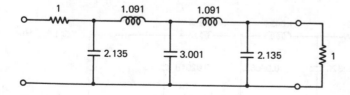

Figure P6.21

6.22 Show the signal-flow graph of a
(a) Seventh-order Chebyshev lowpass filter

(b) Seventh-order inverse Chebyshev lowpass filter

(c) Seventh-order elliptic lowpass filter

6.23 Work out the details of the technique for all-pole bandpass filters in Section 6.4.1.3, Fig. 6-22, for a fourth-order Chebyshev lowpass prototype; the bandpass filter must have a center frequency $f_0 = 3.4$ kHz and a 3-dB bandwidth $\Delta f = 600$ Hz. Use single-amplifier biquads.

6.24 The Åckerberg-Mossberg (AM) circuit with *inverting* bandpass gain was shown in Fig. 5-29. In the signal-flow graph technique based on second-order bandpass sections, however, both inverting *and* noninverting biquads are required. In order to save the hardware needed for an additional inverter, derive a modified version of the AM circuit that realizes a *noninverting* bandpass gain. *Hint:* Change the polarities of the op amps and change some interconnections in the circuit [11]; see Figs. 6-27 and 6-28 for a generalization of this procedure.

6.25 Repeat Problem 6.23, but use Åckerberg-Mossberg sections. Make use of the results in Problem 6.24.

6.26 Design a leapfrog filter, using Åckerberg-Mossberg biquads, to realize an eighth-order Butterworth bandpass filter with $f_0 = 2$ kHz, $\Delta f = 200$ Hz, and source and load resistors $R = 600$ Ω. The prototype lowpass takes on one of the forms in Fig. P6.26. Use the results of Problem 6.24.

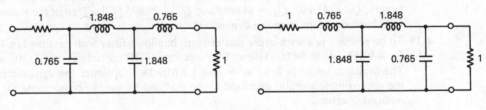

Figure P6.26

6.27 For your circuit of Problem 6.26, calculate as a function of frequency, and evaluate at the bandpass center frequency, the total sensitivity of the transfer function magnitude to changes in the pole frequencies ω_{0i} of the second-order sections. Assume that the sections track, i.e., $\omega_{0i} = \omega_0$ for all i.

6.28 The *LC* ladder realization of the Thomson filter of Problem 2.24 is given in Fig. P6.28. Implement this circuit in active form by the signal-flow graph method. Maximize the dynamic range.

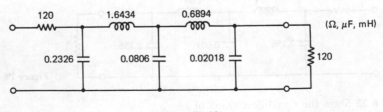

Figure P6.28

6.29 Repeat Problem 6.28 for the circuit in Fig. P6.29, which is the *LC* realization of the bandpass filter of Problem 2.27.

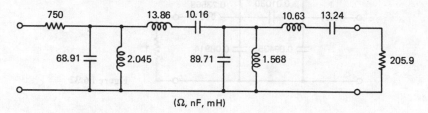

(Ω, nF, mH)

Figure P6.29

6.30 The *LC* realization of the specifications of Problem 1.23 is shown in Fig. P6.30.
(a) Realize this circuit by the signal-flow graph method.
(b) Realize the circuit by Bruton's technique.

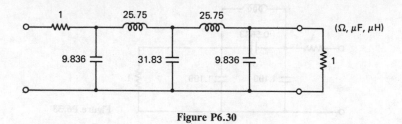

Figure P6.30

6.31 Consider the lowpass filter in Fig. P6.31. Develop an active realization based on Bruton's transformation using GIC-based FDNR elements. Be sure to take care of the biasing problem and to correct or minimize the effect of any biasing resistors.

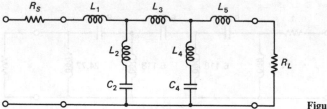

Figure P6.31

6.32 The transfer function

$$H(s) = \frac{s^2 + 49}{(s^2 + \sqrt{2}s + 1)(s^2 + 3.5s + 49)}$$

$$R_S = R_L = 1$$

can be realized as shown in Fig. P6.32 with normalized elements. The normalizing frequency is $f_n = 1.923$ kHz; source and load terminations are 3.2 kΩ. Implement the filter by the signal-flow graph method.

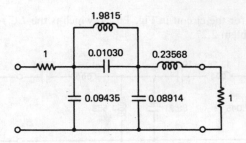

Figure P6.32

6.33 The circuit in Fig. P6.33 is the *LC* realization of the specifications in Problem 1.41.
   (a) Find a realization by the signal-flow graph method.
   (b) Find a realization by Gorski-Popiel's technique.

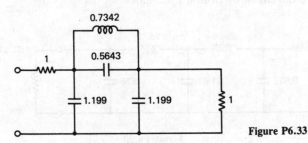

Figure P6.33

6.34 The highpass ladder in Fig. P6.34 meets the requirements of Problem 2.30.   Implement the filter by simulating the inductors by GIC circuits.

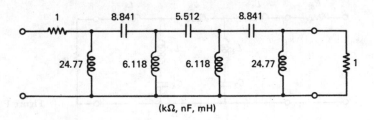

(kΩ, nF, mH)

Figure P6.34

6.35 The specifications of Problem 2.34 are implemented by the *LC* ladder in Fig. P6.35. Realize the ladder
   (a) Via a signal-flow graph simulation
   (b) Via Gorski-Popiel's method

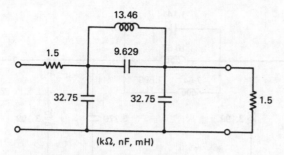

13.46

9.629

1.5

32.75   32.75   1.5

(kΩ, nF, mH)

Figure P6.35

6.36 A highpass filter is to be built as an active circuit with $R_S = R_L = 600\ \Omega$, maximally flat passband in $f \geq 3.2$ kHz, $A_p \leq 2$ dB, transmission zero at $f_z = 1$ kHz. A passive realization of the highpass circuit is shown in Fig. P6.36. Find an active realization using a suitable component-simulation procedure.

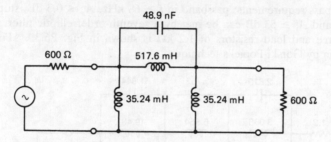

48.9 nF

600 Ω   517.6 mH

35.24 mH   35.24 mH   600 Ω

Figure P6.36

6.37 A realization of the requirements of Problem 1.55, using the lowpass-to-bandreject transformation, is shown in Fig. P6.37. Convert the passive design into an active RC circuit

(a) By the signal-flow graph method

(b) By Gorski-Popiel's method

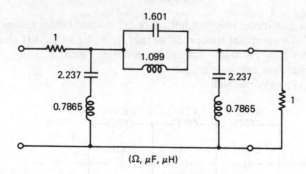

1.601

1

1.099

2.237   2.237

0.7865   0.7865   1

(Ω, μF, μH)

Figure P6.37

6.38 An implementation of the bandpass function specified in Problem 1.54 is shown in Fig. P6.38. Realize the circuit

(a) By the signal-flow graph technique

(b) By Gorski-Popiel's method

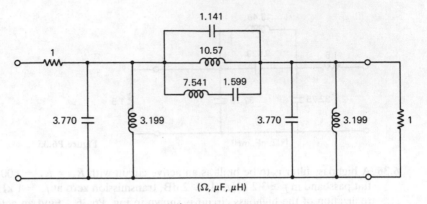

$(\Omega, \mu F, \mu H)$

Figure P6.38

6.39 The highpass requirements: passband in $f \geq 68$ kHz, $A_p \leq 0.3$ dB, stopband in $f \leq$
55 kHz, and $A_s \geq 53$ dB can be met by a seventh-order elliptic filter. The circuit
with source and load resistors of 1.2 kΩ is shown in Fig. P6.39. Find an active
realization by Gorski-Popiel's technique.

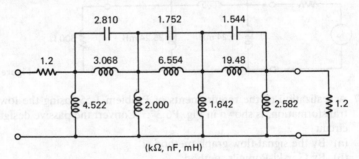

$(k\Omega, nF, mH)$

Figure P6.39

6.40 A normalized eighth-order inverse Chebyshev LC lowpass ladder realization is shown
in Fig. P6.40. Denormalize this circuit so that $R_S = R_L = 1.2$ kΩ and the lowpass
cutoff frequency $f_c = 7.3$ kHz. Realize the resulting circuit
(a) By the signal-flow graph technique
(b) By Gorski-Popiel's method.

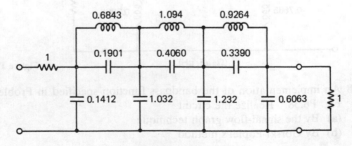

Figure P6.40

6.41 The normalized prototype lowpass circuit shown in Fig. P6.41 results by realizing the transfer function obtained after applying the bandpass-to-lowpass transformation to a bandpass with the specifications:

Equiripple passband in 12 kHz $\leq f \leq$ 18.75 kHz, $A_p \leq$ 0.6 dB
Stopbands with $A_s \geq$ 43 dB in $0 \leq f \leq$ 10 kHz and, approximately, $f \geq$ 22 kHz
A readily tunable transmission zero at $f =$ 8.6 kHz
Source and load resistors 5.5 kΩ

Derive the final form of the LC bandpass ladder and realize the circuit by Gorski-Popiel's method.

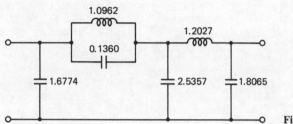

Figure P6.41

Fully Integrated
Continuous-Time Filters

SEVEN

7.1 INTRODUCTION

As mentioned at the end of Chapter 6, with the arrival of LSI or VLSI (large-scale or very large-scale integration), the designer of continuous-time filters has many powerful incentives for trying to develop the filter circuitry in a form that is compatible with the appropriate IC technology (MOS or bipolar). In many instances, a filter, be it active or passive, built with discrete components will be unacceptably bulky when compared to the remaining parts of a system that are increasingly being implemented in fully integrated form. Also, counting hand assembly, manual tuning, and wire interconnections, a discrete design may be too expensive or unreliable. Therefore, in recent years, considerable effort is being devoted to the design of filter circuits which consist only of components that are conveniently realizable on an integrated circuit (IC) chip. In this context we note in passing, that, of course, LC filters are not directly integrable, because there exists no method that permits building an integrated high-quality inductor.

At this point the reader will correctly observe that implementing a monolithic filter in principle ought to be no problem, because all components used in our designs of *active* filters, electronic operational or transconductance amplifiers along with capacitors and resistors, can readily be integrated together on one IC chip. We found, after all, in Chapters 5 and 6 that there was no need for inductors; we simply employed *gain* together with resistors, capacitors, and feedback to implement large pole quality factors and transmission zeros on the $j\omega$-axis. Although this observation is correct, it neglects some difficulties which the engineer must keep in mind when designing a practical integrated filter circuit.

The first problem arises because the *range of available passive elements is very limited*. Not considering special processing techniques, such as resistive thin films on a silicon IC, the maximum resistance of the layers available for building integrated resistors is of the order of 1 kΩ/$\square$.[1] Thus, resistors are normally limited in value to $R \leq 40$ kΩ, i.e., approximately 40 squares in length. Exceeding this limit yields resistors which tend to be physically too large and, in addition, will be seriously affected by parasitic capacitors. Similarly, the maximum value of capacitors is limited to approximately 50 pF because of restrictions on their physical size. Note that an integrated circuit capacitor is described by

$$C = \varepsilon_r \varepsilon_0 \frac{A}{t}$$

where A is the area, t the thickness of the dielectric (typically $t = 600$ Å $= 6 \cdot 10^{-8}$ m), ε_0 the permittivity ($\varepsilon_0 = 8.854 \cdot 10^{-12}$ F/m), and ε_r the relative dielectric constant ($\varepsilon_r = 3.78$ for silicon dioxide). With these numbers, the area of a 50 pF capacitor becomes approximately $(300 \ \mu m)^2$, which is very large compared to all other electronic components.

These restrictions make it quite difficult or at least inconvenient to use standard active *RC* techniques when realizing monolithic active filters at audio frequencies: the frequency parameters are set by *RC* time constants as $\omega_0 = 1/(RC)$; assuming $C = 50$ pF and $f_0 = 3$ kHz, the necessary resistor takes on the value

$$R = \frac{1}{\omega_0 C} \simeq 1 \ \text{M}\Omega$$

requiring a length of the order of 1000 squares to implement! Thus, alternative techniques are needed for low-frequency filters. The methods proposed in the literature employ either MOS transistors biased in the ohmic region [146–150] (see Section 7.4, below) or they simulate a resistor via $1/g_m$ with a transconductance value of the order of μS (compare Fig. 4-13). This technique will be explained further in Section 7.5.

The second difficulty arises because *for accurate performance a filter requires accurate and stable component values*. To set precise frequency parameters, $\omega_0 = 1/(RC)$, evidently requires accurate *absolute values* of resistors and capacitors. But, although *ratios of like components* can be realized very well in integrated-circuit technologies, absolute-value tolerances are generally quite poor: fabrication tolerances ε_r of resistors and ε_c of capacitors of the order of 40% or more are not

---

[1]The resistor realized by a piece of thin resistive material with resistivity ρ, thickness t, length L, and width W has the value

$$R = \frac{\rho L}{t W} = R_s \frac{L}{W}$$

where R_s is called the "sheet resistance" in Ω/$\square$ (ohms per square). R_s is the resistor of one square (equal length and width) of the material of thickness t regardless of the size of $L = W$.

uncommon. Consequently, time constant errors of 80% or higher must be expected:

$$\tau = RC = R_0(1 + \varepsilon_r)C_0(1 + \varepsilon_c) \simeq R_0C_0(1 + \varepsilon_r + \varepsilon_c) = \tau_0(1 + 0.8)$$

where R_0, C_0, and τ_0 are the nominal design values.

Because the second important filter parameter, the quality factor Q, is a *dimensionless* quantity and, therefore, depends only on ratios of like components, it may appear initially that Q must be realizable quite accurately in IC form. The reader will recall though from the discussion in Chapters 5 and 6 that small phase errors in the feedback loops can give rise to significant deviations in quality factor. These phase errors, caused by parasitic components or parasitic effects in the active devices, are generally too difficult to predict and cannot be eliminated by design. Consequently, at least in critical high-frequency requirements, even the quality factor will require tuning.

In discrete form, one can eliminate many of the tolerance problems by an initial tuning step and thereafter attempt to keep the filter within specifications by selecting stable and drift-free circuit components. In a realization in IC form, however, where all components are on a silicon chip and where we have to contend with large processing tolerances and changes caused by temperature drifts, aging, and power supply variations, tuning in the usual sense is clearly not a viable proposition. Although in some technologies initial tuning of elements at additional cost via a laser trimming procedure could be undertaken in order to eliminate the effect of processing tolerances, it is generally not a satisfactory solution, because circuit parameters will rarely stay stable when operating conditions, such as temperature, change. Even a low-sensitivity design, although helpful, will normally not solve the problem, because the expected element variations can be quite large.

To make an integrated filter with its large component tolerances practical, some tuning is clearly quite unavoidable; on the other hand, any "mechanical adjustments" of elements on an IC are, of course, impossible. The generally adopted solution for this dilemma is *automatic electronic tuning*. It is obtained by integrating a tuning or control system along with the filter on the IC such that *the filter tunes itself* automatically and continually against any variations, regardless of their origin if possible. Because of the importance of the tuning problem to practical integrated filters, we shall devote a separate section, Section 7.3, to a detailed discussion of automatic tuning.

Finally, let us point out that, although advances in this direction are being made, the design of IC filters has not yet been reduced to the simple assembly of building blocks. Therefore, the successful design of a filter in integrated form requires a reasonable knowledge of the limitations of IC processing technologies and device modeling. A few elementary concepts will be presented when pertinent in our further discussions; a detailed treatment of these topics, however, is beyond the scope of this text. The reader is urged to consult the appropriate references [151–153].

7.2 SINGLE-ENDED VERSUS FULLY BALANCED DESIGNS

We mentioned before that the main motivation for implementing analog filters in fully integrated form is the intent to place them together with other parts of the system on a single IC chip. These parts will often perform digital or sampled-data signal-processing tasks and thereby generate switching noise, for example. Also, as we shall see later in this chapter, the required control or tuning circuitry frequently includes digital or switching circuits, and, especially, it always contains a time-continuous or clocked reference signal as a necessary part of the tuning scheme. Because of unavoidable parasitics on the IC, there is a real danger that these signals and the switching noise from the digital system parts will tend to be injected into the analog signal-processing circuitry, where they may cause a serious deterioration of the signal-to-noise ratio.[2] This "noise" may be coupled into the filter either directly or via the substrate, the power supply, or the ground lines. To reduce these problems, designers of analog ICs usually build their circuits as *differential* rather than single-ended structures [68, 153]. A further improvement is obtained if the circuitry is not simply differential but *fully balanced* [154], with completely symmetrical layout so that all parasitic injections couple equally into the inverting and the noninverting signal paths as common mode signals. The differential nature of the circuits along with good power supply and common mode rejection then assures that extraneous "noise" voltages contaminate the main signal only minimally. Of course, this design approach does not come without cost, because, as we shall see, it requires duplication of much of the circuitry.

In spite of the practical advantages of balanced designs, we shall in the sections to follow always assume single-ended, i.e., single-input–single-output, structures. We take this approach because we believe that the derivations of a single-ended filter and its operation are easier to understand than those of a balanced differential design. Moreover, as will be pointed out next, once a single-ended design has been found, it is quite easy to convert it into a fully balanced configuration if required.

A general rule to follow when converting a given single-ended circuit into a balanced one is the following:

> Draw the single-ended circuit and identify the ground node(s). Mirror the whole circuit at ground, duplicating all the elements, and divide the gain of all active devices by 2. Change the sign of the gain of all mirrored active elements and merge any so resulting pair with inverting-noninverting gains into one balanced differential input–differential output device. Because signals of both polarities are now available, realize any devices whose sole effect in the original circuit is a sign inversion by a simple crossing of wires [and thereby save the component(s) in question].

---

[2]"Noise" here is to be understood as *any* unwanted signal and not just as the usual electrical (such as thermal) noise.

Before illustrating the application of this simple rule with a few examples which are useful as building blocks for active filters, let us emphasize an important point to remember: in many op amp circuits, additional op amps are used for active compensation or to achieve a phase change; the high-Q integrator (Fig. 4-22) and the noninverting phase-lead integrator (Fig. 4-24a or 7-1b) are good examples. If the corrective functions of these additional op amps are to be retained in the conversion to balanced form, the seemingly "redundant" op amps must remain in the circuit. For instance, in the example below, the fully balanced circuit of Fig. 7-2b is *not* a phase-lead integrator, in contrast to the circuit in Fig. 7-1b, because we proceeded as if the two op amp inverters were ideal and replaced their function simply by crossed wires! There is nothing in Fig. 7-2b that could compensate for the phase lag contributed by the single remaining op amp (Problem 7.2).

7.2.1 Fully Differential Op Amp Integrators

Inverting and noninverting integrators were discussed in Chapter 4, Figs. 4-20 and 4-24a. The circuits are redrawn in Fig. 7-1, together with their mirror images as

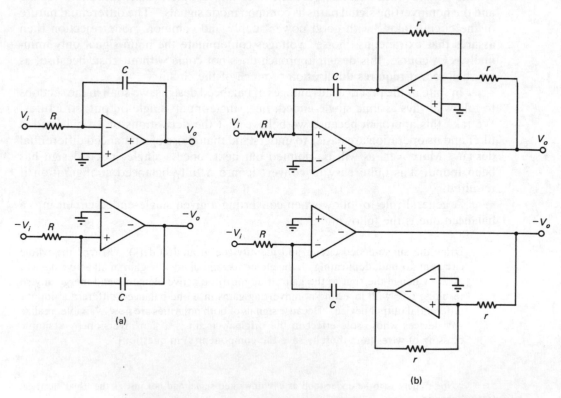

Figure 7-1 Inverting and noninverting op amp integrators with mirror images for conversion to balanced form.

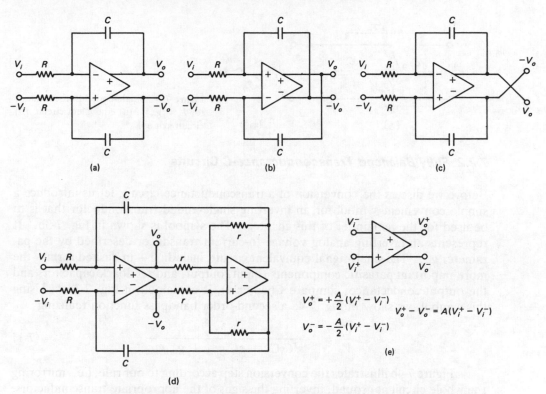

Figure 7-2   Fully balanced versions of the integrators in Fig. 7-1; all opamp output voltages are referenced to ground.   (a–c) Inverting and noninverting integrators; (d) noninverting phase-lead integrator; (e) symbol and defining equations of balanced op amp.

specified in our conversion rule.   Figure 7-2 contains the final converted circuits with differential inputs and outputs.   Note that the gain of the two inverters in Fig. 7-1b is not modified and that the function of the corresponding two op amps is implemented by crossed wires.   We also point out that a noninverting integration can be obtained simply from the balanced Miller integrator in Fig. 7-2a by crossing its output wires, because output voltages of both polarities are now available! Figure 7-2c shows the circuit.   Formally, it is obtained from a Miller-inverter cascade after replacing the inverters by cross connections.   Alternatively, we observe that the circuits in Fig. 7-2b and c are, of course, identical; we have only interchanged the op amp connections.   In Fig. 7-2d, we show a noninverting phase-lead integrator; for this circuit, the function of the *nonideal* op amp inverters of Fig. 7-1b is retained, as discussed earlier.   Of course, these two op amps are again merged into one differential input–differential output device.   In Problem 7.2 the student is asked to verify that the circuits in Fig. 7-2 are indeed balanced versions of those in Fig. 7-1.

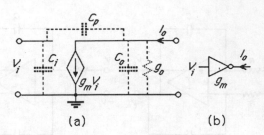

Figure 7-3 Single-ended transconductor $I_o = g_m V_i$: (a) equivalent circuit; (b) circuit symbol.

7.2.2 Fully Balanced Transconductance-C Circuits

Before we discuss the conversion of a transconductance circuit, let us introduce a simple convenient symbol for an inverting single-ended transconductor that is to be used for the remainder of this chapter. The symbol is shown in Fig. 7-3b. It represents an inverting analog voltage-to-current transducer described by the parameter g_m. The small-signal equivalent circuit, including—in dashed form—the more important parasitic components (input, output, and feedback capacitors and the output conductance; compare Chapter 4), is contained in Fig. 7-3a.[3] Using this symbol, we show in Fig. 7-4a a second-order bandpass function realizing

$$\frac{V_o}{V_i} = -\frac{g_{m0}\, sC_2}{s^2 C_1 C_2 + sC_2 g_{m3} + g_{m1} g_{m2}} \tag{7-1}$$

Figure 7-4b illustrates the conversion step according to our rule, i.e., mirroring the whole circuit at ground, inverting the signs of the appropriate transconductors, and replacing the function of inverters by crossed connections. Finally, Fig. 7-4c contains the fully differential circuit. Observe that the total capacitance has been doubled! Three quarters of this capacitance area on the silicon chip can be saved, if instead we connect the capacitors as shown in Fig. 7-4d; but note that in this case the capacitors must be *floating*, which results in more complicated processing, and especially, introduces additional parasitics (the so-called bottom-plate parasitic capacitors [155]; see also Fig. 8-4) that are not present when the capacitors are grounded.

We shall present a formal development for the derivation of the circuit in Fig. 7-4a in Section 7.5, below; in the meantime, the student may wish to verify that Eq. (7-1) is indeed realized by Fig. 7-4a, c, and d (Problem 7.4). It is also instructive to compare this circuit with the one in Fig. 5-37, where OTAs with differential inputs but single-ended outputs were used! The two circuits are really identical, but compared to Fig. 7-4a, OTAs afforded us a certain economy in the number of active devices: no inverter is required, and the availability of two inputs allows us to merge g_{m0} and g_{m3} in Fig. 7-4a into one OTA.

Observe from the above discussion that the conversion from a single-ended

---

[3]An additional parasitic effect that should be considered at very high frequencies is the frequency dependence of the transconductance parameter itself; it can usually be approximated as a phase or delay term via $g_m(j\omega)e^{-j\phi(\omega)}$; ϕ is a phase term that is used to model all high-order parasitic effects.

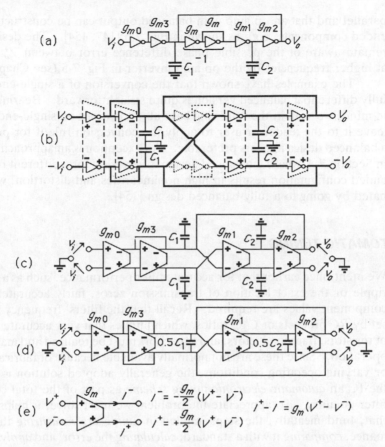

Figure 7-4 (a) Second-order bandpass filter built with single-ended inverting transconductors; (b) the circuit mirrored at ground; (c) the final fully differential configuration with *grounded* capacitors and differential input and output transconductors; (d) the circuit of part c with floating capacitors; (e) symbol and defining equations of balanced transconductance.

to a balanced design generally *doubles* the number of passive components but that the number of active devices, op amps, or transconductors for our case does not increase and, indeed, often decreases. We should bear in mind, though, that active components with differential inputs and differential, balanced outputs frequently consist of more complicated circuitry than their single-ended counterparts. Also, we remind the reader that this chapter is aimed at *fully integrated* filters; the possible lack of commercial availability of op amps and OTAs with differential or balanced outputs is, therefore, of little concern, because the whole IC filter is presumably custom-designed, including the active components. Even if for convenience single-ended active circuits from an existing IC design library are to be used, the conversion is still quite straightforward. Recall that a transconductor with differential output consists of two matched single-ended devices connected in parallel and that an op amp with balanced output can be constructed from single-ended components, e.g., as shown in Fig. 7-5 [147, 154]. The designer must only remain aware of the possible phase difference error between V_o^+ and V_o^- caused at higher frequencies by the op amp inverter in Fig. 7-5 (see Chapter 4).

The examples have shown that the conversion of a single-ended circuit to a fully differential balanced version is quite straightforward. Bearing this simplicity in mind, we shall in the following base all derivations on single-ended designs and leave it to the reader or user to apply the conversion rule if for practical reasons a balanced design appears preferable. An exception is an approach to be discussed in Section 7.4, where a balanced design is *necessary* for a different reason: a single-ended configuration results in high nonlinearities and distortion, which are eliminated by going to a fully balanced design [154].

7.3 AUTOMATIC TUNING

We mentioned earlier that for accurate filter performance, such as a small passband ripple or the exact location of transmission zeros, fairly accurate, low-tolerance component values are required. Recall that the filters' frequency parameters are set by RC products or C/g_m ratios, which implies that very accurate *absolute* values of resistors, transconductors, and capacitors must be realized and maintained during operation. Since these are not normally available because of fabrication tolerances or varying operating conditions, the generally adopted solution is to design onto the IC an *automatic electronic tuning scheme* as part of the total continuous-time filter circuitry. To appreciate the problem we must solve, it helps to understand that, fundamentally, the steps implied in *tuning* are *measuring* the filter performance, *comparing* it with a standard, *calculating* the error, and *applying a correction* to the filter to reduce the error. A review of the literature [113, 145, 147, 154, 156–163] shows that an accurate *reference frequency*, e.g., a system clock (V_{ref} in Fig. 7-6), has been agreed upon among designers as the most reliable standard. From the filter's response to the reference signal at this known frequency, the tuning circuitry must then detect and identify any mistuning, compute the appropriate corrections, and apply them via a suitable control circuit to the filter. Before adopting this intuitively appealing approach blindly, however, we must remember that the filter's purpose is to process the *main information-carrying signal*, with which the *reference signal* must not be allowed to interfere. Applying both signals simultaneously to the filter will likely cause undesirable cross talk or intermodulation.

The problems are avoided or at least minimized by constructing on the IC a

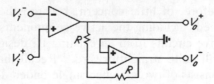

Figure 7-5 Implementation of an op amp with balanced outputs from single-ended devices.

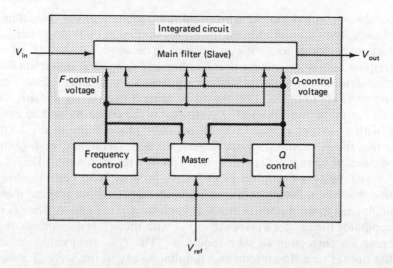

Figure 7-6 Block diagram of master-slave control system for integrated continuous-time filters.

so-called *master*, which represents, i.e., models, all relevant performance criteria of the main filter, the *slave*, with adequate accuracy. The proposed tuning strategy then applies the reference signal to the master and relies on matching and tracking between this master and the sufficiently identical slave on the same chip to tune the slave, i.e., the main filter. Tuning is accomplished by applying the correction voltage simultaneously to both master and slave. To help the student understand the operation without having to consider the actual circuitry, we show in Fig. 7-6 a block diagram of this scheme as it is implemented in almost all cases that have appeared in the literature to date.

We have identified the *main filter* (the slave) that performs the required signal-processing task and the master whose circuitry is designed to be *of sufficient complexity to model the slave's behavior that is relevant for tuning*.[4] The system contains a *frequency-control* block that compares the master's response to the reference signal V_{ref} with the reference signal itself and generates a frequency (F) control voltage which is applied to the master in a closed-loop control scheme such that any detected errors are minimized. Since master and slave are on the same chip and the master is constructed to model the slave's behavior well, we can assume that also the slave filter *errors* match and track those of the master. Consequently, the F-control voltage applied at one or more locations of the slave, as appropriate, can be expected also to correct any frequency errors in the main filter.

In the diagram of Fig. 7-6, we have shown an additional block, labeled *Q-control*. The quality factors Q_i, as we know from earlier chapters, are parameters whose correct values determine among other things the filter's pole locations, its

---

[4]In many cases the master is an appropriately designed voltage-controlled oscillator (VCO); in general, it may consist of any subsection of the slave's circuitry (up to a full duplicate of the slave).

bandwidth, the passband shape, and the steepness of the transition regions. Obviously, the Q_i are important quantities that must be realized correctly for proper system performance. We mentioned in Section 7.1 that Q, as a ratio of two frequencies, is a *dimensionless* number and as such is determined in principle by a ratio of like components (resistor, capacitor, or g_m ratio). Experience has indeed shown [147, 154, 158, 159] this to be correct in practice at fairly low frequencies and for moderate values of Q. However, the concerns we had in Chapters 5 and 6 with Q errors caused by small parasitic phase shifts in the feedback loops of active filters should convince us that for high-frequency, high-Q designs it is unreasonable to expect Q to turn out correctly without tuning. Therefore, we include for generality a Q-control block whose purpose is to permit automatic tuning of the *shape* of a filter's transmission characteristic by a scheme that is completely analogous to the F-control method, as illustrated in Fig. 7-6: The Q-control circuitry compares the master's response to V_{ref} with the reference voltage itself and thereby creates a correction signal if required. The Q-control voltage is then applied to the master in a closed loop and simultaneously to the slave in a way that corrects any detected errors.

As mentioned, all currently proposed automatic tuning schemes, with few exceptions [163], follow the approach of Fig. 7-6. The actual implementation is naturally very circuit-specific; it depends on such items as the technology used, the type of circuit, and the kind of tunable elements available. The details of these control methods will be described later in this chapter in connection with some examples. Also, a discussion of possible implementations of the frequency- and the Q-control blocks along with a few design guidelines will follow below. At this point, let us first present a number of observations that the designer of such systems should keep in mind:

1. It is a key assumption for the correct operation of the control scheme that master and slave are matched in their behavior and track. Therefore, the whole strategy can be only as accurate as the achievable matching of circuit components across a silicon chip or wafer. It would obviously be preferable to tune the main filter directly and thereby eliminate the otherwise unavoidable matching errors between master and slave, but no direct tuning method has been successful to date, since reference signal and main signal cannot be applied *simultaneously* to the filter because of the resulting danger of interference.

   As shown by the material in Chapters 5 and 6, frequency parameters are not very sensitive to parasitic effects so that matching does not appear to be a problem; Q, however, turns out to be *extremely* sensitive to parasites and small phase errors so that great care must be taken when "modeling" the slave's behavior by a master circuit. This problem becomes particularly critical in high-frequency and high-Q designs.

2. Filter control is to be *completely automatic*, including adjusting the circuit elements. Obviously, therefore, the designer must make the important filter

parameters dependent on *electronically variable components*. For instance, variable capacitors can be realized via reverse-biased *pn* junctions [152, 153; see Eq. (7-4)], and an MOS transistor in the triode (nonsaturation) region gives a bias-controllable resistor [153, 154][5]. In the g_m-C approach to be discussed in Section 7.5, the transconductance value is tunable by changing gate-bias voltages or bias currents.

Apart from this self-evident factor, however, the reader is invited to reflect upon the generally quite difficult process and concept of *self-tuning*. A signal is applied to the filter and an error, magnitude and/or phase, is measured (automatically) *at a given frequency*. The tuning circuitry, which for practical reasons should be relatively simple, must then decide on the basis of this error, *which* of the many filter component(s) to vary in order to reduce the error. The problem is made more difficult by the fact that we have only one or at best very few test values[6] and no (on-chip!) computing power. Because of this difficulty it cannot be emphasized strongly enough that the *designer of an IC filter chip must consider at the outset whether the contemplated filter structure is tunable*, that is, whether an observed error can be corrected by varying the value of a dominant tunable component. Considerations of tunability may, for example, dictate cascade designs in spite of the superiority, in principle, of ladder simulations: in cascade circuits the errors can readily be attributed to errors in a second-order section where in turn they are easy to correct; see Chapter 6 and references 113 and 156.

3. The IC contains not only the information signal of interest that is to be processed by the main filter, but also the reference voltage; moreover, quite often there are digital signals that may be needed for switching, time sharing, or multiplexing the filter and/or control circuitry, or that may arise in other parts on the IC chip. The *utmost care* must be taken to *shield* the main filtering path from these extraneous signals, which are likely to interfere with the signal-processing job or at least cause the signal-to-"noise" ratio to deteriorate.

4. The frequency-control loop is designed to reduce to zero any difference between the reference frequency and some characteristic frequency parameter of the master filter. The frequency parameters of the slave are related to those of the master by a multiplying constant that is close but generally not equal to unity.[7] If the reference frequency changes, the master and the

---

[5]Note that the "capacitor" [152, 153] and the "resistor" will be nonlinear and are, therefore, useful only for small signals. A method for obtaining linear filters in spite of the nonlinear MOS "resistors" is discussed in Section 7.4.

[6]One test signal, V_R, at a given frequency gives us just two pieces of information: *magnitude* and *phase* of the filter's response.

[7]In order to minimize the interference problems (see item 3), one generally selects a reference frequency outside the filter's passband. However, the reference frequency should not be chosen "too far" from the filter's critical pole frequencies, because it would complicate the required matching and tracking between master and slave.

matching slave filters, within the tracking range of the control loop, will simply follow suit. In this sense, the filter system operates as a *tracking filter*.

5. Although the filters can be tuned, the engineer is still well-advised to derive the best possible *nominal* design by use of such tools as careful modeling, statistical yield analysis, or design centering. Electronic tuning is almost always realized by changing the dc bias conditions; if very wide tuning ranges are required, caused, for example, by large initial tolerances, often other important filter aspects, such as linearity and dynamic range, are affected. Also, wide tuning requirements tend to result in more complicated filter and control circuitry.

After this introduction to the difficult and crucial problem of automatic tuning of integrated filters, let us now proceed to formalize our objective more carefully, point out extensions and special difficulties, and illustrate along the way some of the concepts with a particularly simple example that nevertheless has been used to design a commercial self-tuned continuous-time filter [160].[8]

Assume we wish to realize a transfer function $H(s, \mathbf{t}, \mathbf{u})$ where s is the complex frequency and $\mathbf{t}$ and $\mathbf{u}$ are the *vectors* of *tuned* and *untuned* circuit components or parameters, respectively.[9] Both $\mathbf{t}$ and $\mathbf{u}$ are subject to errors and tolerances, so that we may write

$$\mathbf{t} = \mathbf{t}_0 + \Delta\mathbf{t} \qquad \mathbf{u} = \mathbf{u}_0 + \Delta\mathbf{u} \tag{7-2}$$

where $\mathbf{t}_0$ and $\mathbf{u}_0$ are the nominal values. Our goal is then to design a control system which adjusts the values of $\Delta\mathbf{t}$ such that

$$H(s, \mathbf{t}, \mathbf{u}) = H(s, \mathbf{t}_0 + \Delta\mathbf{t}, \mathbf{u}_0 + \Delta\mathbf{u})$$
$$= \frac{N_m(s, \mathbf{t}_0 + \Delta\mathbf{t}, \mathbf{u}_0 + \Delta\mathbf{u})}{D_n(s, \mathbf{t}_0 + \Delta\mathbf{t}, \mathbf{u}_0 + \Delta\mathbf{u})} \Rightarrow H(s, \mathbf{t}_0, \mathbf{u}_0) \tag{7-3}$$

i.e., by varying $\Delta\mathbf{t}$, the inaccurate transfer function $H(s, \mathbf{t}_0 + \Delta\mathbf{t}, \mathbf{u}_0 + \Delta\mathbf{u})$ is to be tuned to its nominal value $H(s, \mathbf{t}_0, \mathbf{u}_0)$. Note that $\Delta\mathbf{t}$ may consist of the "tuning changes" $\Delta\mathbf{t}_t$ and the "errors" $\Delta\mathbf{t}_\varepsilon$ caused by processing tolerances, temperature variations, and so forth:

$$\Delta\mathbf{t} = \Delta\mathbf{t}_t + \Delta\mathbf{t}_\varepsilon$$

In other words, tuning the components in the vector $\mathbf{t}$ will not only compensate for the errors in $\mathbf{u}$ but also for errors in $\mathbf{t}$ itself. We have indicated in Eq. (7-3) that $H(s)$ is a ratio of polynomials $N_m(s)$ of order m and $D_n(s)$ of order n; that is, it is determined by $n + m - 1$ coefficients. This fact implies that if Eq. (7-3) holds at $n + m - 1$ frequency points, then it will hold over the entire frequency

---

[8]This discussion follows the treatment in reference 145.

[9]For example, MOS capacitors are not tunable and belong to the vector $\mathbf{u}$; on the other hand, junction capacitors can be tuned by changing the reverse bias voltage: they would belong to $\mathbf{t}$ if they are used for tuning purposes, but to the set $\mathbf{u}$ if they are not tuned.

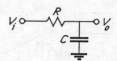

Figure 7-7 First order RC lowpass filter.

domain. We observe, therefore, the first practical difficulty: complete tuning, in general, requires application of at least $n + m - 1$ test frequencies with two possible measurements each (magnitude and phase). In practice, however, only very few reference signals, customarily just one, are applied to the filter chip.

Let us now turn to a simple example to illustrate how an inaccurate RC product would be tuned automatically. Consider the first-order RC lowpass filter in Fig. 7-7 and assume that the resistor is untuned (belonging to **u**) and that the capacitor, a reverse-biased junction, is the tunable element (out of the set **t**) described by

$$C = \frac{C_n}{(1 - V_c/\phi)^m} \tag{7-4}$$

V_c is the reverse bias voltage, ϕ the built-in junction potential, and C_n the device capacitance with no reverse bias applied. The constant m depends on the doping profile of the pn junction ($\frac{1}{3} \le m \le \frac{1}{2}$). The transfer function of this circuit equals

$$\frac{V_o}{V_i} = H(s, C(V_c), R) = \frac{1}{sCR + 1} = \frac{1}{s\dfrac{C_n}{(1 - V_c/\phi)^m}R + 1} \tag{7-5}$$

In this simple example, an error in R (and any error in the value of C) can clearly be corrected by adjusting V_c such that the RC product remains unchanged:

$$(C_0 + \Delta C)(R_0 + \Delta R) = C_0 R_0$$

i.e.,

$$\frac{\Delta C(V_c)}{C_0} = \frac{R_0}{R_0 + \Delta R} - 1 \tag{7-6}$$

The question remains how to implement a circuit that accomplishes this adjustment. Among the various possibilities, a particularly simple *magnitude-locking loop* approach was proposed in reference 160. It is shown in Fig. 7-8, where the PDs are peak detectors and K is a differential amplifier with gain K. V_B is a nominal (for no error) dc reverse bias voltage that may be required to generate the necessary capacitor value by Eq. (7-4); we shall consider it to be a dc offset of $v_c(t)$. The output voltages of the two peak detectors are

$$V_{o1} \simeq \frac{R_2}{R_1 + R_2} V_M \quad \text{and} \quad V_{o2} \simeq \frac{V_M}{\sqrt{(\omega CR)^2 + 1}} \tag{7-7}$$

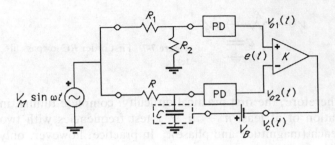

Figure 7-8 Tuning system for the *RC* lowpass filter of Fig. 7-7.

The function of the feedback control system is easiest to understand from the block diagram in Fig. 7-9, which represents the operation of the circuit in Fig. 7-8 in the frequency domain. The system is described by the expression

$$E(s) = V_{o1}(s) - V_{o2}(s) = \frac{1}{1 + KH(s, V_c)}V_{o1}(s) \qquad (7\text{-}8)$$

where $H(s, V_c)$ is the transfer function of the lowpass filter, the right-hand side of Eq. (7-5), at the given frequency ω for a specific value of the control voltage input V_c. It follows from Eq. (7-8) that $E(s)$ goes to zero if the loop gain KH is sufficiently large; thus, in steady state, $V_{o1} \simeq V_{o2}$, or, by Eq. (7-7),

$$CR \simeq \frac{1}{\omega}\sqrt{\left(1 + \frac{R_1}{R_2}\right)^2 - 1} \qquad (7\text{-}9)$$

The equation points out a result that is typical for all automatic tuning schemes:

> The accuracy and long-term stability of the *RC* product, which sets the frequency parameter of the filter, are now as good as those of the applied reference frequency ω and of a ratio of like components on the IC.

In this way all tuning schemes rely on an accurate and stable reference frequency, such as a system clock, and on matched and tracking components on the IC chip.

The tuning strategy until now has ignored the effects of the always present parasitic components. For instance, an IC implementation of the simple lowpass filter in Fig. 7-7 may have, among others a *lossy* capacitor and a *parasitic* shunt

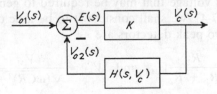

Figure 7-9 Block diagram for the tuning system in Fig. 7-8.

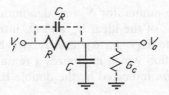

Figure 7-10 First-order lowpass filter with parasitic components.

capacitor across the resistor as is shown in Fig. 7-10. In that case the transfer function equals, with $G = 1/R$,

$$\frac{V_o}{V_i} = \frac{sC_R + G}{s(C + C_R) + G + G_c} \tag{7-10}$$

and we can observe two effects that again are true in general:

1. The presence of parasitic components changes the *values* of the critical frequencies (poles and zeros). For the case of this example, the nominal pole value has moved from $1/(RC)$ to the new position

$$\omega_p = \frac{G + G_c}{C + C_R} = \frac{1}{RC} \frac{1 + R/R_c}{1 + C_R/C} \tag{7-11}$$

2. The presence of parasitic components *may* cause *new* critical frequencies to appear. In the example, the parasitic capacitor C_R gave rise to a new zero at

$$\omega_z = \frac{1}{C_R R}$$

The first of these two effects usually has no major consequences, because in this case the parasitic components lead to the same result as changes in the nominal component values, that is, they look like tolerances. In our example, as far as ω_p is concerned [see Eq. (7-11)], the presence of C_R appears as an error in the value of C equaling $\Delta C = C_R$. Similarly, the capacitor loss, modeled as a conductance G_c, looks like an error in the conductor value $\Delta G = G_c$. Assuming there are *sufficient tunable elements* in the circuit, errors caused by shifted critical frequencies can be corrected by tuning.

The second effect, the creation of new poles and/or zeros, is more serious: there exists no tuning method that reduces the "expanded" transfer function [e.g., Eq. (7-10)] to the ideal one [Eq. (7-5)], because the new parasitic poles and/or zeros cannot be canceled. Thus, if parasitic effects are of concern—as they always are—then there exists no choice of adjustments Δt to the set of tunable components t such that the tuning operation defined in Eq. (7-3) can be satisfied *exactly*. Rather, if there exists a vector of parasitic components p, the transfer function $H(s, t, u)$ becomes

$$H_p(s, t, u, p) = \frac{N_p(s, t, u, p)}{D_p(s, t, u, p)} \tag{7-12}$$

where the degrees of the numerator N_p and denominator D_p, respectively, are in general higher than those of the ideal N_m and D_n introduced in Eq. (7-3). Since H_p cannot be equated to $H(s, t_0, u_0)$ *exactly*, we have to be content with an approximate tuning solution which minimizes a remaining tuning error ε defined via a suitably chosen *norm* indicated by the double bars, as

$$\varepsilon(\Delta t) = \|H_p(s, t_0 + \Delta t, u_0 + \Delta u, p) - H(s, t_0, u_0)\| \Rightarrow \min \qquad (7\text{-}13)$$

For example, by choice of Δt one may wish to minimize the maximum difference between the magnitudes of H_p and H over a frequency range Ω of interest:

$$\varepsilon(\Delta t) = \max_{\omega \in \Omega} \left| |H_p(j\omega, t_0 + \Delta t, u_0 + \Delta u, p)| - |H(j\omega, t_0, u_0)| \right| \Rightarrow \min \qquad (7\text{-}14)$$

Although conceptually the tuning operation described in Eq. (7-13) or (7-14) is easy to understand, the reader will appreciate that in practice it is generally very difficult to perform *automatically, on chip, and with reasonably simple control-system circuitry*. As mentioned in the introduction to this section, the technique involves *measuring* filter performance $|H_p|$ as a function of all parameters u, t, p, and frequency ω, *comparing* it with a standard $|H|$, *computing* a suitable error norm, and *determining* an appropriate correction (i.e., tuning) to perform the minimax operation as required.[10] Clearly, then, the designer is well-advised to search for simplifications to the tuning procedure.

One obvious simplification is achieved via a reduction in the complexity of the function $H_p(s, t, u, p)$, an approach referred to as *partitioning* [145]. As pointed out earlier, it is obtained, for example, through a *cascade* design where any effects within a second-order section are isolated from all the others and where the influence of section errors on the total transfer function is easy to identify and to correct. Of course, the disadvantage of a cascade design when compared, for example, to a ladder is the higher sensitivity to component tolerances in the first place (see Chapter 6).

A second important simplification of the tuning scheme is obtained from topologies in which *parasitics are absorbed* in the main circuit components, as is G_c in the example of Fig. 7-10. In that case the parasitics look like element tolerances, do not increase the order of the system, and therefore do not generate new poles and/or zeros. We shall recognize below as one advantage of the *trans-conductance-C* approach to be discussed in Section 7.5 that most parasitic elements can be absorbed. To anticipate this point, consider the circuit in Fig. 7-4c: the dominant circuit parasitics (see Fig. 7-3), the input capacitors, and for that matter

---

[10]The process as described, finding a correction for the measured error in filter performance $|H_p|$ as a function of *all* parameters, identifies the method as *functional tuning*.

the output capacitors of the g_m elements, *all* appear in parallel with either C_1 or C_2, or they load the input voltage source, where they are relatively harmless. Similarly, all output conductances are in parallel with the "resistor" $1/g_{m3}$ except the output conductance of g_{m1}, which shunts C_2. With the aid of Eq. (7-1) the student can readily verify that the system stays of order 2 and no new poles are created although there is an unimportant negative real high-frequency zero (at $-g_o/C_2$). The parasitics-induced pole variations can readily be tuned by varying the bias conditions of g_{m1} or g_{m2} and g_{m3}.

The third property to look for when designing a system for automatic tuning is *good matching* of like components, which is fortunately a strong point of integrated-circuit implementations. Better matching is generally obtained as the components get closer in value and characteristic, with the best match found for identical components (ratio of value 1) [e.g., 164]. It will be found as an important advantage of the *transconductance-C* approach that most if not all transconductors can be made identical for certain types of filters (ladder simulations). Since all active subblocks—the g_m's—are identical, including intrinsic and layout parasitics, tuning is greatly simplified because it can be reduced to tuning one subcircuit with one sample of these devices and then sending the tuning signal to all other identical devices on the chip.

If parasitics cannot be absorbed as discussed above, the engineer should attempt other *parasitics immunization* techniques with the goal of reducing the transfer function error as far as possible at the design stage.[11] In addition to *absorption*, immunization techniques include *compensation* and *predistortion*. Both rely on a well-defined and well-characterized IC process which via *parasitics extraction* permits the designer to predict—of course, with an amount of uncertainty—at least the nominal parasitic components to be expected. The effect of a particular parasitic parameter may then be *compensated* by strategically placing the opposite parasitic into the circuit such that the two effects just cancel. This is the technique that was used in Chapter 5 when connecting a resistor R_c in parallel to the GIC to cancel the apparent negative loss resistor R_L of Eq. (5-110b). Also, the method that pairs phase-lag with phase-lead integrators in the Åckerberg-Mossberg biquad (Section 5.3.3) or in SFG filters (Section 6.4) may be regarded as parasitic compensation. *Predistortion* simply attempts to counter the errors resulting from parasitic components by design: if, for instance, a parasitic effect is known to result in a 30% increase in Q value, the circuit might be designed with a nominal Q value that is 30% too small. The approach taken in Example 5-8 is a case in point.

To evaluate the operation of the functional block diagram of the master-slave tuning system in Fig. 7-6, it is helpful to develop a mathematical model. Assuming that we can Laplace-transform all signals into the frequency domain and that all

---

[11]"Parasitics-insensitive" designs in switched-capacitor circuits discussed in Chapter 8 are just such techniques.

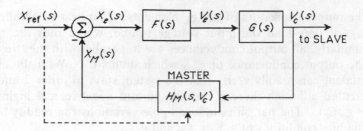

Figure 7-11 Operational diagram of an on-chip tuning system.

blocks can be represented by their transfer functions, we obtain the diagram in Fig. 7-11, where

$X_{ref}(s)$ is an externally applied stable reference signal. As will be explained below, depending on the design of the master H_M, X_{ref} may (dashed line) or may not be applied to H_M.

$X_M(s)$ is an output quantity from the master that represents the relevant performance measure.

$X_e(s)$ is the error in X_M obtained by comparing X_M to X_{ref}.

$V_e(s)$ is the error voltage (or current) produced by the block $F(s)$ from $X_e(s)$.

$V_c(s)$ is the tuning control voltage (or current).

$F(s)$ converts the error information X_e into a suitable electrical signal V_e.

$G(s)$ is the controller which converts V_e into a tuning or control voltage (or current) V_c appropriate for the tunable components used in the design.

$H_M(s, V_c)$ is the transfer characteristic of the master as a function of frequency for a specific value of the parameter V_c.

Operational diagrams, such as the one in Fig. 7-11, apply separately for the frequency- and the Q-control loops identified in Fig. 7-6. Note that the main filter (slave) circuitry which processes the main signal is not shown; V_c will, of course, also be applied to the slave to assure main filter tuning.

The function described by the control system in Fig. 7-11 is

$$X_e(s) = \frac{1}{1 + F(s)G(s)H_M(s, V_c)} X_{ref}(s) \tag{7-15}$$

from which we find with the final-value theorem of the Laplace transformation [4], assuming $x_{ref}(\infty)$ is defined and finite,

$$x_e(\infty) = \frac{1}{1 + F(0)G(0)H_M(0, V_c)} x_{ref}(\infty) \tag{7-16}$$

It may be helpful to relate Eq. (7-15) to the control circuit in Fig. 7-8: we have

$$x_{\text{ref}}(t) = v_{o1}(t) = V_M \frac{R_2}{R_1 + R_2} \qquad x_M(t) = v_{o2}(t) \qquad x_e(t) = e(t)$$

$$F(s) = 1 \qquad G(s) = K \qquad H(s, V_c) = \frac{1}{s \dfrac{C_n}{(1 - V_c/\phi)^m} R + 1}$$

and from Eq. (7-16), with $H(0) = 1$,

$$e(\infty) = \frac{1}{1 + K} v_{o1}(\infty)$$

in agreement with Eq. (7-8). From Eq. (7-16), the steady-state value of the filter error, $x_e(\infty)$, can be made arbitrarily small if the low-frequency loop gain $F(0)G(0)H_M(0,V_c)$ is sufficiently large; i.e., the tuning strategy will converge. The specific time-domain properties of the tuning system, such as the *residual error*, *settling time*, and *tuning range*, depend on the characteristics of the controller [165]. We shall look at the most commonly used schemes in the following.

7.3.1 The Frequency-Control Block

In almost all cases that have appeared in the literature, the frequency-control block identified in Fig. 7-6 is realized by using a phase-locked loop (PLL) approach with a general scheme as shown in Fig. 7-12. The operation proceeds as follows. The phase detector compares the reference signal v_{ref} with the signal v_M generated by the master to produce a raw error signal $\hat{y}(t)$, the lowpass filter LPF eliminates all undesirable high-frequency components, and the cleaned error signal $y(t)$ is applied to the master and the matched and tracking slave in a way that corrects the detected errors. Operational details depend on the particulars of the phase detection and the master circuitry. The most frequently found methods are described in the following subsections.

7.3.1.1 Phase detector is an analog multiplier, master is a voltage-controlled oscillator (VCO). We explain the first standard PLL strategy briefly in the following. It is assumed that the VCO was designed such that its oscillation frequency matches and tracks all frequency parameters in the slave circuitry. With

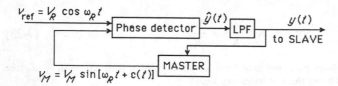

Figure 7-12 Phase-locked loop frequency tuning scheme.

the signals identified in Fig. 7-12, the analog multiplier with gain $\hat{K}$ produces the output

$$\hat{y} = \hat{K}v_{ref}v_M = \hat{K}V_{ref}V_M \cos \omega_R t \sin [\omega_R t + \varepsilon(t)]$$

$$= \tfrac{1}{2}\hat{K}V_{ref}V_M\{\sin \varepsilon(t) + \sin [2\omega_R(t) + \varepsilon(t)]\}$$

Assuming that the lowpass filter LPF passes[12] the low-frequency component $\sin \varepsilon(t)$ with a gain H_0 and rejects the component at $2\omega_R$ completely, we obtain

$$y(t) = \frac{H_0\hat{K}}{2} V_{ref}V_M \sin \varepsilon(t) \triangleq K_y \sin \varepsilon(t) \qquad (7\text{-}17)$$

K_y is a gain constant used to indicate how the phase error $\varepsilon(t)$ is converted into the error signal $y(t)$. We have to assume that, in general, the free-running VCO output frequency ω_F is different from the reference frequency ω_R and that, in addition, there is a phase difference $\phi(t)$; therefore, the total instantaneous phase error can be expressed as

$$\varepsilon(t) = (\omega_F - \omega_R)t - \phi(t) = \Delta\omega_F t - \phi(t) \qquad (7\text{-}18)$$

In usual VCO operation, the instantaneous change of phase, the frequency, is made proportional to the applied signal $y(t)$; thus

$$\frac{d\phi(t)}{dt} = K_F y(t) \qquad (7\text{-}19)$$

An equation for the instantaneous value of the angle error $\varepsilon(t)$ is then obtained by differentiating Eq. (7-18) and using Eqs. (7-19) and (7-17):

$$\frac{d\varepsilon(t)}{dt} + K \sin \varepsilon(t) = \Delta\omega_F \qquad (7\text{-}20)$$

where we have called the loop gain $K = K_F K_y$. Note that $\Delta\omega_F$ is the initial open-loop frequency error of the VCO. Without attempting to solve this nonlinear differential equation, we observe that in steady state, $d\varepsilon(t)/dt = 0$, so that the residual steady-state error becomes

$$\varepsilon_{ss} = \sin^{-1} \frac{\Delta\omega_F}{K} \qquad (7\text{-}21)$$

with the convergence condition

$$K > |\Delta\omega_F| \qquad (7\text{-}22)$$

because $\sin^{-1}x$ is defined only for $|x| < 1$. With these expressions, the steady-state error signal is

$$y_{ss} = K_y \frac{\Delta\omega_F}{K} = \frac{\Delta\omega_F}{K_F} \qquad (7\text{-}23)$$

---

[12]Note that the lowpass filter passband must have a bandwidth of at least $\Delta\omega = \omega_F - \omega_R$ [see Eq. (7-18)], wide enough to transmit the low-frequency signal.

and the VCO output equals

$$v_M(t) = V_M \sin\left(\omega_R t + \sin^{-1}\frac{\Delta\omega_F}{K}\right) \tag{7-24}$$

i.e., within the *tuning range* [note Eq. (7-22)]

$$\omega_R - K < \omega_F < \omega_R + K \tag{7-25}$$

the steady-state error signal adjusts itself such that the VCO frequency is locked to the reference frequency ω_R with a small residual phase error.   Also note that the final rate of convergence is exponential: if after a time $t = t_0$ we can assume $\varepsilon \ll 1$, then the differential equation [Eq. (7-20)] reduces to

$$\frac{d\varepsilon(t)}{dt} + K\varepsilon(t) \simeq \Delta\omega_F \tag{7-26}$$

with the solution

$$\varepsilon(t) \simeq \frac{\Delta\omega_F}{K} + \left[\varepsilon(t_0) - \frac{\Delta\omega_F}{K}\right] e^{-K(t-t_0)} \tag{7-27}$$

Note that all the equations indicate the desirability of a large loop gain K.

Recall our earlier emphasis in this chapter of the importance of close matching and tracking between master and slave.   Anticipating our need for *Q-control* with its high sensitivity to parasitics and fearing that the performance of a *VCO as master* might be sufficiently different from that of a *filter as slave*, it has been proposed [113] to use a lowpass *filter* section as the master and thereby to insure a better match.[13]   The operation is somewhat different from the one discussed above and will be explained in the following.

7.3.1.2 Phase detector is an analog multiplier, master is a filter.   If the master is a lowpass filter section with transfer function

$$H_M(s) = \frac{H_{M0}\omega_F^2}{s^2 + s\dfrac{\omega_F}{Q_M} + \omega_F^2} \tag{7-28}$$

its phase at an applied frequency ω can readily be calculated to equal

$$\varepsilon(\omega) = -\frac{\pi}{2} + \tan^{-1} Q_M \frac{\omega_F^2 - \omega^2}{\omega_F\omega} \tag{7-29}$$

To generate an output which is usable for our purposes, we inject the reference signal with *known* frequency ω_R both into the phase detector *and* into the master

---

[13]If the designer prefers to stay with a VCO, the best possible match between VCO and "filter" may be obtained by constructing an oscillator from a filter by using a biquad section of such high Q (i.e., $Q \rightarrow \infty$) that the biquad oscillates (see Example 7-1).

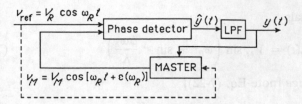

Figure 7-13 Frequency control using a master filter.

as shown dashed in Fig. 7-13 (see also Figs. 7-6 and 7-11). The filter output

$$v_M(t) = V_M \cos [\omega_R t + \varepsilon(\omega_R)] \tag{7-30}$$

where $V_M = |H_M(j\omega_R)|V_R$, is then compared with (multiplied by) the reference voltage; the result is filtered to yield an error signal

$$y(t) = \frac{\hat{K} H_0 |H(j\omega_R)|}{2} V_R^2 \cos \left[\frac{\pi}{2} - \tan^{-1} Q_M \frac{(\omega_F - \omega_R)(\omega_F + \omega_R)}{\omega_F \omega_R} \right] \tag{7-31}$$

$$= K_y \sin \left[\tan^{-1} Q_M \frac{(\omega_F - \omega_R)(\omega_F + \omega_R)}{\omega_F \omega_R} \right]$$

K_y is defined as before, and H_0 is the low-frequency gain of the LPF. With a master *filter*, the applied error voltage $y(t)$ does not change the phase angle directly, but rather it changes the filter's pole frequency such that

$$\omega_{Fa} = \omega_{F0} + K_F y(t) \tag{7-32}$$

where ω_{Fa} and ω_{F0}, respectively, are the *actual* and the *initial untuned* pole frequencies and K_F is the conversion gain from $y(t)$ to $\Delta\omega_F$. In this case, no differential equation can be established for the frequency error, but an estimate of the tuning accuracy can still be found as follows [113]. From Eq. (7-32), the initial open-loop frequency error equals

$$\Delta\omega_{Fi} = \omega_{Fa} - \omega_{F0} = K_F y(t)$$

where, after closing the loop, $y(t)$ is to be equated with Eq. (7-31). Thus, with the residual error

$$\Delta\omega_{Fr} = \omega_{Fa} - \omega_R$$

we have

$$\Delta\omega_{Fi} = K_F K_y \sin \left(\tan^{-1} \left[Q_M \Delta\omega_{Fr} \frac{\omega_{Fa} + \omega_R}{\omega_{Fa}\omega_R} \right] \right) \tag{7-33}$$

which can be solved for $\Delta\omega_{Fr}$. If we label the loop gain $K = K_F K_y$ as before, we find from Eq. (7-33)

$$\frac{\Delta\omega_{Fr}}{\omega_R} = \frac{\omega_{Fa}}{\omega_{Fa} + \omega_R} \frac{1}{Q_M} \tan \left(\sin^{-1} \frac{\Delta\omega_{Fi}}{K} \right) \tag{7-34}$$

The expression can be simplified if we may assume that the error is small so that $\omega_{Fa} \simeq \omega_R$; in that case, with $K \gg \Delta\omega_{Fi}$,

$$\frac{\Delta\omega_{Fr}}{\omega_R} \simeq \frac{1}{2Q_M} \frac{\Delta\omega_{Fi}}{K} \tag{7-35}$$

Observe that a system with a master *filter* results in a residual frequency error, whereas operation with a master *VCO* yields $\Delta\omega_{Fr} = 0$ (with $d\varepsilon/dt = 0$). Note that Eq. (7-22) must again be satisfied so that the tuning range as before equals

$$\omega_R - K < \omega_F < \omega_R + K$$

and—as is intuitively obvious because of the steeper phase slope—that a *high-Q* master filter will lead to a reduced frequency tuning error.

It is apparent from Eq. (7-17) that the gain K_y and, therefore, the loop gain K depend on the amplitudes of the signals applied to the phase detector. To avoid any difficulties with variable gains (if the signal amplitudes should be time-varying) and at the same time achieve simpler and more dependable phase-detection circuitry, many authors have proposed [145, 147, 156] to use instead of an analog multiplier, a *digital* phase detector, such as an Exclusive OR (EXOR) gate. We shall explain this approach next.

7.3.1.3 Phase detector is an EXOR gate, master is a VCO.

To ready the signals for digital processing and to eliminate amplitude information, the phase detector input signals v_{ref} and v_M are first sent through hard limiters, such as comparators or high-gain inverters, to convert them into square waves x_{ref} and x_M, respectively; they are illustrated in Fig. 7-14, along with the phase detector output signal $\hat{y}$. The lowpass loop filter extracts the average value y, which from Fig. 7-14 can readily be computed to equal

$$y = 4MH_0 \frac{\tau}{T_R} = \frac{2}{\pi} MH_0\varepsilon \triangleq K_y\varepsilon \tag{7-36}$$

A plot of $y(\tau)$ is shown in Fig. 7-15; in Eq. (7-36), M and T_R, respectively, are the maximum value and the period of the square wave inputs, H_0 is the low-frequency

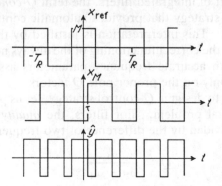

Figure 7-14 Square wave input signals and output of an EXOR gate.

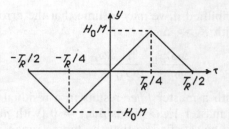

Figure 7-15 Transfer characteristic of an EXOR gate phase detector.

gain of the LPF, and we have converted the time shift τ into a phase ε via $\varepsilon = 2\pi(\tau/T_R)$ in order to have a direct comparison with the analog case in Eq. (7-17). Note that the equations derived for the analog multiplier as phase detector can be used also for the EXOR gate if we replace $\sin \varepsilon$ by ε. Specifically, with $\tau \le T_R/4$, i.e., $\varepsilon \le \pi/2$, Eqs. (7-18), (7-19), (7-23), (7-26), and (7-27) are seen to be valid; the EXOR-based PLL locks ω_F to ω_R with a residual steady-state phase error

$$\varepsilon_{ss} = \left| \frac{\Delta\omega_F}{K} \right| \tag{7-37}$$

so that, because $\varepsilon_{ss} < \pi/2$ from Fig. 7-15, the tuning range is given by

$$\omega_R - \frac{\pi}{2}K < \omega_F < \omega_R + \frac{\pi}{2}K \tag{7-38}$$

Evidently, there is no significant difference in the PLL operation with a digital or an analog phase detector; as a matter of fact, apart from the factor $\pi/2$ in Eq. (7-38), for $\varepsilon \ll 1$, both systems behave identically. The same can be said for a tuning method that pairs an EXOR gate with a master *filter*. We shall leave the investigation of some details of this case to the problems (Problem 7.8).

7.3.2 The Q-Control Block

We mentioned repeatedly that, at least for applications at higher frequencies and for medium to high values of the quality factor Q, some automatic control or tuning of Q will be necessary because of the large sensitivity of Q to parasitic phase shifts on the IC. In the context of integrated filters, the term *Q-control* is to be understood to mean a suitable strategy that provides automatic control of the *shape of the transfer characteristic*. This interpretation is justified by the fact that the Q-control operation implies the correct functioning of the always necessary frequency tuning scheme. But with accurate frequency parameters assured, the transfer function *shape* depends only on the proper pole Q factors.

Before endeavoring to design a Q-control strategy, let us pause for a moment to discuss the fundamental problem. For filters, the *quality factor* is typically defined as a frequency divided by the difference of two frequencies

$$Q \triangleq \frac{\omega_0}{\omega_2 - \omega_1}$$

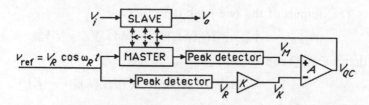

Figure 7-16 Magnitude-locking loop for Q-control.

at specified magnitude levels, such as V_{max} and V_{3dB}. Thus, to determine Q in an automatic tuning system requires us in principle to measure three voltages *at three different frequencies, on chip, with*, we want to stress again, *reasonably simple circuitry*. It implies that, for each Q to be tuned, three different *accurate* test frequencies are to be generated on chip (or supplied externally) for which the master circuit's response must be evaluated. It will be apparent to the reader that this scheme would be exceedingly complicated.

Fortunately, a far simpler approach is available which makes use of the fact that in a transfer function, Q errors and magnitude errors are closely related. To illustrate, consider the simple second-order lowpass function of Eq. (7-28); assuming ω_F has been tuned correctly, at $\omega = \omega_F$ we find

$$|H_M(j\omega_F)| = H_{M0}Q_{M0}$$

where Q_{M0} is the nominal design value. Because the value of dc gain H_{M0} is set by a ratio of like components (resistor or capacitor ratio) and is not subject to parasitic phase errors, it can be assumed to be accurate. Therefore, any Q error, ΔQ_M, reflects itself directly in a gain error,

$$|H_M(j\omega_F)| = |H_M(j\omega_F)|_0 + \Delta H_M = H_{M0}(Q_{M0} + \Delta Q_M) \qquad (7\text{-}39)$$

i.e.,

$$\Delta H_M = H_{M0}\,\Delta Q_M \qquad (7\text{-}40)$$

This proportionality of ΔH_M and ΔQ_M is the clue to a workable Q-control method: we measure ΔH_M at *one* frequency via a simple peak detector and deduce by virtue of the known value of H_{M0} the error ΔQ_M. A diagram for this method, the Q-control block of Fig. 7-6, is given in Fig. 7-16.

The master is built to model all relevant Q errors of the slave correctly. The two peak detectors are identical. K is a dc amplifier and A is a dc differential amplifier (comparator); their gains are K and A, respectively. It is assumed that all frequency parameters of master and slave are tuned correctly by means of a suitable frequency-control loop as discussed earlier.[14] V_{QC} is returned at one or more points to both master and slave in a way that is appropriate for adjusting the quality factors.

---

[14]Note that the frequency parameters *must* be correct for the Q-control scheme to work; this requirement, however, poses no difficulties, because, as seen earlier, the frequency-control system will converge independently of Q.

The outputs of the two peak detectors are

$$V_M = |H_M(j\omega_R)|V_R \quad \text{and} \quad V_K = KV_R \tag{7-41}$$

so that the dc Q-control voltage V_{QC} equals

$$V_{QC} = A(V_M - V_K) = A\left[|H_M(j\omega_R)| - K\right]V_R \tag{7-42}$$

From Eq. (7-42) we have clearly

$$|H_M(j\omega_R)| - K = \frac{1}{A}\left(\frac{V_{QC}}{V_R}\right)\Bigg|_{A\to\infty} \simeq 0 \tag{7-43}$$

which implies $K \simeq |H_M(j\omega_R)|$, or

$$|H_M(j\omega_R)|V_R \simeq KV_R \tag{7-44}$$

i.e., the two amplitudes are locked together. To carry the concept further, let us assume that the slave circuit is built as a cascade of second-order sections (see Chapter 6) and that the master is one representative lowpass biquad[15] [Eq. (7-28)], so that by Eq. (7-39)

$$|H_M(j\omega_R)| = |H_M(j\omega_R)|_0 + \Delta H_M = H_{M0}(Q_{M0} + \Delta Q_M)$$

If we design the dc gain of the amplifier K to equal the nominal value—i.e., if $|H_M(j\omega_R)|_0 = H_{M0}Q_{M0}$—but allow for a (small) design or processing error ε_K, Eq. (7-44) yields

$$H_{M0}(Q_{M0} + \Delta Q_M) = K + \varepsilon_K \tag{7-45}$$

i.e.,

$$\Delta Q_M = \frac{\varepsilon_K}{H_{M0}} \tag{7-46}$$

Because the resistor (or capacitor or transconductance) ratio K can be made very precise, we would normally have $\varepsilon_K \to 0$ and, consequently, obtain ideally a zero Q error. The resulting control voltage V_{QC} depends on the circuit implementation, that is, on the function $\Delta Q(V_{QC})$; assuming that $\Delta Q_M = -K_Q V_{QC}$, we find from Eq. (7-42) with Eq. (7-45)

$$V_{QC} = A(H_{M0}Q_{M0} - K - \varepsilon_K - H_{M0}K_Q V_{QC})V_R$$

which can be solved for

$$V_{QC} = \frac{-A\varepsilon_K V_R}{1 + AK_Q H_{M0}V_R}\Bigg|_{A\to\infty} \simeq -\frac{\varepsilon_K}{K_Q H_{M0}} \simeq 0 \tag{7-47}$$

(Note that K_Q has units of V^{-1}!) We observe that the magnitude-locking loop attempts to reduce the control voltage to zero; if a finite dc control voltage is

---

[15]Bandpass, highpass, or other types of biquads can, of course, also be used with the appropriate modifications in the following equations.

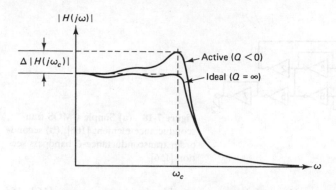

Figure 7-17   Measured errors in a high-order lowpass filter. $\Delta|H(j\omega_c)| > 0$ indicates negative losses ($Q < 0$).

necessary to maintain the required value Q_{M0}, it must be inserted as a dc "offset" into the feedback loop as was indicated in Fig. 7-8, where a similar magnitude-locking scheme was used, but for frequency control [160].

From the above discussion it may appear that this Q-control strategy will work very well for second-order sections and cascade filters, but that it may not be useful for other types of filter topologies, such as ladder simulations. Recall, though, from Fig. 3-11 and the related discussion, that incorrect transfer function shapes can always be attributed to "Q errors," provided they are interpreted appropriately.[16] For example, in Fig. 3-11, partially redrawn in Fig. 7-17, passband errors are caused by finite (here: negative) inductor or capacitor losses, i.e., finite component Q factors. Therefore, from the measured errors, $\Delta|H(j\omega_c)|$, we may deduce the required corrective (tuning) steps in quite the same fashion as in designs based on biquads; only the "tuning algorithm" that identifies which component to vary to effect the correction will generally be less obvious. Careful predesign simulations, however, will usually indicate the most appropriate tuning elements.

Because the practical implementation of a tuning system depends on the specific circuitry and the chosen technology, we feel that the reader may benefit from an example. Thus, before proceeding with the general discussion, let us illustrate some of the tuning concepts by an example which assumes use of a specific CMOS transconductor so that implementation details can be discussed [156].

Example 7-1

An integrated second-order bandpass filter is to be designed in CMOS technology. The center frequency is to be $f_0 = 3.81$ MHz, the quality factor $Q = 9.84$, and the midband gain is to equal 20 dB. Both f_0 and Q must be self-tuned. For the bandpass section, use the circuit in Fig. 7-18b with the simple transconductor circuit in Fig. 7-18a. [The student is encouraged to derive this circuit from the passive prototype: identify the gyrator, the positive and negative resistors,[17] and the buffer (Problem

---

[16]Note that frequency parameters are as always assumed to be correct!

[17]Positive and negative "resistors" are used to avoid the otherwise large transconductance ratio of value Q [see Eq. (7-53b)] and to be able to achieve easy tunability and large values of Q in spite of parasitic output conductances g_o in the g_m elements.

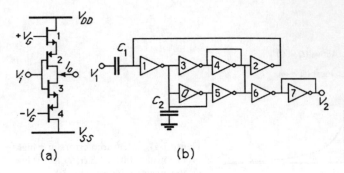

Figure 7-18 (a) Simple CMOS transconductance element [166]; (b) second-order transconductance-C bandpass section [156].

7.10)]. Assume $V_{DD} = -V_{SS} = 5$ V and reasonable device parameters [153, 156, 166]; neglect all parasitics (Fig. 7-3a).

Solution. Assuming that the circuit in Fig. 7-18a consists of two matched n-channel devices and two matched p-channel devices and that all four MOS transistors are in their saturation (constant-current) region, the single-ended four-transistor CMOS transconductor can be shown to realize [166]

$$g_m = \frac{I_0}{V_i} = 4k_{\text{eff}} (V_G - V_{Tc}) \tag{7-48}$$

where

$$V_{Tc} = \tfrac{1}{2}\Sigma V_T = \tfrac{1}{2} (V_{Tn1} + V_{Tn3} + |V_{Tp2}| + |V_{Tp4}|) \tag{7-49}$$

and

$$k_{\text{eff}} = \frac{k_n k_p}{(\sqrt{k_n} + \sqrt{k_p})^2} \tag{7-50}$$

is the effective transconductance parameter of the circuit. k_n and k_p are the transconductance parameters of the individual n-channel and p-channel devices, respectively:

$$k_{n,p} = \tfrac{1}{2} \left(\mu_{\text{eff}} C_{\text{ox}} \frac{W}{L} \right) \Big|_{n,p} \tag{7-51}$$

Here, μ_{eff}, C_{ox}, W, and L are the effective mobility, oxide capacitance per unit area, channel width, and channel length, respectively, and $V_{Tn,p}$ are the threshold voltages of the devices [151]. We shall assume[18]

$$\tfrac{1}{2} \mu_{\text{eff},n} C_{\text{ox}} = 24 \ \mu\text{A/V}^2 \qquad \tfrac{1}{2} \mu_{\text{eff},p} C_{\text{ox}} = 9 \ \mu\text{A/V}^2$$

$$V_{Tn} = -V_{Tp} = 0.95 \text{ V} \qquad L = 5 \ \mu\text{m}$$

The derivation of Eqs. (7-48) through (7-50) made certain simplifying assumptions which are acceptable for the discussion in this text; a more careful analysis can be found in the literature [166], which also shows that with an appropriate choice of design parameters the -3 dB frequency of this transconductor can be > 100 MHz.

[18]The simplifying assumption $V_{Tn} = -V_{Tp}$ is, of course, not realistic, but for this example is not important.

In the circuit in Fig. 7-18b we used the transconductor symbol introduced in Fig. 7-3b. Neglecting all parasitic components, and choosing $C_1 = C_2 = C$ and all g_m elements equal except g_{m6}, g_{m7}, and g_{mQ}, the filter circuit in Fig. 7-18b can be shown to realize (Problem 7.10)

$$H_2(s) = -\frac{g_{m6}}{g_{m7}} \frac{\frac{g_m}{C} s}{s^2 + \frac{g_{mQ} - g_m}{C} s + \left(\frac{g_m}{C}\right)^2} = -H_B \frac{\omega_0 s}{s^2 + \frac{s\omega_0}{Q} + \omega_0^2} \qquad (7\text{-}52)$$

Note from Eq. (7-48) that g_m is tunable by varying the gate bias voltage V_G; thus, in our circuit we can adjust the pole frequency

$$\omega_0 = \frac{g_m}{C} \qquad (7\text{-}53a)$$

the pole equality factor

$$Q = \frac{g_m}{g_{mQ} - g_m} \qquad (7\text{-}53b)$$

and the bandpass gain constant

$$H_B = \frac{g_{m6}}{g_{m7}} \qquad (7\text{-}53c)$$

by varying the appropriate transconductor value. If we now choose $C = 4.2\,pF$, we compute from Eq. (7-53)

$$g_m = \omega_0 C = 100.54 \ \mu S$$

$$g_{mQ} = g_m \left(\frac{1}{Q} + 1\right) = 110.76 \ \mu S$$

$$20 \log \left(\frac{g_{m6}}{g_{m7}} Q\right) = 20 \quad \text{i.e.,} \ \frac{g_{m6}}{g_{m7}} = \frac{10}{Q} = 1.016 \approx 1$$

Let us pick[19] $g_{m6} = g_{m7} = g_m$. A good choice for the channel widths is [166]

$$W_p = 25 \ \mu m \quad \text{and} \quad W_n = 20 \ \mu m$$

so that, by Eqs. (7-51) and (7-50),

$$k_n = 96 \ \mu A/V^2 \qquad k_p = 45 \ \mu A/V^2 \qquad k_{\text{eff}} = 15.86 \ \mu A/V^2$$

and, from Eq. (7-48),

$$V_G = \frac{g_m}{4k_{\text{eff}}} + V_{Tc} = 3.49 \ V \qquad V_{GQ} = \frac{g_{mQ}}{4k_{\text{eff}}} + V_{Tc} = 3.65 \ V$$

This completes the nominal design.

To proceed with the design, we observe that in practice it will undoubtedly be necessary to tune the filter section against initial tolerances and the effect of parasitics

---

[19]This choice implies that g_{m6} and g_{m7} form a unity-gain buffer which prevents any load from modifying the section's quality factor.

and to keep tuning it because of variations caused by changing operating conditions. Note that even with accurate design and processing and with stable operating conditions, the effect of the parasitic components will cause errors in the pole positions. Neglecting C_p and C_o from Fig. 7-3a, it is apparent that the capacitor C_1 in Fig. 7-18b is in parallel with C_i and g_o, C_2 is paralleled by $2C_i$, and the "resistor" $1/g_{mQ}$ is in parallel with $3g_o$. Also, the "resistor" $1/g_{m4}$ is shunted by $2g_o$ and $3C_i$. Consequently, rather than from the ideal denominator of Eq. (7-52), the actual pole positions are determined by the expression

$$s(C_2 + 2C_i) + g_{mQ} + 3g_o$$

$$- \left[g_{m5} - \frac{g_{m1}g_{m2}}{s(C_1 + C_i) + g_o} \right] \frac{g_{m3}}{g_{m4} + 2g_o + s(2C_i)} = 0 \qquad (7\text{-}54)$$

which, with all g_m except g_{mQ} identical and no parasitics, reduces to the ideal denominator polynomial. In reality, ω_0 and Q are changed and a parasitic pole is introduced by the nonideal inverter whose gain equals

$$- \frac{g_{m3}}{g_{m4} + 2g_o + s(2C_i)}$$

We shall, therefore, contemplate a tuning scheme for both ω_0 and Q as outlined in Figs. 7-12 and 7-16, with an EXOR gate as phase detector, a master VCO for frequency control, and a master filter for Q control. Note that we use *two* masters; the reasons will become apparent as we progress in our example. The control diagram is shown in Fig. 7-19. The frequency-control part contains an f-VCO; two hard limiters to obtain square wave inputs for the EXOR gate; and two lowpass filters, the loop-filter LPF1 and the filter LPF2, whose purpose is further removal of high-frequency noise before the control voltage enters the slave filter. Their operation and design will be explained during the following discussion.

Considering the importance of a good match between master and slave, we form an oscillator (f-VCO) by copying the bandpass section, Fig. 7-18b, but grounding the input and setting Q to infinity; i.e., ideally $g_{mQ} = g_{m5}$ according to Eqs. (7-53b) and (7-54).[20] In that case, the resonance frequency equals, nominally,

$$\omega_{VCO} = \frac{\sqrt{g_{m1}g_{m2}}}{C_{VCO}} = \frac{g_m}{C_{VCO}} \qquad (7\text{-}55)$$

as in Eq. (7-53a) and all parasitic effects on ω_{VCO} are essentially the same in master and slave. Recalling the earlier observation that the reference frequency should be outside the passband, but for better matching "not too far" from it, let us choose f_{VCO} = 2.9 MHz, and, as in the filter section, $C_1 = C_2 = C_{VCO}$ with all identical transconductances $g_m = 100.52\ \mu S$ (except for g_{mQ}). We obtain

$$C_{VCO} = \frac{g_m}{\omega_{VCO}} = \frac{100.54}{2\pi \cdot 2.9}\ \text{pF} = 5.52\ \text{pF}$$

---

[20]In practice, the designer would use thorough simulations to determine the value of g_{mQ} appropriate for oscillations. Careful analysis yields $g_{mQ} < g_{m5}$. It may even be necessary to control g_{mQ} by a separate tuning loop or to permit some external adjustments of the Q tuning voltage. These details are not important for this example and shall be omitted.

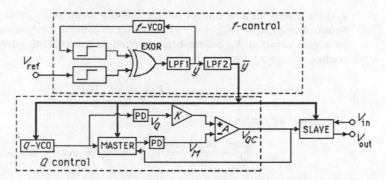

Figure 7-19   Control system for Example 7-1.

The VCO signal and the externally supplied reference signal are then converted to square waves by hard limiters and compared in the EXOR gate as discussed earlier. The limiters can be implemented as shown in Fig. 7-20a, i.e., by having the transconductance of Fig. 7-18a working into its g_0 load (gain equals $-g_m/g_o$), or simply as CMOS inverters (transistors 2 and 3 in Fig. 7-18a).   The EXOR output signal, after cleaning in LPF1, is given in Eq. (7-36), with $M = V_{DD}$ (signal swing approximately from V_{DD} to V_{SS}):

$$y = \frac{2}{\pi} V_{DD} H_0 \varepsilon \tag{7-56}$$

The measured phase error ε is caused by the difference between the reference frequency ω_R and the free-running VCO frequency ω_F and by an actual phase shift ϕ as in Eq. (7-18), so that

$$\frac{d\varepsilon}{dt} = \omega_F - \omega_R - \frac{d\phi}{dt} = \Delta\omega_{VCO} - \frac{d\phi}{dt}$$

$d\phi/dt$ is the instantaneous change in ω_{VCO} that can be related to the oscillator frequency [Eq. (7-55)] via Eq. (7-48):

$$\omega_{VCO} = \frac{4k_{\text{eff}}}{C_{VCO}} (v_G - V_{Tc}) = \frac{4k_{\text{eff}}}{C_{VCO}} (V_{G0} - V_{Tc}) - \frac{4k_{\text{eff}}}{C_{VCO}} v_g = \omega_F - \frac{d\phi}{dt}$$

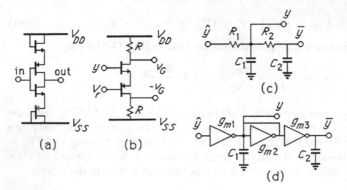

Figure 7-20   (a) Hard limiter; (b) bias generator; (c and d) lowpass filters.

v_g is the change in gate control voltage from the nominal V_{G0} that is generated in the f-control loop: $v_G = V_{G0} - v_g$. Further let us assume that the signal y is multiplied by a gain constant K_G before being applied to the control gates in the VCO as a control voltage v_g; i.e.,

$$v_g = K_G y = \frac{2}{\pi} K_G H_0 V_{DD} \varepsilon \tag{7-57}$$

Therefore, we obtain

$$\frac{d\varepsilon}{dt} + \frac{8 k_{\text{eff}} K_G H_0}{\pi C_{\text{VCO}}} V_{DD} \varepsilon \triangleq \frac{d\varepsilon}{dt} + K\varepsilon = \Delta\omega_{\text{VCO}} \tag{7-58}$$

with the steady-state error ε_{ss}, for $d\varepsilon/dt = 0$,

$$\varepsilon_{ss} = \frac{\Delta\omega_{\text{VCO}}}{K} = \frac{1}{K_G H_0} \frac{\Delta\omega_{\text{VCO}}}{\omega_R} \frac{\pi}{8} \frac{\omega_R C_{\text{VCO}}}{k_{\text{eff}} V_{DD}} \tag{7-59}$$

For the numbers in this example, we get

$$\varepsilon_{ss} \simeq 0.5 \frac{1}{K_G H_0} \frac{\Delta\omega_{\text{VCO}}}{\omega_R}$$

For instance, if we choose $K_G = H_0 = 1$, a 40% initial f_{VCO} error would result in $\varepsilon_{ss} \simeq 0.2$ rad.

The value of K_G depends, of course, on the specific choice of bias-generation circuitry; for example, making use of the square-law I-V relationship of MOS devices [151], the current in the circuit of Fig. 7-20b can readily be shown to equal

$$I = k_{\text{eff}}[y - V_r - (V_{Tn} + |V_{Tp}|)]^2$$

where k_{eff} was defined in Eq. (7-50) and V_r is a reference voltage. Consequently, v_G is a nonlinear function of y:

$$v_G = V_{DD} - R k_{\text{eff}}[y - V_r - (V_{Tn} + |V_{Tp}|)]^2$$

which can be linearized via

$$\left.\frac{dv_G}{dy}\right|_{y=0} = -2R k_{\text{eff}}[-V_r - (V_{Tn} + |V_{Tp}|)] \triangleq -K_G \tag{7-60}$$

into

$$v_G \simeq V_{DD} - R k_{\text{eff}}[-V_r - (V_{Tn} + |V_{Tp}|)]^2 - K_G y \tag{7-61}$$

With Eq. (7-59), the steady-state gate correction voltage [Eq. (7-57)] equals

$$v_{gss} = \frac{\omega_R C_{\text{VCO}}}{4 k_{\text{eff}}} \frac{\Delta\omega_{\text{VCO}}}{\omega_R} \simeq 1.59 \frac{\Delta\omega_{\text{VCO}}}{\omega_R} \text{ V}$$

$\pm v_{gss}$ is sent to the control gates (Fig. 7-18a) of g_{m1} and g_{m2} in the f-VCO. Because the f-VCO is matched to the slave filter section (the two frequencies differ only by a capacitor ratio 3.81 MHz/2.9 MHz = 1.31), v_{gss}, after further filtering in LPF2 to obtain additional attenuation of high-frequency components, is also applied via identical bias generation circuits to the corresponding transconductors in the slave.

The required reference voltage V_r and the resistor R must be computed from Eqs. (7-60) and (7-61). We find

$$K_G = 2Rk_{\text{eff}}[-V_r - (V_{Tn} + |V_{Tp}|)]$$

and

$$V_{G0} = V_{DD} - Rk_{\text{eff}}[-V_r - (V_{Tn} + |V_{Tp}|)]^2$$

$$= V_{DD} - 0.5K_G[-V_r - (V_{Tn} + |V_{Tp}|)]$$

where from earlier calculations we have $V_{G0m} = 3.49\ V$ for the nominal gate voltage of the transconductance g_m. If we choose for our example $K_G = 2$, we can compute

$$V_{rm} = V_{G0m} - V_{DD} - (V_{Tn} + |V_{Tp}|) = -3.46\ \text{V}$$

which leads to

$$R_m = \frac{1}{k_{\text{eff}}[-V_{rm} - (V_{Tn} + |V_{Tp}|)]} = 41.76\ \text{k}\Omega$$

The subscript m refers to the parameters for g_m control.

Finally, let us pay attention to the design of the loop and lowpass filters LPF1 and LPF2. Both filters are quite uncritical as far as their tolerances are concerned and can be implemented as simple passive RC lowpass blocks as shown in Fig. 7-20c. The circuit realizes

$$\frac{Y}{\tilde{Y}} = H_{\text{LPF1}}(s) = \frac{1 + sC_2R_2}{s^2C_1C_2R_1R_2 + s[C_1R_1 + C_2(R_1 + R_2)] + 1} \tag{7-62a}$$

$$\frac{\overline{Y}}{\tilde{Y}} = H_{\text{LPF2}}(s) = \frac{1}{s^2C_1C_2R_1R_2 + s[C_1R_1 + C_2(R_1 + R_2)] + 1} \tag{7-62b}$$

Assume we want a tuning range $\Delta\omega_{\text{VCO}}$ of 1 Mrad/s (corresponding to initial ω_{VCO} tolerances of approximately $\pm 25\%$); by Eq. (7-37), this implies

$$K > \frac{2\Delta\omega_{\text{VCO}}}{\pi} = 4 \cdot 10^6/\text{s}$$

K is given in Eq. (7-58), and for the numbers in this example,

$$K = K_GH_0\frac{8}{\pi}\frac{k_{\text{eff}}V_{DD}}{C_{\text{VCO}}} \simeq 36.6K_GH_0\ 10^6\ \text{s}^{-1}$$

Therefore, with $K_G = 2$ per our earlier design, we need to implement

$$|H_{\text{LPF1}}(j\Delta\omega_{\text{VCO}})| \triangleq H_0 > \tfrac{1}{18}$$

so that the loop gain is larger than $4 \cdot 10^6\ \text{s}^{-1}$ over the 1-MHz bandwidth.

Concentrate for now on Eq. (7-62a) and assume that the second-order term in the denominator can be neglected in the range $\omega \le 2\pi \cdot 1$ Mrad/s. Let us impose $\simeq 30$ dB attenuation at $2f_R \simeq 6$ MHz. A few trials show that these requirements are satisfied safely by

$$C_1 = C_2 = 20\ \text{pF} \qquad R_1 = 42\ \text{k}\Omega \qquad R_2 = 20\ \text{k}\Omega$$

As mentioned earlier, LPF2 serves to reduce the PLL noise before tying the control voltage to the slave filter: at the output $\overline{Y}$, the 6-MHz signal is attenuated by ≈ 54 dB, an additional 24 dB over the output Y. The necessary filtering can also be achieved with an active g_m-C filter as shown in Fig. 7-20d. We leave to the problems (Problem 7.11) a design of this circuit such that the requirements of this example are met.

Let us now turn our attention to Q tuning, where we use the magnitude-locking technique discussed earlier. Evidently, from Eq (7-52), the gain at $\omega = \omega_0$ equals

$$|H_2(j\omega_0)| = H_B Q$$

where H_B, given in Eq. (7-53c) as a ratio of transconductors, is nominally equal to unity. Any errors in $|H_2(j\omega_0)|$, therefore, are contributed dominantly by errors in quality factor, regardless of origin, and can be detected by measuring $\Delta|H_2(j\omega_0)|$:

$$\Delta Q = \frac{\Delta|H_2(j\omega_0)|}{H_B} = \Delta|H_2(j\omega_0)|$$

As illustrated in the Q-control system, lower part of Fig. 7-19, the master, a *copy* of the slave circuit,[21] is stimulated by a signal from the QVCO, which also is a copy of the slave but with input grounded and Q set to infinity (nominally, $g_{mQ} = g_m$ as discussed before). The two *matched* peak detectors PD can be built, for example, as shown in Fig. 7-21a and b; the FET in Fig. 7-21a permits the capacitor to discharge, depending on the gate voltage V_D; it also serves via the appropriate bias voltage V_{ref} to overcome the necessary diode forward bias of ≈ 0.6 V. The OTA circuit in Fig. 7-21b [145] acts by closing and opening the switch when $V_o < V_i$ and $V_o > V_i$, respectively; the capacitor can discharge via a reverse-biased junction in the switch circuit which can be as simple as an FET or may consist of any of the more sophisticated transmission gates available in the literature [79, 80]. Note that the OTA is a "high-gain amplifier" with gain equal to $-g_m/g_o$. Finally, simple circuits for the dc amplifier K and the dc differential amplifier A can also be found in the literature. As an example, a transconductance implementation of both functions, realizing

$$V_{QC} = \frac{g_m}{g_o}\left(\frac{g_{m1}}{g_{m2}}V_Q - V_M\right) \tag{7-63}$$

is shown in Fig. 7-21c.

As seen in Fig. 7-19, the f-control signal $\overline{y}$ is applied to the QVCO, master, and slave so that, as required, all frequency parameters may be assumed to be tuned correctly. In particular, the QVCO's output frequency[22] is 3.81 MHz (the master is a copy of the slave!), at which the master's response has a *nominal* amplitude

$$V_{M,nom} = KV_Q = Q_{nom}V_Q = 9.84V_Q \tag{7-64}$$

V_Q is the amplitude of the oscillator output voltage. Note from Eq. (7-64) that we chose the amplifier gain K (see Fig. 7-19) equal to Q_{nom}; i.e., by Eq. (7-63), we set

---

[21]It is difficult and unreliable to detect Q errors of the slave *filter* from an infinite-Q master *oscillator*. Therefore, for Q control, the master is a copy of the slave filter.

[22]Note that for this application we preferred to choose a reference frequency (ω_{QVCO}) *inside* the slave filter's passband; this choice permits easy and accurate measuring of Q errors but necessitates that very careful attention be paid to the problem of shielding the slave from the QVCO output signal.

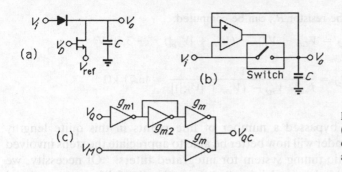

Figure 7-21 (a) Diode peak detector; (b) OTA-C peak detector; (c) transconductance implementation of amplifier K and comparator A.

$g_{m1} = 9.84\, g_{m2}$. This implies that for an incorrect quality factor $Q_{nom} + \Delta Q$, the Q-control signal V_{QC} is

$$V_{QC} = A(KV_Q - V_M) = A[(K + \varepsilon_K) - Q_{nom} - \Delta Q]V_Q = A(\varepsilon_K - \Delta Q)V_Q$$

i.e., for large A,

$$\Delta Q \simeq \varepsilon_K \to 0$$

as in Eq. (7-46). ε_K is a small design or processing error of the dc gain K. Just as for the f-control scheme (Fig. 7-20b), V_{QC} is applied to a bias-generation circuit whose output is sent to the control gates of g_{mQ} in master and slave. From Eq. (7-53b), we find

$$Q = \frac{g_m}{g_{mQ} - g_m} = \frac{g_m}{g_{mQ0} - g_m + \Delta g_{mQ}} = \frac{Q_{nom}}{1 + \dfrac{\Delta g_{mQ}}{g_m} Q_{nom}} \tag{7-65}$$

so that

$$\Delta Q = Q - Q_{nom} = -\frac{Q_{nom}^2 \dfrac{\Delta g_{mQ}}{g_m}}{1 + Q_{nom} \dfrac{\Delta g_{mQ}}{g_m}} \simeq -Q_{nom}^2 \frac{\Delta g_{mQ}}{g_m} \tag{7-66}$$

Further, making use of Eq. (7-48),

$$g_{mQ} = 4k_{eff}(V_{GQ0} - V_{Tc}) + 4k_{eff}v_{gQ} = g_{mQ0} + \Delta g_{mQ}$$

we find from Eq. (7-66)

$$\Delta Q \simeq -\frac{Q_{nom}^2}{V_{GQ0} - V_{Tc}} v_{gQ} = -K_Q v_{gQ} \tag{7-67}$$

The total Q-control voltage v_{GQ} is given by an expression analogous to Eq. (7-61),

$$v_{GQ} \simeq V_{DD} - R_Q k_{eff}[-V_{rQ} - (V_{Tn} + |V_{Tp}|)]^2 - K_G v_{QC} \tag{7-68}$$

with K_G defined as in Eq. (7-60). We found earlier the nominal gate bias voltage of the transconductance g_{mQ} to equal $V_{GQ} = 3.65$ V. Thus, the values of the reference voltage V_{rQ} and of the resistor R_Q can be computed:

$$V_{rQ} = V_{GQ} - V_{DD} - (V_{Tn} + |V_{Tp}|) = -3.30 \text{ V}$$

$$R_Q = \frac{1}{k_{eff}[-V_{rQ} - (V_{Tn} + |V_{Tp}|)]} = 46.71 \text{ k}\Omega$$

Although we still bypassed a number of fine points in this quite lengthy example, we hope the reader will now better be able to appreciate the steps involved in designing an automatic tuning system for integrated filters. Of necessity, we had to use specific circuits when explaining the design details of different functional blocks. These details will, of course, change with different circuit implementations or technologies. Nevertheless, the principles of operation stay the same and can be modified appropriately.

In the remainder of this chapter we shall present a few procedures that are useful and convenient for the implementation of integrated continuous-time filters. The so-called MOSFET-C technique discussed in Section 7.4 is based on relatively standard active op amp–RC filter design methods as discussed in Chapters 5 and 6; however, resistors are replaced by MOSFETs biased in their ohmic region. Because the values of these "resistors" are bias-dependent, they can be used directly as tunable elements with the automatic tuning methods introduced earlier. The main advantage of the technique is its simplicity, which permits the engineer to take advantage of well-developed active RC filter design methods and to use standard "components" (op amps, capacitors, and MOS transistors) that are available in most IC design libraries. Its disadvantages are the restricted frequency range caused by the use of op amps (see Chapter 4) and the inherent nonlinearities of "MOSFET resistors" that must be eliminated by clever design strategies. The transconductance-C technique presented in Section 7.5 appears to emerge as the dominant approach for designing integrated continuous-time filters because transconductors require much simpler circuitry than op amps (e.g., see Fig. 7-18a) and generally have a superior frequency response. Thus, video-frequency filters can be implemented with relative ease as illustrated in Example 7-1. The example also shows that the tunable parameters are now the bias-dependent transconductances. After the detailed discussion of automatic tuning systems in this section, we shall in the following treat the design of suitable tuning circuitry only very briefly and leave it to the reader to develop and adapt the methods of his or her designs.

7.4 MONOLITHIC FILTERS BASED ON OP AMPS [154]

We remarked earlier that, in principle, our op amp–based active RC filters are integrable if we have *electronically tunable* components of a size that can be realized in monolithic form. As an additional requirement, we shall insist that all com-

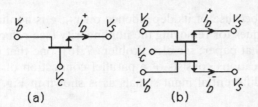

Figure 7-22 (a) n-channel transistor; (b) differentially driven parallel connection.

ponents must be implemented in MOS technology in order to insure compatibility with digital VLSI systems which may reside on the same IC chip.[23] Starting from discrete active RC designs, the engineer can then use all the accumulated knowledge of this field and "construct" the circuit by wiring together (on chip) predesigned op amps that are available in the literature or in integrated-circuit CAD libraries along with resistors and capacitors. Because experts in filter design are frequently not also experts in the design of integrated devices and circuits, this building block approach has considerable advantages for the rapid development of monolithic filters.

Readily available in MOS technology are capacitors and op amps; because normally, neither of these components is tunable,[24] we recognize, therefore, the need for bias-variable resistors. They can be implemented via MOS transistors operating in the ohmic (nonsaturation) region: staying for this text with a simple device model[25] yields for the drain current in the ohmic region of an n-channel[26] MOS transistor (Fig. 7-22a),

$$i_D^+ = k_n[2(V_C - v_S - V_{Tn})(v_D - v_S) - (v_D - v_S)^2] \qquad (7\text{-}69a)$$

so that the incremental conductance equals

$$g = \left.\frac{di_D}{d(v_G - v_S)}\right|_{v_S = \text{const}} = 2k_n[(V_C - v_S - V_{Tn}) - (v_D - v_S)]$$

Note that we have labeled the gate potential V_C to indicate that it will be used as the tuning control voltage. The transconductance parameter k_n was defined in

---

[23]In this text we shall discuss integrated filters only in MOS technology, which increasingly dominates both analog and digital VLSI systems. Of course, with minor modifications of the design details, the principles and techniques can also be applied to bipolar circuits if system requirements should so dictate [113, 161, 167, 168].

[24]One can, of course, vary the gain of op amps by changing the bias currents or voltages; this approach, however, will also tend to change other op amp parameters, such as phase and output impedance, and thereby complicate the design of filters. Further, the development of tunable op amps will involve a certain amount of custom electronic circuit design and detract from our intended fast building block approach.

[25]Simple transistor models are adequate for the discussion in this text. Careful investigations of operation and nonlinearities of the simulated resistors require more refined device models, which can be found in the literature [151].

[26]The discussion holds equally well for p-channel devices with appropriate changes in the signs of voltages and currents.

Eq. (7-51). Evidently, because of its dependence on v_D, g is nonlinear. There exist a number of techniques for removing the nonlinearity (see reference 154 and the many references in that paper; see also Problem 7.16); one that is simple and convenient for our applications consists of a parallel connection of two identical n-channel devices with differential input signals as is shown in Fig. 7-22b. The current i_D^- is

$$i_D^- = k_n[2(V_C - v_S - V_{Tn})(-v_D - v_S) - (-v_D - v_S)^2] \qquad (7\text{-}69\text{b})$$

The nonlinearity is removed by taking the difference between these two currents:

$$i = i_D^+ - i_D^- = 2k_n(V_C - V_{Tn})2v_D \triangleq G \cdot 2v_D \qquad (7\text{-}69\text{c})$$

Thus, in the ohmic region, i.e.,

$$v_D < V_C - V_{Tn} \qquad (7\text{-}70)$$

the configuration in Fig. 7-22b can implement a *linear voltage-tunable resistor* of value

$$R = \frac{1}{G} = \frac{1}{\mu_n C_{ox}(V_C - V_{Tn})} \frac{L}{W} = R_s \frac{L}{W} \qquad (7\text{-}71)$$

that can be varied by adjusting the gate control voltage V_C. Note that the "sheet resistance" R_s can be quite large: with typical values of $\mu C_{ox} = 25$ $\mu\text{A/V}^2$ and $V_C - V_{Tn} = 1$ V bias, we have $R_s = 40$ kΩ.

It remains to be explored how this linear "resistor" obtained through current differencing can be implemented and used in integrated active filters. The solution is found in the balanced designs discussed in Section 7.2. We illustrate the principle on the balanced integrator of Fig. 7-2a; it is redrawn in Fig. 7-23b with additional capacitive inputs for generality and with resistors replaced by MOSFETs. Simple analysis yields for the upper output voltage

$$v_o(t) = v^- - \frac{1}{C_F} \int_{-\infty}^{t} i_F^+(\lambda) \, d\lambda$$

$$= v^- - \frac{1}{C_F} \int_{-\infty}^{t} \left[C \frac{d(v_b - v^-)}{d\lambda} + i_R^+ \right] d\lambda \qquad (7\text{-}72\text{a})$$

$$= v^- - \frac{C}{C_F}(v_b - v^-) - \frac{1}{C_F} \int_{-\infty}^{t} i_R^+(\lambda) \, d\lambda$$

Similarly, we obtain for the lower output

$$-v_o(t) = v^+ - \frac{C}{C_F}(-v_b - v^+) - \frac{1}{C_F} \int_{-\infty}^{t} i_R^-(\lambda) \, d\lambda \qquad (7\text{-}72\text{b})$$

Subtracting Eq. (7-72b) from Eq. (7-72a) yields

$$2v_o(t) = v^- - v^+ - \frac{C}{C_F}(2v_b - v^- + v^+) - \frac{1}{C_F} \int_{-\infty}^{t} (i_R^+ - i_R^-) \, d\lambda \qquad (7\text{-}73)$$

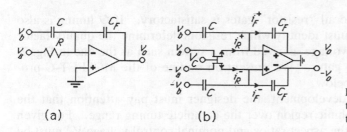

(a)   (b)

Figure 7-23 (a) Op amp–RC integrator; (b) equivalent MOSFET-C integrator.

If the op amp gain is sufficiently large, we have $v^+ = v^-$; further, from Eq. (7-69c),

$$i_R^+ - i_R^- = 2k_n(V_C - V_{Tn})2v_a \triangleq \frac{2v_a}{R}$$

Thus, we obtain finally the *linear* relationship between the output and input signals, v_o, v_a, and v_b,

$$v_o(t) = -\frac{C}{C_F} v_b - \frac{1}{RC_F} \int_{-\infty}^{t} v_a(\lambda)\, d\lambda \tag{7-74a}$$

or, in the Laplace domain,

$$V_o = -\frac{C}{C_F} V_b - \frac{1}{sC_F R} V_a \tag{7-74b}$$

where R is given in Eq. (7-71). Note, however, that the *internal* currents and voltages in the balanced integrator are *not* linear. The expressions of Eqs. (7-74a) and (7-74b) are identical to those that would be obtained for the single-ended active RC integrator in Fig. 7-23a!

Note the simplicity with which this technique permits the conversion of an active RC prototype into an integrated MOSFET-C filter:

1. Starting from a *suitable* active RC prototype, use the steps described in Section 7.2 to convert the single-ended design into a fully balanced structure. The discrete RC topology must satisfy a few restrictions which are caused by the required current subtraction to eliminate nonlinearities: all resistors must be connected between voltage sources (signal or op amp output terminals) and op amp summing nodes.

2. Replace all resistors in the resulting circuit by MOS transistors with their gates connected to the appropriate control voltage. Compute the aspect ratios L/W such that for given voltages V_C the resistor values are realized correctly.

As we discussed in Section 7.3, the control voltages are derived from appropriate tuning loops. If frequency tuning alone is satisfactory (in applications at low frequencies and low to medium Q), a phase-locked loop (Fig. 7-12 or 7-13) with control voltage applied to all "resistor" gates is satisfactory. If Q tuning is also necessary, the designer must identify which resistors determine the quality factor and apply the Q-control voltage, derived from a system such as the one in Fig. 7-16, to the corresponding gates. We present an outline of the MOSFET-C procedure in Example 7-2, below.

During the circuit development, the designer must pay attention that the transistors stay in their ohmic region over the complete tuning range. For given values of V_{Tn} and μC_{ox}, the aspect ratios and nominal control voltage V_C must be chosen such that the "resistors" can be tuned over all expected fabrication tolerances and, for example, temperature drifts without violating the condition in Eq. (7-70). Also, we point out again that, in general, floating differential op amp output voltages are not satisfactory; rather, both V_o and $-V_o$ must be individually defined with respect to ground, as is indicated in Fig. 7-23b.

Before proceeding to the example, let us make a few comments about the effects of parasitic components associated with our MOSFET "resistors" [154]. Top and side views of the integrated MOS "resistor" are shown in Fig. 7-24a; from this schematic it is clear that the "resistor" is really a *distributed RC line* [169], i.e., it exhibits significant distributed capacitance between gate and channel (C_q) and between channel and substrate (C_b). Both are proportional to the gate area WL and add to give the total distributed parasitic capacitor C_p:

$$C_p = C_q + C_b = (C_{ox} + C_d)(WL) \tag{7-75}$$

C_{ox} was defined in Eq. (7-51), and C_d is the nonlinear depletion-region capacitance per unit area. C_d is approximately 10% of C_{ox} and decreases with increasing bulk bias $|V_B|$ [151]. A small-signal model of the distributed structure, assuming that the dc gate and substrate bias potentials are at ac ground, is given in Fig. 7-24b;

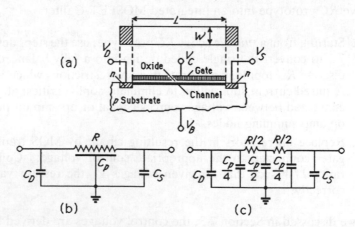

Figure 7-24 (a) Schematic top and side views of MOS "resistor"; (b) small-signal model with distributed capacitor; (c) approximate discrete model.

it also shows the source and drain capacitances, C_S and C_D, respectively, which are contributed by reverse-biased junctions at source and drain and by gate overlap. Fortunately, in our topologies (see rule 1 above and Fig. 7-23b), C_S and C_D have little or no effect because they are either driven by voltage sources or are connected to op amp input terminals. As long as the op amp gain is high, i.e., $v^+ = v^-$, they add equal current to i_F^+ and i_F^- which cancel in the equations when taking the difference. The remaining capacitor, C_p, however, can affect filter performance significantly and should be taken into account, when designing the circuit, by careful modeling and predistortion if necessary. Because many analysis programs do not easily handle distributed elements, we have in Fig. 7-24c given an approximate discrete model that permits simulating the effects with sufficient accuracy.

Example 7-2

Design an Åckerberg-Mossberg bandpass biquad as a fully balanced MOSFET-C circuit to realize the filter parameters $\omega_0 = 75$ krad/s, $Q = 12$, and midband gain $H_M = 3.5$. Use p-channel MOSFETs for the "resistors" and assume that the device parameters are the same as those in Example 7-1. The maximum signal amplitude that must be processed is 0.8 V. A frequency-tolerance analysis indicates that the "resistors" must be tuned over a range of $\pm 25\%$. Assume operation with a ± 5 V supply and op amps with $\omega_t = 15$ Mrad/s. Considering the good performance of the AM biquad (see Section 5.3.3) and the relatively low operating frequency, no tuning is required for Q and gain.

Solution. The Åckerberg-Mossberg circuit was introduced in Fig. 5-29; it is repeated along with its mirror image in Fig. 7-25a. The balanced active RC prototype is obtained by merging the appropriate op amps; the resulting circuit is shown in Fig. 7-25b. Note that we have retained the op amp inverters from the feedback loops of the original circuit as in Fig. 7-2d; had we deleted these op amps and replaced their ideal function by crossing wires as in Fig. 7-2b, the circuit would have been a balanced version of the Tow-Thomas biquad, Fig. 5-28.

The prototype RC circuit realizes the bandpass function [Eq. (5-115a)]

$$\frac{V_b}{V_i} = -\frac{k\omega_0 s}{s^2 + s\omega_0/Q + \omega_0^2}$$

where $\omega_0 = 1/(RC)$. Choosing $C = 12$ pF gives for the nominal design value of the resistor

$$R_{\text{nom}} = \frac{1}{\omega_0 C} = \frac{10^9}{75 \cdot 12}\ \Omega = 1.111\ \text{M}\Omega$$

By Eq. (7-71), this resistor is realized by a MOSFET in the ohmic region as

$$R = \frac{1}{\mu_p C_{\text{ox}}(V_C - V_{Tp})}\frac{L}{W} = \frac{10^6\ \Omega}{18(V_C - V_{Tn})/(1\ \text{V})}\frac{L}{W} \tag{7-76}$$

where we use the fact that $0.5\ \mu_{\text{eff},p} C_{\text{ox}} = 9\ \mu\text{A/V}^2$ from Example 7-1. The expected $\pm 25\%$ tolerances imply that R must be varied in the range

$$833.33\ \text{k}\Omega \leq R \leq 1.389\ \text{M}\Omega$$

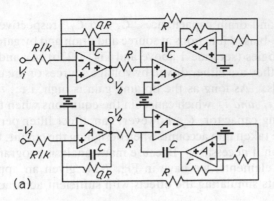

(a)

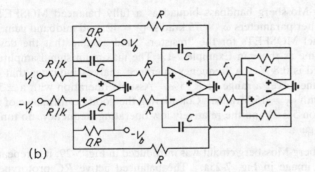

(b)

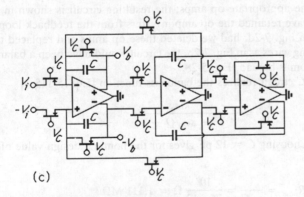

(c)

Figure 7-25 (a) Åckerberg-Mossberg circuit and its mirror image; (b) conversion to a balanced *RC* prototype structure; (c) MOSFET-*C* realization.

If we choose $V_{C,nom} = 2.5\ V$, we obtain from Eq. (7-76) with $V_{Tp} = 0.95$ V the aspect ratio

$$\frac{L}{W} = 18 \cdot 1.111 \cdot (2.5 - 0.95) = 31$$

so that

$$R_{nom} = \frac{1.7222\ M\Omega}{[V_C/(1\ V) - 0.95]}$$

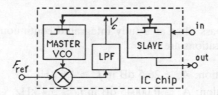

Figure 7-26 Block diagram of the tuning system.

and

$$V_{C,\min} = \left(\frac{1.722}{1.389} + 0.95\right) V = 2.190\,V \qquad V_{C.\max} = \left(\frac{1.722}{0.833} + 0.95\right) V = 3.016\,V$$

According to Eq. (7-70), the maximum linear signal range is then given by $|V_D| \le$ (2.19 − 0.95) V = 1.24 V, which satisfies the specifications. Choosing now $W = $ 5 μm results in the following values for the lengths of the MOSFETs:

$$L = 155\ \mu m \qquad L_Q = LQ = 1860\ \mu m \qquad L_k = \frac{L}{k} = 44.29\ \mu m$$

L_Q and L_k are the gate lengths of the Q- and gain-determining MOSFETs, respectively.

Let us conclude the example with a brief comment about the design of the tuning system. Having agreed initially that no Q-tuning is required, we need only to concern ourselves with frequency tuning. A suitable system is the one shown in the upper part of Fig. 7-19 with the f-VCO designed[27] such that $\omega_{vCO} = 1/(RC_{vCO})$; R is given in Eq. (7-71), i.e., it is realized as a MOSFET identical to those which determine the pole frequency of the filter in Fig. 7-25c. As mentioned, we place ω_{vCO} slightly outside the filter's passband by choosing $C_{vCO} = (\omega_{pole}/\omega_{vCO})C_{filter}$. The gate control voltage V_C generated in the phase-locked loop is then applied simultaneously to all "resistor" gates in the master oscillator and in the slave filter. Figure 7-26 shows a schematic diagram of the system.

As a further illustration of MOSFET-C design, we present in the following an additional example which deals with the signal-flow graph simulation of an LC ladder. Having discussed the SFG method and how to determine the component values in detail in Section 6.4, we shall here only give an outline of the procedure and leave it to the student to fill in the final details (Problem 7.15). Note carefully, though, that the derivation of the signal-flow graph is somewhat different from the general procedure discussed in Chapter 6. It is useful for many important practical special cases, such as the ladder in this example. As we shall see presently, the principle of the SFG simulation does not change; we only interpret and implement the node equations differently, in a way that leads generally to fewer active devices (op amps for our case).

---

[27]Considering the low frequency, any simple oscillator circuit will be adequate. If good matching of parasitics is of concern, an oscillator may be designed by duplicating the filter section of Fig. 7-25c with the Q-determining MOSFETs biased off or removed (i.e., with $QR = \infty$).

Example 7-3

Realize an elliptic lowpass filter as a fully integrated continuous-time MOSFET-C circuit; the filter specifications are:

Stopband attenuation: $A_s \geq 26.5$ dB in $f_s \geq 4.6$ kHz

Passband attenuation: $A_p \leq 0.00045$ dB in $f_p \leq 3.4$ kHz

Source and load resistors: $R_S = R_L = 5.5$ kΩ

Use p-channel MOSFETs with parameters as in Example 7-1; you may consider the integrated balanced op amps to be ideal, i.e., $\omega_s/\omega_t \ll 1$. Derive the passive LC prototype and use a signal-flow graph simulation to develop the active realization. Assume that the tolerances require the "resistors" to be varied by $\pm 30\%$ and that frequency tuning alone will be sufficient to keep the filter within specifications.

Solution. From filter design software or tables [18], we find that a fifth-order elliptic lowpass filter, as shown in Fig. 7-27a, with the normalized component values

$$R_S = R_L = 1 \qquad L_2 = 0.9176 \qquad L_4 = 0.5740$$

$$C_1 = 0.3829 \qquad C_2 = 0.1485 \qquad C_3 = 1.0602 \qquad C_4 = 0.5546 \qquad C_5 = 0.1387$$

will satisfy the specifications. In order to derive the signal-flow graph for this circuit, we perform a source transformation and arrive at the following equations:

$$V_1[s(C_1 + C_2) + G_S] = V_S G_S + sC_2 V_3 - I_{L2}$$

$$I_{L2} = \frac{1}{sL_2}(V_1 - V_3)$$

$$V_3[s(C_2 + C_3 + C_4)] = I_{L2} - I_{L4} + sC_2 V_1 + sC_4 V_5$$

$$I_{L4} = \frac{1}{sL_4}(V_3 - V_5)$$

$$V_5[s(C_4 + C_5) + G_L] = I_{L4} + sC_4 V_3$$

At this point, the student may wish to review the principles of the SFG method in Section 6.4, where we used one integrator for each reactance. However, remembering that capacitors are permissible components, we interpret such terms as $sC_2 V_3$, after impedance level scaling (!), as an input signal to the integrators. Then, rewriting these equations in the following different format to obtain the integrators needed for an SFG simulation, with a saving of several op amps, we obtain:

$$V_1 = \frac{-1}{s(C_1 + C_2) + G_S}(-V_S G_S - sC_2 V_3 + I_{L2}) \tag{7-77a}$$

$$I_{L2} = \frac{1}{sL_2}(V_1 - V_3) \tag{7-77b}$$

$$-V_3 = \frac{-1}{s(C_2 + C_3 + C_4)}(I_{L2} - I_{L4} + sC_2 V_1 + sC_4 V_5) \tag{7-77c}$$

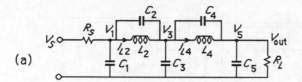

(a)

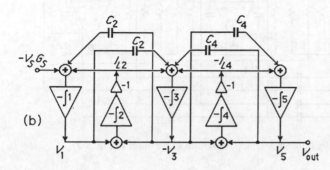

(b)

Figure 7-27 (a) Fifth-order elliptic LC lowpass filter; (b) its signal-flow graph representation.

$$-I_{L4} = \frac{1}{sL_4}(-V_3 + V_5) \qquad (7\text{-}77\text{d})$$

$$V_5 = \frac{-1}{s(C_4 + C_5) + G_L}(-I_{L4} - sC_4V_3) \qquad (7\text{-}77\text{e})$$

Evidently, the integrator functions are

$$-\int_1 \triangleq \frac{-1}{s(C_1 + C_2) + G_S} \qquad \int_2 \triangleq \frac{1}{sL_2} \qquad -\int_3 \triangleq \frac{-1}{s(C_2 + C_3 + C_4)}$$

$$\int_4 \triangleq \frac{1}{sL_4} \qquad -\int_5 \triangleq \frac{-1}{s(C_4 + C_5) + G_L} \qquad (7\text{-}78)$$

The signs were chosen according to the discussion in Section 6.4 in order to avoid having to take differences. The student can easily verify that Eqs. (7-77a) through (7-77e) are realized by the signal-flow graph in Fig. 7-27b. Recall that we need inverting and noninverting integrators to keep the loop gains positive, but that inverting integrators are easier to realize; thus we have labeled all five integrators as $-\int$ and taken care of the signs by separate inverters (conceptually as a "Miller-inverter cascade"). Because our final goal is a fully balanced version of the circuit, however, these inverters will not be necessary, since output voltages of both signs are available.

We have stated earlier our conviction that single-ended circuits are easier to develop and to understand; we have, therefore, derived the SFG realization as a single-ended circuit which permits us to use the concepts of Chapter 6 unmodified. The conversion to balanced form is undertaken by following the rules given in Section 7.2. The student is strongly urged to follow the implementation step by step in order to gain some practice in the conversion process. The final balanced filter is shown in Fig. 7-28.

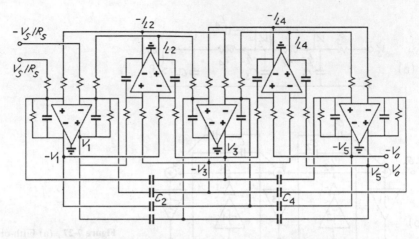

Figure 7-28 Fully differential implementation of the filter in Fig. 7-27.

Note that all loop gains are negative, that inverters are realized by crossing wires, and that, as expected, the first and last integrators are lossy to realize the resistive loads.   After denormalization to obtain the correct frequency range and impedance level, the element values are obtained by following the method discussed in Chapter 6.   Finally, we replace all resistors by properly dimensioned MOSFETs biased in the ohmic region and apply the frequency-control voltage to all gates.   The *f*-control circuit is the same as that for the previous example (Fig. 7-26).   In Problem 7.15, the student is asked to complete the design, following the steps given in Section 6.4 and in this example.

A MOSFET-*C* filter of quite similar specifications, including a frequency-control circuit, has been implemented with excellent performance.   For a detailed discussion of the experimental results, the reader is urged to consult the literature [147, 154].

We have pointed out a number of times—for instance, in Example 7-2—that *saving op amp inverters* by implementing their function by crossed wires, although advantageous from the point of view of component count, power consumption, and noise, nevertheless carries a penalty if these inverters are used for active compensation or phase correction.   Recall the discussion following the general conversion rule in Section 7.2!   Therefore, although the simple design procedure adopted in Example 7-3 has been demonstrated to be quite adequate for low-frequency designs [154], the reader should be cautious before extending the method to higher frequencies, where the two lagging op amp phase shifts become more significant and result in serious *Q* enhancement.   Refer to the discussion in Section 5.3.3 leading from a Tow-Thomas to an Åckerberg-Mossberg biquad!   Note that in the SFG development in Chapter 6 we carefully paired a phase-lead with a phase-lag integrator in each loop of the ladder simulation, but that the realization in Fig. 7-28 contains only phase-lag elements.   This factor, in addition to other parasitic phase shifts, was found to cause the expected passband peaking [96], which had to be corrected with *ad hoc* predistortion via careful modeling.   Such simple predistortion techniques cannot be expected to lead to satisfactory performance as operating frequencies increase; rather, the designer should resort to proven active compensation techniques and use noninverting phase-lead integrator designs such as the one in Figs. 7-2d and 7-25b and c. In any event, although for low-frequency applications the MOSFET-*C* technique is fast and intuitively simple and has been shown to have excellent performance, for higher operating frequencies we believe that op amp–based circuits will continue to suffer from all the problems that have been discussed for discrete realizations in Chapters 4 and 5. For such applications, *transconductance-C* methods appear to become the preferred design approach because transconductors consist of simpler circuits with much better high-frequency performance. We shall devote the next section to a discussion of these procedures.

7.5 MONOLITHIC FILTERS BASED ON TRANSCONDUCTANCES

We have illustrated several times throughout this text that the design of transconductance-*C* filters generally is not only easier and more systematic than that based on op amps, but also that high-performance transconductors can be obtained from simpler circuitry. For instance, a four-transistor single-ended device was shown in Fig. 7-18a; similarly, Fig. 7-29 contains a fairly simple differential CMOS transconductor [170]. A number of additional high-quality voltage-to-current transducers are available in the literature [152, 153, 157, 164, 167, 171]; generally, the first stage of an operational amplifier may be considered to be a voltage-to-current converter, i.e., a transconductance element. As stated earlier, in this text we shall assume that a tunable transconductor of suitable performance is available; for the actual design of the electronic circuitry with low distortion, high power supply rejection, low noise, and good frequency response, the reader is referred to the literature [152, 153, 157]. The following discussion is, therefore, concerned only with the design of g_m-*C* filters, given appropriate g_m elements, such as those in Figs. 7-18a or 7-29.

The development of active filters with transconductors must, fundamentally, follow the same concepts as op amp–based designs. Frequency parameters are determined by *RC* products or, in this case, by C/g_m ratios, and all other filter parameters—gains or quality factors—are set by ratios of capacitors, resistors, or

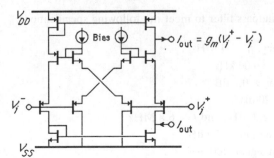

Figure 7-29 Differential CMOS transconductance element.

transconductors. The necessary gain required for *complex* poles is obtained from the g_m elements. Now note that for these filters, there is no need for resistors, because transconductors can be used to simulate the function of resistors (see Figs. 4-13 and 4-34). Conveniently, we may thereby not only save one type of component in our active g_m-RC filters by "merging" the functions of g_m and R, but we obtain "resistors" that are readily tunable by varying the transconductors' bias voltages or currents. Thus, it is quite easy to design g_m-C filters where all relevant parameters are electronically tunable.

Having established this basic fact, we can then develop the elementary building blocks required for analog filters and divide the design approaches of g_m-C filters into the same categories that we found useful for op amp–based active RC circuits: *cascade, multiple-loop feedback*, and the different ways of *ladder simulation*. We shall in the following give design examples for a cascade circuit and a ladder simulation based on *element substitution using gyrators*, and finally, we shall present a *signal-flow graph* ladder simulation method that uses all *identical* transconductors (except possibly one) and only grounded capacitors. We pointed out before that this latter approach proves very convenient for a systematic design in even the simplest IC technology. As in our earlier developments, we shall derive all circuits with single-ended transconductors and refer the student to Section 7.2 for the conversion procedures to differential balanced form.

7.5.1 Cascade Design

We discussed the cascade design of active RC filters in detail in Section 6.2; its use in integrated filters, in spite of the known superiority of, for instance, simulations of LC ladders, is motivated mainly by their easy and transparent tuning methods which permit observed transfer function deviations to be identified with ω or Q errors. Recall that we emphasized in Section 7.3 the importance of being able first to detect and then to correct performance errors with relatively simple control circuitry. Since, apart from the continuous tuning problem, nothing in the development is fundamentally new or different from a discrete approach, we shall illustrate the design directly by an example. The reader is encouraged to review Example 7-1 before studying the following Example 7-4.

Example 7-4

Design an all-pole bandpass filter to meet the following specifications:

> Center frequency: f_0 = 4 MHz
> Bandwidth: Δf = 800 kHz
> Passband ripple: $\leq$ 0.5 dB
> Passband gain: 20 dB
> Stopbands: $f < 2.3$ MHz and $f > 6.8$ MHz
> Stopband attenuation: $\geq$ 70 dB
> Maximum input signal: 100 mV

The circuit is to be implemented in integrated form as a transconductance-C cascade filter in CMOS technology. Considering the specified high-frequency operation, both frequency parameters and quality factors should be tuned automatically. Assume the same MOSFET parameters as in Example 7.1.

Solution [157]. Using the approximation method in Section 1.6 or the appropriate software [11, 29], it is found that an eighth-order Chebyshev bandpass function will meet the specifications. The pole frequency (in MHz) and quality factor of the ith second-order section are

i	1	2	3	4
f_i	3.5807	3.8090	4.1586	4.4237
Q_i	23.858	9.8368	9.8368	23.858

Now note that biquad 2 was designed successfully in Example 7-1, so we shall use it as a basis for the complete filter. A suitable circuit was given in Fig. 7-18b, using the single-ended transconductors of Fig. 7-18a. The second-order transfer function realized is that of Eq. (7-52). We shall continue with a single-ended design; but just for reference, Fig. 7-30 shows a fully differential version of the same circuit, obtained from Fig. 7-18b by the steps given in Section 7.2. The numbers in the transconductors refer to those in Fig. 7-18b.

As in Example 7-1, we shall set $g_{mi} = g_m$, $i = 1, \ldots, 6$, with $g_m = 100.54 \, \mu S$, in order to build the total system with as many identical subcircuits as possible. This choice simplifies design, IC layout, and tuning. Also, *individually* for each section, we set $C_1 = C_2 = C$ to obtain for

Section 1: $C = \dfrac{g_m}{\omega_{01}} = 4.467 \text{ pF}$ $g_{mQ1} = g_m\left(\dfrac{1}{Q_1} + 1\right) = 104.75 \, \mu S$

Section 2: $C = 4.20 \text{ pF}$ $g_{mQ2} = 110.76 \, \mu S$ (Example 7-1)

Section 3: $C = 3.848 \text{ pF}$ $g_{mQ3} = 110.76 \, \mu S$

Section 4: $C = 3.617 \text{ pF}$ $g_{mQ4} = 104.75 \, \mu S$

Let us connect the sections in the order 3, 1, 2, 4 (see reference 156 and Section 6.4.1.2); for this ordering, it is relatively easy to calculate the gain constants such that the maximum signal at each section output equals 1 V (specified gain is 20 dB). We find from Eqs. (7-52) and 7-53c)

$$H_{B1} = \frac{g_m}{g_{m71}} = 0.133 \qquad g_{m71} = 755.9 \, \mu S$$

$$H_{B2} = \frac{g_m}{g_{m72}} = 0.162 \qquad g_{m72} = 620.6 \, \mu S$$

$$H_{B3} = \frac{g_m}{g_{m73}} = 1 \qquad g_{m73} = 100.5 \, \mu S$$

$$H_{B4} = \frac{g_m}{g_{m74}} = 0.414 \qquad g_{m74} = 242.9 \, \mu S$$

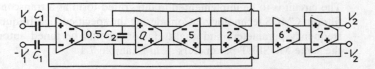

Figure 7-30   Differential balanced version of the circuit in Fig. 7-18b.

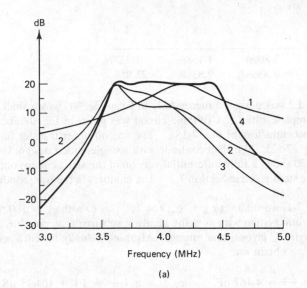

(a)

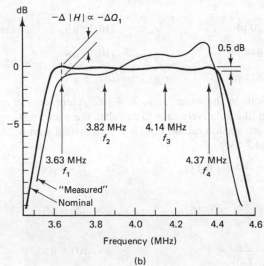

(b)

Figure 7-31 (a) Simulated performance at the biquad outputs after 1, 2, 3, and 4 sections, respectively. (b) Sketch of the nominal and a "measured" transfer function passband with "support" frequencies at the reflection zeros.

With all transconductances computed and the parameters $V_{Tn,p}$ and $0.5\mu_{\text{eff},n,p}C_{\text{ox}}$ known from the process (Example 7-1), the gate aspect ratios and nominal bias voltages can be determined as in Example 7-1. The simulated performance (SPICE [172, 173]) is shown in Fig. 7-31a; it indicates that the maximum signal levels are indeed equalized.

The last question we have to address is that of tuning the frequency parameters and the quality factors. Again, the principle of the operation was discussed in considerable detail in Example 7-1 in connection with the block diagram in Fig. 7-19. However, in the present case we have *four* Q factors, two of them fairly high (≈ 23.9), and operation at video frequencies where parasitic phase shifts can be assumed to be significant. With very careful layout and modeling, the designer may attempt to rely on tracking, and tune all four Q factors from the same control voltage (V_{QC} in Fig. 7-19). Taking a more conservative approach, let us indicate how the four quality factors can be controlled *separately* with the system shown in Fig. 7-32, which is a direct extension of the control circuitry in Fig. 7-19. An explanation of the operation follows.

The frequency-control system is identical to the one in Fig. 7-19; it is an EXOR-based phase-locked loop that locks the frequency of the f-VCO (a copy of one of the slave sections as discussed in Example 7-1) to the reference frequency. The cleaned control voltage $\bar{y}$ is applied to the appropriate transconductance gates in slave, master, and the QVCOs. As was discussed in Example 7-1, the four master and QVCO sections are copies of the corresponding slave circuitry, of course, with the QVCOs' inputs grounded and their quality factors increased to the point of oscillation.[28] Because all frequency parameters on the chip differ only by capacitor ratios, we may, with careful modeling and design, then assume that by tuning the transconductance values, *all* frequencies in the whole system are tuned correctly. (Recall that accurate frequencies are essential for the Q control to perform as required.) For tuning the quality factors of *each section individually*, the Q-control circuitry works exactly as described in Example 7-1; a difficulty arises because the *four* errors of the *four* quality factors cannot be separated by looking at the response of the eighth-order master filter to *one* stimulus. By the same token, we cannot apply signals at four frequencies *simultaneously* to the filter and hope to be able to interpret the response for convenient error correction.

A possible solution is found in applying the required Q-control signals *consecutively*. To this end we construct a digital clock generator (CG in Fig. 7-32) with four nonoverlapping clock phases and a bank of four sample-and-hold capacitors (S/H in Fig. 7-32). During phase 1, the switches connecting QVCO1 and the first of the S/H capacitors, $C_{S/H1}$, are closed, thereby completing the Q-control loop. The

---

[28]Since the oscillation frequencies are set as g_m/C, instead of building *four* complete Q-control oscillators one may contemplate to build only *one* but with capacitors that are switched for the four different required frequencies. However, the designer should note that the oscillation frequencies must be very accurate for the Q-control operation to function with the necessary precision! Connecting different values of C through switches tends to add so many parasitic capacitors to the VCOs that the oscillator frequencies are likely to be incorrect, which in turn results in large Q-tuning errors. Constructing the QVCO essentially as a duplicate of the master-slave circuitry with minimally changed capacitors to account for the slightly different nominal ω_0 values copies also all relevant parasitics which affect the pole frequencies of master and slave.

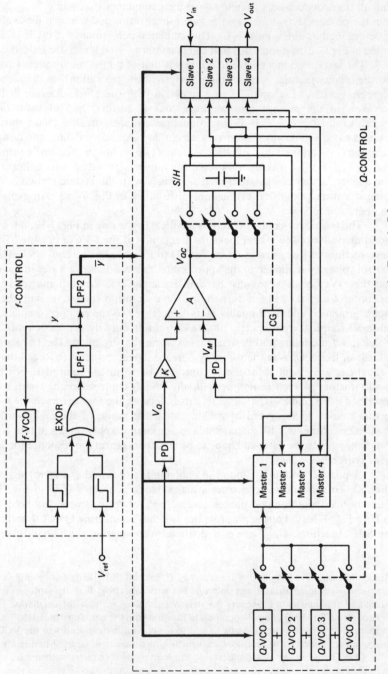

Figure 7-32 Four-point tuning system for Example 7-3.

remaining switches stay open. QVCO1 is set to 3.63 MHz (see Fig. 7-31b), the frequency (the first reflection zero) where the passband response has a maximum, nominally 1 V, a gain of 20 dB. With K set equal to 10 (20 dB!), we compare the master's response at 3.63 MHz to KV_{QVCO1} and generate a Q-control voltage V_{QC} that is sent to $C_{\mathrm{S/H1}}$ and held there until the next update. If the response level at 3.63 MHz is incorrect as we have indicated in Fig. 7-31b by sketching a nominal and a typical "measured" response curve, the error $\Delta|\mathrm{H}|$ can be attributed *dominantly* to an error in the quality factor ΔQ_1 of section 1 and corrected by applying V_{QC} from $C_{\mathrm{S/H1}}$ to the appropriate control gates in sections 1 of the master and the matched slave. Note that this process is identical to the one described in Example 7-1. During the next clock phase, QVCO$_2$, with $f_{\mathrm{QVCO2}} = 3.82$ MHz, and $C_{\mathrm{S/H2}}$ are connected into the control loop; the remaining switches stay open. Again, the master's response at the second reflection zero frequency, 3.82 MHz, is compared to KV_{QVCO2} and the corresponding Q-control voltage V_{QC} is generated and held on $C_{\mathrm{S/H2}}$ until the next update. V_{QC} in this case reflects the error in the master's output level at 3.82 MHz, which arises *dominantly* from a Q error of section 2. Applying V_{QC} to the control gates of the Q-determining transconductors in sections 2 of master and slave will tune the corresponding quality factors. In this way we proceed to the remaining reflection zero frequencies, at 4.14 MHz and at 4.37 MHz, so that all four Q factors are tuned. Note that the *amplitude* of the VCOs is unimportant because it is "compared to itself" [see Eq. (7-44)].

We emphasized in our description that the output level error at one of the test frequencies is only *dominantly*, but not entirely, determined by just *one* of the Q factors. One can show, however, that the method converges after a few (four or five) iterations [174]. Thus, after one cycle, the process is repeated with a chosen clock frequency of a few kilohertz, well outside the desired passband but fast enough to reach the final Q values in milliseconds and, especially, fast enough to prevent drooping of the control voltages stored on $C_{\mathrm{S/H}i}$, $i = 1, \ldots, 4$.[29]

Since this method tunes the four pole frequencies and, implicitly, the four section gain constants and the four section Q factors, we have now tuned *all* available parameters of the cascade realization. It follows, therefore, that such a conservative control system theoretically results in a filter that is *precisely* and *continuously* tuned against fabrication tolerances and against deviations caused by drifts in operating conditions (e.g., temperature) and aging, *as long as* the requirement, as always, of good matching and tracking of the relevant performance parameters in master, slave, and VCOs is met. The main error sources in this tuning system are inaccurate QVCO frequencies [175], which result in Q and gain errors in the biquads.

The filter along with the control circuitry as presented in this example and Example 7-1 was implemented in integrated form in a 3-μm CMOS process. The experimental performance was found to be within the prescribed specifications (see Fig. 7-33); the effects of fabrication tolerances and temperature drifts within the range $0°\mathrm{C} \le T \le 65°\mathrm{C}$ were eliminated by the control circuitry. For further details of design and performance, the reader is referred to the references [156, 174].

---

[29]Note that $C_{\mathrm{S/H}}$ of the order of picofarads and leakage currents through switch diodes of the order of picoamperes results in time constants measured in only seconds.

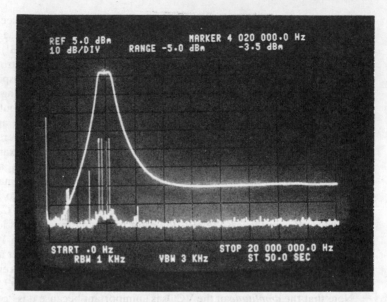

Figure 7-33 Experimental performance of the circuit of Example 7-4. The five discrete spectral lines are the reference signal V_{ref} and the four support frequencies identified in Fig. 7-31b.

7.5.2 Ladder Simulation Using Gyrators

We discussed in Chapters 3 and 6 that *LC* ladders result generally in filters with better performance because of their low passband sensitivities to element tolerances. It stands to reason, therefore, that this type of design should be advantageous for integrated implementations, where, as we have seen, component tolerances of various origins are of considerable concern. For instance, the effects of parasitics can be expected to be much less severe, particularly in those cases where parasitics are *absorbed* and manifest themselves in the form of component tolerances (see Section 7.3). Several monolithic active *LC* ladder simulations of continuous-time filters have been reported in the literature [145, 147, 154, 157–162, 167] with very good or at least promising experimental performance; Example 7-3 is a case in point. At the time of this writing, the main difficulty with ladder simulations is the lack of well-understood, reliable Q-tuning algorithms. The frequency parameters are as always and quite easily stabilized by "slaving" them to a master *RC* time constant. But in contrast to cascade circuits (Example 7-4), the problem arises in identifying which *tunable components* should be varied in order to correct an observed error in the transmission characteristic. Although some progress is being made in this area [145], the solutions are not general, but very much *ad hoc* and not as transparent as in cascade realizations. Nevertheless, the student should be aware of the design possibilities, which follow in principle those of discrete active *RC* filters discussed in Section 6.4.   The following example, again taken from the current literature, will illustrate one of the methods.

Example 7-5

Design an integrated lowpass filter in CMOS technology to meet the following specifications:

Simulated *LC* ladder, resistive terminations
Passband: $0 \leq f \leq 4.5$ MHz
Passband ripple: $\alpha_p \leq 0.5$ dB
Stopband: $f \geq 7.7$ MHz
Stopband attenuation: $\alpha_s \geq 23$ dB

Solution [157].   The methods of Section 1.6, along with Table III-36 in Appendix III, indicate that a third-order elliptic filter function will satisfy the requirements.   The transfer function is found to be

$$H(s) = \frac{0.28163(s^2 + 3.2236)}{(s + 0.7732)(s^2 + 0.4916s + 1.1742)}$$

$H(s)$ can be realized as a resistively terminated *LC* ladder by the methods discussed in Chapter 2; the circuit with normalized elements is shown in Fig. 7-34.   Leaving the impedance level open for now but rescaling the components for the required frequency, $\omega_p = 2\pi \cdot 4.5 \cdot 10^6$ rad/s, results in

$$R_S = R_L = 1 \qquad \hat{C}_1 = \hat{C}_3 = \frac{1.293}{2\pi \cdot 4.5} \ \mu\text{s} = 45.73 \ \text{ns}$$

$$\hat{C}_2 = 13.104 \ \text{ns} \qquad \hat{L} = 29.613 \ \text{ns}$$

Let us choose a ladder implementation which simulates the components (rather than the equations); i.e., we must use two gyrators to obtain a simulated floating inductor as was shown in Fig. 4-26b.   The high operating frequency indicates that the active elements must be transconductors; thus, a gyrator is built from an inverting and a noninverting transconductor (Fig. 4-27).   Using the symbol introduced in Fig. 7-3b, the complete ladder simulation is given in Fig. 7-35a.   The student should have no difficulties in identifying the two gyrators, the "floating inductor," the two "load resistors," and the input voltage-to-current converter.   The fully differential version of the circuit, derived by the method given in Section 7.2, is shown in Fig. 7-35b.

Let us proceed with the differential circuit.   If we choose $g_m = 200$ μS and for convenience make g_m^{-1} the normalizing resistor, the required elements become

$$\frac{1}{R_S} = \frac{1}{R_L} = 200 \ \mu\text{S}$$

$$C_1 = C_3 = \hat{C}_1 g_m = 9.146 \ \text{pF} \qquad C_2 = \hat{C}_2 g_m = 2.621 \ \text{pF} \qquad C_L = \hat{L} g_m = 5.923 \ \text{pF}$$

Considering that these capacitors are very small and that three of them will further be divided by 2 (see Fig. 7-35b), the designer is well-advised to take into account the effect of the unavoidable parasitic input and output capacitors of the transconductance elements.   Observe from the circuit diagram in Fig. 7-35b that $2(C_i + C_o)$ appears

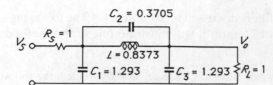

Figure 7-34 Elliptic lowpass ladder for Example 7-5.

in parallel with $0.5C_L$ and $0.5C_3$, and that $2C_i + 3C_o$ shunts $0.5C_1$. Note that all parasitic capacitors can be taken care of by *absorption* without increasing the degree of the transfer function! The values of these parasitics depend, of course, on the details of the transconductance design, which is beyond the scope of this text. A suitable differential input–differential output device is given in reference 157, where C_i and C_o values of 0.42 pF and 0.22 pF, respectively, are claimed. If we use these values, the final capacitors for the design are computed as follows:

$$0.5C_1 = 3.073 \text{ pF} \qquad C_2 = 2.621 \text{ pF}$$

$$0.5C_3 = 3.292 \text{ pF} \qquad 0.5C_L = 1.682 \text{ pF}$$

Finally note that an LC ladder has an intrinsic gain factor of $\frac{1}{2}$, i.e., a 6-dB loss, built into the transfer function. If desired, we can eliminate this loss in the active realization by choosing the input transconductance value equal to $2g_m = 400 \mu S$.

Our remaining task is the tuning circuit. As master we shall use an oscillator that models as closely as possible the frequency parameters of the filter. A convenient circuit is an LC oscillator where L is, of course, realized via the *same* gyrator (with the same parasitics) that is used in the filter. Figure 7-36 shows the circuit principle. It uses a differential phase detector in the phase-locked loop to detect the difference between the oscillation frequency and the reference frequency. The phase detector output voltage, after cleaning from high-frequency components (LPF), is used to adjust the g_m bias in master and slave. Choosing as reference $f_R = 8.5$ MHz, a frequency in the stopband (in the vicinity of the transmission zero) but close to the critical

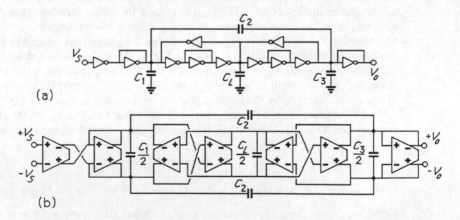

Figure 7-35   (a) Single-ended transconductance-C simulation of Fig. 7-34; (b) differential version of the circuit in part a. All transconductors have the value g_m, and $C_L = Lg_m^2$. Note definitions in Fig. 7-4e.

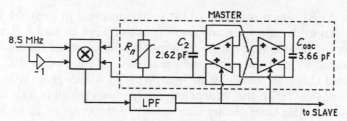

Figure 7-36 Frequency-control loop for the filter of Fig. 7-35b.

passband corner, lets us compute the required capacitor values. Let C_2 be the same value as in the filter (2.621 pF); then the oscillation frequency equals (see Fig. 7-36)

$$f_{osc} = \frac{1}{2\pi}\sqrt{\frac{g_m^2}{(C_2 + C_i + C_o)(C_{osc} + C_i + C_o)}}$$

which may be solved for $C_{osc} = 3.660$ pF. R_n is a nonlinear resistor [157] used to control the oscillation amplitude. It should be apparent, that good modeling of the circuitry is necessary in order to insure that the master really models the slave's behavior. For example, the 2.62-pF capacitor C_2 will have to be further reduced if the input capacitances of R_n and of the phase detector are not negligible! The circuit along with the control loop was implemented in a 3-μm CMOS process and performed essentially according to specifications, within the temperature range $-10°C \leq T \leq 60°C$ (see Fig. 7-37).

The student may wonder at this point why no Q-control loop was used in spite of our earlier warnings about high frequencies and medium- to high-Q designs. Note that although the filter operates at video frequencies, it has very low quality factors: the highest pole Q in the given third-order elliptic filter is only about 2.2. Nevertheless, looking carefully at the experimental performance of the chip, we still observe toward the bandedge of the transfer characteristic a noticeable droop

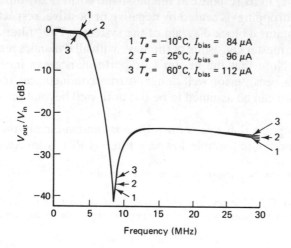

1 $T_a = -10°C$, $I_{bias} = 84\ \mu A$
2 $T_a = 25°C$, $I_{bias} = 96\ \mu A$
3 $T_a = 60°C$, $I_{bias} = 112\ \mu A$

Figure 7-37 Experimental performance of the integrated filter of Fig. 7-35b [157]. (© IEEE, June 1988.)

(by $\gtrsim 1$ dB), which is typical of the behavior found in lossy LC filters (see Fig. 3-11). An explanation for this result is not difficult to find. Remember from our earlier discussions that the finite output conductances of the transconductors (Fig. 7-3) result in lossy inductors and capacitors. Both $C_L/2$ and $C_3/2$ are shunted by $2g_o$, and $C_1/2$ is in parallel with $3g_o$. With $g_o = 2\mu S$ given from the design of the active devices [157], the student can easily show that all reactances, apart from C_2, at $f = 4.5$ MHz have quality factors only of the order of 20 to 30! The effects of these low Q factors are at least partially offset by other Q-enhancing parasitic phase shifts, as determined by careful modeling on programs such as SPICE, and are further compensated by the designers' ad hoc changes of component values. Details can be found in reference 157.

The student will appreciate that such ad hoc system optimization is very sensitive to process and layout, and is not reliable enough for commercial designs. A possible automatic tuning solution to the problem was outlined in connection with Fig. 7-17; it proceeds as follows [145]—however, at a cost of additional circuitry. Having identified the circuit's (positive or negative) "losses" as the cause of the (drooping or peaking) deviations,[30] we need a method that permits us to cancel the losses of either sign. It can be achieved by inserting in parallel with each capacitor (apart from C_2) large *tunable* compensation resistors $R_c = 1/G_c$ that can take on *positive or negative values* such that, nominally,

$$G_c + 2g_o - G_\phi = 0 \quad \text{or} \quad G_c + 3g_o - G_\phi = 0$$

depending on the case.[31] $-G_\phi$ is used to represent the effect of Q-enhancing phase shifts. Figure 7-38a is a suitable simple circuit which realizes the differential conductance

$$G_c = g_m - g_m(V_{QC})$$

Evidently, G_c can be positive, negative, or zero, depending on the applied value of V_{QC}. A Q-control loop is then constructed which compares the filter's "performance at dc" ($V_o/V_i = 1$) to its response at the passband corner; any difference in signal level (peaking or drooping) is caused by negative or positive, respectively, "losses." Figure 7-39 contains a block diagram of the system. The "filter at dc" is obtained by building a dc master consisting of the slave with all dynamics removed as shown in Fig. 7-38b. Note that the master's dc performance can indeed be measured at $f = 4.5$ MHz because for well-designed transconductances the gain of the dc master (Fig. 7-38b) can be assumed to be flat until well beyond the filter's cutoff frequency.

The student will recognize that this control loop is fundamentally the same as that in Fig. 7-19; an additional simple lowpass filter (LPF) is inserted as a

---

[30]Q-enhancing phase shifts can be interpreted via *negative* resistors, $-R_\phi = -1/G_\phi$, i.e., as negative losses.

[31]Returning to Example 7-1, Fig. 7-18b, and Eqs. (7-52) and (7-54), we note that it is the function of the transconductance g_{m5} to permit Q tuning over positive or negative deviations, $\pm\Delta Q$, i.e., Q may be increased or decreased.

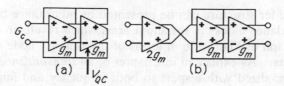

Figure 7-38 (a) Negative "resistor";
(b) dc master.

precaution before the Q-control signal V_{QC} enters the slave because the VCO frequency is in the passband. Observe that the system forces the level of the transfer characteristic at the passband corner to be equal to unity *regardless* of whether peaking or drooping occurs and independent of the error sources. As long as matching across the chip can be relied upon, the effects of positive or negative losses and of circuit, process, and layout parasitics on the filter will be eliminated automatically.

7.5.3 Ladder Simulation by Signal-Flow Graph Methods

As the last implementation method of transconductance-C integrated filters, let us discuss the signal-flow graph (SFG) simulation of general parameter ladder filters. From our work in Chapter 6 we can expect that this procedure is very powerful; furthermore, it is easy and methodical to apply. We recall that the SFG method needs only inverting and noninverting integrators in addition to summers, all of which we have seen to be realizable readily with transconductances and capacitors. The procedure to be derived in the following requires only *grounded* capacitors, and in many cases all transconductors can be *identical*. Consequently, the IC designer may concentrate his or her efforts primarily on the best possible transconductance element. Once such an optimized device is available, it is placed in an IC design library and called upon as a fundamental building block which can be wired together with grounded capacitors to form the filter with prescribed specifications. In this respect the procedure is comparable to the one based on op amps in Section 7.4, but it is usable at much higher frequencies: filters operating in the low megahertz range have been demonstrated (see Examples 7-4 and 7-5), and higher frequency (up to 100 MHz) bipolar transconductance-C designs are under investigation.

The design method [145] is based on the simulation of ladder filters by their "highest-level" signal-flow graphs as illustrated in connection with Fig. 6-11. As

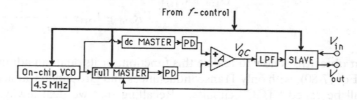

Figure 7-39 Q-control scheme for transconductance-C ladder filter.

shown in Fig. 6-12a, a ladder structure can be presented by admittance blocks in the series arms and impedance blocks in the shunt arms which results in the corresponding signal-flow graph of Fig. 6-12b in terms of the primary node voltages and primary mesh currents. As explained in Chapter 6, all *transmittances* z_i and y_j are assumed to be normalized with respect to both frequency and impedance level. Therefore the parameters and variables in Fig. 6-12b are dimensionless, z_i and y_j are voltage transfer functions, and i_k is to be interpreted as a voltage signal, regardless of notation. Nevertheless, in the following we shall refer to the intermediate functions as *admittances* or *transfer functions* in order to be able to keep track of the repeated required conversions.

In a practical ladder filter, the circuit branches shown in Fig. 6-12a typically consist of series and parallel combinations of inductors, capacitors, and possibly resistors. Therefore, in general, the branch transmittance functions z_i or y_j of the *RLC* oneports under consideration are mathematically of the form of a continuous fraction,

$$F(s) = G_{01}(s) + \cfrac{1}{G_{02}(s)}$$

$$+ \cfrac{1}{G_{11}(s) + \cfrac{1}{G_{12}(s) + \cfrac{1}{G_{21}(s) + \cfrac{1}{G_{22}(s) + \cfrac{1}{G_{31}(s) + \dots}}}}} \tag{7-79a}$$

where

$$G_{ij}(s) = k_{ij0} + k_{ij1}s \tag{7-79b}$$

and where k_{ij0} and k_{ij1} are real constants that correspond to resistor-conductor and inductor-capacitor, respectively, depending on the situation in the original ladder. Note that Eq (7-79) is general and some coefficients may be zero or infinite depending on the particular case. For instance, the impedance function of the oneport in Fig. 7-40 is

$$Z(s) = R_{010} + L_{011}s + \cfrac{1}{G_{020} + C_{021}s}$$

$$+ \cfrac{1}{G_{110} + C_{111}s + \cfrac{1}{R_{210} + L_{211}s + \cfrac{1}{G_{220} + C_{221}s}}} \tag{7-80}$$

Our objective is now to realize this function, or its dual, an admittance of the form of Eq. (7-80), with only Transconductances and Grounded Capacitors, i.e., by what shall be called "TGC" circuits. Recalling that we deal always with *normalized* quantities, the total design is then reduced to the synthesis of TGC subcircuits that

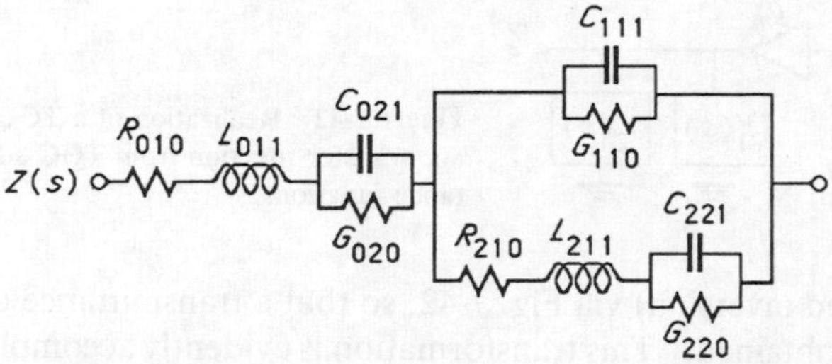

Figure 7-40 Typical branch of an *RLC* ladder.

simulate the normalized driving-point functions of *RLC* circuits (the *transmittances*) by voltage transfer functions.

We shall derive two methods for accomplishing this simulation; they are analogous to the procedure for op amp–based SFG simulations in Chapter 6, which the student may want to review before proceeding further. We shall continue with our agreement to develop single-ended circuits, using the transconductance symbol of Fig. 7-3b, and rely on the rules of Section 7.2 if differential, balanced designs are desired for a final implementation.

Method 1. A glance at Eq. (7-79) shows that the realization of $F(s)$ requires fundamental immittances, either constant or proportional to s, on which the repeated operations of *addition* and *inversion* have to be performed. Therefore, apart from transconductance elements, the synthesis requires grounded capacitors and constants as shown in Fig. 7-41. Note that the constant is a grounded resistor of value $1/K$; it is, of course, implemented with the aid of a transconductor of value K so that, as pointed out earlier, really only two fundamental elements are needed. *Addition* of two or more admittances and *inversion* can be obtained quite simply by the circuit in Fig. 7-42, which implements the *voltage transfer function*

$$F_1(s) = \frac{V_o(s)}{V_i(s)} = -\frac{g_m}{Y_1(s) + Y_2(s)} \tag{7-81}$$

A simple example of the application of Fig. 7-42 is illustrated in Fig. 7-43a, which shows a fundamental inverting lossy integrator together with its special cases of either G or C equal to zero in Fig. 7-43b and c. Note that Fig. 7-43c for $K = g_m$ is a unity-gain inverter. Figure 7-43d illustrates how the summing operation required in the signal-flow graph can be obtained with only transconductors.

To proceed further with the general synthesis, we recognize that a voltage transfer function, e.g., Eq. (7-81), must be converted back to an admittance for

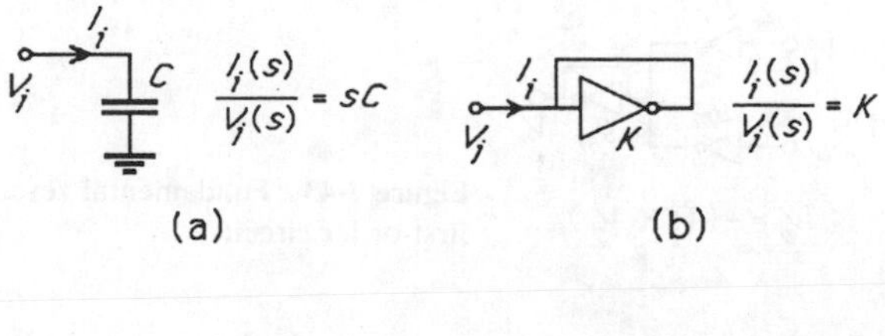

Figure 7-41 Fundamental circuit structures for method 1.

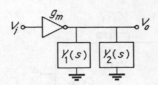

Figure 7-42 Realization of a TGC voltage transfer function from TGC admittance functions.

further processing (renewed inversion) via Fig. 7-42, so that a transmittance of the form of Eq. (7-79a) can be obtained. This transformation is evidently accomplished by the circuit of Fig. 7-44, which converts the output voltage of the block $F_1(s) = V_{out}/V_i$ into a current and thereby realizes

$$Y(s) = \frac{I_i}{V_i} = g_m F_1(s) \tag{7-82}$$

For Eq. (7-82) to hold, it is, of course, assumed that in Fig. 7-44 the input current of the block realizing the voltage transfer function $F_1(s)$ is zero; an exception is the trivial case of a short-circuit connection to implement a unity transfer function (Fig. 7-41b). Note also that a *negative* admittance along with potential instabilities is obtained if $F_1(s)$ is *inverting*, i.e., if $F_1(s) = -V_{out}/V_i$.

These two simple steps, *alternating between admittances and voltage transfer functions*, complete the synthesis process. Thus, the design starts at the "lowest" or "innermost" level of the immittance [e.g., at $G_{22}(s)$ in Eq. (7-80)] with the input admittance functions sC and/or K (constant), built as shown in Fig. 7-41. It continues by taking the reciprocal (Fig. 7-42), converting back to an admittance (Fig. 7-44), and using repeated additions and inversions as needed until the ladder arm, that is, the signal-flow graph transmittance, is realized. As an illustrative example, consider the ladder shunt arm shown in Fig. 7-45; the desired transmittance function is

$$\frac{V_2}{V_{I1} - V_{I2}} = \frac{g_m}{sC_4 + \dfrac{1}{sL_3 + 1/(sC_3)}} = \frac{g_m}{sC_4 + \dfrac{g_m^2}{sg_m^2 L_3 + g_m^2/(sC_3)}} \tag{7-83}$$

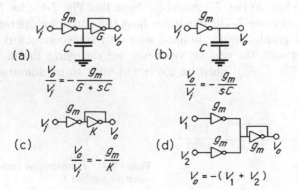

(a) $\dfrac{V_o}{V_i} = -\dfrac{g_m}{G + sC}$

(b) $\dfrac{V_o}{V_i} = -\dfrac{g_m}{sC}$

(c) $\dfrac{V_o}{V_i} = -\dfrac{g_m}{K}$

(d) $V_o = -(V_1 + V_2)$

Figure 7-43 Fundamental zeroth- and first-order circuits.

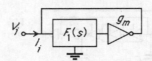

Figure 7-44 Realization of a TGC admittance function from a TGC voltage transfer function.

(Recall that I_k is represented by a voltage, V_{Ik}; i.e., $V_{Ik} = g_m^{-1}I_k$.) Starting from the "innermost" element, C_3, the reciprocal is taken as in Fig. 7-42 (or specifically in Fig. 7-43b) to give $-g_m/(sC_3)$; the minus sign is removed through multiplication by -1 (Fig. 7-43c), the voltage transfer function is transformed back into an admittance [of value $g_m^2/(sC_3)$] via Fig. 7-44, and a capacitor of value $C_{L3} = g_m^2 L_3$ is added in parallel; this combination is inverted, multiplied by -1 as before, and converted into an admittance so that a capacitor of value C_4 can be added in parallel. Finally, application of Fig. 7-43d converts this combination into the desired transmittance which can be driven by $V_{Ii} - V_{I2}$ as required. This process formally leads to the expression on the right-hand side of Eq. (7-83). Figure 7-46 shows the resulting circuit, with g_m normalized to unity. Observe that all transconductors are identical!

To help the student further to appreciate the ease with which these structures are synthesized, consider the following example.

Example 7-6

In Chapter 5, Section 5.3.2, we derived an op amp realization of a second-order bandpass filter from the RLC prototype in Fig. 5-22, repeated here as Fig. 7-47. Use the technique discussed in the present section to derive a fully differential g_m-C implementation to realize $f_o = 1MHz$, $Q = 6$, and midband gain $H_B = 1$. For ease of processing, both capacitors should be grounded; i.e., Fig. 7-18b is not a suitable solution. For this example, neglect all parasitic components (see Problem 7.5 for the effects of parasitics).

Solution. The passive prototype realizes the function

$$\frac{V_o}{V_i} = \frac{sG/C}{s^2 + sG/C + 1/(LC)} \qquad (7\text{-}84a)$$

Following our design procedure, we realize the inductor via Fig. 7-43b, cascaded with Fig. 7-43c ($K = g_m$) to remove the minus sign, and convert the resulting transfer function $+g_m/(sC_L)$ into an admittance $+g_m^2/(sC_L)$ with the help of Fig. 7-44. Note that we have just constructed a gyrator to realize the grounded inductor via $L = C_L/g_m^2$. In parallel with this inductor we connect the capacitor C and the grounded resistor $1/g_m$, obtained from Fig. 7-41b. Because this last step amounts implicitly to a source

Figure 7-45 Typical branch of an LC ladder network.

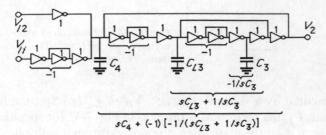

$$-1/sC_3$$

$$\overline{sC_{L3} + 1/sC_3}$$

$$\overline{sC_4 + (-1)\,[-1/(sC_{L3} + 1/sC_3)]}$$

Figure 7-46   TGC subcircuit implementing the transmittance function of Eq. (7-83) by method 1.
Note: To simplify the figure, all transconductors have been normalized to unity, i.e., $g_m = 1$.

transformation, we must drive the circuit from a current source $I = g_m V_i$. The final filter, obtained by method 1, is shown in Fig. 7-4a (with $C_1 = C$, $C_2 = C_L$, and $g_{mi} = g_m$), where it was used to illustrate also the requested conversion to balanced form, Fig. 7-4c. The circuit realizes

$$\frac{V_o}{V_i} = -\frac{g_m s C_L}{s^2 C C_L + s C_L g_m + g_m^2} \tag{7-84b}$$

so that

$$\omega_0 = \frac{g_m}{\sqrt{C C_L}} \qquad Q = \sqrt{\frac{C}{C_L}} \qquad H_B = 1$$

Choosing $C_L = 2$ pF and assuming that the frequency response of g_m is much wider than f_0 results in the nominal values $C = 72$ pF and $g_m = 75.4$ μS.

Method 2. The second realization method is obtained by interpreting Eq. (7-79) from a slightly different point of view. All components $G_{ij}(s)$ of the *scaled* ladder arm are considered to be voltage transfer functions (instead of immittances) from which $F(s)$ *is constructed by repeated summing and inverting.* Like method 1, method 2 starts from the "innermost" fundamental blocks of order 1 or 0 that may be realized as in Fig. 7-43; the operations of summing and inverting are then performed as shown in Figs. 7-48 and 7-49, as the reader may verify by elementary analysis.

Note that Fig. 7-49 is a combination of Figs. 7-44 and 7-42.[32]   The minus signs can always be removed by use of unity-gain inverters (Fig. 7-43c, $K = g_m$) if required.   Observe also that the summing function in Fig. 7-48 is really obtained by *summing the currents* at node n; the output voltage is realized by sending this sum current through the "resistor" $1/g_m$.   The interpretation of *current summing* often allows the designer to save many redundant transconductances.

In this way, starting from the fundamental transfer functions, Eq. (7-79), we proceed sequentially by summing and inverting until the prescribed branch transmittance is realized.

---

[32]Observe also that $F(s)$ must generally be *non*inverting, i.e., $+F(s)$, to assure stability of the feedback loop.   See Probl. 7.19.

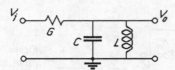

Figure 7-47 Second-order RLC band-pass filter.

A formal example will help to illustrate the process. Realizing again the circuit of Fig. 7-45 described by Eq. (7-83), we rewrite the expression as follows

$$\frac{V_2}{V_{I1} - V_{I2}} = \frac{1}{s\dfrac{C_4}{g_m} + \dfrac{1}{sg_mL_3 + g_m/(sC_3)}} \tag{7-85}$$

and observe that we must realize the *transfer functions*

$$\frac{V_3}{V'} = \frac{g_m}{sC_3} \qquad \frac{V_L}{V'} = sg_mL_3 \qquad \frac{V'}{V_2} = \frac{1}{sg_mL_3 + g_m/(sC_3)} \qquad \frac{V_4}{V_2} = s\frac{C_4}{g_m} \tag{7-86}$$

in order to arrive at the circuit. The different voltages are identified in Fig. 7-50. Considering for the moment the boldface connections *left open*, the reader will have no difficulty verifying that the subcircuits realize

$$\frac{V_3}{V'} = -\frac{g_m}{sC_3} \qquad \frac{V_L}{V'} = -\frac{sC_L}{g_m} \qquad \frac{V_4}{V_2} = +\frac{sC_4}{g_m}$$

Note that the inductor is implemented as $L_3 = C_L/g_m^2$. The two transconductors to the right of V_3 and V_L convert V_3 and V_L into currents; they are summed into

$$I_1 = -g_m(V_3 + V_L) = g_m\left(\frac{g_m}{sC_3} + \frac{sC_L}{g_m}\right)V'$$

which, from the leftmost device connection, equals g_mV_2. Consequently, the upper part of Fig. 7-50 realizes

$$V' = +\frac{1}{g_m/(sC_3) + sC_L/g_m}V_2$$

The separate circuits in Fig. 7-50 are now combined as before by connecting the

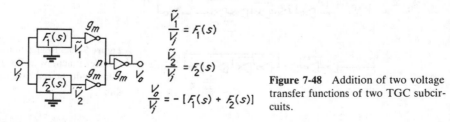

$$\frac{\tilde{V}_1}{V_i} = F_1(s)$$

$$\frac{\tilde{V}_2}{V_i} = F_2(s)$$

$$\frac{V_o}{V_i} = -[F_1(s) + F_2(s)]$$

Figure 7-48 Addition of two voltage transfer functions of two TGC subcircuits.

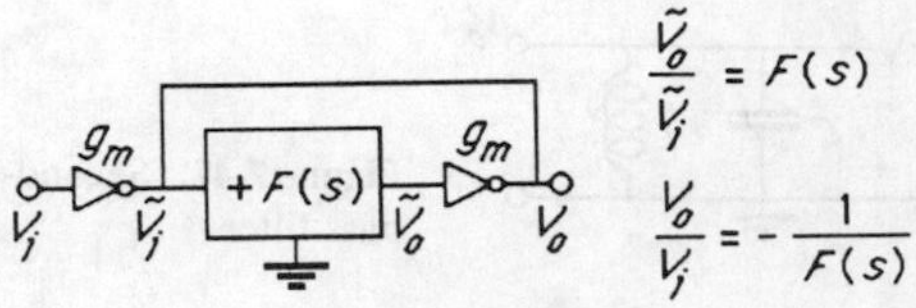

$$\frac{\tilde{V}_o}{\tilde{V}_i} = F(s)$$

$$\frac{V_o}{V_i} = -\frac{1}{F(s)}$$

Figure 7-49 Realization of the reciprocal of a voltage transfer function of a TGC circuit.

two inputs V_2 and summing the output voltages V' and V_4 via a current conversion as

$$I_2 = -g_m(V' + V_4) = -g_m\left(\frac{1}{g_m/(sC_3) + sC_L/g_m} + \frac{sC_4}{g_m}\right)V_2$$

i.e.,

$$\frac{V_2}{I_2} = -\frac{1}{g_m\left(\dfrac{1}{g_m/(sC_3) + sC_L/g_m} + \dfrac{sC_4}{g_m}\right)}$$

The circuit connections are shown in boldface on the right side in Fig. 7-51. As further indicated in the figure, we finish the wiring by forcing I_2 to equal $-I_3$, which, in turn, is made proportional to the input voltage difference:

$$-I_3 = g_m(V_{I2} - V_{I1})$$

Thus we have realized

$$\frac{V_2}{V_{I1} - V_{I2}} = \frac{1}{\dfrac{sC_4}{g_m} + \dfrac{1}{g_m/(sC_3) + sC_L/g_m}} \tag{7-87}$$

as was specified in Eq. (7-85). Note the simple and systematic procedure and observe again that all transconductances are identical. To gain additional practice, the student may want to solve the following problem.

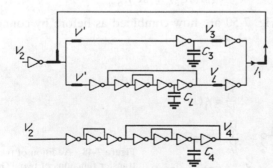

Figure 7-50 Realization of the three transfer functions in Eq. (7-86). All transconductances have the same value g_m.

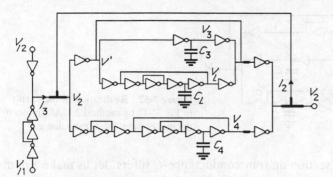

Figure 7-51 g_m-C circuit realizing Eq. (7-87). All transconductances have the same value g_m.

Example 7-7

Realize the LC circuit of Fig. 7-47, Example 7-6, as a transconductance-C filter by method 2.

Solution. The transfer function to be realized was given in Eq. (7-84a); it can be rewritten in a form more convenient for this case as

$$\frac{V_o}{V_i} = \frac{1}{1 + sC/G + R/(sL)} \tag{7-88a}$$

i.e., we must realize the TGC transfer functions

$$1 \qquad \frac{sC}{G} \qquad \frac{R}{sL}$$

which then must be added (Fig. 7-48) and the sum inverted (Fig. 7-49). After our detailed derivation of the circuit in Fig. 7-51, the student should have no problem in identifying the constituent branches and in assembling the overall topology given in Fig. 7-52. Specifically, we recognize the relationships

$$\hat{V} = -V' \qquad V_L = -\frac{g_m}{sC_L} V' \qquad V_C = -\frac{sC}{g_m} V'$$

with $V' = V_o$. The three voltages are summed via

$$I = -g_m(\hat{V} + V_L + V_C) = g_m\left(1 + \frac{g_m}{sC_L} + \frac{sC}{g_m}\right)V_o$$

which is equal to $g_m V_i$. Thus we obtain

$$\frac{V_o}{V_i} = \frac{1}{1 + \dfrac{g_m}{sC_L} + \dfrac{sC}{g_m}} = \frac{g_m s C_L}{s^2 C C_L + s C_L g_m + g_m^2} \tag{7-88b}$$

as required by Eq. (7-88a). Equation (7-88b) is the same as Eq. (7-84b) apart from the unimportant sign inversion; i.e., the element values are those computed in Example 7-6.

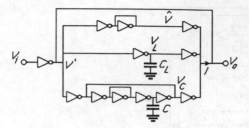

Figure 7-52 Realization of the circuit in Fig. 7-47 by method 2. All transconductances have the same value g_m.

Before leaving this section on transconductance-C filters, let us make a number of observations that relate to the practicality of this approach.

1. The reader who has followed our discussion up to this point will appreciate that the development of a fully integrated continuous-time filter is not a trivial matter: the engineer must be experienced not only in circuit design (filter and control loop) but also in the design of the electronic active devices (op amps, transconductors, and so forth) and IC layout.[33]   Any simplification of the design task is, therefore, very desirable.   In this respect, the op amp–based procedure of Section 7.4 [154] and, for higher frequencies, the transconductance-C method [145, 180] are quite advantageous because of the systematic approach which removes from the filter designer at least the burden of device design.   Instead, the engineer may take an existing, presumably optimized, op amp or transconductance *building block* from an IC design library and assemble the system by systematically "wiring together" many identical subblocks.

2. The device designer in turn can concentrate on developing a gain element that is optimized with respect to such factors as wide bandwidth, good linearity, low noise with wide dynamic range, low power consumption, good power supply rejection, and tunability.

3. We note from the examples that the transconductance-C methods use relatively many active devices.   The designs are clearly impractical or at least inconvenient for discrete filter implementations.   On an integrated circuit, however, *gain*, "a transistor," within reason is relatively cheap;[34] i.e., the number of transconductances is less important.   Furthermore, let us point out that the large number of gain cells is caused by three factors:
   a) The desire for a simple and systematic design methodology.
   b) The insistence on only identical transconductors and grounded capacitors, the advantages of which are ease of fabrication, immunity to many critical parasitics (bottom-plate capacitors), and better matching and tracking

---

[33]Both these latter aspects are beyond the scope of this text, and the reader is referred to the literature [153, 177, 178].

[34]But note that active devices contribute noise and consume power!

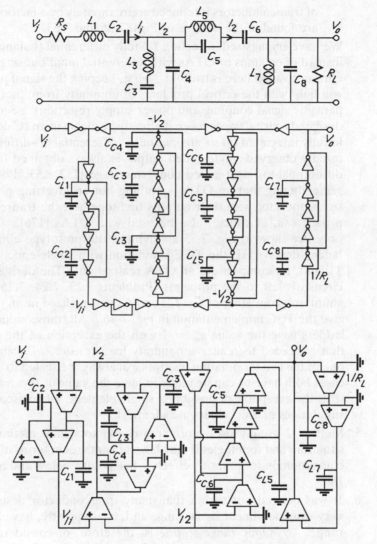

Figure 7-53 Eighth-order elliptic *LC* bandpass ladder and its TGC and OTA realizations [180]. All unlabeled transconductors and OTAs have $g_m = 1$.

with easier electronic tuning. If these restrictions are relaxed, e.g., if floating capacitors are permitted, a number of active devices (transconductors) can often be saved in certain situations and with special topologies (see Example 7-5 and references 154, 157, and 176).

c) The derivation, for pedagogic reasons, is based on single-ended transconductors. If, in practice, devices with differential inputs (OTAs) or with both differential inputs and differential outputs are used, the number of transconductors is reduced approximately by a factor of 2.[35]   Examples are found in the designs of Figs. 7-35 and 7-53.

4. We have emphasized in Section 7.4 fully differential (balanced) active devices instead of op amps or OTAs with differential input but single-ended outputs. Our motivation here is twofold.   First, keeping the signal path fully balanced will help with the critical problems of immunity from "noise" (such as stray parasitic signal coupling and power supply rejection).   Second, the custom-designed active devices (even if already available in an IC design library) used in fully integrated filters are *naturally* "differential in–differential out."   The usually observed single-ended output is always obtained from an additional differential-to-single-ended converter stage [152, 153, 179].   Nevertheless, remembering that an OTA is nothing but an inverting plus a noninverting transconductor with their outputs tied together, the transconductance-C approach can, of course, also be used with OTAs [176].   As an illustration only, we show in Fig. 7-53 an eighth-order prototype elliptic LC bandpass ladder design [11], the TGC simulation with single-ended devices (method 1), and, for comparison, an OTA realization.   The derivation of the active circuits is left to the problems (Problems 7.23, 7.24, 7.25).   Note that the shunt arm consisting of C_3, L_3, and C_4 was realized in an earlier example to give the TGC implementation in Fig. 7-46.   All transconductors in the active ladders have the value $g_m = 1$ with the exception of the one labeled $1/R_L$ that is needed to realize a nonunity load resistor.   Because in LC ladders one of the two terminating resistors can always be scaled to unity and in many cases both have or can be made to have the value 1, it can be concluded that with the exception of possibly a single element, all transconductors in TGC simulations of LC ladders are identical.

5. Method 2 usually uses more transconductors than method 1 but in some situations and topologies it may have advantages in terms of sensitivity to and compensation for parasitic effects.   See point 7 below and references 145 and 180.

6. It was pointed out earlier that many transconductor designs suffer from a very restricted linear signal range and, consequently, have a limited dynamic range.   *Dynamic range scaling* is, therefore, of considerable practical importance and should always be performed in transconductance-C filters.   Fortunately, for the signal-flow graph ladder simulations discussed in this section, such scaling is quite easy; it follows exactly the procedure discussed in Chapter 6, Section 6.4 [181]: as was illustrated in Fig. 6-18, all signal levels can be

---

[35]Note, though, that this smaller number of transconductors will generally not reduce the thermal noise level in the circuit: OTAs and "differential in–differential out" transconductors require for their implementation more transistors, which tend to generate more noise.

equalized by multiplying them with the appropriate constants. An all-transconductance realization for such scale factors was shown in Fig. 7-43c.[36]

7. *Excess phase* problems in transconductance-C filters are, in principle, the same as those in op amp–based designs. The phase errors arise from parasitics inside the devices (Fig. 7-3), modeled as

$$g_m(j\omega) \simeq |g_m(j\omega)|e^{-j\phi(\omega)} \simeq |g_m(j\omega)|e^{-j\omega\tau(\omega)} \tag{7-89}$$

where $\tau(\omega)$ is an intrinsic transconductance delay, and from parasitic elements in the layout of the circuit. Due to the better frequency response of transconductors, the effects of $\tau(\omega)$ are hardly noticeable at low frequencies; still, since g_m-C circuits are aimed at high-frequency operation, excess phase effects—mainly Q enhancement resulting in the slanted responses illustrated in Fig. 3-11 or 7-17—will rarely be completely negligible. For instance, the uncompensated and untuned TGC circuit in Fig. 7-53, built with the transconductors in Fig. 7-18a, was simulated on SPICE for a 3-μm CMOS process; the response is shown in Fig. 7-54 [180]. The deviations typical of Q enhancement, caused by uncompensated excess phase errors, are clearly visible.

We have pointed out earlier in both discussion and examples that these problems necessitate an automatic Q-tuning scheme; however, for high-order filters, the Q-control circuitry tends to become increasingly complicated (see Fig. 7-32), so that minimizing the errors in the design stage is desirable.

Again, the scaling transformation of signal-flow graphs [181], as illustrated in Fig. 6-16, comes to our rescue [145]. To repeat, the scaling transformation states that the *overall* transfer function of a system does not change[37] if all signals entering a subgraph (the contour C in Fig. 6-16) are multiplied by a factor M, provided that all signals leaving the subgraph are multiplied by $1/M$. At the same time, of course, the signals *within* the subgraph are multiplied by M. The clue is then to choose $M(s)$ as a pure delay of appropriate sign and magnitude and thereby furnish the relevant signals in the filter with the desired phase. Thus, if we can find TGC circuits to realize

$$M^+ = e^{-sT} \quad \text{and} \quad M^- = \frac{1}{M^+} = e^{+sT} \tag{7-90}$$

---

[36]Note that implementing arbitrary scaling factors would require that the g_m's can no longer be identical throughout the filter. Depending on the details of the IC design library, this will cause no major difficulties, because it can be achieved by a simple scaling of the W/L ratios of the CMOS devices [see Eqs. (7-48) through (7-51)]. However, if desired we can retain the identical g_m building block approach and achieve *approximate* dynamic range optimization by scaling signal levels by *integer* multipliers. For instance, implementing K in Fig. 7-43c from two transconductances g_m in parallel realizes $V_o/V_i = -\frac{1}{2}$; similarly, replacing g_m in Fig. 7-43c by two transconductances in parallel and setting $K = g_m$ results in $V_o/V_i = -2$. Mostly, a scale factor no larger than 3 or smaller than $\frac{1}{3}$ will be necessary.

[37]Apart from possibly a multiplication by M.

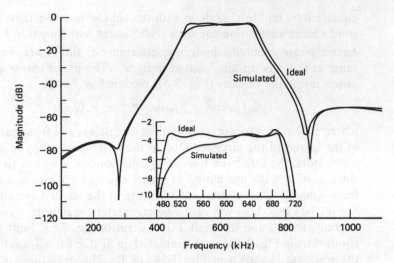

Figure 7-54 Simulated magnitude response of the TGC bandpass filter of Fig. 7-53 for 500 kHz center frequency and 0.5 dB–ripple bandwidth of 200 kHz. Ideal: simulation for *ideal* transconductors. Simulated: simulation for *real* transconductors (Fig. 7-18a).

any phase errors throughout a signal-flow graph TGC filter can be corrected in principle. A detailed discussion of the method and a procedure that yields the minimum number of delay elements can be found in reference 145. After the nominal phase errors are compensated in the design stage, relying on careful modeling, the remaining deviations must be eliminated by automatic tuning.

7.6 SUMMARY

In this chapter, we have discussed the tools required for the development of fully integrated continuous-time active filters. We have seen that in the design of such circuits, one uses in principle the same methods as for discrete designs; that is, we need the functions, if not the elements, of "resistors" and "capacitors" in order to set time constants (frequency parameters) in addition to *gain* to be able to realize *complex* poles. We found as the main difference from discrete filter design that attention must be paid to the demands of automatic tuning. Because tuning will be performed electronically, the designer must provide for a sufficient number of electrically variable components in the circuit whose values can be changed by altering bias conditions. Thus, "capacitors" are often realized as reverse-biased *pn* junctions as long as signals are not so large as to cause unacceptable nonlinearities.   Mostly, designers have opted for fixed MOS capacitors and delegated the tunability requirement to the "resistors," which are implemented either as MOSFETs in the triode region or as "reciprocal transconductances," $1/g_m$, both of which are readily variable via a bias current or voltage.   The third "component," gain, either is obtained from op amps permitting, with minor modifications, the use of classical active RC design methods, or is derived from transconductors.   In that case the functions of "R" and "gain" are merged into one type of device so that only transconductors and capacitors are needed for the implementation of continuous-time integrated filters.   Additional advantages of transconductors, when compared to op amps, are their simpler circuitry and their much wider bandwidth.   Thus, transconductors are emerging as the elements of choice for implementing "gain," in particular for high-frequency communications filters, where the operating frequencies extend to hundreds of kilohertz or even megahertz.

We have stressed repeatedly the importance and difficulty of automatic tuning in the design of IC continuous-time filters.   Indeed, unless it lends itself to automatic tuning by reasonably simple control circuitry, the filter will be quite useless in practice.   Because of the wide tolerances of absolute values of RC products attainable in IC processing, automatic control of a filter's frequency parameters will *always* be necessary.   In addition, for any but the most uncritical applications, some tuning of the transfer function shape, termed Q control in our discussion, is generally unavoidable because of the sensitivity to parasitic effects.   We have discussed the theoretical background of several tuning methods and have illustrated their implementation in a number of examples which, we hope, the student can use as a base for new designs.

The design of an integrated filter is a demanding task, requiring that the engineer be familiar with filter theory and with design, modeling, and layout of integrated circuits.   Evidently, it is very much a custom approach that does not lend itself readily to automation.   However, the g_m-C design method presented in Section 7.5 points the way toward implementing integrated filters out of mostly identical "analog gates," the transconductance building blocks, and capacitors. Note that many if not all the elements of the control circuitry, such as master filter, oscillators, limiters, comparators, peak detectors, and high-gain and low-gain amplifiers, can also be implemented from these same components.   With further development, we expect that the design of a monolithic filter chip will become considerably simpler and require less custom work.

The primary applications of *IC continuous-time filters* are expected to lie in the higher-frequency domain; at lower frequencies, such as in the audio range, better and more dependable results appear to be attainable by a different type of circuit technique, *switched-capacitor filters*.   These *SC filters* are well developed and have found wide commercial applications; they will be our topic in the remaining chapter of this text.

PROBLEMS

7.1 In transconductance-C filters (as in switched-capacitor filters; see Chapter 8), many filter parameters depend on or are related by *capacitor ratios*. Thus, capacitor arrays should be designed for best possible *ratio matching*. Show that one of the error sources in the implementation of capacitor ratios, the poor *edge definition* Δx caused by the optical resolution in the photomasking process, can be eliminated by constructing a large capacitor from a parallel connection of minimum-size unit capacitors C instead of just scaling the aspect ratios (see Fig. P7.1, where it is assumed that, nominally, $L_6 W_6 = 6 L_1 W_1$).

Calculate the capacitor ratios c (nominally, $c = 6$) for the two designs in Fig. P7.1a and b and determine the effect of Δx in both cases. A common ground plane is assumed.

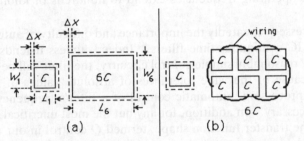

(a)   (b)   **Figure P7.1**

7.2 Analyze the circuits in Fig. 7-2 and compare their performance with that of the corresponding single-ended ones in (the top halves of) Fig. 7-1.

7.3 Determine the intrinsic (i.e., with no load) frequency response of the transconductance model in Fig. 7-3a. Assume $g_m = 100$ μS, $C_i = 0.2$ pF, $C_p = 0.01$ pF, $C_o = 0.05$ pF, and $g_o = 2$ μS and compute the 3-dB frequency of the circuit. Determine the delay τ used in the approximate transconductance model $|g_m(j\omega)| e^{-j\omega\tau}$. Over which frequency range is the model valid?

7.4 Verify that Eq. (7-1) is realized by the circuits in Fig. 7-4a, c, and d.

7.5 Investigate the effects of parasitic input capacitances and output conductances of the transconductances in Figs. 7-4b and c on the realized bandpass function.

7.6 **(a)** If the resistor ratio R_1/R_2 in Fig. 7-8 has an error of ρ percent, what is the error of the RC product?

(b) Repeat part (a) for a frequency error $\Delta\omega$ of Ω percent.

(c) Assume that the two peak detectors in Fig. 7-8 are mismatched by ε percent. How does this design error affect the tuning accuracy of the RC product?

7.7 Verify Eq. (7-36).

7.8 Analogously to the development in Section 7.3.1, investigate performance and operation of a frequency-tuning system that is implemented with an EXOR gate and a master *filter*.

7.9 Derive a block diagram for a tuning system to correct an error in the operation of a ladder filter as depicted in Fig. 7-17. Discuss the block diagram circuit in sufficient detail to make the operation clear to a nonspecialist designer. In particular, think about *which* ladder elements could be varied in order to achieve tuning.

7.10 Derive the circuit in Fig. 7-18b by identifying the building blocks from which it is assembled. Verify Eqs. (7-52) and (7-53). Determine the effects of the finite input capacitors and output resistors of the transconductors.

7.11 Analyze the circuit in Fig. 7-20d and determine component values such that the filtering requirements of Example 7-1 are realized.

7.12 Give a circuit diagram for a MOSFET-*C* implementation of the general AM biquad of Fig. 5-31.

7.13 Give a circuit diagram for a MOSFET-*C* realization of the active *R* circuit in Fig. 5-32 with $d = 1$. Discuss the practicality of the resulting circuit in terms of tuning, op amp implementation, and frequency response. "Invent" a control system (block diagram) for tuning ω_0 and Q for this bandpass filter.

7.14 Starting from Fig. 7-27, derive the circuit in Fig. 7-28.

7.15 Using the PMOS parameters of Example 7-1, complete the design of the filter in Example 7-3.

7.16 A MOSFET circuit for resistor implementation to achieve complete nonlinearity cancellation along with some applications has been discussed in reference 182. It is based on the configuration of four matched transistors in Fig. P7.16b; all four devices are biased in the ohmic region. Using a careful model [151], the drain current of each MOSFET can be represented as [182]

$$i_D = G(v_1 - v_2) - g(v_1) + g(v_2) \qquad \text{(P7-16)}$$

where G is given in Eq. (7-71) and $g(\ )$ is a *nonlinear* function that can be shown to be independent of the gate voltage V_C.

Show that the four-transistor topology in Fig. P7.16b realizes the linear relationship

$$i_1 - i_2 = G(v_1 - v_2)$$

which is the same as that implemented by the resistor *pair* in Fig. P7.16c. Show how this concept can be used to realize *linear* differential input–*single*-ended output MOSFET-*C* integrators and summers [182] and, from there, by the usual methods, higher-order active MOSFET-*C* filters.

(a)   (b)   (c)   **Figure P7.16**

7.17 Derive the circuit in Fig. 7-30.

7.18 Derive the circuits of Fig. 7-35 a and b. Analyze the circuits to obtain the realized transfer function(s) in terms of the components. Compare the function(s) with $H(s)$ given in Example 7-5.

7.19 Show that $F(s)$ in Fig. 7-49 must be noninverting to avoid stability problems.

Hint: Assume an (unavoidable) parasitic component is connected to the input of $F(s)$.

7.20 Convert the circuit in Fig. 7-51 to balanced form.

7.21 Convert the circuit in Fig. 7-52 to balanced form.

7.22 Derive an OTA (single-ended-output) version of the circuit in Fig. 7-52.

7.23 (a) Derive the TGC circuit in Fig. 7-53. Determine the component values if the *impedance-normalized* elements of the LC ladder are

$$R_S = 1 \qquad L_1 = 626.0\mu \; C_2 = 29.49\mu \; L_3 = 63.79\mu \; C_3 = 1269\mu \; C_4 = 361.8\mu$$

$$L_5 = 231.3\mu \; C_5 = 35.75\mu \; C_6 = 37.99\mu \; L_7 = 54.71\mu \; C_8 = 314.1\mu \; R_L = 1.888$$

(b) Develop a balanced version of the circuit.

7.24 (a) Derive a TGC simulation of the LC ladder in Fig. 7-53 by method 2.

(b) Develop a balanced version of the circuit.

7.25 (a) Derive the OTA circuit in Fig. 7-53.

(b) Construct an OTA version of the circuit derived in Problem 7.24(a).

7.26 [145] Design a TGC simulation of an LC lowpass ladder to realize:

Maximum attenuation in passband: $4.343 \cdot 10^{-4}$ dB
Minimum attenuation in stopband: 26.83 dB
Passband edge: $\omega_p = 1$
Stopband edge: $\omega_s = 1.7$

The normalized component values are obtained from the design tables [18]; they are

$$R_S = R_L = 1 \qquad L_2 = 0.9176 \qquad L_4 = 0.5740$$

$$C_1 = 0.3829 \qquad C_2 = 0.1485 \qquad C_3 = 1.0602 \qquad C_4 = 0.5546 \qquad C_5 = 0.1387$$

The LC prototype is shown in Fig. P7.26.

Choose the denormalizing frequency as $\omega_n = 2\pi \cdot 363$ krad/s.

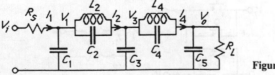

Figure P7.26

Switched-Capacitor Filters

EIGHT

8.1 INTRODUCTION

We have seen in Chapter 7 that there are several techniques that enable us to realize continuous-time active filters in fully integrated form. However, regardless of the approach taken, the implementation of practical circuits requires that an on-chip automatic tuning system is designed which guarantees accurate and drift-free operation of the filter. In particular, we found that Q control is advisable for operation at high frequencies and for medium to high quality factors, and that tuning the frequency parameters is *always* necessary because of the poor control over absolute values in semiconductor processing. Recall that a frequency is set by an RC product,

$$\omega_0 = \frac{1}{RC} \tag{8-1a}$$

so that precise *absolute values* of resistors and capacitors are necessary for accurate designs. For predictable IC processing, it would be preferable to set ω_0 by a *ratio* of elements, say, a capacitor ratio c, such that

$$\omega_0 = cf_c = \frac{C_R}{C} f_c \tag{8-1b}$$

Note that we had to multiply c by a frequency, f_c, in order to obtain the correct units, $1/time$. A comparison of Eqs. (8-1a) and (8-1b) indicates that this goal can be accomplished if we find a circuit that simulates the function of a resistor as

$$R = \frac{1}{f_c C_R} \qquad (8\text{-}2)$$

The design of such a circuit is not difficult (see, for example, references 47, 153, 165, and 183): we need only remember that capacitors store charge and that current is transport of charge. Thus, if the capacitor C_R is connected to a node with voltage V_1, it stores $Q_1 = C_R V_1$; connecting it thereafter to a node with voltage V_2 recharges the capacitor to $Q_2 = C_R V_2$. The charge packet transferred from V_1 to V_2 is, therefore,

$$\Delta Q = Q_1 - Q_2 = C_R(V_1 - V_2) \qquad (8\text{-}3)$$

Figure 8-1a shows the configuration. Let now the switch S be flipped periodically, with a *clock period T*, such that the *clock frequency $f_c = 1/T$* is so large compared to the signal frequency f of the two voltage "sources" V_1 and V_2,

$$f_c \gg 2\pi f \qquad (8\text{-}4)$$

that these two signals can be assumed to be constant over the period T: the flowing charge packets on the average can then be considered as a current

$$I \simeq \frac{\Delta Q}{T} = \Delta Q f_c = f_c C_R(V_1 - V_2) \qquad (8\text{-}5)$$

which indicates that the *switched capacitor* in Fig. 8-1a behaves *approximately* like the resistor

$$R \simeq \frac{V_1 - V_2}{I} = \frac{1}{f_c C_R} \qquad (8\text{-}6)$$

required in Eq. (8-2) [47, 153]. The approximate equivalence of Fig. 8-1a and b is thereby established. Figure 8-1c shows how the switch can be implemented in MOS technology. We use two transistors which are biased on and off in an alternating fashion with two *nonoverlapping* clock signals ϕ_1 and ϕ_2 as shown in Fig. 8-2. Assuming n-channel enhancement devices, the transistors are on (current flows, switch is closed) when the clock signals ϕ_i are high. We shall assume in this book that the input signals of the SC filters are sampled only during ϕ_1 and then held over the full period T.

Let us point out here that this approach of "resistor implementation" solves the *two* main problems which we found troublesome in our discussion of continuous-time integrated filters in Chapter 7:

1. Frequency parameters, i.e., *RC* time constants, are set by *ratios* of MOS capacitors and by a *clock frequency*:

$$\tau_0 = \frac{1}{f_c} \frac{C}{C_R}$$

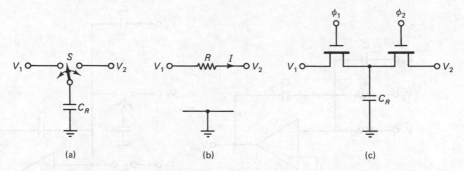

Figure 8-1 (a) Switched capacitor; (b) approximate resistor equivalent; (c) MOS implementation.

MOS capacitor ratios can be implemented repeatably with about 0.1% accuracy; further, clock frequencies are obtained from a crystal-controlled clock generator and are very precise and stable. Consequently, frequency parameters can be *designed* accurately, and *no postdesign tuning is required.*

2. By Eq. (8-6), a 1pF capacitor clocked at 100 kHz results in $R = 10$ MΩ; the required silicon area is only approximately $(50 \ \mu m)^2$. Nominal *unit capacitors* with values as small as a fraction of 1 pF are routinely used in filters designed by the switched-capacitor (SC) approach. Thus, *large "resistors"* can be built easily on a very small area of silicon.

Based on these very preliminary considerations, we are led to conclude that the SC method looks promising for the design of integrated filters [195], provided that suitable filter circuits can be developed. To support our motivation further, we recall that, for example, an integrating summer was identified as a fundamental

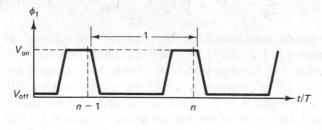

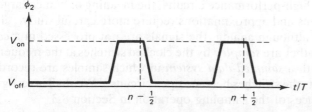

Figure 8-2 Nonoverlapping clock signals ϕ_1 and ϕ_2; note that the time axis is normalized with respect to the clock period.

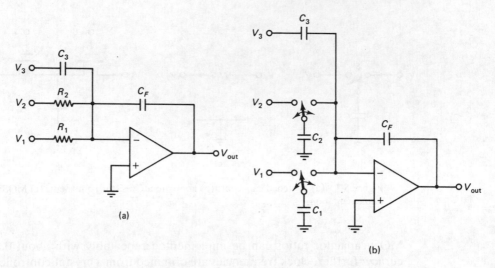

Figure 8-3   (a) Active RC building block; (b) approximate SC equivalent.

active filter building block in Chapters 4 through 7. Such a basic circuit can readily be implemented in the SC approach as is illustrated in Fig. 8-3: Fig. 8-3a shows the active RC circuit, and, using our approximate resistor equivalent, Fig. 8-3b shows its SC implementation. Extensions to additional capacitive or resistive inputs are obvious. The circuits realize

$$V_{out} = -\frac{1}{j2\pi f C_F}(G_1V_1 + G_2V_2) - \frac{C_3}{C_F}V_3 \qquad (8\text{-}7a)$$

for Fig. 8-3a and, assuming all switches are clocked at f_c with $f_c \gg 2\pi f$,

$$V_{out} \simeq -\frac{f_c}{j2\pi f C_F}(C_1V_1 + C_2V_2) - \frac{C_3}{C_F}V_3 \qquad (8\text{-}7b)$$

for Fig. 8-3b. The one-to-one correspondence between the two realizations is apparent. The approximation in Eq. (8-7b) becomes an equality only for $f_c \rightarrow \infty$.

Relatively undemanding low-frequency and low-Q switched-capacitor filters can be designed from active RC prototypes by use of the simple analogy of Eq. (8-6) *provided only that the condition in Eq. (8-4) is satisfied*; that is, the clock frequency f_c must be "much larger" than the highest signal frequency f. But in order to be able to design high-performance circuits, the meaning of "much larger" and the resulting limitations and approximations require more careful study. The difficulties arise because, although analog, the signals are not processed in a *continuous-time* fashion but rather are *sampled* by the clocked switches at the frequency f_c; i.e., we are dealing with a *sampled-data system* in which samples are recorded on capacitors at the instants the switches become open circuits. We shall investigate the details and consequences of the sampling operation in Section 8.3.

An additional difficulty arises because of our need for accuracy and our desire to implement the circuits with the smallest possible capacitors in order to save silicon area.   Note from Fig. 8-1c that C_R is in parallel with parasitic (drain-source diffusion) capacitors C_p from the switches and from routing metalization so that not C_R but rather $C_R + 2C_p$ is converted to a "resistor," resulting in incorrect time constants.   To solve this problem and make the circuits *insensitive to parasitics*, we need to modify the simple topology in Fig. 8-1c.   We shall consider the details in Section 8.2.

8.2 THE EFFECT OF PARASITICS—PARASITICS-INSENSITIVE BUILDING BLOCKS

To understand the origin of parasitic capacitors, we refer the student to Fig. 8-4 which illustrates the various parasitic capacitances associated with the silicon realization of the capacitor C_R.   The sketch shows a double polysilicon CMOS capacitor: C_R is measured between the two polysilicon layers (terminals a–b); in addition there is a "bottom-plate" parasitic, C_b, between the Poly 1 bottom-plate of C_R and the substrate, and a top-plate parasitic, C_t, between the top-plate of C_R and the substrate. C_t is comprised of two components: C_j, voltage dependent nonlinear capacitances associated with source-drain diffusions of the switches and C_m, the accumulated capacitance formed by all the metal routing which connects the top-plate of C_R to various components in the filter.   The presence of C_j can cause noticeable distortion in the filter signals.   Typically, C_b is about 10–20% of the value of C_R and, depending on the switch sizes and the amount of routing, C_t is 1–5% of C_R.   Because of the potentially large size of C_b, the conventional

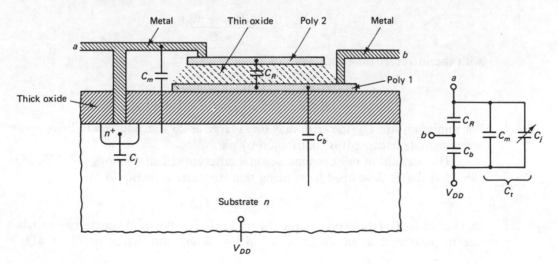

Figure 8-4 (a) Implementation of an MOS double polysilicon capacitor; (b) small-signal equivalent circuit. *Note*: $C_t = C_m + C_j$.

wisdom is to connect the bottom-plate of C_R always to either an independent voltage source or a op amp output. This connection rule tends the reduce the accumulated parasitic capacitance at the op amp virtual ground. This is perhaps more important for power supply rejection than for the accuracy of C_R.

It is important to recognize that all these parasitics are connected to the substrate bias (V_{DD} for CMOS), which for ac signal considerations is the equivalent of ground. Unfortunately, due to the use of switching power converters, V_{DD} is not a "clean" dc supply and contains "noise" which enters the SC filter through these parasitics. The transfer ratio V_{out}/V_{dd} (where we assume the noise V_{dd} to be an ac signal) measures the ability of the filter to reject this power supply injected noise. This transfer ratio, called the power supply rejection ratio or PSRR, is proportional to the size of the parasitic capacitances. Additional parasitics to be concerned with are the gate-source, C_S, and gate-drain, C_D, capacitors, which are illustrated in the MOS transistor model in Fig. 7-24. Since the signal passes through a switch from source-to-drain, C_S and C_D do not get involved with signal transmission, but allow the clock signals ϕ_1, ϕ_2 to couple into the filter, resulting in clock feedthrough. One consequence of the sampled-data nature of SC filters is that clock feedthrough appears at the output as additional dc offset.

Figure 8-5a illustrates how all these parasitics enter an SC integrator. Note that all capacitors at the inverting input node of the op amp are connected to virtual ground; thus, they and the bottom-plate capacitor of C_R are always shorted. The bottom-plate capacitor of C_F and, the switched-capacitor C_{sw}, at the left side of the left switch (ϕ_1) are voltage-driven and, therefore, inconsequential.[1] Thus, the relevant circuit is that of Fig. 8-5b, which shows the capacitors that actually contribute to the circuit's operation. Clearly, in Fig. 8-5b the capacitor $C_R + C_p$ would be "converted" into an equivalent resistor so that the integrator realizes

$$\frac{V_{out}}{V_{in}} \simeq -\frac{f_c(C_R + C_p)}{j\omega C_F}$$

with the incorrect time constant

$$\tau = \frac{1}{f_c}\frac{C_F}{C_R}\frac{1}{1 + C_p/C_R} = \tau_0\frac{1}{1 + C_p/C_R}$$

For minimum-size C_R, the error may be as large as 20% in addition to being quite unpredictable because it is contributed by parasitics.

The way out of this dilemma lies in a different circuit topology [47, 155, 185, 186, 194] that is developed by noticing that the current of Eq. (8-5),

$$I \simeq f_c C_R(V_1 - V_2)$$

can be realized by the two-switch configuration in Fig. 8-6a. When the two switches are in position 1 as shown ($\phi = \phi_1$ in Fig. 8-6b), the charge on C_R is $\Delta Q =$

---

[1]We assumed here that the op amp gain is sufficiently high and that V_{in} and V_{out} are ideal voltage sources. If these assumptions are not true, further error sources must be considered [184, 185].

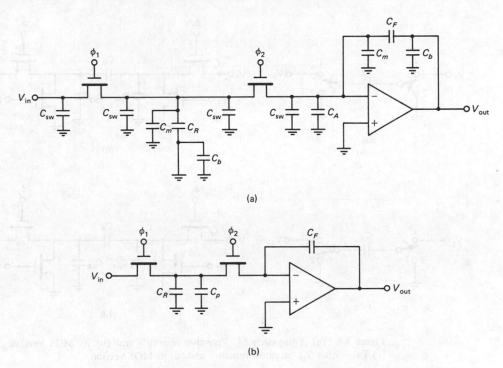

Figure 8-5   (a) SC integrator with all parasitic capacitors; (b) SC integrator with all operative parasitic capacitors that are active lumped into C_p; typically, $C_p =$ 0.01 − 0.05 pF for $C_R =$ 1pF. Symmetrical transmission switches were assumed. C_A is the input capacitor of the op amp.

$C_R(V_1 - V_2)$; when the switches are in position 2 ($\phi = \phi_2$), C_R discharges completely. Therefore, when the switching frequency is f_c, the average current is that given in Eq. (8-5). The MOS implementation of Fig. 8-6a is shown in Fig. 8-6b along with the relevant parasitics C_{p1} and C_{p2}, which, as always, consist of switch and routing parasitics; in addition, depending on the connection of C_R, either C_{p1} or C_{p2} will have an additional contribution from C_b. Note that although during phase ϕ_1 these two parasitic capacitors are charged to V_1 and V_2, respectively, during phase ϕ_2 they both discharge harmlessly to ground without transferring any charges to nodes V_1 or V_2. A minor but very useful modification of this topology is shown in Fig. 8-6c and d: we observe that during phase ϕ_1, C_R is charged with $Q_1 = C_R V_1$ and during ϕ_2 with $Q_2 = C_R V_2$. But by changing the switch phase, the polarity of C_R is inverted, so that the total charge packet flowing from V_1 to V_2 is now $\Delta Q = C_R(-V_1 - V_2) = -C_R(V_1 + V_2)$; in effect, the sign of V_1 is changed. The same result can be obtained by interchanging the switch phases, i.e., $\phi_1 \to \phi_2$ and $\phi_2 \to \phi_1$ in Fig. 8-6d. Note again that the presence of C_{p1} and C_{p2} has no effect on circuit operation. The utility of this scheme [47, 185, 194] becomes readily apparent when we remember that these capacitors, switched at the frequency f_c, are to simulate resistors connected to the virtual ground inputs

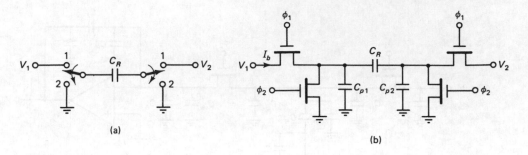

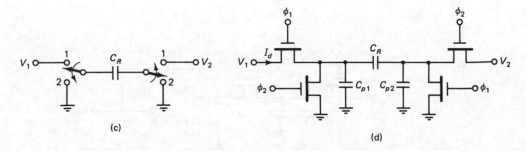

Figure 8-6 (a) Two-switch SC "positive resistor" and (b) its MOS version. (c) Two-switch SC "negative resistor" and (d) its MOS version.

of op amps (Fig. 8-7), so that $V_2 = 0$. Thus, in Fig. 8-6b and d, respectively, we have

$$I_b \simeq f_c C_R V_1 \quad \text{and} \quad I_d \simeq -f_c C_R V_1$$

Consequently, Figure 8-7a realizes a *parasitics-insensitive inverting integrator*,

$$\frac{V_{\text{out}}}{V_{\text{in}}} \simeq -\frac{f_c C_R / C_F}{j\omega} \tag{8-8a}$$

and, by the simple step of a change in switch phasing, Fig. 8-7b yields a *parasitics-insensitive noninverting integrator*,

$$\frac{V_{\text{out}}}{V_{\text{in}}} \simeq +\frac{f_c C_R / C_F}{j\omega} \tag{8-8b}$$

Let us emphasize that insensitivity to parasitics is important not only for reasons of accuracy and reduced nonlinearities but that it also results in reduced silicon area: the filters may be built with smaller capacitors which no longer have to dominate the parasitics. We note that a crucial assumption in the above arguments is that V_1 and V_2 are either ideal voltage sources (an input or the output of an ideal op amp) or an ideal op amp virtual ground. Although these assumptions are only realized approximately in practice, parasitic insensitivity is nearly complete.

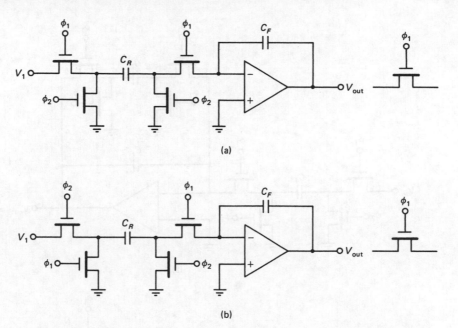

(a)

(b)

Figure 8-7   (a) Parasitics-insensitive inverting integrator; (b) parasitics-insensitive noninverting integrator.   The trailing switch symbolizes that V_{out} is to be sampled during the clock phase ϕ_1.

Having obtained both inverting and noninverting parasitics-insensitive *lossless* integrators, we recall from Chapters 4 through 7 that *lossy* integrators, or, more generally, *summing lossy* integrators, are also needed for the design of active filters. A lossless integrator is made lossy by placing a resistor into the feedback path around the op amp.   Thus, the circuit is obtained by adding an SC resistor to the integrator in Fig. 8-7a as is shown in Fig. 8-8a.   Note that this SC resistor has to be positive; the switch phasing must, therefore, be chosen correctly to maintain stability!   The circuit realizes

$$\frac{V_{out}}{V_{in}} \simeq -\frac{f_c C_1}{j\omega C_F + f_c C_2} = -\frac{C_1}{C_F}\frac{f_c}{j\omega + f_c C_2/C_F} \tag{8-9}$$

We can realize a non-inverting lossy integrator by simply inverting the phases (i.e. $\phi_2 \to \phi_1$ and $\phi_1 \to \phi_2$) of the two switches on the input or left side of C_1 in Fig. 8-8a.

Notice that during phase ϕ_1 the left plate of C_2 and the right plate of C_1 in Fig. 8-8a are both connected to the inverting input terminal of the op amp, whereas during phase ϕ_2 they are both connected to ground.   This indicates that the corresponding switches may be shared, as we have illustrated in the redrawn version of the lossy integrator in Fig. 8-8b.   Such *switch sharing* is possible in many SC situations and can be used profitably to simplify the circuitry.

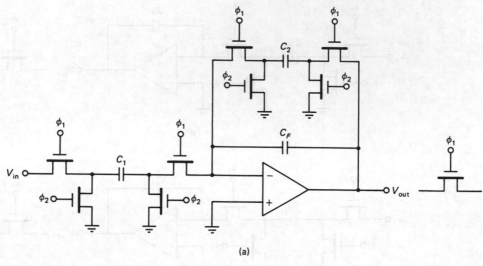

(a)

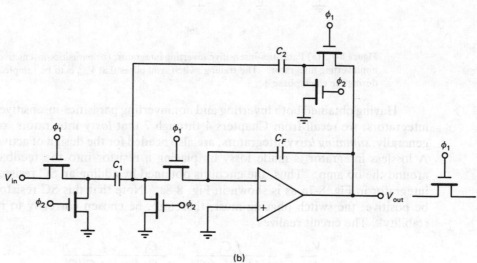

(b)

Figure 8-8   (a) Lossy inverting integrator; (b) illustrating switch sharing. The trailing switch at the output symbolizes that V_{out} is to be sampled during ϕ_1.

A summing integrator was shown conceptually in Fig. 8-3b; adding loss, we obtain the general first-order SC building block in Fig. 8-9.

Using again our approximate resistor equivalent [Eq. (8-6)], we obtain

$$V_{out} \simeq -\frac{1}{j\omega C_F + f_c C_4}[f_c(C_1 V_1 - C_2 V_2) + j\omega C_3 V_3] \qquad (8\text{-}10)$$

Note that the switch phasing at the left plate of C_2 resulted in our integrating the *difference* between (scaled versions of) V_1 and V_2.

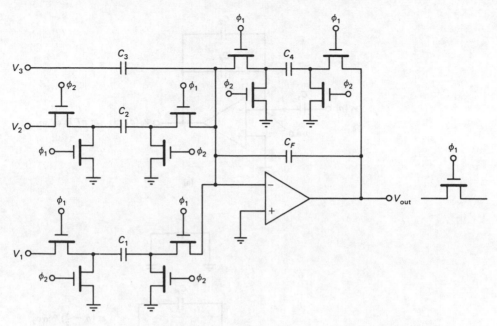

Figure 8-9   General first-order SC building block. V_{out} is sampled during ϕ_1.

Let us emphasize again at this point that all transfer characteristics derived for the SC circuits are based on the assumption of Eq. (8-4), namely, that the switching or clock frequency is much larger than the signal frequency ($f_c \gg 2\pi f$). Under these conditions, the SC circuits in Fig. 8-6 approximate positive or negative resistors, $R \simeq \pm 1/(f_c C_R)$, and the circuits' operation is almost continuous. We mentioned earlier that for undemanding applications, workable circuits could be designed based on this simple equivalence starting from active RC prototypes. However, as operating frequencies increase relative to the clock frequency, one notices increasing deviations in the filters' characteristics from the specified performance. Thus, a more careful analysis is needed. Referring to the waveforms of the switching functions in Fig. 8-2, let us look at the operation of the inverting lossy integrator. For this discussion, and through much of this chapter, let us assume that the sampled input $v_1(n)$ changes value "instantaneously" at only one time instant per clock period T, say during the ϕ_1 phase. The input $v_1(n)$ then remains constant through the next ϕ_2 phase such that $v_1(n - \tfrac{1}{2}) = v_1(n - 1)$. Consider the integrator in Fig. 8-8a. The circuit is redrawn in Fig. 8-10a and b for the two phases ϕ_1 and ϕ_2, respectively. It is perhaps easiest to obtain an exact analysis of the circuit by calculating the current $i(t)$ through the capacitor C_1 during ϕ_1:

$$i(t) = C_1 \frac{dv_1}{dt} = -(C_F + C_2) \frac{dv_2}{dt} \tag{8-11}$$

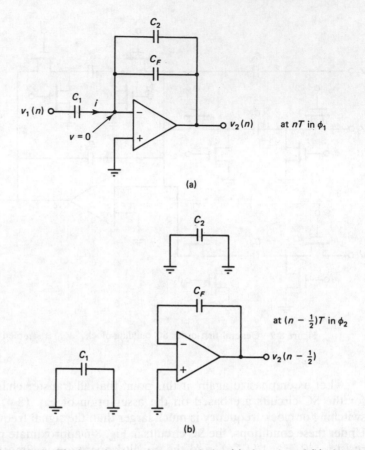

Figure 8-10 The inverting lossy integrator during (a) phase ϕ_1 and (b) phase ϕ_2.

In order to obtain an equation for the charges, we integrate over the *normalized* time axis from the previous switch status $(n - \frac{1}{2})$ during ϕ_2 until the present status (n) in ϕ_1: we find

$$(C_F + C_2)v_2(n) - (C_F + C_2)v_2(n - \tfrac{1}{2}) = -C_1[v_1(n) - v_1(n - \tfrac{1}{2})] \quad (8\text{-}12a)$$

The initial conditions at $t = (n - \frac{1}{2})T$ can be found from Fig. 8-10b: the capacitors C_1 and C_2 are discharged to ground; i.e., their charge is zero:

$$C_1 v_1(n - \tfrac{1}{2}) = 0 \quad \text{and} \quad C_2 v_2(n - \tfrac{1}{2}) = 0$$

Thus

$$(C_F + C_2)v_2(n) - C_F v_2(n - \tfrac{1}{2}) = -C_1 v_1(n) \quad (8\text{-}12b)$$

Further, the op amp and C_F have been isolated since the time $(n - 1)T$; thus v_2 has maintained its value and $v_2(n - \frac{1}{2}) = v_2(n - 1)$; note also that disconnecting

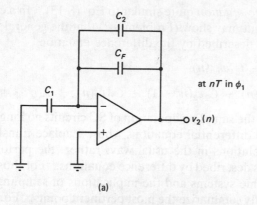

at nT in ϕ_1

(a)

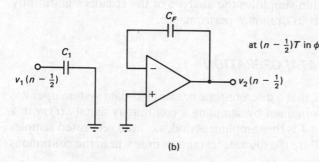

at $(n - \frac{1}{2})T$ in ϕ_2

(b)

Figure 8-11 The noninverting lossy integrator during (a) phase ϕ_1 and (b) phase ϕ_2.

C_2 does not change v_2! Thus, Eq. (8-12b) can be rewritten as

$$C_F[v_2(n) - v_2(n - 1)] + C_2 v_2(n) = -C_1 v_1(n) \qquad (8-13)$$

which is a *difference equation* for the relationship between v_1 and v_2.

In a similar fashion, we can analyze the noninverting damped integrator, Fig. 8-7b with added loss: it is shown for the two switch phases in Fig. 8-11a and b. Note that at time $t = (n - \frac{1}{2})T$, the charge $+C_1 v_1(n - \frac{1}{2})$ is recorded on C_1. Then at $t = nT$, the charge on C_1 is inverted and $-C_1 v_1(n - \frac{1}{2})$ is added to the charge on $(C_F + C_2)$. Thus $(C_F + C_2)v_2(n) - C_F v_2(n - \frac{1}{2}) = C_1 v_1(n - \frac{1}{2})$. Once again, since the op amp and C_F are isolated during the ϕ_2 phase,[2] $v_2(n - \frac{1}{2}) = v_2(n - 1)$. Also due to our assumption regarding $v_1(n)$ we have $v_1(n - \frac{1}{2}) = v_1(n - 1)$. Hence,

$$C_F[v_2(n) - v_2(n - 1)] + C_2 v_2(n) = C_1 v_1(n - 1) \qquad (8-14)$$

[2] v_2 does *not* change by switching C_2 out of the circuit at $t = (n - \frac{1}{2})T$.

This expression is a *difference equation* quite similar to Eq. (8-13). In a completely analogous fashion, the student may show (Problem 8.2) that the general first-order building block in Fig. 8-9 is described by the difference equation

$$C_F[v_{out}(n) - v_{out}(n - 1)] + C_4 v_{out}(n)$$

$$= -C_1 v_1(n) + C_2 v_2(n - 1) - C_3[v_3(n) - v_3(n - 1)] \quad (8\text{-}15)$$

We observe, therefore, that the sampling operation of SC circuits no longer permits us to write continuous-time differential equations whose Laplace transforms give us the frequency-domain solutions in the usual way; rather, the performance of these discrete-time circuits is described by difference equations. For those students not familiar with discrete-time systems and the implications of sampling, we shall in the following section briefly summarize the most pertinent points. For a detailed discussion the student should consult the references [1–5].

It is more than a coincidence that the switched capacitors at the op amp input, for our building block circuits in Figs. 8-8 and 8-9, all were chosen to connect to the op amp in the same phase (i.e., in these cases ϕ_1). During the other phase (ϕ_2 in these cases) the op amp and C_F are isolated and do not receive any new charges. This type of operation simplifies the analysis of the circuits significantly and, more importantly, results in optimum performance.

8.3 DISCRETE-TIME (SAMPLED-DATA) OPERATION

We saw in Sections 1.2 and 8.2 that a discrete-time or sampled-data system operates on samples $x(nT)$ which are obtained by sampling a continuous signal $x(t)$ with a sampling frequency $f_c = 1/T$. T is the sampling period, i.e., the generated samples are T seconds apart. Thus nT are the discrete instants in time where the continuous signal is sampled:

$$x(nT) = x(t)|_{t = nT} \quad (8\text{-}16)$$

The output of a sampled-data system is a similar *sequence* $y(nT)$ which then in many applications may have to be converted back into the continuous-time analog signal $y(t)$ by circuitry which *reconstructs* $y(t)$ from the given samples $y(nT)$.

We shall in the following consider the process of sampling a signal $x(t)$ in some detail and investigate its effect on the frequency spectrum of $x(t)$.

8.3.1 The Sampling Process

The generation of the sampled signal $x^{\#}(t)$ with sample values $x(nT)$ may be understood as a multiplication of $x(t)$ by a *periodic* sampling signal $s(t)$ obtained from the conceptual circuit in Fig. 8-12a. At the times $t = nT$, switch S_1 is closed momentarily to charge the storage capacitor C instantaneously to the values $x(nT)$; it is discharged to zero τ seconds later by closing switch S_2. The capacitor voltage

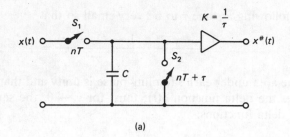

(a)

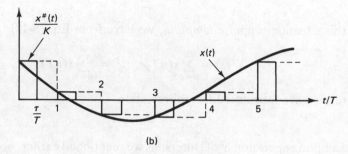

(b)

Figure 8-12   (a) Sampling stage; (b) continuous signal $x(t)$ and its sampled version $x^{\#}(t)$.

is viewed through a buffer amplifier of gain $K = 1/\tau$. Mathematically, this operation can be expressed as

$$x^{\#}(t) = x(t)s(t) = K\sum_{n=0}^{\infty} x(nT)[u(t - nT) - u(t - nT - \tau)] \quad (8\text{-}17)$$

where we have made use of the unit step function $u(t - t_0)$, which changes its value instantaneously from 0 to 1 at $t = t_0$. Also, we have assumed that $x(t) = 0$ for $t < 0$. $x^{\#}(t)$ is shown in Fig. 8-12b. The (single-sided) Laplace transform of the sampled signal is obtained by recalling the transform pairs

$$\mathcal{L}\{u(t)\} = \frac{1}{s} \qquad \mathcal{L}\{u(t - t_0)\} = \frac{1}{s}\, e^{-st_0}$$

Thus,

$$\begin{aligned}
X^{\#}(s) &= \mathcal{L}\{x^{\#}(t)\} = K\sum_{n=0}^{\infty} x(nT)\left(\frac{1}{s}\, e^{-snT} - \frac{1}{s}\, e^{-s(nT + \tau)}\right) \\
&= \frac{1}{\tau}\frac{1 - e^{-s\tau}}{s}\sum_{n=0}^{\infty} x(nT)e^{-snT}
\end{aligned} \quad (8\text{-}18)$$

The multiplier $(1 - e^{-s\tau})/(s\tau)$ depends on the width τ of the sampling pulse; we shall in the following assume τ to be very small so that

$$\frac{1}{\tau}\frac{1 - e^{-s\tau}}{s} \simeq \frac{1 - (1 - s\tau)}{s\tau} = 1$$

Note that the area under each sampling pulse is unity and that for $\tau \to 0$ the pulse approximates the delta function $\delta(t)$; thus, for $\tau \to 0$, the sampling function is a sequence of delta functions:

$$s(t) = \sum_{n=0}^{\infty} \delta(t - nT) \tag{8-19}$$

For this situation, *impulse sampling*, we have from Eq. (8-18)

$$X^{\#}(s) = \sum_{n=0}^{\infty} x(nT)e^{-snT} = \sum_{n=0}^{\infty} x(nT)z^{-n} \tag{8-20}$$

where we have introduced the abbreviation

$$z = e^{sT} \tag{8-21}$$

The sampling operation in SC filters, as we mentioned earlier, occurs at the instants the switches become open circuits. Hence, sampling in SC circuits is very nearly ideal so that the impulse sampling in Eq. (8-18) characterizes the process accurately.

The right-hand side of Eq. (8-20) is called the (single-sided) *z-transform* $X(z)$ of the time sequence $x(nT)$:

$$\mathcal{Z}\{x(nT)\} = X(z) = \sum_{n=0}^{\infty} x(nT)z^{-n} \tag{8-22a}$$

Note that, apart from the notation introduced in Eq. (8-21), the z-transform is identical to the Laplace transform and, therefore, obeys all the same properties [1–5]. On the $j\omega$-axis, the reciprocal of the operator z represents a delay of T seconds:

$$z^{-1} = e^{-j\omega T}$$

i.e., by Eq. (8-22a), $X(z)$ is the sum of the samples or *weights* $x(nT)$, each delayed by nT seconds. It is customary to drop the explicit reference to T in the z-transform. Thus in the following discussion we shall often set $T = 1$; i.e.,

$$\mathcal{Z}\{x(n)\} = X(z) = \sum_{n=0}^{\infty} x(n)z^{-n} \tag{8-22b}$$

which is equivalent to normalizing the time axis with respect to T as was done in Figs. 8-2 and 8-12b. Equation (8-22) is often written as a transform pair

$$\mathcal{Z}\{x(n)\} \leftrightarrow X(z) \quad \text{with} \quad \mathcal{Z}\{x(n - k)\} \leftrightarrow z^{-k}X(z) \tag{8-22c}$$

where n and k are integers. The second pair in Eq. (8-22c) reflects the time shift property of the z-transformation; remember that for $s = j\omega$, z^{-k} is simply a delay of k units of time.   Additional properties of the z-transform are given in Appendix II.

Since in this book we are interested in spectrum-shaping filters, it is of interest to determine the effect of sampling the signal $x(t)$ on its spectrum

$$\mathscr{F}\{x(t)\} = X(j\omega)$$

$\mathscr{F}\{\;\}$ denotes the Fourier transform.   To this end we note that $s(t)$ in Eq. (8-19) is a periodic function that can be represented by a Fourier series

$$s(t) = \sum_{k=-\infty}^{+\infty} C_k e^{jk\omega_c t} \tag{8-23}$$

where $\omega_c = 2\pi/T$ is the sampling frequency and

$$C_k = \frac{1}{T}\int_{-T/2}^{+T/2} s(t) e^{-jk\omega_c t} dt = \frac{1}{T}\int_{-T/2}^{+T/2} \delta(t)\, e^{-jk\omega_c t}\, dt = \frac{1}{T} \tag{8-24}$$

are the Fourier coefficients.   Substituting Eqs. (8-23) and (8-24) in Eq. (8-17) yields

$$x^{\#}(t) = x(t)s(t) = \frac{1}{T} \sum_{k=-\infty}^{+\infty} x(t) e^{jk\omega_c t} \tag{8-25}$$

and

$$\mathscr{F}\{x^{\#}(t)\} = X^{\#}(j\omega) = \frac{1}{T} \sum_{k=-\infty}^{+\infty} \mathscr{F}\{x(t) e^{jk\omega_c t}\}$$

which may be solved to give the important result

$$X^{\#}(j\omega) = \frac{1}{T} \sum_{k=-\infty}^{+\infty} X[j(\omega - k\omega_c)] \tag{8-26}$$

From Eq. (8-26) we see that the sampling operation has introduced infinitely many new spectrum components which are obtained as translations of the spectrum $X(j\omega)$ of the unsampled original signal $x(t)$ by integer multiples of the sampling (or clock) frequency ω_c.   Additionally, for ideal impulse sampling, all spectra are multiplied by an unimportant constant factor $1/T$.   The situation is illustrated in Fig. 8-13b for a band-limited spectrum $X(j\omega)$ with highest frequency ω_B and $\omega_c < 2\omega_B$.   Evidently, the individual translations (dashed lines) overlap so that the sum spectrum (solid line) will permit no easy way of recovering the original spectrum (Fig. 8-13a).   Spectrum components at frequencies higher than $f_c/2$ are said to be *aliased* into the baseband; i.e., they *appear as if they were lower than $f_c/2$*.   To remedy the situation, no overlap of the translated components (no aliasing) is permitted

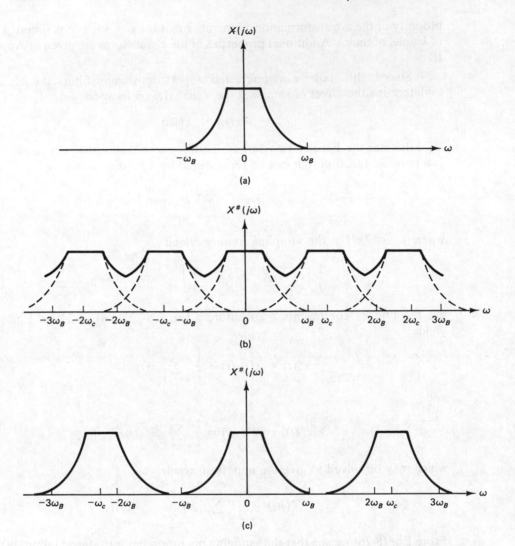

Figure 8-13 Spectra for (a) continuous baseband signal $x(t)$, (b) $x(t)$ sampled at $\omega_c < 2\omega_B$, and (c) $x(t)$ sampled at $\omega_c > 2\omega_B$.

to occur; this requires clearly that

$$f_c > 2f_B \qquad (8\text{-}27)$$

as is shown in Fig. 8-13c. In that case a simple continuous-time lowpass filter with transfer characteristic given in Fig. 8-14 will extract the baseband and eliminate all the translated spectral components. Since the step of lowpass-filtering the spectrum $X^{\#}(j\omega)$ of the sampled signal permits us to recover the spectrum $X(j\omega)$ of

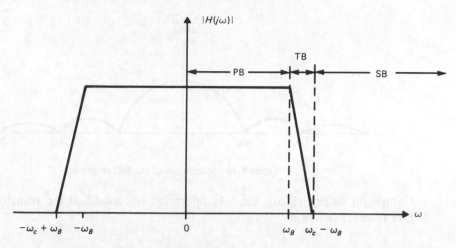

Figure 8-14   Reconstruction filter.

the original continuous-time signal, $x(t)$ can be fully *reconstructed* from just the sample values $x(nT)$ at the instants nT.   This statement along with Eq. (8-27) is a way of expressing the important *sampling theorem* attributed to *Shannon* [8]:

If a function $x(t)$ has a band-limited spectrum $X(j\omega)$ such that $X(j\omega) = 0$ for $|\omega| > \omega_B$ with $|\omega_B| \le \omega_c/2$, then $x(t)$ is uniquely described by its values at uniformly spaced time instants T seconds apart, where $T = 2\pi/\omega_c$.

In the communications literature, the minimum sampling frequency $\omega_c = 2\omega_B$ is also called the *Nyquist* rate.

The filter characterized in Fig. 8-14 is often referred to as a *reconstruction filter*.   Clearly, the greater the separation between ω_B and ω_c, the wider can be the transition band (TB) between passband (PB) and stopband (SB) for the reconstruction filter.   We have seen in earlier chapters in this book that widening the transition band results in lower-order and, therefore, less complicated and cheaper filter realizations.

In many sampled-data systems, the signals are not just sampled but they are *sampled and held* until the next sampling period.   This means $x(t)$ maintains the value $x(nT)$ until the next update at time $(n + 1)T$, as is indicated with dashed lines in Fig. 8-12b.   In fact when we derived Eq. (8-14) we assumed that input signal v_1 and, correspondingly with the switch arrangements chosen in Fig. 8-9, output signal v_2 were sampled and held in this manner.   The sample-and-hold (S/H) operation can be viewed as a special case of our conceptual circuit in Fig. 8-12a with S_2 removed.   Mathematically, this means that in Eqs. (8-17) and (8-18) we can set $\tau = T$ so that we obtain

$$X_{\text{S/H}}^{\#}(s) = \frac{1 - e^{-sT}}{sT} \sum_{n=0}^{\infty} x(nT)e^{-snT} \qquad (8\text{-}28)$$

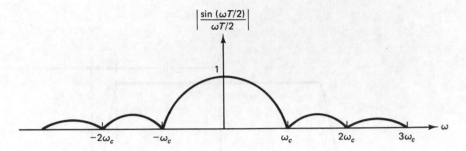

Figure 8-15 Magnitude of the S/H response.

Compared to Eq. (8-20), the S/H operation has modified the transform by the S/H transfer function

$$H_{\text{S/H}}(s) = \frac{1 - e^{-sT}}{sT} \tag{8-29a}$$

or, on the $j\omega$-axis,

$$H_{\text{S/H}}(j\omega) = \frac{1 - e^{-j\omega T}}{j\omega T} = \frac{e^{j\omega T/2} - e^{-j\omega T/2}}{j\omega T} e^{-j\omega T/2} = \frac{\sin(\omega T/2)}{\omega T/2} e^{-j\omega T/2} \tag{8-29b}$$

Thus, apart from a delay $T/2$, the S/H operation multiplies ("filters") the magnitude of the spectrum of the sampled signal by the so-called $(\sin x)/x$ factor sketched in Fig. 8-15:

$$X_{\text{S/H}}^{\#}(s) = \frac{1 - e^{-sT}}{sT} X^{\#}(s) \tag{8-30}$$

If the signals are sampled and held, the magnitude spectrum sketched in Fig. 8-13c, as obtained by ideal impulse sampling, must be multiplied by the $(\sin x)/x$ factor.

Based on our foregoing discussion, we can now sketch a diagram of all filter blocks required in a complete sampled-data system (Fig. 8-16). Because real signals are generally not band-limited, a continuous-time so-called *anti-aliasing filter* is required at the input in order to attenuate any unwanted high-frequency components beyond $f = f_B$; it is followed by a sampling or an S/H stage to convert the

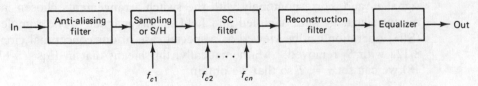

Figure 8-16 Block diagram of a complete sampled-data system.

continuous band-limited signal $x(t)$ into discrete form $x(nT)$ with $T \leq 1/(2f_B)$; at this point the signal is ready for the discrete-time (in our case: switched-capacitor) filter which generates the desired discrete output samples $y(nT)$ with the spectrum $Y^\#(s)$. The translated components of this spectrum are eliminated (sufficiently attenuated) by the continuous-time reconstruction or *smoothing* lowpass filter, and finally, if unacceptable amplitude distortion was introduced by an S/H operation, a correction stage (an amplitude equalizer) with transfer function $H_{eq}(s) = 1/H_{S/H}(s)$ must be used. The description of Fig. 8-16 assumed that input and output are continuous-time analog signals; if the input signal is already in discrete form and the output can be in discrete form for further processing, then of course only the switched-capacitor block in Fig. 8-16 is needed for filtering. Although in this text we shall always assume a single clock-frequency, there are several applications where multiple clock frequencies are useful [155, 187–190]. Thus, for generality, we have indicated in Fig. 8-16 different clock frequencies.

8.3.2 Application of the z-Transform in Sampled-Data Systems

We shall see presently that employing the z-transform in the analysis of discrete-time systems yields many of the same benefits that resulted from using the Laplace transform in continuous-time systems. Most important, just as the Laplace transform converts continuous-time differential equations into algebraic equations, the difference equations that we saw arising in the analysis of discrete-time systems are converted into algebraic equations by the z-transform. That is, applying the z-transformation to a difference equation such as Eq. (8-15) gives us a rational function in z which describes the behavior of the discrete-time circuit. Before we consider this process and its interpretation in some detail, let us look at a few examples for finding the z-transform of common discrete-time series and discuss how the time-domain samples are found from a given z-transform.

Example 8-1

Determine the z-transform for the sampled versions of the unit step $x_1(t) = u(t)$ and of the exponential function $x_2(t) = e^{-at}u(t)$.

Solution. Sampled at intervals T, we have $x_1(n) = 0$ for $n < 0$ and $x_1(n) = 1$ for all $n \geq 0$. Thus, using Eq. (8-22), we find

$$X_1(z) = \sum_{n=0}^{\infty} z^{-n} \tag{8-31a}$$

which is recognized as a geometric series that converges for $|z| > 1$ to

$$X_1(z) = \frac{1}{1 - z^{-1}} \tag{8-31b}$$

Similarly, we find for $x_2(t)$ the sampled version $x_2(n) = e^{-anT}$ for $n \geq 0$ so that by

Eq. (8-22) we have

$$X_2(z) = \sum_{n=0}^{\infty} e^{-naT} z^{-n} = \frac{1}{1 - e^{-aT} z^{-1}} \qquad |z| > e^{-aT} \qquad (8\text{-}32)$$

Both time series are seen to result in rational functions in z. The z-transforms of additional commonly encountered time functions are found in Appendix II.

To find the time-domain samples from a given rational z-transform $X(z)$, it is most convenient to expand $X(z)$ into partial fractions such that the first- or second-order terms given in Appendix II can be identified. Since all transforms [except that of $\delta(nT)$] are seen to have a factor z in the numerator, it is convenient to expand $X(z)/z$ into partial fractions and then multiply the result by z. Partial fraction expansions are discussed in the references [1–5], but an example will illustrate the method.

Example 8-2

Find the time sequence represented by the transform

$$X(z) = \frac{1 - 2z^{-1}}{1 - 0.3z^{-1} + 0.42z^{-2} - 0.08z^{-3}}$$

Solution. The first step is to write $X(z)$ in the form of a ratio of polynomials in z rather than in z^{-1} and factor the polynomials to find their roots; the result is

$$X(z) = \frac{z^2(z - 2)}{(z - 0.2)(z - 0.05 + j0.630764)(z - 0.05 - j0.630764)}$$

Next, we form

$$\frac{X(z)}{z} = \frac{A}{z - 0.2} + \frac{B}{z - 0.05 + j0.630764} + \frac{B^*}{z - 0.05 - j0.630764}$$

where

$$A = (z - 0.2)\frac{X(z)}{z}\Big|_{z = 0.2} = -0.857143$$

and

$$B = (z - 0.05 + j0.630764)\frac{X(z)}{z}\Big|_{z = 0.05 - j0.630764} = 1.58610e^{-j0.945375}$$

B^* is the conjugate of B. Thus, we have found the required expansion of $X(z)$ as

$$X(z) = -0.857143\,\frac{z}{z - 0.2}$$

$$+ 1.58610\left(\frac{ze^{-j0.945375}}{z - 0.05 + j0.630764} + \frac{ze^{+j0.945375}}{z - 0.05 - j0.630764}\right)$$

which, with Appendix II, results in

$$x(n) = -0.857143(0.2)^n + 1.58610e^{-j0.945375}(0.05 - j0.630764)^n$$

$$+ 1.58610e^{+j0.945375}(0.05 + j0.630764)^n$$

Using further that

$$(0.05 + j0.630764)^n = 0.63274^n e^{j1.49169n}$$

we find finally

$$x(n) = -0.85714(0.2)^n + 2 \cdot 1.5861 \cdot (0.6327)^n \cos(1.4917n + 0.94538)$$

Let us now consider the transformation of a difference equation into the z-domain. A single-input–single-output difference equation has the form

$$y(n) + \sum_{k=1}^{N} b_k y(n-k) = \sum_{k=0}^{M} a_k x(n-k) \tag{8-33}$$

where $x(n)$ and $y(n)$, respectively, represent the input and output sequences. M and N are two finite integers, and the sampling interval T has been set equal to unity. Applying the definition of the z-transform, Eq. (8-22), to Eq. (8-33) results in

$$Y(z)\left(1 + \sum_{k=1}^{N} b_k z^{-k}\right) = X(z) \sum_{k=0}^{M} a_k z^{-k} \tag{8-34a}$$

The ratio of output transform to input transform is the z-domain transfer function

$$H(z) = \frac{Y(z)}{X(z)} = \frac{\displaystyle\sum_{k=0}^{M} a_k z^{-k}}{1 + \displaystyle\sum_{k=1}^{N} b_k z^{-k}} \tag{8-34b}$$

$H(z)$ with $b_k \neq 0$ and $N > 1$ represents a so-called *recursive* or *infinite impulse response (IIR)* filter; if $b_k = 0$ for all k, $H(z)$ is called a *nonrecursive* or *finite impulse response (FIR)* filter [1–5, 47, 155].

When the input is a unit pulse function, $x(n) = \delta(n)$, we have, from Appendix II, $X(z) = 1$ and, by Eq. (8-34b), $Y(z) = H(z)$. Thus, $H(z)$ serves a role analogous to that of the transfer function $H(s)$ in the continuous-time domain, and the inverse z-transform of $H(z)$ is the unit pulse response $h(n)$.

$H(z)$ is a ratio of polynomials in z^{-1} which can be factored as

$$H(z) = \frac{a_0 \displaystyle\prod_{k=1}^{M} (1 - \alpha_k z^{-1})}{\displaystyle\prod_{k=1}^{N} (1 - \beta_k z^{-1})} \tag{8-34c}$$

where $z = \alpha_k$ and $z = \beta_k$, respectively, are the zeros and the poles of $H(z)$. We discussed earlier that the inverse z-transform via a partial fraction expansion is used to identify the time sequence. Thus, with $X(z) = 1$, a typical term of the pulse response is obtained via

$$Y(z) = H(z) = \frac{A}{1 - \beta z^{-1}} = \frac{Az}{z - \beta} \rightarrow y(n) = h(n) = A\beta^n$$

It shows quite clearly that the samples are diverging (increasing in magnitude with increasing time nT) if $|\beta| > 1$. Consequently, we conclude that:

1. If $|\beta| > 1$, the system is unstable; it has a pole *outside the unit circle* in the z-domain: $|z_p| = \beta > 1$.
2. If $|\beta| = 1$, the system is unstable; it has a pole *on the unit circle* in the z-domain: $|z_p| = \beta = 1$. The samples $y(n)$ do not go to zero as n increases; if $\beta = +1$, $y(n) = A$; if $\beta = -1$, the samples are a sequence of numbers with alternating signs at the rate $\omega_c/2$, $y(n) = (-A)^n$.
3. If $|\beta| < 1$, the system is stable; it has a pole *inside the unit circle* in the z-domain: $|z_p| = \beta < 1$.

We can further visualize the situation if we also investigate how the discrete-time z-domain relates to the continuous-time s-domain. This is done most readily by recalling the definition [Eq. (8-21)], which for $s = \sigma + j\omega$ results in

$$z = e^{sT} = e^{\sigma T} e^{j\omega T} \tag{8-35}$$

Clearly, then, we have

$$|z| = e^{\sigma T}$$

with $|z| < 1$ for $\sigma < 0$ (s is in the left half-plane); also, for physical frequencies ($s = j\omega$),

$$|z| = |e^{j\omega T}| \equiv 1$$

Thus, we conclude that Eq. (8-35) maps the $j\omega$-axis of the s-plane onto the unit circle in the z-plane, and the left half of the s-plane (where stable continuous-time filters have their poles) onto the interior of the unit circle in the z-plane (where stable discrete-time filters have their poles). Also, transmission zeros on the $j\omega$-axis in the s-plane are translated into points on the unit circle in the z-plane. It follows further that the continuous-time filter specifications which are given along the $j\omega$-axis should be transformed into discrete-time specifications *on the unit circle*. The mapping property is illustrated in Fig. 8-17. Note also that z is *periodic*: adding $j2\pi m$ to the exponent in Eq. (8-35) does not change the value of z for integer values of m. Therefore, only the horizontal strip of width $|\omega T| \leq \pi$, i.e., $|\omega| \leq \omega_c/2$, in the s-plane is needed to fill the entire z-plane. Additional horizontal strips of width $|\omega| = \omega_c$ in the s-plane map over (on top of) the "baseband z-domain," or as is said sometimes, they fill additional z-domain sheets.

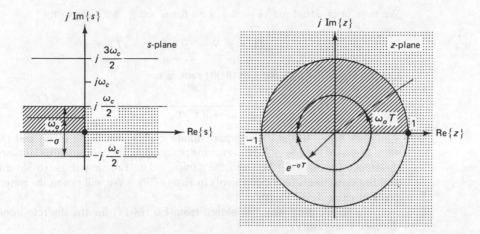

Figure 8-17   Mapping between the s-plane and z-plane: $z = e^{sT}$.

Let us illustrate some of these points with a simple example.

Example 8-3

Apply the z-transformation to the difference equations, Eqs. (8-13) and (8-14). Discuss the stability of the corresponding circuits.

Solution.   Using Eq. (8-22) and denoting

$$\mathcal{L}\{v(n)\} = V(z)$$

the difference equation, Eq. (8-13), becomes

$$C_F[V_2(z) - z^{-1}V_2(z)] + C_2V_2(z) = -C_1V_1(z)$$

so that the z-domain transfer function of the *inverting damped* integrator equals

$$H_{i,d}(z) = \frac{V_2(z)}{V_1(z)} = -\frac{C_1}{C_F(1 - z^{-1}) + C_2} = -\frac{C_1}{C_F}\frac{1}{1 + C_2/C_F - z^{-1}} \quad (8\text{-}36)$$

where $z^{-1} = e^{-sT}$.   The pole of this function is seen to be at

$$z_p = \frac{1}{1 + C_2/C_F} < 1 \quad (8\text{-}37)$$

so that the damped integrator is stable.

We reasoned earlier that an *inverting lossless* integrator is obtained by removing from Fig. 8-8a the switched capacitor C_2, i.e., the damping.   For $C_2 = 0$, we obtain from Eq. (8-36)

$$H_{i,l}(z) = -\frac{C_1}{C_F}\frac{1}{1 - z^{-1}} = -\frac{C_1}{C_FT}\frac{T}{1 - z^{-1}} \quad (8\text{-}38)$$

We note briefly that for $|sT| \ll 1$, i.e., for $\omega \ll f_c$, we have with $e^{-sT} \simeq 1-sT$

$$\frac{1}{T}(1 - z^{-1}) \simeq \frac{1}{T}(1 - 1 + sT) = s$$

Consequently, Eqs. (8-36) and (8-38) reduce to

$$H_{i,d}(e^{j\omega T}) \simeq -\frac{C_1}{C_F}\frac{f_c}{j\omega + f_c C_2/C_F} \quad \text{and} \quad H_{i,l}(e^{j\omega T}) \simeq -\frac{f_c C_1}{j\omega C_F} \qquad (8\text{-}39)$$

which are equal to the earlier approximate expressions [Eqs. (8-8a) and (8-9)] that were obtained for $f_c \to \infty$. However for practical f_c, there is an important measurable difference (or "error") between the sampled-data functions Eqs. (8-36) and (8-38) and the continuous-time equivalents in Eq. (8-39). We will probe the nature of this error in more detail shortly.

In an identical way, we obtain from Eq. (8-14) for the discrete *n*oninverting *d*amped integrator

$$[C_F(1 - z^{-1}) + C_2]V_2(z) = z^{-1}C_1 V_1(z)$$

i.e.,

$$H_{n,d}(z) = \frac{V_2(z)}{V_1(z)} = +\frac{z^{-1}C_1}{C_F(1 - z^{-1}) + C_2} = +\frac{C_1}{C_F}\frac{z^{-1}}{(1 - z^{-1}) + C_2/C_F} \qquad (8\text{-}40)$$

Notice that, apart from the expected sign inversion and a factor z^{-1} in the numerator $H_{n,d}(z)$ looks just like $H_{i,d}(z)$; thus it has the same pole and is stable. Again, if $f_c \gg 2\pi f$, Eq. (8-40) reduces to the "almost continuous" function

$$H_{n,d}(e^{j\omega T}) \simeq +\frac{C_1}{C_F}\frac{f_c}{j\omega + f_c C_2/C_F}$$

and if $C_2 = 0$, we get the *n*oninverting *l*ossless SC integrator

$$H_{n,l}(z) = +\frac{C_1}{C_F}\frac{z^{-1}}{1 - z^{-1}} \qquad (8\text{-}41)$$

It remains for us to investigate the *frequency response* of a filter characterized by the discrete-time transfer function $H(z)$; i.e., we need to determine filter behavior along the $j\omega$-axis, which now means, of course, that $H(z)$ is to be evaluated on the unit circle. Looking at gain in dB and phase, we find

$$G(\omega) = 20 \log |H(z)|_{z=e^{j\omega T}} \qquad (8\text{-}42a)$$

$$\phi(\omega) = \tan^{-1}\frac{\text{Im }\{H(z)\}}{\text{Re }\{H(z)\}}\bigg|_{z=e^{j\omega T}} \qquad (8\text{-}42b)$$

Example 8-4

Find the magnitude and phase responses of the damped inverting and noninverting integrators discussed in Example 8-3.

Solution. The exact z-domain transfer function of the inverting integrator is given in Eq. (8-36); it can be written as

$$H_{i,d}(e^{j\omega T}) = -\frac{C_1}{C_F} \frac{z^{1/2}}{z^{1/2} - z^{-1/2} + \frac{C_2}{C_F} z^{1/2}}\bigg|_{z=e^{j\omega T}}$$

$$= -\frac{C_1}{C_F} \frac{1}{2j\sin\frac{\omega T}{2} + \frac{C_2}{C_F}\left(\cos\frac{\omega T}{2} + j\sin\frac{\omega T}{2}\right)} e^{+j\omega T/2} \qquad (8\text{-}43)$$

$$= -\frac{C_1}{C_F} \frac{1}{j\sin\frac{\omega T}{2}\left(2 + \frac{C_2}{C_F}\right) + \frac{C_2}{C_F}\cos\frac{\omega T}{2}} e^{+j\omega T/2}$$

Note that the real part of the denominator, the damping factor $(C_2/C_F)\cos(\omega T/2)$, is frequency-dependent and becomes approximately constant only for very small values of ωT. Letting $\alpha = C_2/C_F$, the magnitude is

$$G_{i,d}(\omega) = 20 \log \frac{C_1/C_F}{\sqrt{4(1 + \alpha)\sin^2\frac{\omega T}{2} + \alpha^2}} \qquad (8\text{-}44a)$$

and the phase becomes

$$\phi_{i,d}(\omega) = \pi - \tan^{-1}\frac{(2 + \alpha)\sin(\omega T/2)}{\alpha\cos(\omega T/2)} + \frac{\omega T}{2} \qquad (8\text{-}44b)$$

The term π in Eq. (8-44b) represents the minus sign in Eq. (8-43). Proceeding along similar lines, we find for the transfer function of the *noninverting damped* integrator, Eq. (8-40),

$$H_{n,d}(e^{j\omega T}) = +\frac{C_1}{C_F} \frac{z^{-1/2}}{z^{1/2} - z^{-1/2} + \frac{C_2}{C_F} z^{1/2}}\bigg|_{z=e^{j\omega T}}$$

$$= +\frac{C_1}{C_F} \frac{1}{j\sin\frac{\omega T}{2}\left(2 + \frac{C_2}{C_F}\right) + \frac{C_2}{C_F}\cos\frac{\omega T}{2}} e^{-j\omega T/2} \qquad (8\text{-}45)$$

with $G_{n,d}(\omega) = G_{i,d}(\omega)$ given in Eq. (8-44a) and

$$\phi_{n,d}(\omega) = -\tan^{-1}\frac{(2 + \alpha)\sin(\omega T/2)}{\alpha\cos(\omega T/2)} - \frac{\omega T}{2} \qquad (8\text{-}46)$$

We leave it to the student to sketch these frequency responses versus ω (Problem 8-11). Let us, however, still point out the exact transfer functions for the *lossless* integrators. With $C_2 = 0$, we obtain from Eq. (8-43)

$$H_{i,l}(e^{j\omega T}) = -\frac{C_1}{C_F} \frac{z^{1/2}}{z^{1/2} - z^{-1/2}}\bigg|_{z=e^{j\omega T}} = -\frac{C_1}{TC_F} \frac{T}{2j\sin(\omega T/2)} e^{+j\omega T/2} \qquad (8\text{-}47)$$

and from Eq. (8-45)

$$H_{n,l}(e^{j\omega T}) = +\frac{C_1}{C_F}\frac{z^{-1/2}}{z^{1/2} - z^{-1/2}}\bigg|_{z=e^{j\omega T}} = +\frac{C_1}{TC_F}\frac{T}{2j\sin(\omega T/2)}e^{-j\omega T/2} \quad (8-48)$$

In a similar manner, the z-domain description of the general first-order block in Fig. 8-9, obtained from Eq. (8-15), can be shown to be

$$[C_F(1 - z^{-1}) + C_4]V_{out}(z) = -C_1V_1(z) + z^{-1}C_2V_2(z) - C_3(1 - z^{-1})V_3 \quad (8-49)$$

We leave its derivation and a discussion of magnitude and phase responses to the problems (Problem 8.12).

At this point, three observations are in order:

First, it is somewhat reassuring to note that for $2\pi f \ll f_c$ both Eq. (8-47) and Eq. (8-48) reduce again to the approximate integrator functions obtained earlier from the simple "resistor switched-capacitor equivalence". However, for practical f_c, there is a measurable error which we can determine by expressing the inverting and noninverting integrator functions, Eqs. (8-47) and (8-48), in the form

$$H_{i,l}(e^{j\omega T}) = -\frac{C_1}{C_F T}\frac{1}{j\omega}[1 - \Delta G(\omega)]e^{j\Delta\phi}$$

$$H_{n,l}(e^{j\omega T}) = +\frac{C_1}{C_F T}\frac{1}{j\omega}[1 - \Delta G(\omega)]e^{-j\Delta\phi}$$

where

$$\Delta G = 1 - \frac{\omega T/2}{\sin(\omega T/2)} \quad \text{and} \quad \Delta\phi = \omega T/2$$

Second, we observe that the inverting integrator has a *phase lead error* $\Delta\phi = +\omega T/2$, whereas the noninverting integrator has a *phase-lag error* $\Delta\phi = -\omega T/2$. Thus, when we endeavor to assemble higher-order switched-capacitor filters from loops of inverting and noninverting integrators as was discussed in Chapters 5 through 7 for the continuous-time case, the phase errors in the loops cancel, because in $H_{i,l}(z)H_{n,l}(z)$ the product $z^{1/2} z^{-1/2} = 1$.

Third, we again call the reader's attention to the fact that the z-domain transfer functions are *periodic*! This feature is quite evident from Eqs. (8-42a) and (8-42b), or, for example, Eqs. (8-47) and (8-48). In practice, this means that the frequency response of $H(z)$ is uniquely determined only in the baseband, i.e., in the frequency range $0 \le \omega \le \omega_c/2$, with the response in $\omega_c/2 < \omega < \omega_c$ being a mirror image of the baseband. Further periods of $H(z)$ are obtained when increasing ω beyond ω_c, which means by repeatedly traversing the unit circle in the z-plane.

   The periodicity of $H(z)$ will clearly give rise to potential problems and errors because in the continuous-time domain, a transfer characteristic $H(j\omega)$ is prescribed along the *whole* frequency axis $0 \le \omega < \infty$. We shall present below a suitable frequency transformation which maps the frequency interval $[0, \omega_c/2]$ of the discrete-time domain into the range $0 \le \omega \le \infty$ of the continuous-time domain. First, however, we illustrate gain calculations and the periodic nature of the z-domain functions with the following example.

Example 8-5

Determine the magnitude of the transfer function whose poles and zeros are sketched in Fig. 8-18. Draw $G(\omega)$ qualitatively for the baseband and illustrate the periodicity of $G(\omega)$ as ω increases.

Solution.   There are several ways for writing the transfer function from the pole-zero plot:

$$H(z) = K\frac{(z - z_1)(z - z_1^*)}{(z - z_2)(z - z_2^*)} = K\frac{z^2 - (2\cos\omega_1 T)z + 1}{z^2 - (2r\cos\omega_2 T)z + r^2}$$

$$= K\frac{1 - (2\cos\omega_1 T)z^{-1} + z^{-2}}{1 - (2r\cos\omega_2 T)z^{-1} + r^2 z^{-2}}$$

(8-50a)

r is the distance of the pole location from the origin in the z-plane. On the $j\omega$-axis, we find

$$H(z)\bigg|_{z = e^{j\omega T}} = K\frac{(e^{j\omega T} - e^{j\omega_1 T})(e^{j\omega T} - e^{-j\omega_1 T})}{(e^{j\omega T} - re^{j\omega_2 T})(e^{j\omega T} - re^{-j\omega_2 T})}$$

(8-50b)

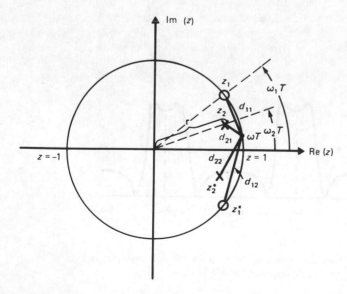

Figure 8-18   Pole-zero plot for Example 8-5.

so that, from Eq. (8-42a), the gain is computed as

$$G(\omega) = 20 \log \left(K \frac{|e^{j\omega T} - e^{j\omega_1 T}| \, |e^{j\omega T} - e^{-j\omega_1 T}|}{|e^{j\omega T} - re^{j\omega_2 T}| \, |e^{j\omega T} - re^{-j\omega_2 T}|} \right)$$

$$= 20 \log \left(K \frac{d_{11} d_{12}}{d_{21} d_{22}} \right)$$

where the d_{ij}, $i, j = 1, 2$, are the distances of the poles and zeros from the point $e^{j\omega T}$ as is indicated in Fig. 8-18. These distances are measured as functions of ωT as the point $e^{j\omega T}$ travels around the unit circle from $\omega = 0$ (i.e., $z = 1$) to $\omega = \omega_c/2$ (i.e., $z = -1$). Observe that when ωT comes close to $\omega_2 T$, d_{21} becomes a minimum and $G(\omega)$ will have a peak; also, when $\omega T = \omega_1 T$, $G(\omega) = 0$, which means we have a transmission zero at $\omega = \omega_1$. Since $\omega_1 > \omega_2$, the pole-zero plot is that of a lowpass filter; a qualitative sketch of the baseband response is shown in Fig. 8-19a. As we

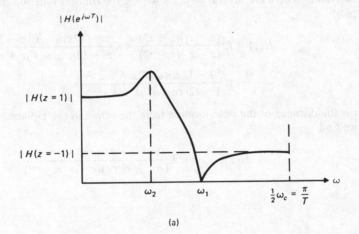

(a)

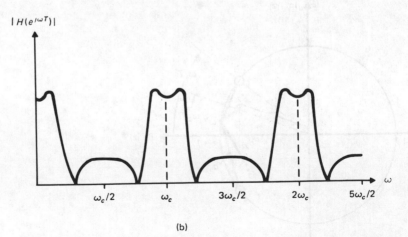

(b)

Figure 8-19 (a) Baseband gain response $|H(e^{j\omega T})|$ for the pole-zero plot in Fig. 8-18; (b) repeated baseband gain response for $\omega > \omega_c/2$.

travel around the unit circle from $\omega_c/2$ to ω_c (i.e., $z = 1$), the baseband response is repeated (mirrored at the line $\omega_c/2$). Continued travel around the unit circle will then result in the periodic response shown in Fig. 8-19b.

We develop next the so-called *bilinear s-to-z transformation*

$$s_{ct} = \frac{2}{T} \frac{z - 1}{z + 1} \tag{8-51}$$

which maps the top half of the unit circle in the z-plane, $\exp(j\omega T)$ with $0 \le \omega \le \omega_c/2$, into the positive ω_{ct}-axis, $0 \le \omega_{ct} \le \infty$, in the s_{ct}-plane. In order to keep the different frequency parameters separate in the following equations, we shall label $s_{ct} = \sigma_{ct} + j\omega_{ct}$ the frequency in the *continuous-time* domain and refer to the frequency in the sampled-data or discrete-time domain as $s = \sigma + j\omega$.

Our goal, a periodic transformation that maps an angle ζ between 0 and π into the infinite interval $0 \le \omega_{ct} \le \infty$, is accomplished by the trigonometric function

$$\omega_{ct} = \frac{2}{T} \tan \frac{\zeta}{2} = \frac{2}{T} \tan \left(\frac{\pi}{2} \frac{f}{f_c/2} \right) = \frac{2}{T} \tan \frac{\omega T}{2} \tag{8-52}$$

as shown in Fig. 8-20. The factor $2/T$ was introduced to assure that for $\omega T \ll 1$ the continuous-time frequency ω_{ct} is equal to the sampled-data frequency ω. This choice permits us to use the approximate equivalences employed earlier for the case of very fast sampling ($\omega \ll \omega_c$). Notice also from Eq. (8-52) or from Fig. 8-20 that, although the transformation is approximately linear for $|\omega| \ll \omega_c$,

$$\omega_{ct} = \frac{2}{T} \tan \frac{\omega T}{2} \bigg|_{\omega T \ll 1} \simeq \frac{2}{T} \frac{\omega T}{2} = \omega$$

the overall transformation is *nonlinear* because the infinite frequency interval $0 \le \omega_{ct} \le \infty$ is compressed into the finite range $0 \le \omega \le \omega_c/2$. The axis is said to be "*warped*."

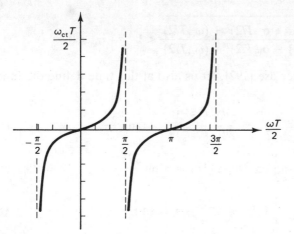

Figure 8-20 The nonlinear relationship between ω and ω_{ct}.

The desired relationship between s_{ct} and z can be obtained from Eq. (8-52) by noting that:

$$\tan\frac{\omega T}{2} = \frac{\sin\dfrac{\omega T}{2}}{\cos\dfrac{\omega T}{2}} = \frac{\dfrac{1}{2j}(e^{j\omega T/2} - e^{-j\omega T/2})}{\dfrac{1}{2}(e^{j\omega T/2} + e^{-j\omega T/2})} = \frac{1}{j}\frac{z^{1/2} - z^{-1/2}}{z^{1/2} + z^{-1/2}}\Bigg|_{s=j\omega}$$

Therefore, from Eq. (8-52) we find

$$j\omega_{ct} = \frac{2}{T}\frac{z^{1/2} - z^{-1/2}}{z^{1/2} + z^{-1/2}}\Bigg|_{s=j\omega}$$

and, after *analytic continuation* from the unit circle into the z-plane and from the $j\omega_{ct}$-axis into the s_{ct}-plane,

$$s_{ct} = \frac{2}{T}\frac{z - 1}{z + 1} \quad \text{or} \quad z = \frac{1 + s_{ct}T/2}{1 - s_{ct}T/2} \tag{8-53a}$$

Equations (8-53a) give the frequency transformations between the continuous domain, s_{ct}, and the discrete domain, z. Specifically, the physical frequencies are related by the nonlinear expressions

$$\frac{\omega_{ct}T}{2} = \tan\frac{\omega T}{2} \quad \text{and} \quad \frac{\omega T}{2} = \tan^{-1}\frac{\omega_{ct}T}{2} \tag{8-53b}$$

Notice from Eq. (8-53a) that the bilinear transformation is *rational*, so that a rational function in s_{ct} will give a rational function in z. Also note that the left half s_{ct}-plane is mapped onto the inside of the unit circle in the z-plane so that stable s_{ct}-plane poles are transformed into stable z-plane poles:

$$|z| = \left|\frac{1 + \sigma_{ct}T/2 + j\omega_{ct}T/2}{1 - \sigma_{ct}T/2 - j\omega_{ct}T/2}\right|$$

$$= \sqrt{\frac{(1 + \sigma_{ct}T/2)^2 + (\omega_{ct}T/2)^2}{(1 - \sigma_{ct}T/2)^2 + (\omega_{ct}T/2)^2}} < 1 \qquad \sigma_{ct} < 0 \tag{8-54}$$

In anticipation of their later use [192], let us also at this time define the functions [196]

$$\lambda \triangleq \frac{s_{ct}T}{2} = \frac{z^{1/2} - z^{-1/2}}{z^{1/2} + z^{-1/2}} = \tanh\frac{sT}{2} \tag{8-55a}$$

$$\gamma = \frac{1}{2}(z^{1/2} - z^{-1/2}) = \sinh\frac{sT}{2} \tag{8-55b}$$

$$\mu = \frac{1}{2}(z^{1/2} + z^{-1/2}) = \cosh\frac{sT}{2} \tag{8-55c}$$

and note from the relationships between the hyperbolic sine, cosine, and tangent functions [23] that

$$\lambda = \frac{\gamma}{\mu} \qquad \mu^2 - \gamma^2 = 1 \qquad z^{1/2} = \mu + \gamma \qquad \text{(8-55d)}$$

μ, γ, and λ are simply different transformed frequency variables which turn out to be convenient in the SC synthesis procedures to be discussed below.   An indication of their use is illustrated in Example 8-6.

Example 8-6

Express the transfer functions of the lossless and damped SC integrators discussed in Examples 8-3 and 8-4 in terms of the frequency parameters in Eq. (8-55).   Also express the characteristic of the general first-order block in Fig. 8-9 in terms of γ and μ.

Solution.   From Eqs. (8-43) and (8-45) we have

$$H_{i,d}(z) = -\frac{C_1 z^{1/2}}{C_F(z^{1/2} - z^{-1/2}) + C_2 z^{1/2}} = -\frac{C_1 z^{1/2}}{\gamma(2C_F + C_2) + \mu C_2} \qquad \text{(8-56a)}$$

$$H_{n,d}(z) = +\frac{C_1 z^{-1/2}}{C_F(z^{1/2} - z^{-1/2}) + C_2 z^{1/2}} = +\frac{C_1 z^{-1/2}}{\gamma(2C_F + C_2) + \mu C_2} \qquad \text{(8-56b)}$$

The lossless inverting and noninverting integrators are obtained by setting $C_2 = 0$:

$$H_{i,l}(z) = -\frac{C_1}{C_F} \frac{z^{1/2}}{z^{1/2} - z^{-1/2}} = -\frac{C_1}{2C_F} \frac{1}{\gamma} z^{1/2} \qquad \text{(8-57a)}$$

$$H_{n,l}(z) = +\frac{C_1}{C_F} \frac{z^{-1/2}}{z^{1/2} - z^{-1/2}} = +\frac{C_1}{2C_F} \frac{1}{\gamma} z^{-1/2} \qquad \text{(8-57b)}$$

We observe that $H_{i,l}(z)$ and $H_{n,l}(z)$ are *integrators in the γ-plane* in addition to having a leading or lagging phase error $\pm \omega T/2$ and, furthermore, that the damping component C_2 of $H_{i,d}(z)$ and $H_{n,d}(z)$ affects the integrator time constants in addition to contributing a *frequency-dependent* loss term: μC_2.   Also note, that in the γ-plane the value of the integrating capacitor C_F is effectively doubled.

For the general first-order block in Fig. 8-9, we find from Eq. (8-49)

$$z^{-1/2}[C_F(z^{1/2} - z^{-1/2}) + z^{1/2}C_4]V_{\text{out}}(z)$$

$$= -C_1 V_1(z) + z^{-1}C_2 V_2(z) - z^{-1/2}C_3(z^{1/2} - z^{-1/2})V.$$

$$[(2C_F + C_4)\gamma + C_4\mu]V_{\text{out}} = -(\mu + \gamma)C_1 V_1 + z^{-1/2}C_2 V_2 - 2C_3\gamma V_3 \qquad \text{(8-58)}$$

8.4 *DESIGN OF SWITCHED-CAPACITOR FILTERS*

In this section, we shall derive a few design methods for switched-capacitor (SC) filters.   To this end, we may start from a prescribed z-domain transfer characteristic and match to it the z-domain transfer function realized by a given SC filter topology.

This is the customary procedure followed in s-domain active RC designs, and it has also been adopted in much of the switched-capacitor literature. However, in order to facilitate the reader's understanding of the derivations and design procedures, we shall follow in our discussion as far as possible the analogous development for passive and active continuous-time filters presented in earlier chapters of this book. We mentioned in the beginning of this chapter that uncritical SC filters could be obtained by simply replacing each resistor in an active RC realization by its SC equivalent (Fig. 8-6). However, as we have seen, these designs are *only approximate* and are based on the assumption that the sampling frequency is *much higher* than the largest signal frequency, so that $\omega T \ll 1$. Because recent research [155, 191, 192] has shown that the exact SC implementation of a prescribed transfer characteristic is not more difficult than an approximate realization, we shall discuss only exact methods. Generally speaking, the SC design procedures can be understood as developing a predistorted continuous-time active RC model circuit in a transformed frequency domain. This circuit is then translated into the SC domain by replacing each resistor by the appropriate switched capacitor of Fig. 8-6b or d. For an excellent overview of, and the reasoning for, many of the approximate design procedures, the reader is referred to reference 155.

Because SC filters implement a required transfer characteristic in the z-domain [Eq. (8-34b)] but approximation procedures and extensive design tables for given transmission specifications may be more readily available for filters in the s_{ct}-domain, we discuss first the transformation of the s_{ct}-domain transfer function into the z-domain. With a suitable z-domain transfer function known, we shall then present SC realizations in the two most popular forms: just as in the continuous-time domain, these are filters realized as a cascade of second-order sections (biquads) and ladder simulations.

8.4.1 The z-Domain Transfer Function

Let us assume that the desired filter characteristics in the *sampled-data domain* are specified in the range $0 \leq \omega \leq \omega_c/2$. For example, a lowpass filter would be specified as having a passband in $0 \leq \omega \leq \omega_p$ and a stopband in $\omega_s \leq \omega \leq \omega_c/2$. Also, certain transmission zeros may be prescribed, such as ω_z indicated in Fig. 8-21a. The approximation problem, however, is normally handled in the continuous-time domain, which means that we must first apply the frequency transformation, Eq. (8-53b), to the sampled-data specifications in order to translate the requirements into the ω_{ct}-axis. Thus the passband corner would become

$$\omega_{ct,p} = \frac{2}{T} \tan \frac{\omega_p T}{2}$$

Similarly, the stopband corner and the transmission zero are transformed into

$$\omega_{ct,s} = \frac{2}{T} \tan \frac{\omega_s T}{2} \quad \text{and} \quad \omega_{ct,z} = \frac{2}{T} \tan \frac{\omega_z T}{2}$$

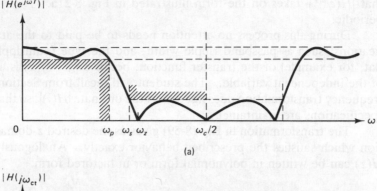

(a)

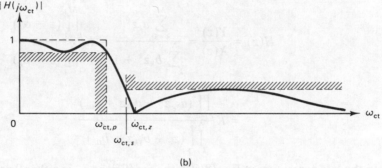

(b)

Figure 8-21   Filter specifications (a) in the sampled-data domain and (b) in the continuous-time domain obtained by prewarping.

and the point $\omega_c/2$ is translated into

$$\omega_{ct} = \frac{T}{2} \tan \frac{2\pi f_c T}{4} = \frac{T}{2} \tan \frac{\pi}{2} = \infty$$

This process, referred to as *prewarping*, is analogous to the frequency transformations discussed in Chapter 1.   Figure 8-21b shows the result of prewarping in the continuous-time domain.   Next, using any of the available procedures or tables, e.g., those discussed in Section 1.6 or in references 11 to 13, 18, 24, and 29, we find a continuous-time transfer function $H(s_{ct})$ whose magnitude $|H(j\omega_{ct})|$ satisfies the prescribed specifications in $0 \le \omega_{ct} < \infty$.   An example of a suitable lowpass characteristic is also shown in Fig. 8-21b.   The continuous-time plane s_{ct} is then transformed into the z-plane by substituting Eq. (8-51) in $H(s_{ct})$:

$$H(z) = H(s_{ct})|_{s_{ct} = (2/T)(z-1)/(z+1)} \tag{8-59}$$

Note that $H(z)$ and $H(s_{ct})$ are *not* the same functions!   The notation $H(\ )$ has been retained only for convenience of reference.   The result of the transformation in Eq. (8-59) is that the positive frequency axis $0 \le \omega_{ct} \le \infty$ is transformed into the angle $0 \le \omega T \le \pi$ along the circumference of the unit circle in the z-plane such that $|H(e^{j\omega T})|$ takes on the form illustrated in Fig. 8-21a. Note that $|H(e^{j\omega T})|$ is periodic!

During this process no attention needs to be paid to the attenuation specifications, such as passband ripple width, and the type (equiripple or maximally flat, for example) of the transfer function, because $s_{ct} = f(z)$ is a transformation of the independent variable. The student will recall from Section 1.6 that such a frequency transformation does not distort the ordinate $|H(\ )|$, so that all attenuation specifications are maintained.

The transformation in Eq. (8-59) gives us the desired z-domain transfer function which satisfies the prescribed behavior exactly. Analogously to Eq. (8-34), $H(z)$ can be written in polynomial form or in factored form:

$$H(z) = \frac{Y(z)}{X(z)} = \frac{\sum\limits_{k=0}^{M} a_k z^k}{\sum\limits_{k=1}^{N} b_k z^k + 1} = K \frac{\prod\limits_{k=1}^{M} (z - \alpha_k)}{\prod\limits_{k=1}^{N} (z - \beta_k)}$$

$$= K \frac{\prod\limits_{k=1}^{M/2} (a_{2k} z^2 + a_{1k} z + a_{0k})}{\prod\limits_{k=1}^{N/2} (z^2 + b_{1k} z + b_{0k})} \tag{8-60}$$

In the last expression in Eq. (8-60), the coefficient a_{2k} is either 0 or 1. We assumed that M and N are even, and we have combined the (usually) conjugate complex roots α_k and β_k (see Example 8-5) into second-order factors with real coefficients. Naturally, if M and/or N are odd, a first-order factor will appear in numerator and/or denominator.

The z-domain transfer function must then be realized by a suitable SC filter. The procedures will be discussed in Sections 8.4.2 and 8.4.3. First, however, we shall illustrate in Example 8-7 the steps used to derive $H(z)$ from prescribed filter requirements.

Example 8-7

To be realized is a switched-capacitor lowpass filter with the specifications

Passband attenuation: $A_p \leq 0.9$ dB in $0 \leq f \leq 3.2$ kHz $= f_p$
Stopband attenuation: $A_s \geq 22$ dB in 4.3 kHz $= f_s \leq f \leq f_c/2$

The clock frequency $f_c = 24$ kHz. Find the z-domain transfer function.

Solution. First, we prewarp the specifications into the continuous-time domain by applying Eq. (8-53b):

$$2\pi \cdot f_{ct,p} = \frac{2}{T} \tan \frac{\omega_p T}{2} = 48 \cdot 10^3 \tan \pi \frac{f_p}{f_c} = 48 \cdot 10^3 \tan \pi \frac{3.2}{24} = 2\pi \cdot 3401 \, \text{rad/s}$$

$$2\pi \cdot f_{ct,s} = \frac{2}{T} \tan \frac{\omega_s T}{2} = 48 \cdot 10^3 \tan \pi \frac{f_s}{f_c} = 48 \cdot 10^3 \tan \pi \frac{4.3}{24} = 2\pi \cdot 4820 \, \text{rad/s}$$

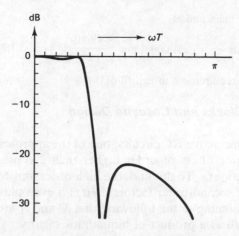

Figure 8-22 Plot of Eq. (8-63) in dB versus ωT.

Thus, along the ω_{ct} axis, the specifications are

Passband attenuation: $A_p \leq 0.9$ dB in $0 \leq f_{ct} \leq 3.401$ kHz $= f_{ct,p}$
Stopband attenuation: $A_s \geq 22$ dB in 4.820 kHz $= f_{ct,s} \leq f_{ct} \leq f_c/2$

To keep the filter order as low as possible, we choose an elliptic approximation function; from Table III-3b in Appendix III we find

$$H(\bar{s}_{ct}) = \frac{0.2745(\bar{s}_{ct}^2 + 2.4136)}{(\bar{s}_{ct} + 0.6358)(\bar{s}_{ct}^2 + 0.3614\,\bar{s}_{ct} + 1.0418)} \tag{8-61}$$

where $\bar{s}_{ct}$ is normalized with respect to the passband corner, i.e., with Eq. (8-51),

$$\bar{s}_{ct} = \frac{s_{ct}}{2\pi \cdot f_{ct,p}} = \frac{2}{\omega_{ct,p}T}\frac{z-1}{z+1} = \frac{f_c}{\pi f_{ct,p}}\frac{z-1}{z+1} = 2.2469\frac{z-1}{z+1} \tag{8-62}$$

Substituting Eq. (8-62) in Eq. (8-61) gives the desired z-domain transfer function which has the prescribed behavior (see Fig. 8-22) along the unit circle in the z-plane:

$$H(z) = \frac{0.2745\left[\left(2.2469\dfrac{z-1}{z+1}\right)^2 + 2.4136\right]}{\left[\left(2.2469\dfrac{z-1}{z+1}\right) + 0.6358\right]\left[\left(2.2469\dfrac{z-1}{z+1}\right)^2 + 0.3614\left(2.2469\dfrac{z-1}{z+1}\right) + 1.0418\right]}$$

$$= \frac{0.10285(z+1)(z^2 - 0.70621z + 1)}{(z - 0.55889)(z^2 - 1.1579z + 0.76494)} \tag{8-63}$$

Notice that $H(z)$ has a zero at $z_1 = -1$, corresponding to $\omega_{ct} = \infty$, and a pair of complex conjugate zeros at $z_{2,3} = e^{\pm j\omega T} = e^{\pm j1.2100}$. This gives transmission zeros at

$$\omega = \pm\frac{1.2100}{T} = \pm 1.2100 \cdot 24 \text{ krad/s} = \pm 2\pi \cdot 4.6219 \text{ krad/s}$$

in the sampled-data domain and at

$$\omega_{ct} = \pm\frac{2}{T}\tan\frac{\omega T}{2} = \pm 48{,}000\left(\tan\pi\,\frac{4.6219}{24}\right) = \pm 33.189 \text{ krad/s}$$

in agreement with the requirement in Eq. (8-61)!

8.4.2 Biquad Building Blocks and Cascade Design

Just as with continuous-time active RC circuits, one of the approaches taken in the design of switched-capacitor filters of order higher than 2 consists of cascading first- and second-order sections. To this end, the high-order transfer function $H(z)$ is split into the product of second-order factors if $H(z)$ is even plus one first-order factor if $H(z)$ is odd. Assuming in the following that M and N are even and $N \geq M$, we can write Eq. (8-60) as a product of biquadratic factors:

$$H(z) = K\prod_{k=1}^{N/2}\frac{a_{2k}z^2 + a_{1k}z + a_{0k}}{z^2 + b_{1k}z + b_{0k}} \tag{8-64}$$

As before, the coefficient a_{2k} is either 0 or 1. Naturally, if N is odd, $H(z)$ will contain a first-order factor

$$H_1(z) = \frac{a_1 z + a_0}{b_1 z + 1} \tag{8-65}$$

which is easy to realize (Problem 8.19). To implement Eq. (8-64), we need an SC circuit that realizes the general biquadratic transfer function

$$H_2(z) = K_2\frac{a_2 z^2 + a_1 z + a_0}{z^2 + b_1 z + b_0} \tag{8-66}$$

Several such circuits exist in the literature [47, 155, 185, 191, 194]; the most intuitive derivation starts from a known active RC biquad and makes use of the approximate equivalence [Eq. (8-6)] and the fact that both positive and negative "resistors" can be obtained by appropriate switch phasing (Fig. 8-6). For example, a convenient starting point is the general *two-integrator loop* of Fig. 5-27 leading to the Tow-Thomas biquad of Fig. 5-28 or the Åckerberg-Mossberg biquad of Fig. 5-31. The noninverting integrator is realized by choosing a *negative* SC resistor at the input of the second op amp so that the op amp inverters of active RC realizations can be eliminated. Let us further generalize the two-integrator loop by employing feed-in capacitors and, since the implementation of both positive and negative resistors poses no difficulties, feed-in resistors of both signs; it leads to the greatest flexibility in realizing arbitrary numerator polynomials (transmission zeros). The resulting circuit is shown in Fig. 8-23. As indicated by the broken line, it consists of two general first-order building blocks. Having verified by routine analysis that this circuit realizes a general biquadratic transfer function (Problem 8.20), we replace each resistor by a switched capacitor with switch phasing chosen to implement the appropriate signs. Figure 8-24 shows the circuit, where we have labeled

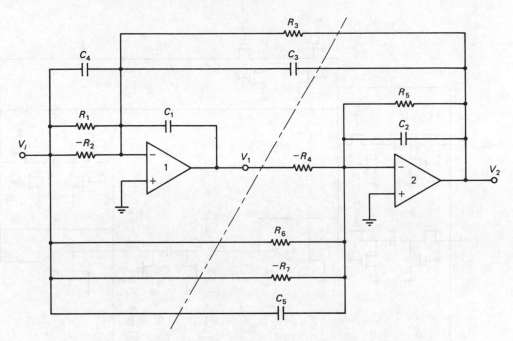

Figure 8-23 Active *RC* prototype circuit for the SC biquad in Fig. 8-24.

the capacitors $A, B, \ldots, L$ [47, 194] to simplify notation. Analysis of the two general first-order summing integrators indicated by the broken line[3] proceeds with the help of Fig. 8-9 and Eq. (8-49); we obtain:

$$D(1 - z^{-1})V_1 = [-G + z^{-1}H - L(1 - z^{-1})]V_i - [C + E(1 - z^{-1})]V_2 \qquad (8\text{-}67a)$$

$$[B(1 - z^{-1}) + F]V_2 = Az^{-1}V_1 + [z^{-1}J - I - K(1 - z^{-1})]V_i \qquad (8\text{-}67b)$$

Eliminating V_2 and V_1, respectively, from these two equations results in the z-domain transfer functions realized at the outputs of the two op amps:

$$H_{2,1}(z) = \frac{V_1}{V_i}$$

$$= \frac{[zC + E(z-1)][zI - J + K(z-1)] - [zF + B(z-1)][zG - H + L(z-1)]}{A[zC + E(z-1)] + D(z-1)[zF + B(z-1)]}$$

$$(8\text{-}68a)$$

$$H_{2,2}(z) = \frac{V_2}{V_i}$$

$$= -\frac{A[zG - H + (z-1)L] + D(z-1)[zI - J + (z-1)K]}{A[zC + E(z-1)] + D(z-1)[zF + B(z-1)]} \qquad (8\text{-}68b)$$

[3]Since by assumption the signals are sampled only once per clock period T, we may combine two of these blocks to obtain the SC biquad in Fig. 8-24.

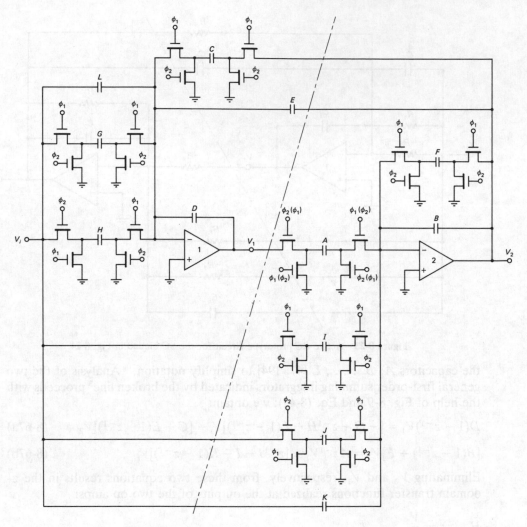

Figure 8-24 SC biquad circuit derived from Fig. 8-23. As indicated, the output is sampled at ϕ_1.

Evidently, $H_{2,1}(z)$ and $H_{2,2}(z)$ are two second-order transfer functions which can be equated to Eq. (8-66) by an appropriate choice of the capacitors A through L. We shall consider the details of this process presently. First, however, let us make an interesting observation which relates to the dynamic range of our SC circuit. As discussed earlier in this book, in order to optimize the dynamic range, we need to equalize the signal levels that the two op amps must process. Suppose, then, analysis of the circuit shows that for a particular design $|V_{1,\max}|$ is m times as large as $|V_{2,\max}|$. A study of Eqs. (8-68a) and (8-68b) shows that replacing capacitors

A and D by mA and mD, respectively, i.e.,

$$A \rightarrow mA \quad \text{and} \quad D \rightarrow mD \tag{8-69a}$$

will reduce the signal level of V_1 by a factor m without changing V_2. Note that A and D are the capacitors connected to the output of op amp 1. Alternatively, we could have replaced capacitors B, C, F, and E as follows:

$$B \rightarrow \frac{1}{m}B \qquad C \rightarrow \frac{1}{m}C \qquad F \rightarrow \frac{1}{m}F \qquad E \rightarrow \frac{1}{m}E \tag{8-69b}$$

to make V_2 m times larger without changing the level of V_1. Note that these capacitors are connected to the output of op amp 2. Further observe that the capacitors A, B, F, I, J, and K, and the capacitors C, D, G, H, and L, respectively, of the two summing integrators can be scaled independently to obtain convenient values.

Returning now to Eq. (8-68b), whose zeros are formed in a simpler way than those of $H_{2,1}(z)$, we observe that there exists considerable freedom in choosing the capacitors to implement a given biquadratic function. $H_{2,2}(z)$ can be brought into the form

$H_{2,2}(z)$

$$= -\frac{D(I+K)z^2 + [A(G+L) - D(I+J+2K)]z + [D(J+K) - A(H+L)]}{D(B+F)z^2 + [A(C+E) - D(F+2B)]z + DB - AE}$$

$$\tag{8-70}$$

which is to be compared with a given expression of the form of Eq. (8-66).

A special case of the configuration in Fig. 8-24 is obtained by setting both feed-in capacitors L and K to zero. This choice leads to one of the circuits proposed in references 47 and 194, where the resulting popular biquad, shown in Fig. 8-25 using *switch sharing*, was analyzed and discussed in considerable detail. Switch sharing, i.e., the elimination of redundant switches, was discussed earlier in connection with Fig. 8-8. Setting $K = L = 0$ and normalizing B and D to unity reduces Eq. (8-70) to

$$H_{2,2}(z) = -\frac{Iz^2 + [AG - (I+J)]z + (J - AH)}{(1+F)z^2 + [A(C+E) - (F+2)]z + 1 - AE} \tag{8-71a}$$

which shows that the value of the capacitor A is a free parameter that can be chosen arbitrarily. Setting also $A = 1$ yields

$$H_{2,2}(z) = -\frac{Iz^2 + (G - I - J)z + (J - H)}{(1+F)z^2 + (C + E - F + 2)z + 1 - E} \tag{8-71b}$$

Equation (8-71b) shows clearly that the poles of this SC biquad are determined by the feedback capacitors C, E, and F, whereas the zeros depend on the feed-in capacitors I, J, G, and H. Thus, poles and zeros are separately adjustable and, in addition, the circuit allows independent tuning of the three numerator coeffi-

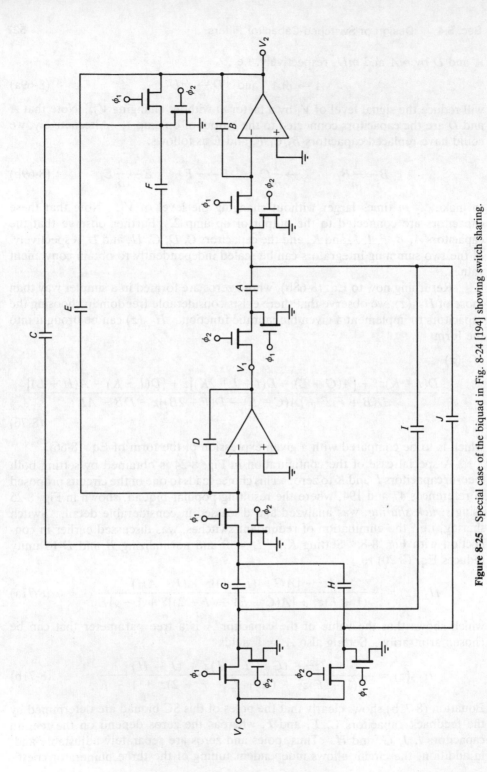

Figure 8-25 Special case of the biquad in Fig. 8-24 showing switch sharing.

cients, permitting arbitrary zeros to be realized. Equation (8-71b) indicates further that pole Q can be implemented either with E damping (i.e., $F = 0$) or with F damping ($E = 0$). One can show that all stable poles can be realized [194].[4] The derivation of the design equations will be left to the problems (Problem 8.23).

A different, frequently used biquad that has appeared in the literature with a number of modifications [185, 191, 193, 194] can be derived from the general topology, in Fig. 8-24 by setting $F = H = I = 0$ and again $B = D = 1$. Figure 8-26 shows this circuit, again using switch sharing. From Eq. (8-70), the transfer function is

$$H_{2.2}(z) = -\frac{Kz^2 + [A(G + L) - (J + 2K)]z + [(J + K) - AL]}{z^2 + [A(C + E) - 2]z + 1 - AE} \tag{8-72}$$

The choice of the remaining capacitors, computed by comparing Eq. (8-72) with Eq. (8-66), determines the location of the poles and zeros, i.e., the type of filter. The circuit can realize any type of second-order transfer function.

An alternative intuitive method for determining the required components in the SC filter starts from a second-order continuous-time prototype in the s_{ct}-domain. If we *rescale* the s_{ct}-axis by $T/2$, i.e., if we write the active RC prototype function in the λ-domain [$\lambda = s_{ct}T/2$; see Eq. (8-55a)], the most general second-order continuous-time transfer function becomes

$$H_2(\lambda) = -\frac{a\lambda^2 + b\lambda + c}{\lambda^2 + d\lambda + e} \tag{8-73a}$$

Using next the *bilinear transformation* [Eq. (8-55a)], $\lambda = (z - 1)/(z + 1)$, we obtain the function

$$H_2(z) = -\frac{(a + b + c)z^2 + 2(c - a)z + (a - b + c)}{(1 + d + e)z^2 + 2(e - 1)z + (1 - d + e)} \tag{8-73b}$$

which is to be compared with Eq. (8-72). To simplify algebra and notation, we set in addition to $B = D = 1$ also $A = 1$ and abbreviate

$$\frac{1}{1 + d + e} = m$$

Recall that A, B, and D were shown to be free parameters which can be adjusted later to optimize dynamic range and to scale the capacitors at each op amp input. Comparing coefficients between Eqs. (8-72) and (8-73b) then yields for the denominator:

$$1 - E = m(1 - d + e) \quad \text{and} \quad C + E - 2 = 2m(e - 1)$$

---

[4]Both E circuits and F circuits can be used to realize arbitrary transfer functions. Often, high-Q designs are more efficiently realized as E circuits (using less total capacitor area) whereas F circuits result in more efficient low-Q designs [47].

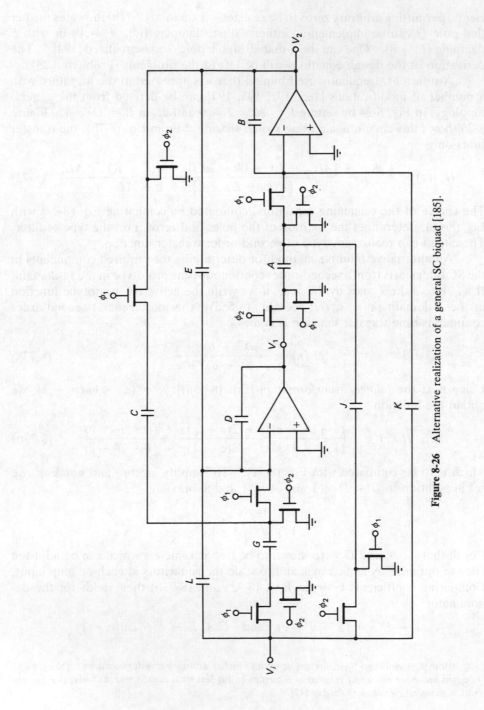

Figure 8-26 Alternative realization of a general SC biquad [185].

These two equations may be solved for C and E to give

$$C = 1 + m(3e - d - 1) \quad \text{and} \quad E = 1 - m(1 - d + e) \qquad (8\text{-}74)$$

for setting the correct pole positions. Similarly, we obtain for the numerator polynomials:

$$K = m(a + b + c) \qquad (8\text{-}75a)$$

$$2m(c - a) = G + L - J - 2K \qquad (8\text{-}75b)$$

and

$$m(a - b + c) = J + K - L \qquad (8\text{-}75c)$$

Subtracting Eq. (8-75c) from Eq. (8-75a) results in

$$L - J = 2mb \qquad (8\text{-}75d)$$

and adding Eq. (8-75c) to Eq. (8-75a) gives

$$2m(c + a) = 2K - L + J$$

which together with Eq. (8-75b) yields

$$G = 4mc \qquad (8\text{-}75e)$$

Equations (8-74) and (8-75) permit us to compute the element values of the SC biquad for arbitrarily prescribed coefficients a, b, c, d, and e in the prototype function in Eq. (8-73a). The pole positions always determine the capacitors A and E by Eq. (8-74). From the remaining equations [Eq. (8-75)], we obtain the capacitors G, J, K, and L for the feed-in components. For instance, the typical biquad functions result in:

Lowpass: $a = b = 0$ → $J = L = 0, K = mc, G = 4mc$

Bandpass: $a = c = 0$ → $G = J = 0, K = mb, L = 2mb$

Highpass: $b = c = 0$ → $G = J = L = 0, K = ma$

Notch: $b = 0$ → $J = L = 0, K = m(a + c), G = 4mc$

Allpass: $a = 1, b = -d, c = e$ → $L = 0, K = m(1 - d + e), J = 2md, G = 4mc$

Note, that either L or J in Eq. (8-75d) is a free parameter; in fact one of them can be set to zero. In practice this choice can be used to reduce sensitivities or capacitor spreads. Once the "raw" values for capacitors A through L have been calculated, A and D can be adjusted for optimum dynamic range and capacitor groups (C, D, E, G, L) and (A, B, J, K) independently scaled to obtain conveniently realizable capacitor values.

The design of a specified high-order transfer function as a cascade of SC biquads follows, then, exactly the procedure outlined in Section 6.2 for continuous-time active RC filters; we shall not repeat the discussion at this point and instead refer the reader to Chapter 6 for the details. In general, the filter requirements, specified along the sampled-data axis ω, are prewarped to the continuous-time axis ω_{ct} so that a transfer function $H(s_{ct})$ can be found (Example 8-7). The questions of factoring, pole-zero pairing, section ordering, and gain assignment are handled in the s_{ct}-domain (Examples 6-1 through 6-3). The resulting second-order sections are frequency-scaled via $s_{ct}T/2 = \lambda$ and further transformed via $\lambda = (z - 1)/(z + 1)$ in order to obtain the z-domain transfer functions which can be compared with the appropriate SC biquad, such as the one in Fig. 8-26 [Eq. (8-72)]. The component values are then obtained by applying Eqs. (8-74) and (8-75). The following example will illustrate the procedure.

Example 8-8

Design an SC cascade bandpass filter with the following specifications:

1 dB–equiripple passband in 6.73 kHz $\leq f \leq$ 9.42 kHz
>60 dB attenuation in 0 $\leq f \leq$ 5.56 kHz
>48 dB attenuation in 18.5 kHz $\leq f \leq f_c/2$

The clock frequency is $f_c = 124$ kHz.

Solution. Applying Eq. (8-53b), the specifications are transformed (prewarped) into the continuous-time domain as demonstrated in Example 8-7; we obtain:

1 dB–equiripple passband in 6.80 kHz $\leq f_{ct} \leq$ 9.60 kHz
>60 dB attenuation in 0 $\leq f_{ct} \leq$ 5.60 kHz
>48 dB attenuation in 20 kHz $\leq f_{ct} < \infty$

These requirements are exactly the ones of Example 6-1, except that the frequency is lower by a factor of 10. It follows that the transfer function is that of Eq. (6-8) but with the frequencies scaled appropriately by a factor of 10:

$$H(s_{ct}) = \frac{Ks_{ct}^2 \, (s_{ct}^2 + 4.100^2)(s_{ct}^2 + 5.4092^2)}{\displaystyle\prod_{i=1}^{4} \left(s^2 + \frac{\omega_{cti}}{Q_i} + \omega_{cti}^2\right)} \tag{8-76}$$

where

$$\omega_{ct1} = 6.8093 \qquad Q_1 = 32.047 \qquad \omega_{ct2} = 7.3094 \qquad Q_2 = 10.404$$

$$\omega_{ct3} = 8.3794 \qquad Q_3 = 7.500 \qquad \omega_{ct4} = 9.5528 \qquad Q_4 = 14.604$$

All frequency parameters are normalized with respect to $\omega_0 = 2\pi$ krad/s.

In Examples 6-2 and 6-3 we found the optimal gain constants and the section ordering (note that they are independent of the frequency scaling!); the result was:

$$H(s_{ct}) = \frac{1.60 \cdot 9.5528 \, s_{ct}}{s_{ct}^2 + \dfrac{9.5528}{14.604} s_{ct} + 9.5528^2} \cdot \frac{0.72(s_{ct}^2 + 5.4092^2)}{s_{ct}^2 + \dfrac{6.8093}{32.047} s_{ct} + 6.8093^2}$$

$$\frac{0.32(s_{ct}^2 + 4.100^2)}{s_{ct}^2 + \dfrac{7.3049}{10.404}s_{ct} + 7.3049^2} \cdot \frac{0.45 \cdot 8.3704\, s_{ct}}{s_{ct}^2 + \dfrac{8.3704}{7.500}s_{ct} + 8.3704^2}$$

$$= T_1(s)T_2(s)T_3(s)T_4(s) \tag{8-77}$$

In the next step, we denormalize the frequency by ω_0 and renormalize it with respect to $T/2$. This amounts to multiplying numerator and denominator of each biquad by the *square* of

$$\frac{\omega_0 T}{2} = \frac{2\pi \cdot 1000}{2 \cdot 124,000} = \frac{\pi}{124} = 0.02533542$$

as is demonstrated in the following equation, where we have used the subscript n to indicate *normalized* quantities:

$$H = \frac{a_n s_n^2 + b_n s_n + c_n}{s_n^2 + d_n s_n + e_n} = \frac{a_n\left(\dfrac{s}{\omega_0}\right)^2 + b_n\dfrac{s}{\omega_0} + c_n}{\left(\dfrac{s}{\omega_0}\right)^2 + d_n\dfrac{s}{\omega_0} + e_n} \times \frac{\left(\dfrac{\omega_0 T}{2}\right)^2}{\left(\dfrac{\omega_0 T}{2}\right)^2}$$

$$= \frac{a_n\left(\dfrac{sT}{2}\right)^2 + b_n\dfrac{\omega_0 T}{2}\dfrac{sT}{2} + c_n\left(\dfrac{\omega_0 T}{2}\right)^2}{\left(\dfrac{sT}{2}\right)^2 + d_n\dfrac{\omega_0 T}{2}\dfrac{sT}{2} + e_n\left(\dfrac{\omega_0 T}{2}\right)^2} = \frac{a\lambda^2 + b\lambda + c}{\lambda^2 + d\lambda + e}$$

Thus, from the function in the normalized s_{ct}-plane we obtain the coefficients in the λ-plane as follows:

$$a = a_n$$

$$b = b_n\frac{\omega_0 T}{2} = 25.33542 \cdot 10^{-3}b_n$$

$$c = c_n\left(\frac{\omega_0 T}{2}\right)^2 = 641.8837 \cdot 10^{-6}c_n$$

$$d = 25.33542 \cdot 10^{-3}d_n$$

$$e = 641.8837 \cdot 10^{-6}e_n$$

and the continuous-time transfer function [Eq. (8-77)] to be realized becomes in the λ-plane:

$$H(\lambda) = \frac{0.38724\lambda}{\lambda^2 + 0.016572\lambda + 0.058576} \frac{0.72\lambda^2 + 0.013552}{\lambda^2 + 0.0053832\lambda + 0.029762}$$

$$\times \frac{0.32\lambda^2 + 0.0034528}{\lambda^2 + 0.017789\lambda + 0.034252} \frac{0.095430\lambda}{\lambda^2 + 0.028276\lambda + 0.044973}$$

$$= T_1(\lambda)T_2(\lambda)T_3(\lambda)T_4(\lambda) \tag{8-78}$$

With these numbers, we obtain from Eqs. (8-74) and (8-75) the following parameters:

T_1: $A = 0.4668$ $E = 0.06604$ $K = 0.3602$ $L = 1.543$ $J = G = 0$
T_2: $A = 0.3391$ $E = 0.03067$ $K = 0.7087$ $L = J = 0$ $G = 0.1544$
T_3: $A = 0.3609$ $E = 0.09371$ $K = 0.3075$ $L = J = 0$ $G = 0.03638$
T_4: $A = 0.4094$ $E = 0.1287$ $K = 0.08892$ $L = 0.4344$ $J = G = 0$

Note that these values are scaled such that $B = D = 1$. If we further choose the smallest capacitor as $C_{min} = 0.5$ pF, the circuit capacitors become:

T_1: $B = D = 7.57$ $A = C = 3.53$ $E = 0.500$ $K = 2.73$ $L = 11.7$
T_2: $B = D = 16.3$ $A = C = 5.53$ $E = 0.500$ $K = 11.6$ $G = 2.52$
T_3: $B = D = 13.7$ $A = C = 4.96$ $E = 1.29$ $K = 4.23$ $G = 0.500$
T_4: $B = D = 5.62$ $A = C = 2.32$ $E = 0.724$ $K = 0.500$ $L = 2.44$

(all values in picofarads). Notice that the largest capacitor ratio in each section is determined approximately by the quality factor of that section and that the total capacitance required to realize the eighth-order filter is less than 100 pF. All values scale directly with C_{min}. For example, if for our design with parasitics-insensitive biquads $C_{min} = 0.2$ pF would be acceptable, all capacitors could be reduced by a factor of 0.4 with the total capacitance being less than 40 pf.

8.4.3 Ladder Simulations

Several times throughout this book we alluded to the fact that in many situations LC ladders form the best possible filter realization because of their low passband sensitivities to component tolerances. Consequently, it is only natural that many designers have worked on developing switched-capacitor filters which simulate the performance of lossless ladders. From the study of SC design methods up to this point in our discussion, the student will correctly anticipate that an SC realization of LC ladder structures should indeed be possible: we have only to note that inverting and noninverting, lossy or lossless, summing SC integrators can be built (see Figs. 8-7 through 8-9) and that these are the only "components" necessary for obtaining a ladder via the signal-flow graph (SFG) method. The SFG method was discussed at great length in Chapter 6 for continuous-time active RC designs. Since it will be used in the following for switched-capacitor ladder implementations in an almost identical fashion, the student is strongly advised to review the relevant material at this time (Section 6.4.1.4). We note already here that the systematic procedure to be discussed has one limitation: it requires that the series arms in the prototype LC ladder have at least one inductor. Thus, all-pole highpass functions, e.g. cannot be realized by this method. For such filters, we use instead the cascade design discussed in Section 8.4.2.

Like all other SC filter designs, ladder simulations start from transfer characteristics specified on the discrete-time frequency axis ω in $0 \leq \omega \leq \omega_c/2$; from the ω-axis, the specifications are transformed (prewarped) onto the continuous-time axis ω_{ct} by applying Eq. (8-53b). The prewarping process was explained in Section 8.4.1 and in Examples 8-7 and 8-8. Further, as was also explained in Example 8-8, the approximation problem is solved on the ω_{ct}-axis, leading to the transfer function $H(s_{ct})$, which is renormalized into the λ-plane by setting $s_{ct}T/2 = \lambda$.

At this point, we have obtained a transfer function $H(\lambda)$ which satisfies the prescribed filter requirements along the λ-axis in $0 \le \lambda \le \infty$ *exactly*. These requirements can then be transformed *exactly* onto the unit circle in the z-plane by Eq. (8-59), i.e., by substituting Eq. (8-55a):

$$\lambda = \frac{z - 1}{z + 1} \tag{8-79}$$

The question of how to realize the resulting transfer function as an SC ladder simulation is addressed most easily and intuitively by staying in the continuous-time λ-domain, i.e., by realizing $H(\lambda)$ instead of $H(z)$.[5] Let us, therefore, realize the transfer function $H(\lambda)$ as an *LC* ladder in the λ-plane by the signal-flow graph method. It was explained in Section 6.4.1.4 that this process leads to the typical ladder branches shown in Fig. 6-26 and further to the transmittances $y(\lambda)$ for the series and $z(\lambda)$ for the shunt ladder arms defined in Eqs. (6-74c) and (6-75c); they are repeated here for convenience in the λ-plane:

$$i_o = y(\lambda)(av_1 + bv_2) = \frac{av_1 + bv_2}{\lambda l_1 + \dfrac{1}{\lambda c_1} + \dfrac{1}{\lambda c_2 + 1/(\lambda l_2)}} \tag{8-80a}$$

$$v_o = -z(\lambda)(ai_1 + bi_2) = -\frac{ai_1 + bi_2}{g_k + \lambda c_1 + \dfrac{1}{\lambda l_1} + \dfrac{1}{\lambda l_2 + 1/(\lambda c_2)}} \tag{8-80b}$$

Clearly, it is irrelevant whether the independent variable is labeled s or λ; the process stays the same. Recall that all lowercase v and i symbols represent *voltages* in the signal-flow graph and that the lowercase component symbols are normalized ladder elements. Note also that we have eliminated the possible resistor in the series branch because we are concerned with *lossless* ladders. Resistors occur only as terminations, where they will be handled as shunt elements. Figure 8-27 contains the normalized series and shunt branches which are relevant for our discussion. Note that the circuits are *in the λ-plane*.

The general SFG ladder simulation, including scaling for dynamic range optimization, was shown in Fig. 6-27. In Chapter 6, we proceeded to realize the transmittances by an appropriate interconnection of s-domain integrators, leading to the active *RC* branch circuits in Fig. 6-30. By complete analogy, we should now endeavor to implement Eqs. (8-80a) and (8-80b) via integrators in the λ-*domain*. To this end, recall that λ eventually is to be the bilinear transform [Eq. (8-79)] when the final circuit is obtained as a switched-capacitor filter; i.e., we require an SC integrator to realize

$$\pm \frac{1}{k\lambda} = \pm \frac{1}{k \tanh (sT/2)}$$

---

[5]Note that for ease of reference we have labeled all transfer functions as $H(\ )$ regardless of their arguments. Actually, as was pointed out earlier, the frequency transformations result, of course, in different functions.

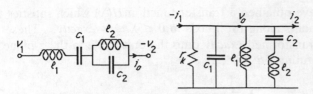

Figure 8-27 Normalized LC ladder branches in the λ-plane.

Circuits which implement various approximations of this expression are available in the literature [e.g., 47, 155], and even an exact strays-insensitive realization of a *bilinear* integrator can be built (see Problem 8-38). However, this circuit is more complicated than the strays-insensitive integrators which realize

$$\pm \frac{1}{k\gamma} = \pm \frac{1}{k \sinh (sT/2)}$$

Apart from the delay terms $e^{\pm sT/2}$, which will be shown below to cancel in this design process, we found that the circuits in Fig. 8-7 [Eq. (8-57)] realize such a function.

A solution for this problem was offered in reference 192, where it was proposed to use *frequency-dependent impedance scaling*, a concept that we encountered already in Chapter 6, Section 6.4.2.2, in Bruton's transformation; note that impedance scaling, frequency-dependent or otherwise, does not alter $H(\lambda)$. As scaling factor in Eq. (8-80), we use $\mu = \cosh (sT/2)$ as defined in Eq. (8-55c): multiplying $y(\lambda)$ and dividing $z(\lambda)$ by μ results in

$$\mu y(\lambda) = \cfrac{1}{\cfrac{\lambda}{\mu}l_1 + \cfrac{1}{\lambda\mu c_1} + \cfrac{1}{\lambda\mu c_2 + 1/\left(\cfrac{\lambda}{\mu}l_2\right)}} \qquad (8\text{-}81\text{a})$$

$$-\frac{z(\gamma)}{\mu} = -\cfrac{1}{\mu g_k + \lambda\mu c_1 + \cfrac{1}{\cfrac{\lambda}{\mu}l_1} + \cfrac{1}{\cfrac{\lambda}{\mu}l_2 + \cfrac{1}{\lambda\mu c_2}}} \qquad (8\text{-}81\text{b})$$

If we now make use of Eq. (8-55d), i.e., if we substitute $\lambda\mu = \gamma$ and $\mu^2 = 1 + \gamma^2$, we obtain

$$\mu y(\lambda) \triangleq y(\gamma) = \cfrac{1}{\cfrac{\gamma}{1 + \gamma^2}l_1 + \cfrac{1}{\gamma c_1} + \cfrac{1}{\gamma c_2 + 1/\left(\cfrac{\gamma}{1 + \gamma^2}l_2\right)}}$$

$$= \cfrac{1}{\cfrac{1}{\gamma c_1} + \cfrac{1}{\gamma/l_1 + 1/(\gamma l_1)} + \cfrac{1}{\gamma(c_2 + 1/l_2) + 1/(\gamma l_2)}} \qquad (8\text{-}82\text{a})$$

$$-\frac{z(\lambda)}{\mu} \triangleq -z(\gamma) = -\cfrac{1}{\mu g_k + \gamma c_1 + \cfrac{1}{\cfrac{\gamma}{1+\gamma^2}l_1} + \cfrac{1}{\cfrac{\gamma}{1+\gamma^2}l_2 + \cfrac{1}{\gamma c_2}}}$$

$$= -\cfrac{1}{\mu g_k + \gamma\left(c_1 + \cfrac{1}{l_1}\right) + \cfrac{1}{\gamma l_1} + \cfrac{c_2}{1 + c_2 l_2}\cfrac{\gamma(\gamma^2 + 1)}{\gamma^2 + 1/(1 + c_2 l_2)}} \qquad (8\text{-}82\text{b})$$

By resynthesizing the admittance (see Chapter 2), the last term in the denominator of Eq. (8-82b) can be brought into the form

$$\frac{c_2}{1 + c_2 l_2}\gamma + \cfrac{1}{\cfrac{(1 + c_2 l_2)^2}{c_2^2 l_2}\gamma + \cfrac{1}{\cfrac{c_2^2 l_2}{1 + c_2 l_2}\gamma}}$$

so that we obtain for the shunt transmittance the expression:

$$-z(\gamma) = -\cfrac{1}{\mu g_k + \gamma\left(c_1 + \cfrac{1}{l_1} + \cfrac{c_2}{1 + c_2 l_2}\right) + \cfrac{1}{\gamma l_1} + \cfrac{1}{\cfrac{(1 + c_2 l_2)^2}{c_2^2 l_2}\gamma + \cfrac{1}{\cfrac{c_2^2 l_2}{1 + c_2 l_2}\gamma}}}$$

$$\triangleq -\cfrac{1}{\mu g_k + \gamma c_a + \cfrac{1}{\gamma l_1} + \cfrac{1}{\gamma l_b + 1/(\gamma c_b)}} \qquad (8\text{-}82\text{c})$$

The definitions of the abbreviations c_a, c_b, and l_b should be apparent. The circuits realizing Eqs. (8-82a) and (8-82c) *in the γ-plane* are shown in Fig. 8-28a and b. We note here especially that the capacitor c_a is always nonzero.

Comparing Figs. 8-27 and 8-28 and referring back to the SFG technique, it becomes apparent that, by analogy, the transformed filter can now be realized as an *active RC circuit in the γ-plane*. Assuming this being accomplished, we can then replace each active *RC* integrator by an SC integrator to obtain the final SC simulation of a lossless ladder, obtained with the *bilinear transformation* as desired for an *exact* realization.

At this point the student will correctly object that the SFG active *RC* ladder realization method provided no procedure for handling "frequency-dependent loss elements" of the form r_k/μ which appear in the input and output shunt branches of our transformed ladder. However, far from posing any difficulties, the occurrence of the loss terms r_k/μ is very fortunate indeed for the intended exact realization, because in the final transformation into switched-capacitor form, lossy

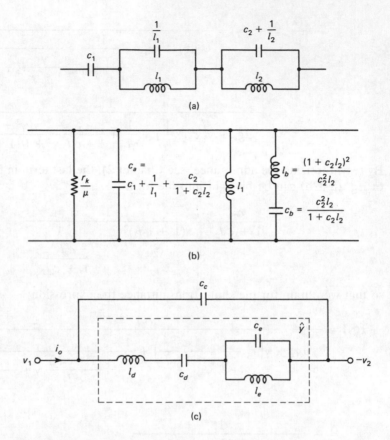

Figure 8-28 The series (a) and shunt (b) branches in the γ-plane after scaling the impedance level by μ. Note that the "resistor" is frequency-dependent. (c) Modified realization of the series branch.

integrators were found in our earlier work to possess just such a loss term [see Eq. (8-56)]!

We are ready, therefore, to transform the active *RC* circuit for the shunt branch (Fig. 8-28b) into the SC domain; it is done formally by replacing each resistor by a switched capacitor, paying attention while doing so that the correct switch phasing is implemented for noninverting integrators. Op amp–based inverters are no longer needed! The resulting circuit is shown in Fig. 8-29a, along with a version in Fig. 8-29b that saves 16 switches by using switch sharing. To obtain the transfer function in the z-domain, we proceed step by step by considering the individual integrators with the aid of Eq. (8-49), adapted to the notation in Fig. 8-29a:

$$C_4(1 - z^{-1})V_c = -C_{c4}V_b$$

$$C_3(1 - z^{-1})V_b = +C_{c3}z^{-1}V_o + C_{R4}z^{-1}V_c$$

$$C_2(1 - z^{-1})V_a = +C_{c2}z^{-1}V_o$$

$$C_1(1 - z^{-1})V_o = -C_{i1}V_{i1} - C_{i2}V_{i2} - C_0V_o - C_{R2}V_a - C_{R3}V_b$$

If these equations are solved for V_o, we obtain (Problem 8.35):

$$V_o = -\cfrac{C_{i1}V_{i1} + C_{i2}V_{i2}}{C_0 + C_1(1 - z^{-1}) + \cfrac{C_{R2}C_{c2}z^{-1}}{C_2(1 - z^{-1})} + \cfrac{C_{R3}C_{c3}z^{-1}}{C_3(1 - z^{-1}) + \cfrac{C_{R4}C_{c4}z^{-1}}{C_4(1 - z^{-1})}}}$$

(8-83a)

Multiplying further numerator and denominator by $z^{1/2}$ and making use of Eq. (8-55) results in

$$V_o = -\cfrac{(C_{i1}V_{i1} + C_{i2}V_{i2})z^{1/2}}{C_0\mu + (2C_1 + C_0)\gamma + \cfrac{C_{R2}C_{c2}}{2C_2\gamma} + \cfrac{C_{R3}C_{c3}}{2C_3\gamma + \cfrac{C_{R4}C_{c4}}{2C_4\gamma}}}$$

(8-83b)

We observe that, apart from the delay term $z^{1/2}$, the correspondence between Eqs. (8-82c) and (8-83b) is complete. We have obtained, therefore, an exact realization of the shunt branch in Fig. 8-28b based on the bilinear transform. We pointed out in connection with Eqs. (8-56) and (8-57) and we emphasize again that in the γ-plane the values of all integrating capacitors are doubled and that the capacitor representing the damping (C_0) is added to the (doubled) integrating capacitor C_1 of the *damped* integrator. We note also that $C_1 \neq 0$ to provide continuous feedback around the op amp because, as mentioned earlier, $c_a \neq 0$ in Eq. (8-82c).

The series transmittance [Eq. (8-82a)] differs from the shunt branch by a minus sign, by the absence of the terms μg_k and γc_a in the denominator, and by the presence of *two* parallel resonance circuits in series. The implementation proceeds along the same lines as that followed for Eq. (8-82c), it is in theory straightforward but the absence of the term γc_a poses a practical problem: it leads to a realization with an op amp that operates always in open loop. (The reader should demonstrate this fact by realizing Eq. (8-82a) directly.) A method that guides us out of this dilemma, at least in most cases of practical interest, lies in re-synthesizing Eq. (8-82a) in the form

$$y(\gamma) = \gamma c_c + \cfrac{1}{\gamma l_d + \cfrac{1}{\gamma c_d + \cfrac{1}{\gamma c_e + \cfrac{1}{\gamma l_e}}}}$$

(8-82d)

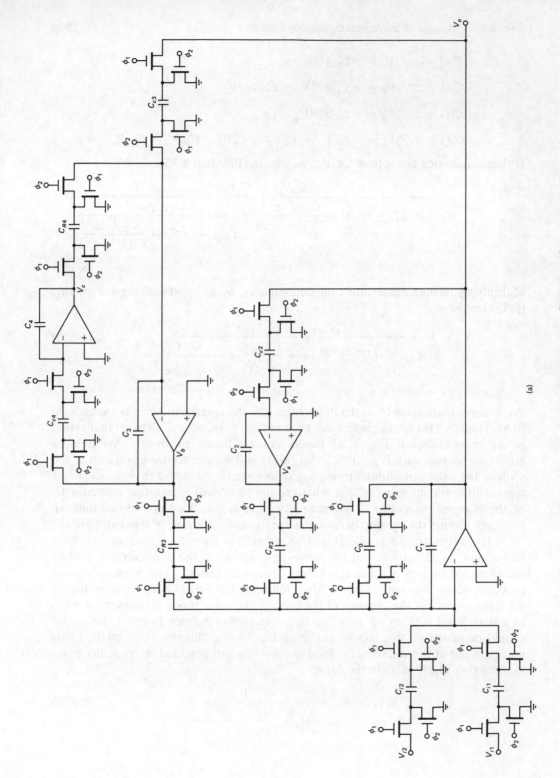

(a)

Figure 8-29 (a) SC circuit realizing the shunt transmittance [Eq. (8-82c)]; (b) same circuit using switch sharing.

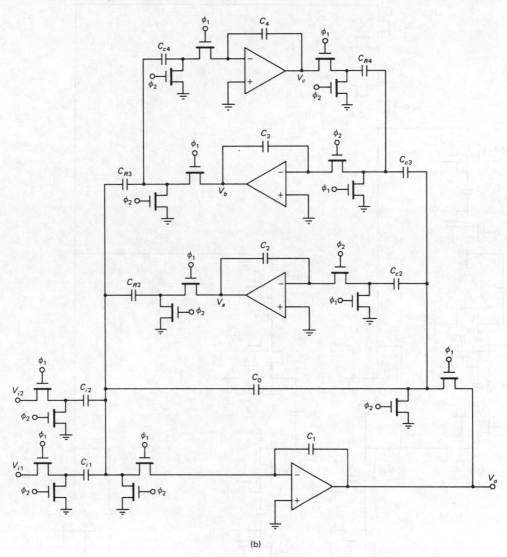

(b)

Figure 8-29 (*Continued*)

The reader will recall from Chapter 2 that all coefficients in Eq. (8-82d) are guaranteed to be positive because $y(\gamma)$ is a realizable *LC* admittance. The corresponding circuit is shown in Fig. 8-28c which can be analyzed, in terms of the quantities identified in the figure, as follows:

$$i_0 = \hat{y}(\gamma) \, (av_1 + bv_2) + \gamma c_c(av_1 + bv_2) \tag{8-84a}$$

$$= \frac{av_1 + bv_2}{\gamma l_d + \dfrac{1}{\gamma c_d} + \dfrac{1}{\gamma c_e + \dfrac{1}{\gamma l_e}}} + \gamma c_c(av_1 + bv_2)$$

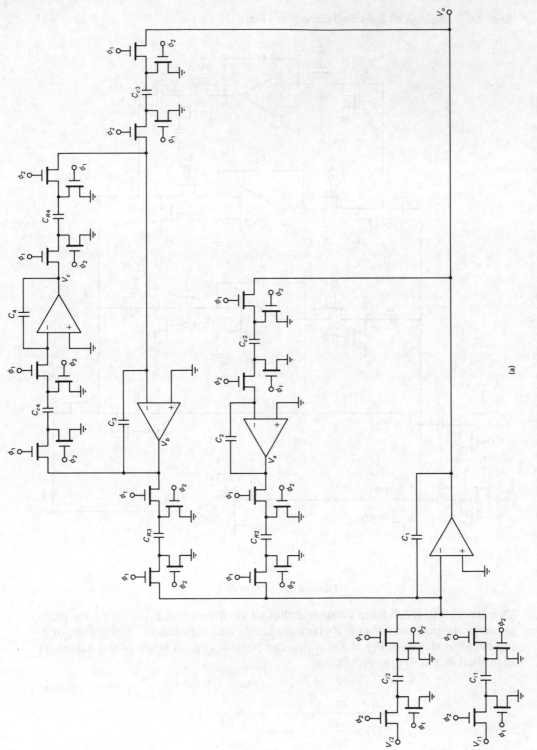

Figure 8-30 (a) Circuit realizing the series transmittance of Eq. (8-84b); (b) same circuit using switch sharing.

(a)

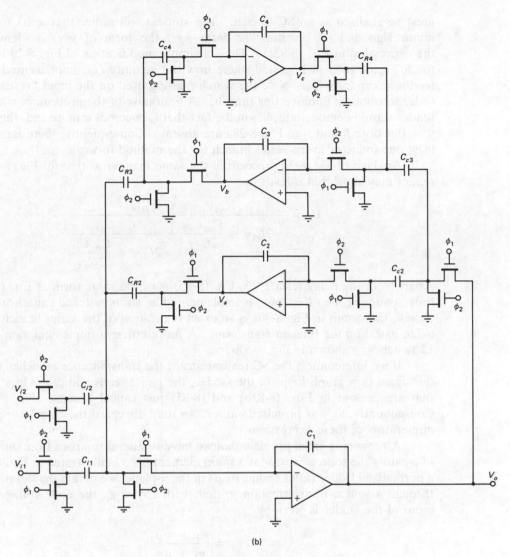

(b)

Figure 8-30 *(Continued)*

As done earlier in Eq. (8-80), in Eq. (8-84) we have again included the scale factors *a* and *b*. Now note that the effect of γc_c can be handled simply as an *unswitched* capacitor c_c connected between the circuit nodes with voltages v_1 and $-v_2$ (compare the method used in Example 7-3) so that only the transmittance

$$\hat{y}(\gamma) = \cfrac{1}{\gamma l_d + \cfrac{1}{\gamma c_d + \cfrac{1}{\gamma c_e + \cfrac{1}{\gamma l_e}}}} \qquad (8\text{-}84b)$$

must be realized as an SC branch. The student will notice that apart from the minus sign and the missing loss term μg_k, the form of $\hat{y}(\gamma)$ is identical to the expression in Eq. (8-82c). Thus, a simple modification of Fig. 8-29 leads to the SC realization in Fig. 8-30 where now the quantity l_d is implemented by the feedback capacitor C_1. Note the switch phasing used on the input "resistors" in order to obtain a noninverting branch. A limitation of the procedure, which the reader should demonstrate, lies in the fact that l_d becomes infinite and, therefore, $\hat{y} = 0$ if both l_1 and l_2 in Fig. 8-28a are absent. Consequently, there must be at least one inductor in the series branch for the method to work.

Analysis of Fig. 8-30 proceeds in the same manner as that for Fig. 8-29; the reader may show that it leads to

$$V_o = +\cfrac{[C_{i1}V_{i1} + C_{i2}V_{i2}]z^{-1/2}}{2C_1\gamma + \cfrac{C_{R2}C_{c2}}{2C_2\gamma + \cfrac{C_{R3}C_{c3}}{2C_3\gamma + \cfrac{C_{R4}C_{c4}}{2C_4\gamma}}}} \tag{8-85}$$

Apart from the delay term $z^{-1/2}$, Eq. (8-85) is of the same form as Eq. (8-84b); thus, assuming the effect of c_c is implemented as an unswitched capacitor as discussed, the circuit in Fig. 8-30a is an exact realization of the series branch in Fig. 8-28c based on the bilinear transform. A more efficient implementation, saving 12 switches, is shown in Fig. 8-30b.

If we interconnect the SC realizations of the transmittance branches to form the signal-flow graph loops of the ladder, the phase leads and phase lags $\pm \omega T/2$ that are present in Eqs. (8-83b) and (8-85) just cancel because $z^{1/2}z^{-1/2} = 1$. Consequently, as was promised earlier, we may disregard the terms $z^{\pm 1/2}$ in the numerators of these expressions.

A remaining small problem that we have to attend to arises from our choice of treating the source resistor as a *shunt* element. As is illustrated in Fig. 8-31 on a normalized ladder (after scaling by μ) in the γ-plane, we obtain the shunt resistor through a source transformation so that, with $1/r_i = g_i$, the node voltage at the input of the ladder is given by

$$v_1 = \frac{g_i\mu v_i + i_2}{y_1 + \mu g_i} \tag{8-86}$$

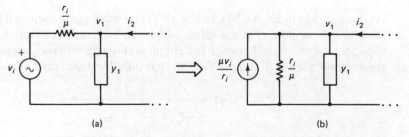

(a)   (b)

Figure 8-31   (a) Ladder input branches; (b) source transformation.

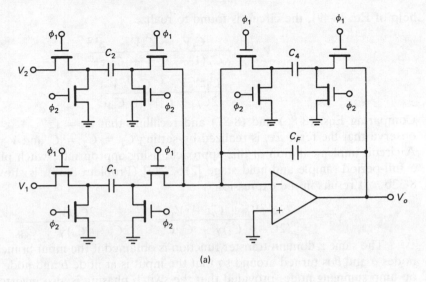

(a)

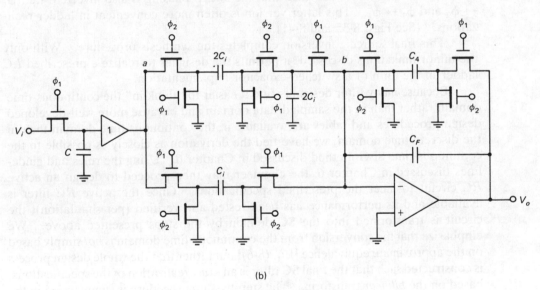

(b)

Figure 8-32   Realization of the input term μv_i: (a) two-input lossy integrator; (b) obtaining μv_i by sample-and-hold circuit and switch phasing.

As mentioned, the term μg_i in the denominator is realized automatically in a lossy SC integrator; the problem arises from the frequency-dependent source, $g_i \mu v_i$, which must be implemented at the ladder input.   To derive a realization of this term, we consult again the general first-order block in Fig. 8-9, which is redrawn in a somewhat modified form appropriate for our case in Fig. 8-32a.   With the help of Eq. (8-49), the circuit is found to realize

$$V_o = -\frac{C_1 V_1 + C_2 V_2}{C_F(1 - z^{-1}) + C_4\, z^{1/2}}\, \frac{z^{1/2}}{}$$

$$= -\frac{C_1 z^{1/2} V_1 + C_2 z^{1/2} V_2}{(2C_F + C_4)\gamma + C_4\mu} \tag{8-87}$$

Comparing Eqs. (8-86) and (8-87) and recalling that $\mu = (z^{1/2} + z^{-1/2})/2$, we observe that the term μv_i is realized by setting $C_1 = C_2 = C_i$ and $V_2 = z^{-1}V_1$. A circuit implementation of this approach, using appropriate switch phasing and a full-period sample-and-hold stage [216, 217] (Problem 8-38), is shown in Fig. 8-32b. It results in the expression

$$V_o = -\frac{C_i z^{1/2} V_1 + C_i z^{-1/2} V_1}{(2C_F + C_4)\gamma + C_4\mu} = -\frac{2C_i \mu V_1}{(2C_F + C_4)\gamma + C_4\mu} \tag{8-88}$$

The same z-domain transfer function is obtained if the input branch between nodes a and b is turned around so that the input is at node b and node a is at the op amp summing node, provided that the switch phasing is also inverted, i.e., $\phi_1 \rightarrow \phi_2$ and $\phi_2 \rightarrow \phi_1$. This latter version is often more convenient in ladder realizations. (See Figs. 8-35 and 8-41).

This final source conversion completes the synthesis procedure. With only the minor limitations discussed, it permits the designer to realize a prescribed LC ladder in the form of a switched-capacitor implementation.

Because many filter designers find it easier to "think in" the continuous-time domain rather than in the sampled-data domain and because more well-developed design procedures and tables are available in the continuous-time domain than in the discrete-time domain, we have tied the derivation as closely as possible to the continuous-time SFG method discussed in Chapter 6. Using the rules and guidelines discussed in Chapter 6, the engineer may then proceed to design an active RC circuit to meet the prescribed specifications. Once the active RC filter is available and its performance has been tested and verified (per simulation), the circuit is transformed into the SC domain by the steps presented above. We emphasize that the conversion from the continuous-time domain is *not* simply based on the approximate equivalence [Eq. (8-6)] but rather that the whole design process is constructed such that the final SC filter is an *exact* realization of the specifications, based on the *bilinear* transform. The step-by-step procedure is summarized in the following:

Step 1. Prewarp the filter specifications from the sampled-data axis ω into the continuous-time axis ω_{ct} by Eq. (8-53b). Use any convenient approximation routine to obtain $H(s_{ct})$. Renormalize $H(s_{ct})$ into $H(\lambda)$ by setting $s_{ct}T/2 = \lambda$.[6]

Step 2. Realize $H(\lambda)$ as an LC ladder in the λ-domain with normalized ad-

---

[6]In this notation, s_{ct} is the actual *denormalized* frequency in radians per second. Because most approximation routines or filter tables work in normalized frequencies s_n, with the upper passband corner ω_0 normalized to unity, we have $s_{ct} = \omega_0 s_n$, and λ is obtained by renormalizing the frequency as $\lambda = (\omega_0 T/2)s_n$.

mittances g_{in}, $1/(\lambda l_k)$, λc_k, and g_{out} such that the series branches contain inductors.

Step 3. Scale the impedance level by μ to obtain the admittances μg_{in}, $(1 + \gamma^2)/(\gamma l_k)$, γ_k^c and μg_{out} in the γ-plane. Perform a source transformation, if necessary, to obtain g_{in} as a shunt element. If convenient or desired, redraw the ladder topology in the γ-plane, showing ladder arms of the forms indicated in Fig. 8-28b and c.

Step 4. Treating now the γ-plane as a normalized s-plane s_n, proceed to realize the γ-plane circuit as an active RC filter in the s_n-plane, using the SFG technique of Section 6.4.1.4 in all details, i.e., topology, component values, scaling for element values and dynamic range, and so forth. Be sure to realize the parallel capacitors of the series transmittance [c_c in Eq. (8-82d) and in Fig. 8-28c] as unswitched capacitors between the corresponding main circuit nodes. Note that negative resistors can be used because they are easy to implement in the final SC conversion. During this process, disregard that the source and load resistors are frequency-dependent; the terminations will become automatically correct in the final transformation to switched-capacitor form.

Step 5. Convert the active RC circuit into the SC domain as follows:

- Replace each conductor G by a switched capacitor of value $C_s = TG$.

  To compensate for the effective doubling of all integrating capacitors and for the effective addition of the damping capacitor to the integrating capacitor of a *lossy* integrator.

  (i)   Subtract the capacitor representing the loss (the termination resistor) from each (*unswitched*) integrating capacitor of a *lossy* integrator and then

  (ii)   Replace each *unswitched* capacitor C_u by $0.5C_u$.

  $f_c = 1/T$ is the clock frequency. For the "SC resistors," use the strays-insensitive realizations of Fig. 8-6b or d, depending on whether inverting or noninverting integrators are to be implemented.

- Realize the input signal branch as shown in Fig. 8-32b.

- To obtain practical component values, all capacitors may be scaled by a convenient factor a. It is illustrated, in the examples below, how in a given design this factor may also be used to minimize the total capacitor area [192].

The justification for some of the details of step 5 is obtained from the following consideration. After prewarping the specifications into the continuous-time domain, assume that a particular active RC branch has been built to realize [compare also Eq. (6-86)]:

$$V_o = -\frac{G_1 V_{i1} + G_2 V_{i2}}{G_0 + s_{ct}C_1 + \dfrac{G_{R2}G_{c2}}{s_{ct}C_2} + \dfrac{G_{R3}G_{c3}}{s_{ct}C_3 + \dfrac{G_{R4}G_{c4}}{s_{ct}C_4}}} \tag{8-89}$$

Replacing each conductor G by f_cC results in

$$V_o = -\frac{f_cC_{i1}V_{i1} + f_cC_{i2}V_{i2}}{f_cC_0 + s_{ct}C_1 + \dfrac{f_cC_{R2}f_cC_{c2}}{s_{ct}C_2} + \dfrac{f_cC_{R3}f_cC_{c3}}{s_{ct}C_3 + \dfrac{f_cC_{R4}f_cC_{c4}}{s_{ct}C_4}}}$$

which, with $f_c = 1/T$, $C_i \to 0.5C_i$, $i = 2, 3, 4$, and $C_1 \to (C_1 - C_0)/2$, can be brought into the form

$$V_o = -\frac{C_{i1}V_{i1} + C_{i2}V_{i2}}{C_0 + \dfrac{s_{ct}T}{2}(C_1 - C_0) + \dfrac{C_{R2}C_{c2}}{\dfrac{s_{ct}T}{2}C_2} + \dfrac{C_{R3}C_{c3}}{\dfrac{s_{ct}T}{2}C_3 + \dfrac{C_{R4}C_{c4}}{\dfrac{s_{ct}T}{2}C_4}}} \qquad (8\text{-}90)$$

Next, we observe that the exact return to the γ-plane replaces $s_{ct}T/2$ by $\gamma = \sinh(sT/2)$ and multiplies C_0 by $(\mu + \gamma)$ [see Eq. (8-55d)]. Finally, we multiply each capacitor by a suitable scaling factor a to obtain:

$$V_o = -\frac{aC_{i1}V_{i1} + aC_{i2}V_{i2}}{aC_0\mu + aC_1\gamma + \dfrac{aC_{R2}aC_{c2}}{aC_2\gamma} + \dfrac{aC_{R3}aC_{c3}}{aC_3\gamma + \dfrac{aC_{R4}aC_{c4}}{aC_4\gamma}}} \qquad (8\text{-}91)$$

Comparing Eq. (8-91) with Eq. (8-83b) shows that the effects of doubling the unswitched capacitors and adding the damping element to the integrating capacitor of the lossy integrator are corrected. The missing delay term $z^{1/2}$ is unimportant because the terms $z^{-1/2}$ and $z^{+1/2}$ will cancel in the two-integrator loops of the SFG implementation.

As an example for scaling by a, assume that in the active RC circuit an integrator time constant $\tau = 5$ μs was realized via $R = 1$ kΩ and $C = 5$ nF. For the SC implementation, $f_c = 144$ kHz is used. It follows that the unswitched integrator capacitor and the switched capacitor equal, respectively,

$$C_{int} = \frac{a}{2}C = 2.5a \text{ nF}$$

$$C_R = \frac{a}{Rf_c} = 6.944a \text{ nF}$$

Choosing $a = 4 \cdot 10^{-3}$ yields $C_{int} = 10$ pF and $C_R = 27.77$ pF.

Let us demonstrate the complete process by guiding the reader through an example. It is kept in literal, symbolic (rather than numerical) form so that the details of the steps can be illustrated more readily.

Example 8-9

A bandpass filter to meet a certain passband ripple requirement in $\omega_2 \leq \omega \leq \omega_3$ is to be designed as a switched-capacitor simulated LC ladder circuit. The stopband attenuation specifications are in $0 \leq \omega \leq \omega_1$ and $\omega_4 \leq \omega \leq \omega_c/2$. The clock frequency equals f_c.

Solution. We start by prewarping the requirements from the ω-axis to the continuous-time domain via Eq. (8-53b):

$$f_{ct,i} = \frac{f_c}{\pi} \tan\left(\pi \frac{f_i}{f_c}\right) \quad \text{with} \quad f_i = \frac{\omega_i}{2\pi} \quad i = 1, 2, 3, 4$$

Using a suitable approximation routine (e.g., Section 1.6, or reference 11 or 29), a transfer function $H(s_{ct})$ is found next which satisfies the prescribed specifications. Assume that a function of order 5 is obtained:

$$H(s_{ct}) = \frac{\hat{b}_4 s_{ct}^4 + \ldots + \hat{b}_0}{s_{ct}^5 + \hat{a}_4 s_{ct}^4 + \ldots + \hat{a}_0} \frac{(T/2)^5}{(T/2)^5}$$

which is renormalized into the λ-plane by multiplying numerator and denominator by $(T/2)^5$ as indicated. The result is the transfer function $H(\lambda)$; let it be of the form

$$H(\lambda) = \frac{b_4 \lambda^2 (\lambda^2 + b)}{\lambda^5 + a_4 \lambda^4 + a_3 \lambda^3 + a_2 \lambda^2 + a_1 \lambda + a_0} \tag{8-92}$$

$H(\lambda)$ can now be realized in the λ-plane by use of the usual LC synthesis procedures. The resulting ladder topology with *normalized* element values (lowercase symbols) is shown in Fig. 8-33a. The components can be denormalized via

$$R_i = r_i R_p \qquad C_i = c_i \frac{T}{2} \frac{1}{R_p} \qquad L_i = l_i \frac{T}{2} R_p \tag{8-93}$$

where R_p is a suitably chosen normalizing resistor.

Staying with the normalized λ-plane ladder, we next perform a source transformation and scale the impedance level by μ, defined in Eq. (8-55c). These steps convert the circuit into the γ-plane ladder shown in Fig. 8-33b. Note that we brought the series branch into the form prescribed in Fig. 8-28c. If not already done so, we next analyze the γ-plane circuit to establish the maxima of all signals throughout the ladder. The details of this process were explained in Section 6.4.1. Let us assume that the maxima of v_1, i_2, and v_o can be equalized by inserting the multipliers (scale factors) α and β into the signal-flow graph of the γ-plane ladder as shown in Fig. 8-33c. Also indicated is a factor $K/(\alpha\beta)$ to achieve a total gain K in the circuit. Following step 4, we analyze Fig. 8-33b with the scale factors of Fig. 8-33c to obtain

$$-v_1 = -\frac{[K/(\alpha\beta)]g_1 v_s + (1/\beta)(-i_2) + [1/(\alpha\beta)]s_n c_a v_0}{g_1 + s_n(c_1 + c_a)} \tag{8-94a}$$

$$-i_2 = \frac{\beta(-v_1) + (1/\alpha)Kv_0}{s_n l_b + 1/(s_n c_b)} \tag{8-94b}$$

$$v_0 = -\frac{\alpha(-i_2) + \alpha\beta s_n c_a(-v_1)}{g_2 + s_n(c_5 + c_a + 1/l_4) + 1/(s_n l_4)} \tag{8-94c}$$

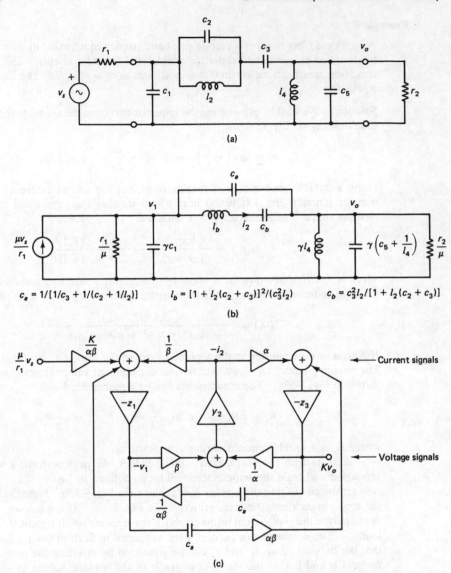

Figure 8-33 (a) LC bandpass ladder in the λ-plane; (b) the same ladder in the γ-plane after source transformation and impedance scaling by μ; (c) signal-flow graph representation of the same ladder, including signal-level scaling.

Notice that we treated the γ-plane as the "s_n-plane" and that the factor μ on the conductors was disregarded when writing Eqs. (8-94).

Our next step consists of developing an active RC signal-flow graph circuit in the s_n-plane to realize Eq. (8-94). The procedure was discussed in Chapter 6; note, however, that we may now use positive and negative resistors as needed. Figure 8-34 shows the result. A simple rule to remember is the following:

Drawing the *current*-signal line on top and the *voltage*-signal line on the bottom of the signal-flow graph (Fig. 8-33c) and positioning all op amps vertical (see Fig. 8-34), then in drawing the active RC implementation

- All resistors located *above* an op amp are positive.
- All resistors located *below* an op amp are negative.
- All resistors connected *directly between* the current and voltage lines are positive.

Analyzing this circuit yields the following equations in which C_{ni}, $i = 1, \ldots, 7$, represent frequency-normalized capacitors, i.e., $C_{ni} = C_i(2/T)$:

$$V_a = -\frac{G_6}{s_n C_{n3}}(-V_2) \tag{8-95a}$$

$$V_b = \frac{G_{10}}{s_n C_{n5}}V_o \tag{8-95b}$$

$$-V_1 = \frac{R_a G_1 V_{in} + R_a G_2(-V_2) + s_n C_{n7} R_a V_o}{R_a G_0 + s_n C_{n1} R_a} \tag{8-95c}$$

$$-V_2 = \frac{R_b G_3(-V_1) + R_b G_4 V_o}{s_n \dfrac{1}{\dfrac{C_{n3}}{R_b G_6 G_7}} + s_n C_{n2} R_b} \tag{8-95d}$$

$$V_o = -\frac{R_c G_5(-V_2) + s_n C_{n6} R_c(-V_1)}{R_c G_8 + s_n C_{n4} R_c + s_n \dfrac{1}{\dfrac{C_{n5}}{G_9 G_{10} R_c}}} \tag{8-95e}$$

Note that we have included scaling resistors R_a, R_b, and R_c for the transmittances in the last three equations to help in obtaining practical element values. Equations (8-95c) through (8-95e) are now to be compared with Eqs. (8-94a) through (8-94c); the comparison yields the following expressions for the denormalized component values:

$$R_0 = \frac{R_a}{g_1} \qquad R_1 = \frac{\alpha\beta}{K}\frac{R_a}{g_1} \qquad R_2 = \beta R_a \qquad R_3 = \frac{R_b}{\beta} \tag{8-96a}$$

$$R_4 = \alpha R_b \qquad R_5 = \frac{R_c}{\alpha} \qquad R_8 = \frac{R_c}{g_2}$$

$$C_1 = \frac{(c_1 + c_a)T}{2}\frac{1}{R_a} \qquad C_2 = \frac{l_b T}{2}\frac{1}{R_b} \qquad C_4 = \frac{(c_5 + c_a + 1/l_4)T}{2}\frac{1}{R_c}$$

$$C_6 = \alpha\beta c_a \frac{T}{2}\frac{1}{R_c} \qquad C_7 = \frac{1}{\alpha\beta}c_a\frac{T}{2}\frac{1}{R_c} \tag{8-96b}$$

$$R_6 R_7 = \frac{c_b T}{2}\frac{R_b}{C_3} \qquad R_9 R_{10} = \frac{l_4 T}{2}\frac{R_c}{C_5} \tag{8-96c}$$

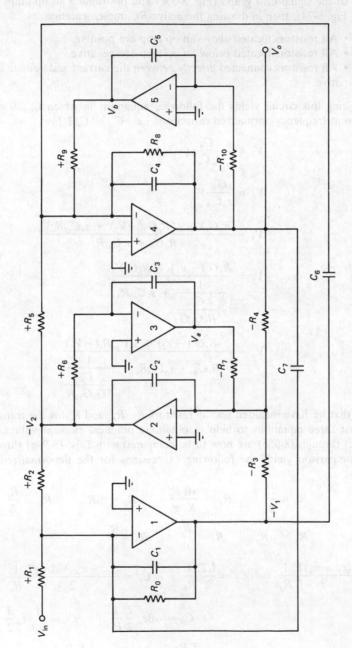

Figure 8-34 Signal-flow graph active *RC* realization of the γ-plane ladder in Fig. 8-33b in the normalized s_n-plane.

Since the last two equations determine only the product of two resistors, the additional degrees of freedom are used to equalize the signal levels as was explained in Chapter 6. For this example, let us assume that circuit analysis showed that

$$|V_a|_{max} = \frac{1}{m_a}|V_2|_{max} \qquad |V_b|_{max} = \frac{1}{m_b}|V_o|_{max}$$

Note that V_a and V_b, respectively, represent the voltage on capacitor c_b, and the current through the inductor l_4 in the γ-plane ladder. Therefore, the necessary scale factors m_a and m_b can be obtained already from the initial analysis of the ladder, performed to yield the factors α and β.

We then consult Eqs. (8-95a) and (8-95b) and observe that the signal levels are equalized by the following substitutions (the symbol $\Rightarrow$ stands for "replace by"):

$$G_6 \Rightarrow m_a G_6 \qquad G_{10} \Rightarrow m_b G_{10} \qquad (8\text{-}97a)$$

Further, from Eqs. (8-95d) and (8-95e), in order not to change the frequency response we have to set:

$$G_7 \Rightarrow \frac{1}{m_a}G_7 \qquad G_9 \Rightarrow \frac{1}{m_b}G_9 \qquad (8\text{-}97b)$$

With Eqs. (8-96) and (8-97), we can compute the *final* resistor values, $R_{f,i}$:

$$R_{f,6}m_a = R_{f,7}\frac{1}{m_a} = \sqrt{\frac{c_b T}{2}\frac{R_b}{C_3}} \qquad R_{f,9}\frac{1}{m_b} = R_{f,10}m_b = \sqrt{\frac{l_4 T}{2}\frac{R_c}{C_5}} \qquad (8\text{-}98)$$

Notice that in the element equations, the capacitors C_1, C_2, C_4 and C_6 can be scaled by an appropriate choice of R_a, R_b and R_c, and C_3 and C_5 are arbitrary.

At this stage, we have expressed all components of the active *RC prototype* or *model* in terms of the known elements of the γ-plane *LC* ladder, the clock frequency $f_c = 1/T$, and some conveniently chosen scaling resistors. The active *RC* design is, therefore, complete and we can transform the model circuit into the SC domain. To this end, we replace in Eqs. (8-96) and (8-98) each resistor $\pm R_i$ by a *switched capacitor* $\pm C_{s,i}$ of the appropriate sign, such that

$$R_i \Rightarrow C_{s,i} = \frac{T}{R_i} \qquad i = 0, 1, \ldots, 10 \qquad (8\text{-}99a)$$

and all capacitors C_j by *unswitched* capacitors $C_{u,j}$ such that

$$C_j \Rightarrow C_{u,j} = \frac{C_j}{2} \qquad j = 2, 3, 5$$
$$\qquad (8\text{-}99b)$$
$$C_1 \Rightarrow C_{u,1} = \frac{1}{2}(C_1 - C_{s,0}) \qquad C_4 \Rightarrow C_{u,4} = \frac{1}{2}(C_4 - C_{s,8})$$

Last, if needed, all capacitors may be scaled by a common scale factor a to obtain capacitor values which are practical for integrated SC filters.

The final circuit, including the version using switch sharing, is shown in Fig. 8-35. Note that the design was performed almost completely in the more familiar continuous-time domain.

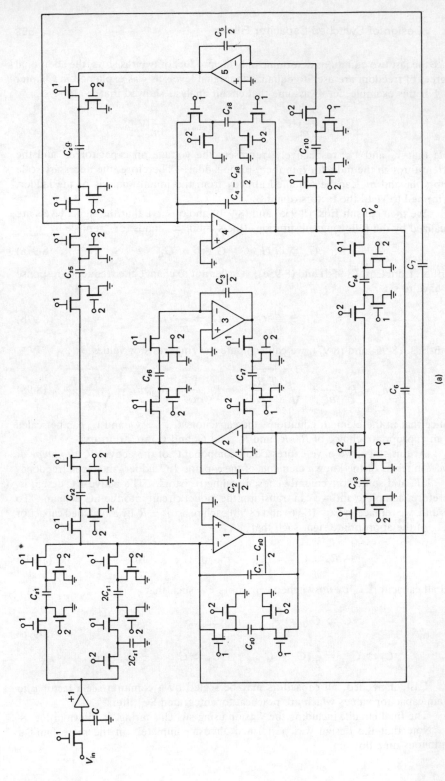

Figure 8-35 (a) SC filter realizing the transfer function of Eq. (8-92); (b) same circuit using switch sharing.

554

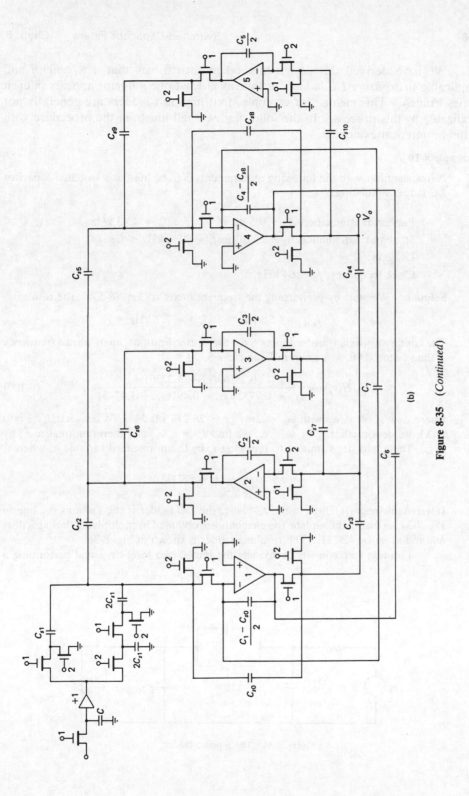

Figure 8-35 (*Continued*)

We have derived this design method in such a way that it is *general* and applicable to *arbitrary LC* ladders as long as at least one inductor appears in each series branch. This means, for example, that highpass ladders are generally not realizable by this process. In the following we shall illustrate the procedure with a final numerical example:

Example 8-10

A lowpass filter with the following requirements is to be built as a switched-capacitor *LC* ladder simulation:

Passband: attenuation $A_p \leq 0.5$ dB in $0 \leq f \leq f_p = 4.43$ kHz
Stopband: attenuation $A_s \geq 23$ dB in $f_s = 7.36$ kHz $\leq f \leq f_c/2$
DC gain: $K = 1$
Clock frequency: $f_c = 64$ kHz

Solution. We start by prewarping the frequency axis via Eq. (8-53b); the result is:

$$f_{p,\text{ct}} = 4.5 \text{ kHz} \qquad f_{s,\text{ct}} = 7.7 \text{ kHz}$$

The continuous-time transfer function for these specifications, apart from a frequency scaling factor 1000, was given in Example 7-5,

$$H(s_n) = \frac{0.28163(s_n^2 + 3.2236)}{(s_n + 0.7732)(s_n^2 + 0.4916s_n + 1.1742)} \tag{8-100}$$

where now $s_n = s_{\text{ct}}/\omega_0$ with $\omega_0 = 2\pi f_{p,\text{ct}} = 28.274$ krad/s. To convert $H(s_n)$ into $H(\lambda)$, we denormalize s_n by ω_0, i.e., we find $s_{\text{ct}} = s_n \omega_0$, and then renormalize s_{ct} by $2/T$. These two steps amount to replacing s_n by the normalized variable λ_n where

$$\lambda_n = \frac{\lambda}{\omega_0 T/2} = 4.5271\lambda$$

(Derive this result!) Thus, in the λ_n-plane, the *LC* ladder is the same as the one in Fig. 7-34, so that in the λ-plane the elements are obtained by multiplying the capacitors and inductors by 4.5271. The resulting ladder is shown in Fig. 8-36.

The next step consists of dividing the impedance level by μ and performing a

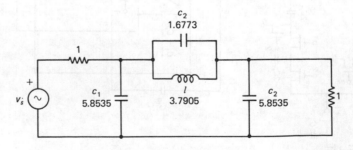

Figure 8-36 The λ-plane ladder.

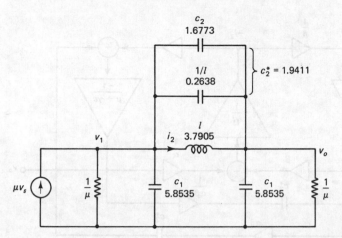

Figure 8-37 The γ-plane ladder.
Note: $c_2 + 1/l = c_2^*$.

source transformation. The resulting γ-plane ladder, shown in Fig. 8-37, can be analyzed to yield

$$-v_1 = -\frac{\mu v_s + (-i_2) + \gamma c_2^* v_o}{\mu + \gamma(c_1 + c_2^*)} \tag{8-101a}$$

$$-i_2 = \frac{-v_1 + v_o}{\gamma l} \tag{8-101b}$$

$$v_o = -\frac{(-i_2) - \gamma c_2^* v_1}{\mu + \gamma(c_1 + c_2^*)} \tag{8-101c}$$

This set of equations is realized by the signal-flow diagram in Fig. 8-38, where we have labeled $c_1 + c_2^* = c$. The indicated scaling factors α, β, and K are to be determined next, such that the signal levels at the integrator outputs are equalized and that the total filter gain equals unity. Analyzing the original ladder or the SFG circuit in Fig. 8-38 for α = β = 1, shows that

$$|v_o|_{max} = 0.5 \qquad |v_1|_{max} = 0.704 \qquad |i_2|_{max} = 6.424$$

Thus,

$$\alpha = \frac{6.424}{0.5} = 12.85 \qquad \beta = \frac{0.704}{6.424} = 0.1096$$

Remember that, at every node, the scaling factors must multiply each incoming signal and divide each outgoing signal in order not to alter the transfer function. Further, to realize the prescribed unity gain, we obviously need

$$K = 2$$

Figure 8-39 shows a plot of the scaled integrator outputs versus the normalized discrete-time frequency axis $\pi f/f_c$.

At this stage it is a simple matter to derive an active RC signal-flow graph circuit that implements the diagram in Fig. 8-38. The circuit is shown in Fig. 8-40; in the

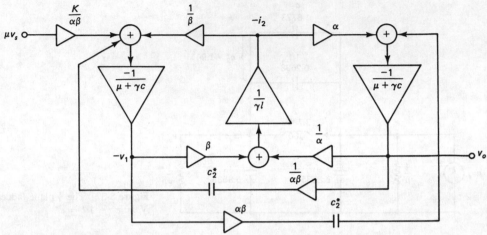

Figure 8-38 Block diagram to realize Eqs. (8-101a) through (8-101c). *Note:* $c_1 + c_2^* = c$.

normalized frequency domain s_n, with frequency-normalized capacitors C_{ni}, it realizes the equations

$$-V_1 = -\frac{R_a G_1 V_{\text{in}} + R_a G_2(-V_2) + s_n C_{n5} R_a V_o}{R_a G_0 + s_n C_{n1} R_a}$$ (8-102a)

$$\rightarrow -\frac{\frac{K}{\alpha\beta}\mu v_s + \frac{1}{\beta}(-i_2) + \frac{1}{\alpha\beta}\gamma c_2^* v_o}{\mu + \gamma(c_1 + c_2^*)}$$

$$-V_2 = \frac{R_b G_3(-V_1) + R_b G_4 V_o}{s_n C_{n2} R_b} \rightarrow \frac{\beta(-v_1) + (1/\alpha)v_o}{\gamma l}$$ (8-102b)

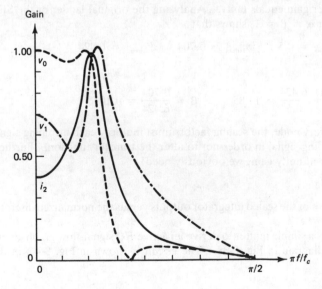

Figure 8-39 The equalized signal levels in the SFG circuit.

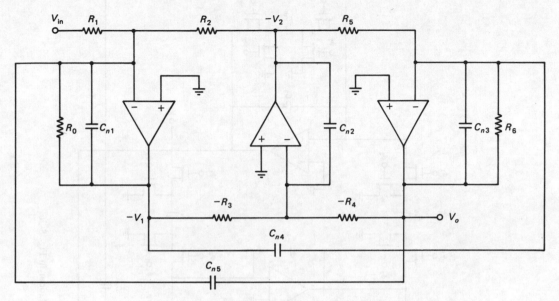

Figure 8-40 Active RC realization of the SFG in Fig. 8-38.

$$V_o = -\frac{R_cG_5(-V_2) + s_nC_{n4}R_c(-V_1)}{G_6R_c + s_nC_{n3}R_c} \rightarrow -\frac{\alpha(-i_2) + \alpha\beta\gamma c_2^*(-v_1)}{\mu + \gamma(c_1 + c_2^*)} \quad (8\text{-}102c)$$

R_a, R_b, and R_c are suitably chosen normalizing resistors. As indicated in Eq. (8-102), the active RC expressions are then to be equated to Eq. (8-101), *including the scaling factors*. Recall that γ is to be treated as a normalized frequency and that the effect of μ is neglected, i.e., we set $\mu = 1$. The comparison yields:

$$R_1 = \frac{\alpha\beta}{K}R_a = 0.7041R_a \qquad R_2 = \beta R_a = 0.1096R_a \qquad R_0 = R_a$$

$$C_1 = (c_1 + c_2^*)\frac{T}{2R_a} = \frac{60.89\ \mu s}{R_a} \qquad C_5 = \frac{c_2^*}{\alpha\beta}\frac{T}{2R_a} = \frac{10.77\ \mu s}{R_a}$$

$$R_3 = \frac{R_b}{\beta} = 9.125R_b \qquad R_4 = \alpha R_b = 12.85R_b$$

$$C_2 = l\,\frac{T}{2R_b} = \frac{29.61\ \mu s}{R_b}$$

$$R_5 = \frac{R_c}{\alpha} = 0.0778R_c \qquad R_6 = R_c$$

$$C_3 = (c_1 + c_2^*)\frac{T}{2R_c} = \frac{60.89\ \mu s}{R_c} \qquad C_4 = \alpha\beta c_2^*\frac{T}{2R_c} = \frac{21.36\ \mu s}{R_c}$$

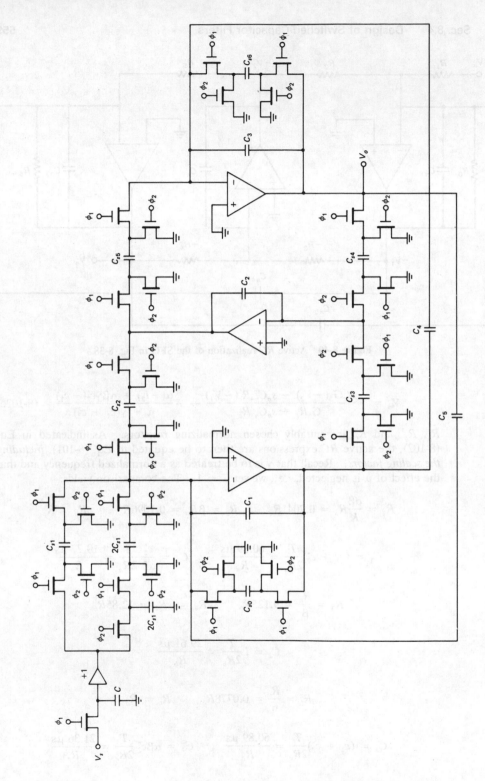

Figure 8-41 The SC realization of the ladder in Fig. 8-36.

The active RC circuit can now be converted into the switched-capacitor domain. To this end, we replace each resistor R by a switched capacitor of value $C_s = T/R$ per Eq. (8-99a) with appropriate switch phasing, and each capacitor by an unswitched capacitor $C_{u,i}$ per Eq. (8-99b). Remember to subtract the loss terms from the integrator capacitors of the lossy integrators! The resulting circuit is shown in Fig. 8-41. The scaling resistors R_a, R_b, and R_c, as well as the global multiplier a, are chosen to get convenient practical capacitor values. We find

$$C_{s1} = \frac{aT}{0.7041R_a} = \frac{22.191a}{R_a} \; \mu s \qquad C_{s2} = \frac{aT}{0.1096R_a} = \frac{142.56a}{R_a} \; \mu s$$

$$C_{s0} = \frac{aT}{R_a} = \frac{15.63a}{R_a} \; \mu s \qquad C_{u1} = \frac{aC_1 - C_{s0}}{2} = \frac{22.63a}{R_a} \; \mu s \qquad C_{u5} = \frac{5.84a}{R_a} \; \mu s$$

$$C_{s3} = \frac{aT}{9.125R_b} = \frac{1.712a}{R_b} \; \mu s \qquad C_{s4} = \frac{aT}{12.85R_b} = \frac{1.216a}{R_b} \; \mu s \qquad C_{u2} = \frac{14.81a}{R_b} \; \mu s$$

$$C_{s5} = \frac{aT}{0.0778R_c} = \frac{200.8a}{R_c} \; \mu s \qquad C_{s6} = \frac{aT}{R_c} = \frac{15.63a}{R_c} \; \mu s$$

$$C_{u3} = \frac{aC_3 - C_{s6}}{2} = \frac{22.63a}{R_c} \; \mu s \qquad C_{u4} = \frac{10.68a}{R_c} \; \mu s$$

In order to bring the capacitors into the picofarad range, we choose $a = 10^{-6}$. Further, let us select arbitrarily 5 pF as the smallest integrator capacitor; this choice leads to $R_a = 4.526 \; \Omega$ and

$$C_{s1} = 4.903 \text{ pF} \qquad C_{s2} = 31.50 \text{ pF} \qquad C_{s0} = 3.453 \text{ pF}$$

$$C_1 = 5 \text{ pF} \qquad C_5 = 1.290 \text{ pF}$$

Similarly, we find $R_b = 2.962 \; \Omega$ and

$$C_{s3} = 0.578 \text{ pF} \qquad C_{s4} = 0.411 \text{ pF} \qquad C_2 = 5 \text{ pF}$$

and finally $R_c = 4.526 \; \Omega$, so that

$$C_{s5} = 44.37 \text{ pF} \qquad C_{s6} = 3.453 \text{ pF} \qquad C_3 = 5 \text{ pF} \qquad C_4 = 2.630 \text{ pF}$$

8.5 SOME PRACTICAL OBSERVATIONS, PRECAUTIONS, AND LIMITS

With the exception of the existence of parasitic capacitors, we have in this chapter considered only the *ideal* operation of SC filters; i.e., we have, for example, assumed capacitor ratios to be exact and switches and op amps to be ideal. These assumptions imply that all capacitors charge and discharge instantaneously with the correct charge values and that the charge signals are processed completely and without delay by the op amps. In reality, these assumptions are, of course, not valid, so that various nonideal effects on circuit performance must be taken into account. Thus, we note, for example, that, since SC networks operate by processing small charge packets, the accumulation of random stray charges must be avoided. These charges may arise from various sources, such as noise, parasitic feed-through of clock signals, leakage through reverse-biased junctions, or op amp offset voltages. Further, just as in continuous-time active *RC* circuits, we must investigate the effects of the op amps' finite bandwidth, gain, slew rate, noise, and output resistance. Also, since the circuits operate in a switched mode, op amp settling time is a pertinent parameter. At a first glance, it may appear that op amp noise is the only noise source of concern because the remaining components are capacitors and switches. But recall that the "on" resistance of MOS switches is by no means zero but depending on transistor size, has values between 5 kΩ and 20 kΩ. In addition, the "on" resistance noise is sampled and held by the switched capacitors, leading to consequences not present in continuous-time filters.

Since all these effects are interconnected, an exact analysis tends to become exceedingly complicated and will be omitted. Instead, we shall in this section only present a qualitative discussion to alert the student to pay attention to these factors when designing practical SC filters. Detailed results can be found in numerous papers where the nonideal behavior of SC operation is investigated. We refer the reader to references 184 and 197 and, in particular, to Chapter 7 of reference 155 and their literature citations.

8.5.1 Design and Layout Precautions

When constructing a practical integrated SC filter circuit, a number of points should be considered in order to maximize the likelihood of satisfactory performance. Many errors that result from violating these rules would be caught by circuit simulation; others, because of lack of adequate models, would be felt only during measurement of the final processed circuit, where corrections are impossible or at least very expensive. In the following, we point out a few of the more important aspects which the designer must bear in mind.

1. Op amps must not be operated in open-loop fashion at any time, in order to avoid nonlinearities and saturation. Therefore, a switched capacitor alone should not be used to complete the feedback path around an op amp, because it does not provide a continuous path; an unswitched capacitor is the minimum circuitry to close the feedback path.

2. No circuit nodes must be completely isolated by capacitors. Because all capacitor plates are subject to charge accumulation from various sources, such as leaking reverse-biased isolation junctions, there must be a path either directly or through a switched capacitor to a voltage source or to ground so that stray charge accumulations can be discharged. This argument implies, for instance, that the feedback path around op amp(s) must be closed by a

switched capacitor (dc feedback) in order to avoid charge accumulation at the summing node(s) leading to saturation of the op amp(s).

3. The bottom plate of every capacitor should be connected either directly or through a switch to ground or to a voltage source. In MOS capacitors, there is always a relatively large and nonlinear parasitic capacitor (approximately 10% to 20% of the intended circuit capacitor) between the bottom plate and the substrate (ac ground). The prescribed connection (see Fig. 8-9) either grounds out the parasitic capacitor or connects it across a voltage source; in any case, it has no effect on circuit performance.

4. The noninverting op amp input should be kept at ac ground. If this input terminal is connected to a signal voltage, the circuit tends to become sensitive to all parasitic capacitors from switches, signal lines, substrate, and so on, which are connected to the inverting input. Consequently, the op amp building blocks are no longer strays-insensitive. In addition, requirements on the op amp's common mode performance become more demanding.

5. Although switched-capacitor filters are analog, they always contain a considerable number of digital circuits because of the needed timing and clocking circuitry or because the filters share the same IC chip with other digital systems. As a result, considerable switching noise caused by clocks and switching transients must be expected to reside on the power supply lines. If these lines are shared by both analog and digital circuits of the system, switching noise will couple into the analog SC filter because of the finite bias line resistance. The situation can be much improved by providing separate bias lines for the analog and the digital circuitry. In that case the digital power line noise travels to the chip pads, where it can be filtered by bypass capacitors. An even better solution, which was also discussed in Chapter 7 for continuous-time designs, consists of a fully differential layout. In that case power supply noise and many parasitic voltages occur as common mode signals and, therefore, do not appear in the output.

6. Lines with digital signals should be kept as far away as possible from lines that carry the analog signal. If layout considerations force analog and digital signal lines into close proximity, noise injection via capacitive coupling along the surface or through the substrate can be minimized by shielding the analog lines via running them between two grounded metal lines and, if feasible, over a grounded polysilicon line (see Fig. 8-42a).[7]

7. A serious problem in SC circuits is clock feed-through. Due to sampling, clock feedthrough usually aliases down to dc and adds to the dc offsets of the op amps. Its origin is found in the gate-to-source or gate-to-drain overlap capacitors C_{ov} of the switches at the inputs to the op amps as illustrated in Fig. 8-42b. It shows that the clock signal couples directly into the op amp

---

[7]The layout precautions of points 5 and 6 are, of course, also recommended for the continuous-time circuits discussed in Chapter 7.

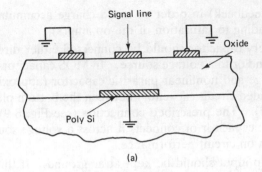

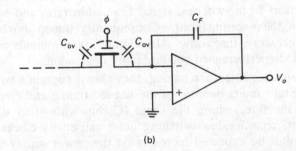

Figure 8-42 (a) Shielding method for signal line; (b) coupling path for the clock signal.

input node and subsequently to the output with a weight factor of C_{ov}/C_F. To minimize this effect the designer can maximize C_F and/or minimize C_{ov} by designing the switches to be as small as practical.[8] Using a large C_F will also serve to increase PSRR and reduce the effect of other spurious signals which couple into the filter via parasitic capacitances. In CMOS design, since feedback confines the voltages at the op amp inputs to very near zero volts, these switches can be minimum sized n-channel devices rather than the complementary p-n switches used in the rest of the circuit where the expected signal swings are larger.

8.5.2 Effects of Nonideal Op Amps

Just as is the case for continuous-time active RC circuits, the performance of switched-capacitor filters depends critically on the operational amplifiers. Important op amp parameters are the finite bandwidth, dc gain, slew rate, and output conductance, especially, because for the simple on-chip amplifiers normally used in SC filters, all these parameters are often one or two orders of magnitude smaller than in the highly developed op amps usually employed in the design of discrete active RC filters. Let us briefly look at some of these effects.

[8]Note, though, the trade-off with the resulting increased switch resistance discussed below.

8.5.2.1 Finite dc gain.

Consider the strays-insensitive integrator[9] of Fig. 8-7a. If the op amp with output voltage V_{out} has a finite dc gain A_0, the op amp input voltage becomes $-V_{out}/A_0$ rather than zero.

The ideal integrator transfer function was given in Eq. (8-38) to be

$$H_{i,I}(z) = -\frac{C_1}{C_F}\frac{1}{1 - z^{-1}} \tag{8-103}$$

We leave it to the problems (Problem 8.47) to prove that for finite dc gain the integrator realizes

$$H(z) = -\frac{C_1}{C_2}\frac{1}{1 + \dfrac{1 + C_1/C_2}{A_0} - z^{-1}\left(1 + \dfrac{1}{A_0}\right)}$$

$$= H_{i,I}(z)\frac{1 - z^{-1}}{1 + \dfrac{1 + C_1/C_2}{A_0} - z^{-1}\left(1 + \dfrac{1}{A_0}\right)} \tag{8-104}$$

Along the unit circle $z = e^{j\omega T}$, $H(z)$ can be brought into the form

$$H(e^{j\omega T}) = H_{i,I}(e^{j\omega T})\frac{1}{1 + \dfrac{1}{A_0}\left\{1 + \dfrac{C_1}{2C_2}\left[1 - j\dfrac{1}{\tan(\omega T/2)}\right]\right\}}$$

$$\simeq H_{i,I}(e^{j\omega T})[1 + m(\omega)]e^{j\theta(\omega)} \tag{8-105}$$

where

$$m(\omega) = -\frac{1}{A_0}\left(1 + \frac{C_1}{2C_2}\right) \quad \text{and} \quad \theta(\omega) = \frac{C_1}{2A_0C_2}\frac{1}{\tan(\omega T/2)} \simeq \frac{C_1/C_2}{A_0\omega T}$$

For the approximations, we assumed reasonable capacitor ratios and values of ωT, so that $|m(\omega)| \ll 1$ and $|\theta(\omega)| \ll 1$. We notice that the finite op amp gain introduced a relative magnitude error $m(\omega)$ and a small[10] phase error $\theta(\omega)$. Completely analogous results can be derived for the noninverting integrator in Fig. 8-7b. In most applications, the magnitude error has negligible effect, but, as in active RC designs, the phase error can be very detrimental in narrowband (high-Q) filters.

8.5.2.2 Finite bandwidth ω_t.

When op amps introduce their own dynamics into SC filter operation, the integrator behavior is best evaluated in the time domain

---

[9]Note that finite gain eliminates the virtual ground at the inverting input terminal. Thus the circuit is no longer really strays-insensitive.

[10]Except as $\omega \to 0$. θ is large unless $\omega > 2/(A_0 T)$.

[198, 199]: at each switching instant, the op amp output voltage is found to approach its final value exponentially or in the manner of a damped oscillation. To avoid errors, the switch "on" time T_{on} of the sampling switch must be large enough to permit the transferred charges to reach their final correct values; i.e., assuming a two-phase clock, the *settling time* of the individual op amps must be less than half the switching period T:

$$t_{settle} < T_{on} \simeq 0.5T$$

However, increasing T, i.e., reducing f_c, too much for a given signal bandwidth entails new problems, such as more complicated and expensive anti-aliasing filters. In many circuits, the acceptable settling time can in effect be increased by a factor of 2 by choice of the appropriate switch phasing. An example is the biquad in Fig. 8-24. As drawn, op amp 1 is sampled during ϕ_1 when its output might not have settled yet. Changing the switch phasing to that indicated in parentheses in the figure does not change the transfer function (Problem 8.48) but permits the op amp to settle until the beginning of time period ϕ_2. The circuit with this particular switch phasing for improved settling behavior was introduced and discussed in references 194 and 218.

In the frequency domain, one finds for both inverting and noninverting integrators after quite lengthy analysis that, in the frequency range of interest, gain and phase errors are proportional to $\exp(-\omega_t T/2)$. To make these errors negligible, we require

$$\omega_t = 2\pi f_t \gg \frac{2}{T} = 2f_c$$

i.e.,

$$f_t \gg \frac{f_c}{\pi}$$

In practice, $f_t \simeq 5f_c$ results in $\exp(-\omega_t T/2) = \exp(-\pi\omega_t/\omega_c) = \exp(-5\pi) = 1.5 \cdot 10^{-7}$; thus, the errors become negligible by choosing the op amp unity gain-bandwidth product of the order of 5 times as large as the clock frequency.

8.5.2.3 Finite slew rate.
As discussed in Chapter 4, slew rate is a measure of the maximum rate of change that the op amp output voltage can sustain. As a consequence, the op amp's response to an input step is slowed down or delayed. The effect of this additional delay is that the output signal reaches its final value only after the *combined* delay and settling times; i.e., we require that

$$t_{settle} + t_{slew} < T_{on}$$

where T_{on} is again the closing time of the switch through which the op amp output is sampled. Note that the shape of the signal while the op amp is slewing does not matter as long as the final correct value is reached before the output is sampled.[11]

8.5.2.4 Finite output resistance.   Simple analysis of SC filters assumes that the op amp output is an ideal voltage source with $R_{out} \simeq 0$. In practice, R_{out} has values of a few kilohms if the op amp has a buffered output stage, and values as high as several hundred kilohms for simple on-chip op amps without a buffered output. Consequently, the op amp must charge the total load capacitance C_L through the resistor R_{out}. If the op amp also has finite ω_t, the analysis becomes relatively complicated; one can show though [155], that the charging time constant is of the order of

$$\tau \simeq 2R_{out}C_L$$

τ must then be much less than T_{on} (say, 5 to 7 times) in order to achieve adequate charge transfer. For instance, if $R_{out} = 100$ kΩ and $C_L = 3$ pF, we obtain $\tau = 0.6$ μs. Using seven as a safety factor results in $T_{on} = 4.2$ μs and results in the maximum switching frequency $f_c = 1/T = 1/(2T_{on}) \simeq 120$ kHz. Because, in practice, f_c is 8 or more times larger than the highest signal frequency, evidently, high-frequency SC operation requires buffered op amp outputs!

8.5.3 Frequency Limitations

It was explained at length in Chapters 4 and 5 for active RC circuits that the maximum signal frequency is limited by the performance of the operational amplifiers. Specifically, we found that the finite slew-rate and gain-bandwidth product ω_t could give rise to serious deviations in the desired filter performance. We pointed out in the preceding sections that the same op amp limitations, along with the finite settling time, also result in errors in the frequency response of switched-capacitor filters, albeit via different mechanisms. For a given op amp, the deviations increase with increasing clock frequency f_c which must be at least twice as large as the highest signal frequency f_B because of the sampling theorem. In practice, f_c is usually chosen to be at least an order of magnitude higher than f_B because of the requirements of practical anti-aliasing and smoothing filters.

An additional cause for errors that may or may not be small compared to limitations imposed by the op amps can be found in the finite switch "on" resistance. Note that each capacitor generally must charge and discharge through two closed MOS switches (e.g., Fig. 8-7). The "on" resistance of a closed MOS transistor switch can be shown to equal [see Eq. (7-76)]

$$R_{on} = \frac{1}{\mu C_{ox}(W/L)(V_{gs} - V_T)}$$

---

[11]Recall that the true instant of sampling is the end of the switch on time.

From this equation we observe that R_{on} is *finite* and, because the gate-to-source voltage varies with the signal being processed, strongly *nonlinear*. In particular, note that R_{on} may become very large depending on the value of the gate-to-source voltage. In CMOS technology, in order to avoid an inadvertent cutoff, the designer may construct the switch as a so-called transmission gate, i.e., as a parallel connection of an NMOS and a PMOS transistor clocked out of phase. For that case, one can show [200] that the maximum through-resistance equals approximately 15 kΩ. The charging time constant for a switched capacitor C becomes

$$\tau \simeq 2R_{on}C$$

a value that should be an order of magnitude times smaller than T_{on} of the switches if charge transfer errors are to be avoided.

In most situations, op amp limitations pose more serious restrictions on the clock frequency, and thereby on the highest signal frequency, than finite switch "on" resistance. In any event, incomplete charge transfer caused by imperfect op amps or switches can be shown to result in severe deviations in the frequency response of SC filters [197].

8.5.4 Noise

In the same way as we defined noise in Chapter 7, noise in SC filters is to be understood as *any unwanted signals interfering with the main signal at the filter output*. Generally, noise in SC filters will be higher than that in equivalent active *RC* designs because of the additional noise sources. As always, we have to contend with the electrical noise [thermal and flicker ($1/f$) noise] contributed by the active devices (op amps and transistor switches). In addition, however, there are the clock feed-through noise mentioned earlier and the noise coupled capacitively over the surface or through the substrate into the signal lines from power and ground lines. The latter two contributions can be reduced significantly by careful bypassing and shielding and by fully balanced symmetrical circuit design, as was discussed earlier. Both $1/f$ and thermal noise have their origin in the physical operation of the MOS devices. As the name implies, $1/f$ noise is large at low frequencies and diminishes as f increases. Its effect on dynamic range can be reduced by clever sampling techniques which transfer the low-frequency noise band away from the signal baseband. Chopper stabilization has been found to be a useful and practical technique [201]. The "white" thermal noise becomes dominant typically at approximately 5 to 10 kHz; using a few realistic assumptions, a switched capacitor C clocked at f_c can be shown to contribute a noise spectral density

$$S \simeq \frac{1}{4}\frac{kT_a}{f_cC}$$

where k is Boltzmann's constant and T_a is the absolute temperature. Note that the noise contribution is inversely proportional to the clock frequency and to C. Considering carefully, further, the noise in an SC integrator [202], one can show that noise power is inversely proportional to C_{min}, i.e., all noise effects are reduced by increasing the minimum circuit capacitors.

8.5.5 Tuning and Programming

Among the advantages of SC filters is the ease with which they can be tuned or programmed to perform different functions. It is evident from Eq. (8-8) that all time constants in SC filters are proportional to $1/f_c$, or conversely, from Eq. (8-1b), that frequency parameters are proportional to f_c. This property implies that the response of a switched-capacitor filter can be shifted along the frequency axis by a simple change in clock frequency, without affecting either Q or gain—for example, for the purpose of building a tracking filter. Even greater versatility can be obtained by switching complete capacitive branches into or out of a given SC filter, such as the general biquad in Fig. 8-25. For example, depending on which of the capacitors I, J, G, and H are in the circuit, we can realize bandpass, lowpass, highpass, notch, or even a general biquadratic transfer function. Also, because all filter coefficients are determined only by capacitor ratios, replacing certain capacitors by suitable *arrays* of capacitors establishes a convenient method for digitally changing—programming—the coefficients of the analog filter. Digitally programmable universal filter chips are, therefore, quite readily realizable.

8.5.6 Comparison with Other Techniques

Analog sampled-data switched-capacitor filters have a number of attractive advantages which led to their rapid acceptance in commercial systems [209–213]. Because of their ability to simulate large resistors and to set accurate frequency parameters via capacitor ratios, they result in low-power, extremely area-efficient realizations of low-frequency fully integrated filters with no need for postdesign tuning or adjusting. SC circuits are as easily programmed and adjusted electronically for different signal-processing requirements as are digital filters, and at the same time they consume less power. The advantage of switched-capacitor analog signal processing, however, comes at a cost which should not be overlooked: if the signals to be processed are in the continuous-time domain, the sampled-data operation necessitates a variety of peripheral circuits which increase chip size, add system noise, and, especially, tend to result in lower operating frequencies than a direct continuous-time approach. First of all, we need continuous-time anti-aliasing and reconstruction filters; then there is the obvious requirement of a system clock, and timing and switching circuitry. Imperfect switches and nonideal op amps lead to incorrect charge transfer, which in turn results in filter transfer function errors.   Since these errors increase with increasing switching frequency, it follows that SC filters are restricted to operating frequencies which are relatively low when compared to a continuous-time approach.   In the continuous-time domain, circuits for operation at frequencies over 100 MHz have been reported [168].   Of course, as was discussed at length in Chapter 7, the advantage of higher operating frequencies also comes with a price: the necessity of generally quite complicated tuning and control circuitry to guarantee accurate filter performance in the face of manufacturing tolerances and changing operating conditions.

Accuracy is generally known or at least assumed to be better in SC filters than in their continuous-time counterparts.   The reported comparisons should be interpreted carefully, however, because they are often based on incomparable assumptions.   There is little doubt that an intrinsic SC filter will have smaller errors in practice than an equivalent continuous-time circuit, because fabricated capacitor ratios and a clock frequency can be maintained more accurately than RC products, even accounting for on-chip tuning.   However, when the *total* performance from analog input to analog output is considered, we have to include in the comparison the *continuous-time* anti-aliasing and reconstruction filters which introduce errors comparable to those of the all–continuous-time analog approach discussed in Chapter 7.   However, as noted earlier the designer can reduce the complexity of the continuous-time antialiasing/reconstruction filters, and reduce the sensitivity of the analog input-to-analog output transfer characteristic to their errors, by increasing f_c.   This is a tradeoff, which obviously becomes more critical as one pushes the SC filter poles and zeros outside the audio range.   Nonetheless, it can be verified that if $f_c > 50f_B$ (where f_B is the highest frequency in the SC filter passband) antialiasing/reconstruction filters, which achieve at least 32 dB of attenuation at $f \geq f_c - f_B$, can be realized with fully integrated second-order active RC lowpass filters with no tuning.   The exact f_c/f_B ratio required for a given design depends on the attenuation desired at $f \geq f_c - f_B$, and the SC filter specifications and the variability of the antialiasing/reconstruction filters at $f \leq f_B$.   Usually the antialiasing specifications are more critical because the reconstruction filter gets a boost from the $(\sin x)/x$ rolloff of the SC filter's inherent sample-and-hold function.   Unity gain Sallen-and-Key lowpass biquads (see Chapter 5) are often used for this application because the dc gain is independent of all external R's and C's.   We leave it to the reader to verify this observation.   Also, we note that the highly accurate capacitor ratios on which SC filter operation depends will at higher frequencies be increasingly perturbed by stray parasitics.   The reason for this experience can be found in the fact that stray-insensitive designs depend on the virtual ground input of op amps (see Fig. 8-7).   But at higher frequencies, op amp gain decreases so that the virtual ground property is no longer available.

Nevertheless, switched-capacitor circuits form a highly desirable economic solution to signal-processing needs over a relatively wide range of frequencies.

8.6 SUMMARY

In this chapter we have introduced the reader to the concept of the *switched-capacitor resistor equivalent*, and we have seen how these SC components can be used to design analog sampled-data filters in integrated form in an area-efficient accurate implementation with no need for on-chip tuning. In this respect, SC designs have a significant advantage over the continuous-time approaches discussed in Chapter 7. On the other hand, the discrete-time nature of SC signal processing necessitates switching, sample-and-hold, anti-aliasing, and reconstruction circuitry, which requires additional silicon area and generates additional noise, especially switching noise and clock feed-through. Furthermore, the required sampling together with technological constraints imposes limits on the frequency range of signals that can safely be handled by SC filters. Although a few impressive laboratory results are available, demonstrating filters at a few hundred kilohertz [204] or even megahertz [205, 206], commercial applications at this time seem to be restricted to signal frequencies below 50 kHz. Unquestionably, these restrictions will be removed in the near future, as advances in technology and new design techniques become available. In this context, we mention in passing some very promising recent developments based on transconductances and current switching [214, 215].

For ease of understanding, and to take advantage of the analog filter design lore in the continuous-time domain, our development of SC filters has drawn upon insightful analogies to continuous-time active-*RC* and *LC* filters. More specifically, we have used frequency transformations which translate the sampled-data filter specifications into a suitable continuous-time frequency plane where an active *RC* filter *model* or *prototype* can be designed. By a simple resistor–switched-capacitor replacement, this predistorted model, in turn, can then be converted readily into an exact SC realization of the desired filter. Apart from an easier, more intuitive understanding of the design methods, the advantage of this approach is that the engineer may make use of the abundance of design algorithms, data, and tables in the literature that deal with continuous-time analog filters. We have also shown that the use of sampled-data theory, difference equations and the *z*-transform results in a generally accurate characterization at all frequencies and provides the designer additional insights into the control of the fine behavior of SC filters which is due to their sampled-data nature. Let us stress the point that "resistor-SC replacements" cannot be used blindly, and they most certainly should only be used after the predistortions and frequency transformations described in this chapter have been applied. One must always remember that even with transformation and predistortion, that the use of continuous-time prototypes is limited to the consideration of frequency domain criteria at low frequencies, $f < f_c/2$. The fine behavior of SC filters, and ultimately their successful realization, depend on time-domain criteria (e.g., the correct phasing of switches and their timing relative to transitions in the sampled-and-held inputs and op amp outputs) and control of sampled-data effects (e.g., antialiasing/reconstruction, aliasing of noise, clock feedthrough and other out-of-band disturbances). Hence, we encourage those readers who intend to pursue the serious design of SC filters to develop a sophisticated working knowledge of the sampled-data techniques introduced in this chapter.

We have discussed only the most popular design methods, i.e., realizations as a cascade of low-order circuits and simulations of *LC* ladder prototypes via signal-flow graphs. However, the active *RC*–SC analogies developed in this chapter will lead the reader to suspect correctly that other schemes, such as element substitution via gyrators or FDNRs (Chapter 6), should also be possible. This is indeed the case: we need only to develop SC inductors, gyrators, or FDNRs. Although one can construct SC circuits which realize these functions, none exist to our knowledge that are suitably parasitic insensitive. One possible explanation for the lack of such SC circuits is that integrator-based designs readily translate into high-quality parasitic insensitive SC structures. The active *RC* structures recommended for gyrators and FDNRs are not integrator-based. The details of these methods will be left to the references [e.g., 47, 207, 208].

In the audio-frequency range, the switched-capacitor technique is an extremely attractive method for placing well-controlled and practical filters onto a silicon chip, integrated along with the remaining parts of a communication system. Testimony to the activity in this area is the rapid emergence of commercial products using SC circuits during recent years [e.g., 209–213].

PROBLEMS

8.1 We have seen in earlier chapters of this book that integration is a fundamental operation in the realization of active filters; in particular, an integrator output voltage v_o is obtained by integrating an input signal v_i, or, expressed differently,

$$\frac{dv_o(t)}{dt} = \pm q v_i(t)$$

q is a constant and time t is normalized with respect to an interval T with $T = 1$. Integrating this equation over the interval $(n - 1)$ to n yields

$$v_o(n) - v_o(n - 1) = \pm q \int_n^{n-1} v_i \, dt \qquad \text{(P8-1)}$$

We shall demonstrate in this problem that the different popular SC integrators simply perform *numerical* integration of signals by the well-known *forward Euler, backward Euler*, or *trapezoidal rules*. Figure P8-1 illustrates these methods.

(a) Find the difference equation obtained by integrating Eq. (P8-1) numerically according to the forward Euler rule. Compare your result with Eq. (8-14) with $C_2 = 0$, i.e., the circuit of Fig. 8-7b.

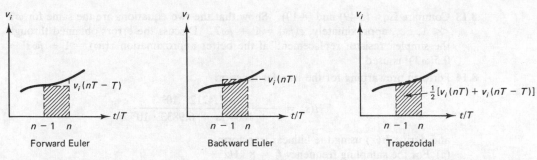

Figure P8-1

(b) Find the difference equation obtained by integrating Eq. (P8-1) numerically according to the backward Euler rule. Compare your result with Eq. (8-13) with $C_2 = 0$, i.e., the circuit of Fig. 8-7a.

(c) Find the difference equation obtained by integrating Eq. (P8-1) numerically according to the trapezoidal rule. It is shown in Section 8.3.2 that this difference equation leads to the important "bilinear transform," Eq. (8-51).

8.2 Verify that the difference equation [Eq. (8-15)] describes the behavior of Fig. 8-9.

8.3 The spectrum of a signal is described by $X(j\omega) = 1 - \omega/\omega_B$ in $|\omega| \le \omega_B$ and $X(j\omega) = 0$ in $|\omega| > \omega_B$. Sketch the spectrum of the sampled signal if $\omega_c = 4\omega_B$, $2\omega_B$, and $1.5\omega_B$.

8.4 A signal has a band-limited spectrum extending to $f_B = 3.5$ kHz. An available reconstruction filter has a transition bandwidth of two octaves (an octave is a factor of 2 in frequency) between passband and stopband. What is the minimum sampling frequency for which the system will operate correctly?

8.5 For the z-domain transfer functions
(a) $H(z) = 1 - z^{-1}$
(b) $H(z) = \dfrac{1}{z^2 - (2\alpha \cos \beta)z + \beta^2}$ $\alpha = 1.3$, $\beta = 0.95$
sketch the magnitude of the frequency response $|H(j\omega)|$.

8.6 Find the z-transform of the sampled version of $x(t) = \cos \omega t$ for $t \ge 0$.

8.7 Repeat Problem 8.6 for $x(t) = e^{at} \cos \omega T$.

8.8 Find the inverse z-transform of $X(z) = Tz^{-1}/(1 - z^{-1})^2$.

8.9 Use the z-transform to solve the difference equations obtained in Problem 8.1; i.e., find $V_o(z)/V_i(z) = H(z)$ for each case. Compare your solution with the results in Example 8-3 and Eq. (8-51).

8.10 Give the z-plane locations, $z = e^{sT}$, into which the poles of a third-order 0.5 dB–ripple Chebyshev lowpass filter are transformed. Do these pole locations indicate a stable filter in the z-domain?

8.11 Sketch Eqs. (8-44a) and (8-44b) versus frequency ωT. Assume $C_1 = 2C_F$ and $C_2 = 0.25C_F$.

8.12 Derive Eq. (8-49). Sketch the gain and phase responses for
(a) $V_1 = V_2 = 0$
(b) $V_1 = V_2 = V_3 = V_i$

8.13 Compare Eqs. (8-49) and (8-10). Show that the two equations are the same for ωT $\ll$ 1, i.e., approximately, $z(j\omega) \simeq 1 + j\omega T$. Discuss the errors obtained through the simple "resistor replacement" if the better approximation $z(j\omega) \simeq 1 + j\omega T - 0.5(\omega T)^2$ is used.

8.14 Perform prewarping for the transfer function

$$H(s) = \frac{s^2 + 1.4212 \cdot 10^5}{s^2 + 1004.2s + 6.9833 \cdot 10^5}$$

and derive $H(z)$ using the bilinear transform
  (a) For the sampling frequency f_c = 8 kHz
  (b) For the sampling frequency f_c = 128 kHz

Discuss the results.

8.15 Prove Eqs. (8-55d).

8.16 A switched-capacitor lowpass filter is to be built which realizes a fifth-order Butterworth characteristic. The passband is in $f \le 4.5$ kHz, and the signal is sampled at f_c = 48 kHz. Using prewarping and the bilinear transform, find the z-domain transfer function $H(z)$ for these specifications. Sketch $|H(e^{j\omega T})|$ versus ωT in $0 \le \omega T \le 2\pi$.

8.17 Repeat Problem 8.16 for a sixth-order geometrically symmetrical Chebyshev bandpass filter with 0.5 dB passband ripple in 12 kHz $\le f \le$ 18.4 kHz. The clock frequency is f_c = 128 kHz.

8.18 Repeat Problem 8.16 for a fourth-order elliptic lowpass filter with a passband ripple of 0.2 dB in $f \le 6.4$ kHz. The filter will be clocked at f_c = 24 kHz.

8.19 Consider the first-order circuit in Fig. P8.19; it is a special case of the first-order block in Fig. 8-9. Note the two different switch phasings indicated in parentheses. Show that the circuit can realize the first-order transfer functions of the form of Eq. (8-65). Find the component values as a function of the coefficients a_0, a_1, and b_1.
(To aid intuition, replace the strays-insensitive switched capacitors by resistors and find the transfer function of the resulting active RC filter.)

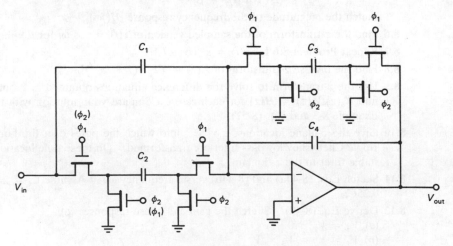

Figure P8-19

8.20 Analyze the circuit in Fig. 8-23 to verify that a general biquadratic transfer function can be realized.

8.21 Analyze the SC biquad in Fig. 8-24 and verify that the transfer functions in Eq. (8-68) are realized.

8.22 Starting from Fig. 8-24, develop the configurations with switch sharing in Figs. 8-25 and 8-26.

8.23 Determine the capacitor values in Fig. 8-25 such that the circuit, Eq. (8-71b), realizes a lowpass, highpass, bandpass, allpass, and band-rejection function, respectively.

  Hint: Start from a biquadratic function in the continuous-time domain s_{ct}, and use the bilinear transformation to obtain the biquadratic z-domain function of the form of Eq. (8-66). The required capacitors are then readily identified.

8.24 Repeat Problem 8.23 for the circuit in Fig. 8-26 [Eq. (8-72)].

8.25 Use the SC circuit in Fig. 8-25 [Eq. (8-71b)] to realize the lowpass notch function

$$H(s_{ct}) = \frac{0.891975 s_{ct}^2 + 1.140926 \cdot 10^8}{s_{ct}^2 + 356.047 s_{ct} + 1.140926 \cdot 10^8}$$

s_{ct} is *not* normalized; i.e., the function indicates a transmission zero at 1800 Hz. The sampling frequency is $f_c = 128$ kHz. Use the bilinear transform and assume pre-warping is not needed due to the relatively high sampling frequency.

  Realize both the circuits with E damping and with F damping. Verify, if possible, that the maximum gain for both circuits at the output V_2 is approximately 10.6 dB and that at the output V_1 for E damping and for F damping the peak gain equals approximately -11 dB. Scale the capacitor values to maximize dynamic range. Scale the capacitors further such that $C_{min} = 1$ pF in both circuits. Compare the two circuits with respect to the total capacitance needed.

8.26 Realize the bandpass function

$$H(s_{ct}) = \frac{2027.9 s_{ct}}{s_{ct}^2 + 641.28 s_{ct} + 1.0528 \cdot 10^8}$$

as a switched-capacitor filter. The sampling frequency is $f_c = 8$ kHz. Use prewarping and the bilinear transformation to derive $H(z)$. Realize $H(z)$ with the circuit in Fig. 8-25 using F damping.

8.27 Realize the function of Problem 8.25 with the circuit in Fig. 8-26.

8.28 Realize the function of Problem 8.26 with the circuit in Fig. 8-26.

8.29 Realize the function

$$H(s_{ct}) = \frac{s_{ct}^2 + 1.4212 \cdot 10^5}{s_{ct}^2 + 1004.2 s_{ct} + 6.9833 \cdot 10^5}$$

using the circuit in Fig. 8-25 with E damping. The sampling frequency is $f_c = 8$ kHz.

8.30 Repeat Problem 8.29 for the circuit in Fig. 8-26 and $f_c = 128$ kHz.

8.31 Use the circuit in Fig. 8-26 to realize a fourth-order cascade switched-capacitor Butterworth lowpass filter with cutoff frequency 5.6 kHz and $f_c = 48$ kHz.

8.32 Use the circuit in Fig. 8-25 to realize a sixth-order all-pole cascade switched-capacitor bandpass filter with an equal-ripple passband in 12 kHz $\leq f_{ct} \leq 36$ kHz. The sampling frequency is 128 kHz.

8.33 Starting from Eq. (8-82b), verify Eq. (8-82c) and synthesize the γ-plane immittances in Fig. 8-28.

8.34 Using the SFG active *RC* techniques discussed in Chapter 6 as a guide, derive the configuration in Fig. 8-29b as a realization of Eq. (8-82c).

8.35 Verify Eq. (8-83b).

8.36 Derive the circuits in Fig. 8-30a and b.

8.37 Verify Eq. (8-85).

8.38 Analyze the circuits in Fig. 8-32b to verify Eq. (8-88). Show that for no damping this circuit actually realizes a bilinear integrator. Show also that the branch containing the capacitors $2C_i$ is parasitics insensitive.

8.39 Analyze the SFG circuit in Fig. 8-34 to verify the set of equations (8-95).

8.40 From the procedure and the equations presented in the text, derive the SC filter in Fig. 8-35a.

8.41 Convert Fig. 8-35a into the minimum switch configuration in Fig. 8-35b.

8.42 Convert the SC filter in Fig. 8-41 into the minimum switch configuration.

8.43 Analyze Fig. 8-41 to obtain $H(\lambda)$. Compare your result with Eq. (8-100).

8.44 Design an SFG switched-capacitor simulation of a fifth-order all-pole *LC* lowpass ladder with maximally flat passband in $0 \leq f \leq 3.4$ kHz and passband attenuation $A_p \leq 0.3$ dB. $f_c = 24$ kHz.

8.45 Realize an SC filter with the specifications of Problem 8.32 as a simulated ladder. The sampling frequency is 128 kHz.

8.46 Design an SFG switched-capacitor simulation of an *LC* lowpass ladder to realize:

  Maximum attenuation in passband: $4.343 \cdot 10^{-4}$ dB

  Minimum attenuation in stopband: 26.83 dB

  Passband edge: $\omega_p = 1$

  Stopband edge: $\omega_s = 1.7$

The normalized component values are obtained from the design tables [18]; they are

$$R_S = R_L = 1 \quad L_2 = 0.9176 \quad L_4 = 0.5740$$

$$C_1 = 0.3829 \quad C_2 = 0.1485 \quad C_3 = 1.0602 \quad C_4 = 0.5546 \quad C_5 = 0.1387$$

The *LC* prototype is shown in Fig. P8-46. Choose the denormalizing frequency as $\omega_n = 2\pi \cdot 36.3$ krad/s; the clock frequency is $f_c = 256$ kHz.

Note: This circuit was designed in Problem 7.26 for a higher operating frequency.

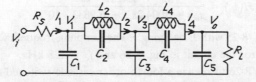

Figure P8-46

8.47 Prove that the SC integrator in Fig. 8-7a realizes the transfer function of Eq. (8-104) if the op amp has a finite dc gain A_0. Also show that this result simplifies to Eq. (8-105) under the assumptions indicated in the text.

8.48 Analyze the circuit in Fig. 8-24 with the switch phasing shown in parentheses and verify that the transfer function is that given in Eq. (8-68).

References

1. Rabiner, L. R., and B. Gold, *Theory and Application of Digital Signal Processing*. Englewood Cliffs, N.J.: Prentice-Hall, 1975.

2. Oppenheim, A. V., and R. W. Schafer, *Digital Signal Processing*. Englewood Cliffs, N.J.: Prentice-Hall, 1975.

3. Bose, N. K., *Digital Filters—Theory and Applications*. New York: Elsevier, 1985.

4. Ziemer, R. E., W. H. Tranter, and D. R. Fannin, *Signals and Systems—Continuous and Discrete*. New York: Macmillan, 1983.

5. Oppenheim, A. V., and A. S. Willsky, *Signals and Systems*. Englewood Cliffs, N.J.: Prentice-Hall, 1983.

6. Lindquist, C., *Active Network Design with Signal Filtering Applications*. Long Beach, Calif.: Steward and Sons, 1977.

7. Ghausi, M. S., *Principles and Design of Linear Active Circuits*. New York: McGraw-Hill, 1965, Chap. 4.

8. Oliver, B. M., J. R. Pierce, and C. E. Shannon, "The Philosophy of PCM," *Proc. IRE*, Vol. 36 (November 1948), 1324–1331.

9. Balabanian, N., and T. Bickart, *Linear Network Theory: Analysis, Properties, Design and Synthesis*. Beaverton, Oreg.: Matrix, 1981.

10. Weinberg, L., *Network Analysis and Synthesis*. New York: McGraw-Hill, 1968, Chap. 11.

11. Sedra, A. S., and P. O. Brackett, *Filter Theory and Design: Active and Passive*. Portland, Oreg.: Matrix, 1978.

12. Daniels, R. W., *Approximation Methods for Electronic Filter Design*. New York: McGraw-Hill, 1974.

13. Christian, E., and E. Eisenmann, *Filter Design Tables and Graphs*. New York: Wiley, 1966.

14. Daryanani, G., *Principles of Active Network Synthesis and Design*. New York: Wiley, 1976.

15. Humphreys, D. S., *The Analysis, Design and Synthesis of Electrical Filters*. Englewood Cliffs, N.J.: Prentice-Hall, 1970.

16. Szentirmai, G., *Computer-Aided Filter Design*. New York: IEEE Press, 1973.

17. Unbehauen, R., "Low-Pass Filters with Predetermined Phase on Delay and Chebyshev Stop-band Attenuation," *IEEE Trans. Circuit Theory*, Vol. CT-15 (December 1968), 337–341.

18. Zverev, A. I., *Handbook of Filter Synthesis*. New York: Wiley, 1967.

19. Su, K. L., *Time Domain Synthesis of Linear Networks*. Englewood Cliffs, N.J.: Prentice-Hall, 1971.

20. Sticht, D. S., and L. P. Huelsman, "Direct Determination of Elliptic Network Functions," *Int. J. Comput. Elec. Eng.*, Vol. 1 (1973), 272–280.

21. Saal, R., *Handbook of Filter Design*. Berlin, West Germany: AEG-Telefunken, 1979.

22. Huelsman, L. P., and P. E. Allen, *Introduction to the Theory and Design of Active Filters*. New York: McGraw-Hill, 1980.

23. Abramowitz, M., and I. A. Stegun, *Handbook of Mathematical Functions*. Washington, D.C.: National Bureau of Standards, 1961.

24. Temes, G. C., and J. W. Lapatra, *Introduction to Circuit Synthesis and Design*. New York: McGraw-Hill, 1977.

25. Balabanian, N., *Network Synthesis*. Englewood Cliffs, N.J.: Prentice-Hall, 1958.

26. Chen, W.-K., *Passive and Active Filters—Theory and Implementation*. New York: Wiley, 1986.

27. Henderson, K. W., and W. H. Kautz, "Transient Response of Conventional Filters," *IRE Trans. Circuit Theory*, Vol. CT-5 (December 1958), 333–347.

28. Christian, E., *LC Filters: Design Testing, & Manufacturing*. New York: Wiley, 1983.

29. *S/FILSYN Software for Filter Analysis and Design*. DGS Associates, 1353 Sarita Way, Santa Clara, CA 95051.

30. Bader, W., "Kopplungsfreie Kettenschaltungen," *Telegr. Ternsprech. Technol.*, Vol. 31 (1942), 177–189.

31. Franco, S., *Design with Operational Amplifiers and Analog Integrated Circuits*. New York: McGraw-Hill, 1988.

32. Moschytz, G. S., "Gain-Sensitivity Product—A Figure of Merit of Hybrid-Integrated Filters Using Single Operational Amplifier," *IEEE J. Solid-State Circuits*, Vol. SC-6 (June 1971), 103–110.

33. Heinlein, W., and H. Holmes, *Active Filters for Integrated Circuits*. Vienna: Oldenbourg Verlag, 1974.

34. Orchard, H. J., "Inductorless Filters," *Electronics Lett.*, Vol. 2 (September 1966), 224–225.

35. Orchard, H. J., G. C. Temes, and T. Cataltepe, "Sensitivity Formulas for Terminated Lossless Two-Ports," *IEEE Trans. Circuits Syst.*, Vol. CAS-32 (May 1985), 459–466.

36. Laker, K. R., R. Schaumann, and M. S. Ghausi, "Multiple-Loop Feedback Topologies for the Design of Low-Sensitivity Active Filters," *IEEE Trans. Circuits Syst.*, Vol. CAS. 26 (January 1979), 1–21. Reprinted in Schaumann et al. [99].

37. Schoeffler, J. D., "The Synthesis of Minimum Sensitivity Networks," *IEEE Trans. Circuit Theory*, Vol. CT-11 (1964), 271–276.

38. Haykin, S. S., and W. J. Butler, "Multiparameter Sensitivity Indexes of Performance for Linear Time-Invariant Networks," *Proc. IEE (Lond.)*, Vol. 117 (1970), 1239–1247.

39. Rosenblum, A. L., and M. S. Ghausi, "Multiparameter Sensitivity in Active RC Networks," *IEEE Trans. Circuit Theory (Special Issue on Active and Digital Networks)*, Vol. CT-18 (November 1971), 592–599. Reprinted in Schaumann et al. [99].

40. Biswas, R. N., and E. S. Kuh, "A Multiparameter Sensitivity Measure for Linear Systems," *IEEE Trans. Circuit Theory*, Vol. CT-18 (November 1971), 718–719.

41. Papoulis, A., *Probability Random Variables and Stochastic Processes*. New York: McGraw-Hill, 1965.

42. Cooper, G. R., and G. D. McGillem, *Probabilistic Methods of Signals and Systems*. New York: Holt, Rinehart & Winston, 1971.

43. Helstrom, C. W., *Probability and Stochastic Processes for Engineers*. New York: Macmillan, 1984.

44. Acar, C., and M. S. Ghausi, "Statistical Multiparameter Sensitivity Measure of Gain and Phase Functions," *Int. J. Circuit Theory Appl.*, Vol. 5 (January 1977), 13–22.

45. Acar, A., K. R. Laker, and M. S. Ghausi, "Statistical Multiparameter Sensitivity Measure in High Q Networks," *J. Franklin Inst.*, Vol. 280 (October 1975), 281–297.

46. Laker, K. R., and M. S. Ghausi, "A Large Change Multiparameter Sensitivity," *J. Franklin Inst.*, Vol. 298 (December 1974), 395–414.

47. Ghausi, M. S., and K. R. Laker, *Modern Filter Design*. Englewood Cliffs, N.J.: Prentice-Hall, 1981.

48. Becker, P. W., and F. Jensen, *Design of Systems and Circuits for Maximum Reliability and Production Yield*. New York: McGraw-Hill, 1977.

49. Pierre, D. A., *Optimization Theory with Applications*. New York: Wiley, 1969; Dover, 1986.

50. Brayton, R. K., and R. Spence, *Sensitivity and Optimization*. New York, Elsevier, 1980.

51. Rosenblum, A. L., and M. S. Ghausi, "Sensitivity Minimization in Active RC Networks," *J. Franklin Inst.*, Vol. 294 (August 1972), 95–111.

52. Fleischer, P. E., "Sensitivity Minimization in a Single Amplifier Biquad Circuit," *IEEE Trans. Circuits Syst.*, Vol. CAS-23 (January 1976), 45–55. Reprinted in Schaumann et al. [99].

53. Laker, K. R., and M. S. Ghausi, "A Comparison of Active Multiple-Loop Feedback Techniques for Realizing High Order Bandpass Filters," *IEEE Trans. Circuits Syst.*, Vol. CAS-21 (November 1974), 774–783.

54. Bandler, J. W., and H. L. Abdel-Malek, "Optimal Centering, Tolerancing, and Yield Determination via Updated Approximations and Cuts," *IEEE Trans. Circuits Syst.*, Vol. CAS-25 (October 1978), 853–870.

55. Director, S. W., and G. H. Hachtel, "The Simplicial Approximation Approach to Design Centering," *IEEE Trans. Circuits Syst.*, Vol. CAS-24 (July 1977), 363–372.

56. Singhal, K., and J. F. Pinel, "Statistical Design Centering and Tolerancing Using Parametric Sampling," *IEEE Proc. IEEE Int. Symp. Circ. Syst.*, 1980, 882–885.

57. Antreich, K., and R. Koblitz, "A New Approach to Design Centering Based on a Multiparameter Yield-Prediction Formula," *Proc. Int. Symp. Circ. Syst.*, 1980, 886–889.

58. Soin, R. S., and R. Spence, "Statistical Exploration Approach to Design Centering," *IEE Proc.*, Vol. 127 (December 1980), 260–269.

59. Wehrhahn, E., "A Cut-Algorithm for Design Centering," *Proc. IEEE Int. Symp. Circ. Syst.*, 1984, 970–973.

60. Wehrhahn, E., and R. Spence, "The Performance of Some Design Centering Methods," *Proc. IEEE Int. Symp. Circ. Syst.*, 1984, 1424–1438.

61. Jones, I., and R. Spence, "The Optimization of Integrated Circuit Cells Subject to a Constraint on Parametric Yield," *Proc. IEEE Int. Symp. Circ. Syst.*, 1984, 974–976.

62. Styblinski, M., and L. J. Opalski, "A Random Perturbation Method for IC Yield Optimization with Deterministic Process Parameters," *Proc. IEEE Int. Symp. Circ. Syst.*, 1984, 977–980.

63. *IEEE Trans. on Computer-Aided Design, Special Issue on Statistical Design of VLSI Circuits*, January 1986.

64. Wehrhahn, E., "A New Simple Heuristic Algorithm for Yield Maximization," *Proc. IEEE Int. Symp. Circ. Syst.*, 1987, 816–819.

65. Geiger, R. L., and E. Sánchez-Sinencio, "Active Filters Using Operational Transconductance Amplifiers: A Tutorial," *IEEE Circuits and Devices Magazine*, Vol. 1 (March 1985), 20–32.

66. Solomon, J. E., "The Monolithic Op. Amp.: A Tutorial Survey," *IEEE J. Solid-State Circuits*, Vol. SC-9 (December 1974), 314–332.

67. Tsividis, K. S., and P. R. Gray, "An Integrated NMOS Operational Amplifier with Internal Compensation," *IEEE J. Solid-State Circuits*, Vol. SC-11 (December 1976), 748–753.

68. Meyer, R. G., Ed., *Integrated Circuit Operational Amplifiers*. New York: IEEE Press, 1978.

69. Sedra, A. S., and K. C. Smith, *Microelectronic Circuits*. New York: Holt, Rinehart & Winston, 1987.

70. Ghausi, M. S., *Electronic Devices and Circuits: Discrete and Integrated*. New York: Holt, Rinehart & Winston, 1985.

71. Wait, F., L. P. Huelsman, and G. Korn, *Introduction to Operational Amplifiers: Theory and Applications*. New York: McGraw-Hill, 1975.

72. Kuo, B., *Automatic Control Systems*. Englewood Cliffs, N.J.: Prentice-Hall, 1962.

73. Rosenstark, S., *Feedback Amplifier Principles*. New York: Macmillan, 1986.

74. Allen, P. E., "Slew Induced Distortion in Operational Amplifiers," *IEEE J. Solid-State Circuits*, Vol. SC-12 (February 1977), 39–44.

75. Allen, P. E., R. K. Cavin, III, and C. G. Kwok, "Frequency Domain Analysis for Operational Amplifier Macromodels," *IEEE Trans. Circuits and Systems*, Vol. CAS-26 (September 1979), 693–699.

76. Bruton, L. T., F. N. Trofimenkoff, and D. H. Treleaven, "Noise Performance of Low-Sensitivity Active Filters," *IEEE J. Solid-State Circuits*, Vol. SC-8 (February 1973), 85–91.

77. Trofimenkoff, F. N., D. H. Treleaven, and L. T. Bruton, "Noise Performance of RC-Active Quadratic Filter Sections," *IEEE Trans. Circuit Theory*, Vol. CT-20 (September 1973), 524–532. Reprinted in [99].

78. Bächler, H. J., and W. Guggenbühl, "Noise and Sensitivity Optimization of a Single-Amplifier Biquad," *IEEE Trans. Circuits and Systems*, Vol. CAS-26 (January 1979), 30–36.

79. National Semiconductor Corporation, *Linear Data Book*, LM 13600, LM 13700, 1982.

80. National Semiconductor Corporation, *Linear Applications Handbook*, 1980.

81. RCA Electronic Components, *Linear Integrated Circuits*, CA 3060, Data File 404, 1970.

82. Geiger, R. L., "Amplifiers with Maximum Bandwidth," *IEEE Trans. Circuits and Systems*, Vol. CAS-24 (September 1977), 510–512.

83. Soliman, A. M., and M. Ismail, "Active Compensation of Operational Amplifiers," *IEEE Trans. Circuits and Systems*, Vol. CAS-26 (February 1979), 112–117.

84. Schaumann, R., "Designing Active RC Biquads with Improved Performance," *IEEE Trans. Circuits and Systems*, Vol. CAS-30 (January 1983), 56–57.

85. Mikhael, W. B., and S. Michael, "Composite Operational Amplifiers: Generation and Finite-Gain Applications," *IEEE Trans. Circuits and Systems*, Vol. CAS-34 (May 1987), 449–460.

86. Michael, S., and W. B. Mikhael, "Inverting Integrator and Active Filter Applications of Composite Operational Amplifiers," *IEEE Trans. Circuits and Systems*, Vol. CAS-34 (May 1987), 461–470.

87. Brand, J. R., and R. Schaumann, "Active *R* Filters: Review of Theory and Practice," *IEE J. Electr. Circuits and Systems*, Vol. 2 (July 1978), 89–101. Reprinted in Schaumann et al. [99].

88. Riordan, R. H. S., "Simulated Inductors Using Differential Amplifier," *Electron. Lett.*, Vol. 3 (February 1967), 50–51.

89. Antoniou, A., "Gyrators Using Operational Amplifiers," *Electron. Lett.*, Vol. 3 (August 1967), 350–352.

90. Antoniou, A., and K. S. Naidu, "Modeling of a Gyrator Circuit," *IEEE Trans. Circuit Theory*, Vol. CT-20 (September 1973), 533–540.

91. Deboo, ,G. J., "A Novel Integrator Results by Grounding Its Capacitor," *Electron. Design*, Vol. 15 (September 1967), 90.

92. Brackett, P. O., and A. S. Sedra, "Active Compensation for High-Frequency Effects in Op-Amp Circuits with Applications to Active *RC* Filters," *IEEE Trans. Circuits and Systems*, Vol. CAS-23 (February 1976), 68–72. Reprinted in Schaumann et al. [99].

93. Temes, G. C., "First-Order Estimation and Precorrection of Parasitic Loss Effects in Ladder Filters," *IEEE Trans. Circuit Theory*, Vol. CT-9 (1967), 385–400.

94. Blostein, M. L., "Sensitivity Analysis of Parasitic Effects in Resistance-Terminated *LC* Two-Ports," *IEEE Trans. Circuit Theory*, Vol. CT-14 (1967), 21–25.

95. Temes, G. C., "Effects of Semiuniform Losses in Reactance 2-Ports," *Electron. Lett.*, Vol. 8, No. 6 (1972), 161–163.

96. Bruton, L. T., and A. I. A. Salama, "Frequency Limitations of Coupled-Biquadratic

Active Ladder Structures," *IEEE J. Solid-State Circuits*, Vol. SC-9 (April 1974), 70–72.

97. Moschytz, G. S., *Linear Integrated Networks—Design*. New York: Van Nostrand Reinhold, 1975.

98. Sedra, A. S., "Generation and Classification of Single-Amplifier Filters," *Int. J. Circuit Theory and Appl.*, Vol. 2 (January 1974), 51–67.

99. Schaumann, R., M. S. Soderstrand, and K. R. Laker, Eds., *Modern Active Filter Design*. New York: IEEE Press Selected Reprint Series/Wiley, 1981.

100. Sedra, A. S., and L. Brown, "A Refined Classification of Single-Amplifier Filters," *Int. J. Circuit Theory and Appl.*, Vol. 7 (January 1979), 127–137. Reprinted in Schaumann et al. [99].

101. Hilberman, D., "Input and Ground as Complements in Active Filters," *IEEE Trans. Circuit Theory*, Vol. CT-20 (September 1973), 540–547.

102. Sallen, R. P., and E. L. Key, "A Practical Method of Designing RC-Active Filters," *IEEE Trans. Circuit Theory*, Vol. CT-2 (March 1955), 74–85.

103. Sedra, A. S., and J. L. Espinosa, "Sensitivity and Frequency Limitations of Biquadratic Active Filters," *IEEE Trans. Circuits Syst.*, Vol. CAS-22 (February 1975), 122–130.

104. Schaumann, R., "Two-Amplifier Active RC Biquads with Minimized Dependence on Op-Amp Parameters," *IEEE Trans. Circuits Syst.*, Vol. CAS-30 (November 1983), 797–803.

105. Fliege, N., "A New Class of Second-Order RC-Active Filters with Two Operational Amplifiers," *Nachrichtentechn. Zeitung*, Vol. 26 (June 1973), 279–282. Reprinted in Schaumann et al. [99].

106. Chiou, C.-F., and R. Schaumann, "Performance of GIC-Derived Active RC Biquads with Variable Gain," *IEE Proc.*, Vol. 128, Pt. G (February 1981), 46–52.

107. Kerwin, W. J., L. P. Huelsman, and R. W. Newcomb, "State-Variable Synthesis for Insensitive Integrated-Circuit Transfer Functions," *IEEE J. Solid-State Circuits*, Vol. SC-2 (January 1967), 87–92.

108. Tow, J., "A Step-by-Step Active Filter Design," *IEEE Spectrum*, Vol. 6 (January 1969), 64–68.

109. Thomas, L. C., "The Biquad: Part I—Some Practical Design Considerations; Part II—A Multipurpose Active Filtering System," *IEEE Trans. Circuit Theory*, Vol. CT-18 (May 1971), 350–357; 358–361.

110. Åckerberg, D., and K. Mossberg, "A Versatile Active RC Building Block with Inherent Compensation for the Finite Bandwidth of the Amplifier," *IEEE Trans. Circuits Syst.*, Vol. CAS-21 (January 1974), 75–78. Reprinted in Schaumann et al. [99].

111. Rao, R. K., and S. Srinivasan, "Low-Sensitivity Active Filters Using the Operational Amplifier Pole," *Proc. IEEE*, 1974, 1713–1714.

112. Rao, R. K., and S. Srinivasan, "A Bandpass Filter Using the Operational Amplifier Pole," *IEEE J. Solid-State Circuits*, Vol. SC-8 (June 1973), 245–246.

113. Chiou, C.-F., and R. Schaumann, "Design and Performance of a Fully Integrated Bipolar 10.7 MHz Analog Bandpass Filter," *IEEE Trans. Circuits Syst.*, Vol. CAS-35 (February 1986), 116–124; Also in *IEEE J. Solid State Circuits*, Vol. SC-21 (February 1986), 6–14.

114. Sánchez-Sinencio, E., R. L. Geiger, and H. Nevárez-Lozano, "Generation of Continuous-Time Two-Integrator Loop OTA Filter Structures," *IEEE Symp. Circuits Syst.*, 1987, 325–328.

115. Sedra, A. S., M. A. Ghorab, and K. Martin, "Optimum Configuration for Single-Amplifier Biquadratic Filters," *IEEE Trans. Circuits Syst.*, Vol. CAS-27 (December 1980), 1155–1163.

116. Geffe, P. R., "A Q-Invariant Active Resonator," *Proc. IEEE*, Vol. 57 (1969), 1442.

117. Tarmy, R., and M. S. Ghausi, "Very High-Q Insensitive Active *RC* Networks," *IEEE Trans. Circuit Theory*, Vol. CT-17 (August 1970), 358–366.

118. Moschytz, G. S., "High Q-Factor Insensitive Active *RC* Network, Similar to the Tarmy-Ghausi Circuit, but Using Single-Ended Operational Amplifiers," *Electron. Lett.*, Vol. 8 (September 1972), 458–459.

119. Revankar, G. N., K. Shankar, and R. B. Datar, "A Modified Moschytz High-Q Filter," *Int. J. Electron.*, Vol. 42 (February 1977), 117–120.

120. Friend, J. J., C. A. Harris, and D. Hilberman, "STAR: An Active Biquadratic Filter Section," *IEEE Trans. Circuits Syst.*, Vol. CAS-22 (February 1975), 115–121. Reprinted in Schaumann et al. [99].

121. Mikhael, W. B., and B. B. Bhattacharyya, "A Practical Design for Insensitive *RC*-Active Filters," *IEEE Trans. Circuits Syst.*, Vol. CAS-22 (May 1975), 407–415.

122. Padukone, P., J. Mulawka, and M. S. Ghausi, "An Active Biquadratic Section with Reduced Sensitivity to Operational Amplifier Imperfections," *J. Franklin Inst.*, Vol. 30 (January 1980), 27–40.

123. Martin, K., and A. S. Sedra, "On the Stability of the Phase-Lead Integrator," *IEEE Trans. Circuits Syst.*, Vol. CAS-24 (June 1977), 321–324.

124. Bowron, P., and A. S. S. Al-Kabbani, "Nonlinear Performance of Composite-Amplifier Filters," *Proc. IEEE Int. Symp. Circ. Syst.*, 1988, 2859–2862.

125. Halfin, S., "An Optimization Method for Cascaded Filters," *Bell Syst. Tech. J.*, Vol. 49 (1970), 185–190.

126. Halfin, S., "Simultaneous Determination of Ordering and Amplifications of Cascaded Subsystems," *J. Optimiz. Theory and Appl.*, Vol. 6 (1970), 356–360.

127. Lüder, E., "Optimization of the Dynamic Range and the Noise Distance of RC-Active Filters by Dynamic Programming," *Int. J. Circuit Theory and Appl.*, Vol. 3 (December 1975), 365–170. Reprinted in Schaumann et al. [99].

128. Snelgrove, W. M., and A. S. Sedra, "Optimization of Dynamic Range in Cascade Active Filters," *Proc. IEEE Int. Symp. Circ. Syst.*, 1978, 151–155.

129. Chiou, C.-F., and R. Schaumann, "Comparison of Dynamic Range Properties of High-Order Active Bandpass Filters," *IEE Proc.*, Vol. 127, Pt. G (June 1980), 101–108.

130. Chiou, C.-F., and R. Schaumann, "Refined Procedure for Optimizing Signal-to-Noise Ratio in Cascade Active Filters," *IEE Proc.*, Vol. 128, Pt. G (August 1981), 189–191.

131. Laker, K. R., and M. S. Ghausi, "Synthesis of a Low-Sensitivity Multiloop Feedback Active Filter," *IEEE Trans. Circuits Syst.*, Vol. CAS-21 (March 1974), 252–259.

132. Laker, K. R., and M. S. Ghausi, "Computer-Aided Analysis and Design of Follow-the-Leader Feedback Active RC Filters," *Int. J. Circuit Theory and Appl.*, Vol. 4 (April 1976), 177–187.

133. Laker, K. R., and M. S. Ghausi, "A Comparison of Active Multiple-Loop Feedback Techniques for Realizing High-Order Bandpass Filters," *IEEE Trans. Circuits Syst.*, Vol. CAS-21 (November 1974), 774–783.

134. Gadenz, R. N., "On Low-Sensitivity Realizations of Band Elimination Active Filters," *IEEE Trans. Circuits Syst.*, Vol. CAS-24 (April 1977), 175–183.

135. Hurtig, G., III, "The Primary Resonator Block Technique of Filter Synthesis," *Proc. Int. Filter Symp.*, 1972, 84.

136. Schaumann, R., W. A. Kinghorn, and K. R. Laker, "Optimal Design Parameters for Minimizing Sensitivity and Distortion in FLF Active Filters," *Proc. 19th Midwest Symp. on Circuits and Systems*, 1976, 390–395.

137. Laker, K. R., and M. S. Ghausi, "Design of Minimum-Sensitivity Multiple-Loop Feedback Bandpass Active Filters," *J. Franklin Inst.*, Vol. 310 (1980), 51–64.

138. Brackett, P. O., and A. S. Sedra, "Direct SFG Simulation of LC Ladder Networks with Applications to Active Filter Design," *IEEE Trans. Circuits Syst.*, Vol. CAS-23 (February 1976), 61–67.

139. Martin, K., and A. S. Sedra, "Design of Signal-Flow Graph (SFG) Active Filters," *IEEE Trans. Circuits Syst.*, Vol. CAS-25 (April 1978), 185–195. Reprinted in Schaumann et al. [99].

140. Martin, K., and A. S. Sedra, "Optimum Design of Active Filters Using the Generalized Immittance Converter," *IEEE Trans. Circuits Syst.*, Vol. CAS-24 (September 1977), 495–503.

141. Antoniou, A., "Realization of Gyrators Using Operational Amplifiers and Their Use in RC-Active Network Synthesis," *Proc. IEE*, Vol. 116 (November 1960), 1838–1850.

142. Gorski-Popiel, J., "RC-Active Synthesis Using Positive-Immittance Converters," *Electron. Lett.*, Vol. 3 (August 1967), 381–382.

143. Bruton, L. T., "Network Transfer Functions Using the Concept of Frequency Dependent Negative Resistance," *IEEE Trans. Circuit Theory.*, Vol. CT-18 (August 1969), 406–408.

144. Bruton, L. T., "Multiple Amplifier RC-Active Filter Design with Emphasis on GIC Realizations," *IEEE Trans. Circuits Syst.*, Vol. CAS-25 (October 1978), 830–845.

145. Tan, M. A., "Design and Automatic Tuning of Fully Integrated Transconductance-Grounded Capacitor Filters," Ph.D. Thesis, University of Minnesota, 1988.

146. Bilotti, A., "Operation of MOS Transistors as a Variable Resistor," *Proc. IEEE*, Vol. 54 (August 1966), 1093–1094.

147. Banu, M., and Y. Tsividis, "An Elliptic Continuous-Time CMOS Filter with On-Chip Automatic Tuning," *IEEE J. Solid-State Circuits*, Vol. SC-20 (December 1985), 1114–1121.

148. Tsividis, Y., Z. Czarnul, and S. C. Fang, "MOS Transconductors and Integrators with High Linearity," *Electron. Lett.*, Vol. 22 (1986), 245–246, 619.

149. Czarnul, Z., "Novel MOS Resistive Circuit for Synthesis of Fully Integrated Continuous-Time Filters," *IEEE Trans. Circuits Syst.*, Vol. CAS-33 (July 1986), 718–721.

150. Han, I. S., and S. B. Park, "Voltage-Controlled Linear Resistor by Two MOS Transistors and Its Application to Active RC Filter MOS Integration," *Proc. IEEE*, Vol. 72 (November 1984), 1655–1657.

151. Tsividis, Y., *Operation and Modeling of the MOS Transistor*. New York: McGraw-Hill, 1986.

152. Gray, P. R., and R. G. Meyer, *Analysis and Design of Analog Integrated Circuits*. New York: Wiley, 1984.

153. Grebene, A. B., *Bipolar and MOS Analog Integrated Circuit Design*. New York: Wiley, 1984.

154. Tsividis, Y., M. Banu, and J. Khoury, "Continuous-Time MOSFET-C Filters in VLSI," *IEEE Trans. Circuits Syst.*, Vol. CAS-33, *Special Issue on VLSI Analog and Digital Signal Processing* (February 1986), 125–140. Also in *IEEE J. Solid-State Circuits*, Vol. SC-21 (February 1986), 15–30.

155. Gregorian, R., and G. C. Temes, *Analog MOS Integrated Circuits for Signal Processing*. New York: Wiley-Interscience, 1986.

156. Park, C. S., and R. Schaumann, "Design of a 4 MHz Analog Integrated CMOS Transconductance C Bandpass Filter," *IEEE J. Solid-State Circuits*, Vol. SC-23 (August 1988), 987–996.

157. Krummenacher, F., and N. Joehl, "A 4-MHz CMOS Continuous-Time Filter with On-Chip Automatic Tuning," *IEEE J. Solid-State Circuits*, Vol. SC-23 (June 1988), 750–758.

158. Khorramabadi, H., and P. R. Gray, "High-Frequency CMOS Continuous-Time Filters," *IEEE J. Solid-State Circuits*, Vol. SC-19 (December 1984), 963–967.

159. Fukahori, K., "A Bipolar Voltage-Controlled Tunable Filter," *IEEE J. Solid-State Circuits*, Vol. SC-16 (December 1981), 729–737.

160. Miura, K., Y. Okada, M. Shiomi, M. Masuda, E. Funaki, Y. Okada, and S. Ogura, "VCR Signal Processing LSIs with Self-Adjusted Integrated Filters," *Proc. 1986 Bipolar Circ. Techn. Mtg.*, 85–86.

161. Moulding, K. W., J. R. Quartly, P. J. Rankin, R. S. Thompson, and G. A. Wilson, "Gyrator Video Filter IC with Automatic Tuning," *IEEE J. Solid-State Circuits*, Vol. SC-15 (December 1980), 963–968.

162. Tan, K. S., and P. R. Gray, "Fully Integrated Analog Filters Using Bipolar-JFET Technology," *IEEE J. Solid-State Circuits*, Vol. SC-12 (December 1978), 814–821.

163. Tsividis, Y., "Self-Tuned Filters," *Electron. Lett.*, Vol. 17 (June 1981), 406–407.

164. McCreary, J. L., "Matching Properties and Voltage and Temperature Dependence of MOS Capacitors," *IEEE J. Solid-State Circuits*, Vol. SC-16 (December 1981), 608–616.

165. Dorf, R. C., *Modern Control Systems*, 4th Ed. Reading, Mass.: Addison-Wesley, 1986.

166. Park, C. S., and R. Schaumann, "A High-Frequency CMOS Linear Transconductance Element," *IEEE Trans. Circuits. Syst.*, Vol. CAS-33 (June 1986), 1132–1138.

167. Voorman, J. O., W. H. A. Bruls, and J. P. Barth, "Integration of Analog Filters in a Bipolar Process," *IEEE J. Solid-State Circuits*, Vol. SC-17 (August 1982), 713–722.

168. Hagiwara, H., M. Kumazawa, S. Takagi, M. Furihata, M. Nagata, and T. Yanagisawa, "A Monolithic Video Frequency Filter Using NIC-Based Gyrators," *IEEE J. Solid-State Circuits*, Vol. SC-23 (February 1988), 175–182.

169. Ghausi, M. S., and J. Kelly, *Introduction to Distributed Parameter Networks*. New York: Holt, Rinehart & Winston, 1968.

170. Seevinck, E., and R. F. Wassenaar, "A Versatile CMOS Linear Transconductor/Square-Law Function Circuit," *IEEE J. Solid-State Circuits*, Vol. SC-22 (June 1987), 366–377.

171. Nedungadi, A., and T. R. Viswanathan, "Design of Linear CMOS Transconductance Elements," *IEEE Trans. Circuits Syst.*, Vol. CAS-31 (October 1984), 891–894.

172. Nagel, L. W., *SPICE2: A Computer Program to Simulate Semiconductor Circuits*, Memorandum No. ERL-M520, University of California, Berkeley, 1975.

173. *SPICE 2G.6 User Manual*.

174. Park, C. S., "A CMOS Linear Transconductance Element and Its Applications in Integrated High-Frequency Filters," Ph.D. Thesis, University of Minnesota, 1988.

175. Park, C. S., and R. Schaumann, "Design of an Eighth-Order Fully Integrated CMOS 4MHz Continuous-Time Bandpass Filter with Digital/Analog Control of Frequency and Quality Factor," *Proc. Int. Symp. Circ. Syst.*, 1987, 754–757.

176. de Queiroz, A. C. M., L. P. Calôba, and E. Sánchez-Sinencio, "Signal Flow Graph OTA-C Integrated Filters," *Proc. Int. Symp. Circ. Syst.*, 1988, 2165–2168.

177. Mead, C. A., and L. A. Conway, *Introduction to VLSI Systems*. Reading, Mass.: Addison-Wesley, 1980.

178. Weste, N., and K. Eshraghian, *Principles of CMOS VLSI Design*. Reading, Mass.: Addison-Wesley, 1985.

179. Allen, P. E., and D. R. Holberg, *CMOS Analog Circuit Design*. New York: Holt, Rinehart & Winston, 1987.

180. Tan, M. A., and R. Schaumann, "Simulating General-Parameter *LC*-Ladder Filters for Monolithic Realizations with Only Transconductance Elements and Grounded Capacitors," *IEEE Trans. Circuits Syst.*, Vol. CAS-36 (February 1989), 299–307.

181. Perry, D. J., "Scaling Transformation of Multiple Feedback Filters," *Proc. IEE*, Vol. 128, Pt. G (1981), 176–179.

182. Ismail, M., S. V. Smith, and R. G. Beale, "A New MOSFET-C Universal Filter Structure for VLSI," *IEEE J. Solid-State Circuits*, Vol. SC-23 (February 1988), 183–194.

183. Fried, D. L., "Analog Sampled-Data Filters," *IEEE J. Solid-State Circuits*, Vol. SC-7 (August 1972), 302–304.

184. Gray, P. R., and R. Costello, "Performance Limitations in Switched-Capacitor Filters," Chap. 10 in Y. Tsividis and P. Antognetti, Eds., *Design of MOS VLSI Circuits for Telecommunications*. Englewood Cliffs, N.J.: Prentice-Hall, 1985.

185. Martin, K., and A. S. Sedra, "Strays-Insensitive Switched-Capacitor Filters Based on the Bilinear z-Transform," *Electron. Lett.*, Vol. 15 (June 1979), 365–366.

186. Gregorian, R., and W. E. Nicholson, "CMOS Switched-Capacitor Filters for a Voice PCM CODEC," *IEEE J. Solid-State Circuits*, Vol. SC-14 (December 1979), 970–980.

187. Ghaderi, M. B., J. A. Nossek, and G. C. Temes, "Narrow-Band Switched-Capacitor Bandpass Filters," *IEEE Trans. Circuits Syst.*, Vol. CAS-29 (August 1982), 557–572.

188. Pandel, J., et al., "Integrated 18th-Order Pseudo N-Path Filter in Integrated VIS SC Technique," *IEEE Trans. Circuits Syst.*, Vol. CAS-33 (February 1986), 158–166.

189. Franca, J. E., and D. G. Haigh, "Design and Applications of Single-Path Frequency-Translated Switched-Capacitor Systems," *IEEE Trans. Circuits Syst.*, Vol. CAS-35 (April 1988), 394–408.

190. Franca, J. E., "Nonrecursive Polyphase Switched-Capacitor Decimators and Inter-polators," *IEEE Trans. Circuits Syst.*, Vol. CAS-32 (September 1985), 877–887.

191. Martin, K., and A. S. Sedra, "Exact Design of Swtiched-Capacitor Bandpass Filters Using Coupled Biquad Structures," *IEEE Trans. Circuits Syst.*, Vol. CAS-27 (June 1980), 469–475.

192. Datar, R. B., and A. S. Sedra, "Exact Design of Strays-Insensitive Switched-Capacitor Ladder Filters," *IEEE Trans. Circuits Syst.*, Vol. CAS-30 (December 1983), 888–898.

193. Sedra, A. S., "Switched-Capacitor Filter Synthesis," Chap. 9 in Y. Tsividis and P. Antognetti, Eds., *Design of MOS VLSI Circuits for Telecommunications.* Englewood Cliffs, N.J.: Prentice-Hall, 1985.

194. Fleischer, P. E., and K. R. Laker, "A Family of Active Switched-Capacitor Biquad Building Blocks," *Bell Syst. Tech. J.*, Vol. 58 (December 1979), 2235–2269.

195. Broderson, R. W., P. R. Gray, and D. A. Hodges, "MOS Switched-Capacitor Filters," *Proc. IEEE*, Vol. 67 (January 1979), 61–75.

196. Scanlan, S. O., "Analysis and Synthesis of Switched-Capacitor State-Variable Filters," *IEEE Trans. Circuits Syst.*, Vol. CAS-28 (February 1981), 85–93.

197. Rudd, E. P., and R. Schaumann, "A Program for the Analysis of High-Frequency Behavior of Switched-Capacitor Filters," *Proc. IEEE Int. Symp. Circ. Syst.*, June 1985, 1173–1176.

198. Temes, G. C., "Finite Amplifier Gain and Bandwidth Effects in Switched-Capacitor Filters," *IEEE J. Solid-State Circuits*, Vol. SC-15 (June 1980), 358–361.

199. Martin, K., and A. S. Sedra, "Effects of the Op Amp Finite Gain and Bandwidth on the Performance of Switched-Capacitor Filters," *IEEE Trans. Circuits Syst.*, Vol. CAS-28 (August 1981), 822–829.

200. Allstot, D. A., and W. C. Black, "Technological Design Considerations for Monolithic MOS Switched-Capacitor Filtering System," *Proc. IEEE*, Vol. 71 (August 1983), 967–986.

201. Hsieh, K.-C., et al., "A Low-Noise Chopper-Stabilized Differential Switched-Capacitor Filter Technique," *IEEE J. Solid-State Circuits*, Vol. SC-16 (December 1981), 708–715.

202. Gobet, C. A., and A. Knob, "Noise Analysis of Switched-Capacitor Networks," *IEEE Trans. Circuits Syst.*, Vol. CAS-30 (January 1983), 37–43.

203. Roberts, G. W., and A. S. Sedra, "Switched-Capacitor Filter Networks Derived from General Parameter Bandpass Ladder Networks," *Proc. IEEE Int. Symp. Circ. Syst.*, May 1988, 1005–1008.

204. Choi, T. C., et al., "High-Frequency CMOS Switched-Capacitor Filters for Commu-nications Applications," *IEEE J. Solid-State Circuits*, Vol. SC-18 (December 1983), 652–664.

205. Masuda, S., et al., "CMOS Sampled Differential Push-Pull Cascade Operational Am-plifier," *Proc. Int. Symp. Circ. Syst.*, 1984, 1211–1214.

206. Song, B. S., "A 10.7 MHz Switched-Capacitor Bandpass Filter," *IEEE J. Solid-State Circuits*, Vol. SC-24 (April 1989), 320–324.

207. Hosticka, B. J., and G. S. Moschytz, "Switched-Capacitor Simulation of Grounded Inductors and Gyrators," *Electron. Lett.*, Vol. 14 (November 1978), 788–790.

208. Temes, G. C., and M. Jahanbegloo, "Switched-Capacitor Circuits Bilinearly Equiv-

alent to the Floating Inductor and FDNR," *Electron. Lett.*, Vol. 15 (November 1979), 87–88.

209. Jacobs, G. M., et al., "Touch-Tone Decoder Chip Mates Analog Filters with Digital Logic," *Electronics* (February 1979), 105–112.

210. White, B. J., G. M. Jacobs, and G. F. Landsburg, "A Monolithic Dual Tone Multi-frequency Receiver," *IEEE J. Solid-State Circuits*, Vol. SC-14 (December 1979), 991–997.

211. Fleischer, P. R., et al., "An NMOS Building Block for Telecommunications Applications," *IEEE Trans. Circuits Syst.*, Vol. CAS-27 (June 1980), 552–559.

212. Kuraishi, Y., et al., "A Single-Chip 20-Channel Speech Spectrum Analyzer Using a Multiplexed Switched-Capacitor Filter Bank," *IEEE J. Solid-State Circuits*, Vol. SC-19 (December 1984), 964–970.

213. Callias, F., F. H. Salchli, and D. Girard, "A Set of Four ICs in CMOS Technology for a Programmable Hearing Aid," *IEEE J. Solid-State Circuits*, Vol. SC-24 (December 1989), 301–312.

214. Haigh, D. G., and J. T. Taylor, "Continuous-Time and Switched-Capacitor Monolithic Filters Based on Current and Charge Simulation," *Proc. IEEE Int. Symp. Circ. Syst.*, May 1989, 1850–1853.

215. Hughes, J. B., N. C. Bird, and I. C. Macbeth, "Switched Currents—A New Technique for Analog Sampled-Data Signal Processing," *Proc. IEEE Int. Symp. Circ. Syst.*, May 1989, 1584–1587.

216. Fleischer, P. E., A. Ganesan and K. R. Laker, "Parasitic Compensated Switched Capacitor Circuits," *Electron. Lett.*, Vol. 17, No. 24 (26 November 1981) pp. 929–931.

217. Laker, K. R., P. E. Fleischer and A. Ganesan, "Parasitic Insensitive, Biphase Switched Capacitor Filters with One Operational Amplifier per Pole Pair," *BSTJ*, Vol. 61 (May 1982) No. 5, pp. 685–707.

218. Laker, K. R., A. Ganesan and P. E. Fleischer, "Design and Implementation of Cascaded Switched Capacitor Delay Equalizers," *IEEE Trans. on Circuits and Systems*, Vol. CAS-32, (July 1985), pp. 700–711.

Laplace Transforms

APPENDIX I

PROPERTIES

Property	Time domain $x(t)$	Frequency domain $X(s)$
Linearity	$ax_1(t) + bx_2(t)$	$aX_1(s) + bX_2(s)$
Differentiation	$\dfrac{d^n x(t)}{dt^n}$	$s^n X(s) - s^{n-1}x(0^-) - \ldots - \dfrac{d^{n-1}x(0^-)}{dt^{n-1}}$
Integration	$\displaystyle\int_{-\infty}^{t} x(\lambda)\,d\lambda$	$\dfrac{X(s)}{s} + \dfrac{1}{s}\dfrac{dx(0^-)}{dt}$
Delay	$x(t - \tau)u(t - \tau)$	$X(s)e^{-s\tau}$
s shift	$x(t)e^{-\sigma t}$	$X(s + \sigma)$
Convolution	$\displaystyle\int_{0}^{\infty} x_1(\lambda)x_2(t - \lambda)\,d\lambda$	$X_1(s)X_2(s)$

TRANSFORM PAIRS

Time signal	Laplace transform
$\dfrac{d^n \delta(t)}{dt^n}$	s^n
$1, u(t)$	$\dfrac{1}{s}$
$\dfrac{t^n e^{-\sigma t}}{n!} u(t)$	$\dfrac{1}{(s + \sigma)^{n+1}}$
$e^{-\sigma t} \cos \omega_0 t\, u(t)$	$\dfrac{s + \sigma}{(s + \sigma)^2 + \omega_0^2}$
$e^{-\sigma t} \sin \omega_0 t\, u(t)$	$\dfrac{\omega_0}{(s + \sigma)^2 + \omega_0^2}$

z-Transforms

APPENDIX II

PROPERTIES

Property	Time sequence $x(n)$	z-transform $X(z)$
Linearity	$ax_1(n) + bx_2(n)$	$aX_1(z) + bX_2(z)$
	$nx(n)$	$-z \dfrac{dX(z)}{dz}$
Delay	$x(n - k), \quad k \geq 0$	$X(z)z^{-k}$
	$x(n + k), \quad k \geq 1$	$X(z)z^k - \displaystyle\sum_{m=0}^{k-1} x(m)z^{-m}$
Convolution	$\displaystyle\sum_{k=0}^{\infty} x_1(k)x_2(n - k)$	$X_1(z)X_2(z)$

TRANSFORM PAIRS ($T = 1$)

Time sequence for $n \geq 0$	z-transform
$\delta(n) \triangleq x(n) = \begin{cases} 1 & n = 0 \\ 0 & n \neq 0 \end{cases}$	1
$1, u(n)$	$\dfrac{1}{1 - z^{-1}}$
n	$\dfrac{z^{-1}}{(1 - z^{-1})^2}$
$e^{-an} = (e^{-a})^n = K^n$	$\dfrac{1}{1 - e^{-a}z^{-1}} = \dfrac{1}{1 - Kz^{-1}}$
ne^{-an}	$\dfrac{e^{-a}z^{-1}}{(1 - e^{-a}z^{-1})^2}$
$e^{-an} \cos bn$	$\dfrac{1 - e^{-a}(\cos b)z^{-1}}{1 - 2e^{-a}(\cos b)z^{-1} + e^{-2a}z^{-2}}$
$e^{-an} \sin bn$	$\dfrac{e^{-a}(\sin b)z^{-1}}{1 - 2e^{-a}(\cos b)z^{-1} + e^{-2a}z^{-2}}$

Tables of Classical Filter Functions

APPENDIX III

TABLE III-1

n	Butterworth polynomials
1	$s + 1$
2	$s^2 + \sqrt{2}\,s + 1$
3	$s^3 + 2s^2 + 2s + 1 = (s + 1)(s^2 + s + 1)$
4	$s^4 + 2.613s^3 + 3.414s^2 + 2.613s + 1 = (s^2 + 0.765s + 1)(s^2 + 1.848s + 1)$
5	$s^5 + 3.2361s^4 + 5.2361s^3 + 5.2361s^2 + 3.2361s + 1$ $(s + 1.0000)[(s + 0.3090)^2 + 0.9511^2][(s + 0.8090)^2 + 0.5878^2]$
6	$s^6 + 3.8637s^5 + 7.4641s^4 + 9.1416s^3 + 7.4641s^2 + 3.8637s + 1$ $[(s + 0.2588)^2 + 0.9659^2][(s + 0.7071)^2 + 0.7071^2][(s + 0.9659)^2 + 0.2588^2]$
7	$s^7 + 4.4940s^6 + 10.0978s^5 + 14.5918s^4 + 14.5918s^3 + 10.0978s^2 + 4.4940s + 1$ $(s + 1.0000)[(s + 0.2225)^2 + 0.9749^2][(s + 0.6235)^2 + 0.7818^2]$ $[(s + 0.9010)^2 + 0.4339^2]$
8	$s^8 + 5.1258s^7 + 13.1317s^6 + 21.8462s^5 + 25.6884s^4 + 21.8462s^3 + 13.1371s^2 +$ $5.1258s + 1$ $[(s + 0.1951)^2 + 0.9808^2][(s + 0.5556)^2 + 0.8315^2][(s + 0.8315)^2 + 0.5556^2]$ $[(s + 0.9808)^2 + 0.1951^2]$
9	$s^9 + 5.7588s^8 + 16.5817s^7 + 31.1634s^6 + 41.9864s^5 + 41.9864s^4 + 31.1634s^3 +$ $16.5817s^2 + 5.7588s + 1$ $(s + 1.0000)[(s + 0.1737)^2 + 0.9848^2][(s + 0.5000)^2 + 0.8660^2][(s + 0.7660)^2$ $+ 0.6428^2][(s + 0.9397)^2 + 0.3420^2]$
10	$s^{10} + 6.3925s^9 + 20.4317s^8 + 42.8021s^7 + 64.8824s^6 + 74.2334s^5 + 64.8824s^4 +$ $42.8021s^3 + 20.4317s^2 + 6.3925s + 1$ $[(s + 0.1564)^2 + 0.9877^2][(s + 0.4540)^2 + 0.8910^2][(s + 0.7071)^2 + 0.7071^2]$ $[(s + 0.8910)^2 + 0.4540^2][(s + 0.9877)^2 + 0.1564^2]$

TABLE III-2a

n	0.5 dB–Ripple Chebyshev filter ($\varepsilon = 0.3493$)
1	$s + 2.863$
2	$s^2 + 1.425s + 1.516$
3	$s^3 + 1.253s^2 + 1.535s + 0.716 = (s + 0.626)(s^2 + 0.626s + 1.142)$
4	$s^4 + 1.197s^3 + 1.717s^2 + 1.025s + 0.379 = (s^2 + 0.351s + 1.064)(s^2 + 0.845s + 0.356)$
5	$s^5 + 1.1725s^4 + 1.9374s^3 + 1.3096s^2 + 0.7525s + 0.1789$ $(s + 0.3623)[(s + 0.1120)^2 + 1.0116^2][(s + 0.2931)^2 + 0.6252^2]$
6	$s^6 + 1.1592s^5 + 2.1718s^4 + 1.5898s^3 + 1.1719s^2 + 0.4324s + 0.0948$ $[(s + 0.0777)^2 + 1.0085^2][(s + 0.2121)^2 + 0.7382^2][(s + 0.2898)^2 + 0.2702^2]$
7	$s^7 + 1.1512s^6 + 2.4126s^5 + 1.8694s^4 + 1.6479s^3 + 0.7556s^2 + 0.2821s + 0.0447$ $(s + 0.2562)[(s + 0.0570)^2 + 1.0064^2][(s + 0.1597)^2 + 0.8001^2][(s + 0.2308)^2 + 0.4479^2]$
8	$s^8 + 1.1461s^7 + 2.6567s^6 + 2.1492s^5 + 2.1840s^4 + 1.1486s^3 + 0.5736s^2 + 0.1525s + 0.0237$ $[(s + 0.0436)^2 + 1.0050^2][(s + 0.1242)^2 + 0.8520^2][(s + 0.1859)^2 + 0.5693^2]$ $[(s + 0.2193)^2 + 0.1999^2]$
9	$s^9 + 1.1426s^8 + 2.9027s^7 + 2.4293s^6 + 2.7815s^5 + 1.6114s^4 + 0.9836s^3 + 0.3408s^2 + 0.0941s + 0.0112$ $(s + 0.1984)[(s + 0.0345)^2 + 1.0040^2][(s + 0.0992)^2 + 0.8829^2][(s + 0.1520)^2 + 0.6553^2][(s + 0.1864)^2 + 0.3487^2]$
10	$s^{10} + 1.1401s^9 + 3.1499s^8 + 2.7097s^7 + 3.4409s^6 + 2.1442s^5 + 1.5274s^4 + 0.6270s^3 + 0.2373s^2 + 0.0493s + 0.0059$ $[(s + 0.0279)^2 + 1.0033^2][(s + 0.0810)^2 + 0.9051^2][(s + 0.1261)^2 + 0.7183^2]$ $[(s + 0.1589)^2 + 0.4612^2][(s + 0.1761)^2 + 0.1589^2]$

TABLE III-2b

n	1.0 dB–Ripple Chebyshev filter ($\varepsilon = 0.5089$)
1	$s + 1.965$
2	$s^2 + 1.098s + 1.103$
3	$s^3 + 0.988s^2 + 1.238s + 0.491 = (s + 0.494)(s^2 + 0.490s + 0.994)$
4	$s^4 + 0.953s^3 + 1.454s^2 + 0.743s + 0.276 = (s^2 + 0.279s + 0.987)(s^2 + 0.674s + 0.279)$
5	$s^5 + 0.9368s^4 + 1.6888s^3 + 0.9744s^2 + 0.5805s + 0.1228$ $(s + 0.2895)[(s + 0.0895)^2 + 0.9901^2][(s + 0.2342)^2 + 0.6119^2]$
6	$s^6 + 0.9282s^5 + 1.9308s^4 + 1.2021s^3 + 0.9393s^2 + 0.3071s + 0.0689$ $[(s + 0.0622)^2 + 0.9934^2][(s + 0.1699)^2 + 0.7272^2][(s + 0.2321)^2 + 0.2662^2]$
7	$s^7 + 0.9231s^6 + 2.1761s^5 + 1.4288s^4 + 1.3575s^3 + 0.5486s^2 + 0.2137s + 0.0307$ $(s + 0.2054)[(s + 0.0457)^2 + 0.9953^2][(s + 0.1281)^2 + 0.7982^2][(s + 0.1851)^2 + 0.4429^2]$
8	$s^8 + 0.9198s^7 + 2.4230s^6 + 1.6552s^5 + 1.8369s^4 + 0.8468s^3 + 0.4478s^2 + 0.1073s + 0.0172$ $[(s + 0.0350)^2 + 0.9965^2][(s + 0.0997)^2 + 0.8448^2][(s + 0.1492)^2 + 0.5644^2]$ $[(s + 0.1759)^2 + 0.1982^2]$
9	$s^9 + 0.9175s^8 + 2.6709s^7 + 1.8815s^6 + 2.3781s^5 + 1.2016s^4 + 0.7863s^3 + 0.2442s^2 + 0.0706s + 0.0077$ $(s + 0.1593)[(s + 0.0277)^2 + 0.9972^2][(s + 0.0797)^2 + 0.8769^2][(s + 0.1221)^2 + 0.6509^2][(s + 0.1497)^2 + 0.3463^2]$
10	$s^{10} + 0.9159s^9 + 2.9195s^8 + 2.1079s^7 + 2.9815s^6 + 1.6830s^5 + 1.2445s^4 + 0.4554s^3 + 0.1825s^2 + 0.0345s + 0.0043$ $[(s + 0.0224)^2 + 0.9978^2][(s + 0.1013)^2 + 0.7143^2][(s + 0.0651)^2 + 0.9001^2]$ $[(s + 0.1277)^2 + 0.4586^2][(s + 0.1415)^2 + 0.1580^2]$

TABLE III-2c

n	2 dB–Ripple Chebyshev filter ($\varepsilon = 0.7648$)

1    $s + 1.308$

2    $s^2 + 0.804s + 0.637$

3    $s^3 + 0.738s^2 + 1.022s + 0.327 = (s + 0.402)(s^2 + 0.369s + 0.886)$

4    $s^4 + 0.716s^3 + 1.256s^2 + 0.517s + 0.206 = (s^2 + 0.210s + 0.928)(s^2 + 0.506s + 0.221)$

5    $s^5 + 0.7065s^4 + 1.4995s^3 + 0.6935s^2 + 0.4593s + 0.0817$
     $(s + 0.2183)[(s + 0.0675)^2 + 0.9735^2][(s + 0.1766)^2 + 0.6016^2]$

6    $s^6 + 0.7012s^5 + 1.7459s^4 + 0.8670s^3 + 0.7715s^2 + 0.2103s + 0.0514$
     $[(s + 0.0470)^2 + 0.9817^2][(s + 0.1283)^2 + 0.7187^2][(s + 0.1753)^2 + 0.2630^2]$

7    $s^7 + 0.6979s^6 + 1.9935s^5 + 1.0392s^4 + 1.1444s^3 + 0.3825s^2 + 0.1661s + 0.0204$
     $(s + 0.1553)[(s + 0.0346)^2 + 0.9867^2][(s + 0.0968)^2 + 0.7912^2][(s + 0.1399)^2 + 0.4391^2]$

8    $s^8 + 0.6961s^7 + 2.2423s^6 + 1.2117s^5 + 1.5796s^4 + 0.5982s^3 + 0.3587s^2 + 0.0729s + 0.0129$
     $[(s + 0.0265)^2 + 0.9898^2][(s + 0.0754)^2 + 0.8391^2][(s + 0.1129)^2 + 0.5607^2]$
     $[(s + 0.1332)^2 + 0.1969^2]$

9    $s^9 + 0.6947s^8 + 2.4913s^7 + 1.3837s^6 + 2.0767s^5 + 0.8569s^4 + 0.6445s^3 + 0.1684s^2 + 0.0544s + 0.0051$
     $(s + 0.1206)[(s + 0.0209)^2 + 0.9919^2][(s + 0.0603)^2 + 0.8723^2] [(s + 0.0924)^2 + 0.6474^2][(s + 0.1134)^2 + 0.3445^2]$

10   $s^{10} + 0.6937s^9 + 2.7406s^8 + 1.5557s^7 + 2.6363s^6 + 1.1585s^5 + 1.0389s^4 + 0.3178s^3 + 0.1440s^2 + 0.0233s + 0.0032$
     $[(s + 0.0170)^2 + 0.9935^2][(s + 0.0767)^2 + 0.7113^2][(s + 0.0493)^2 + 0.8962^2]$
     $[(s + 0.0967)^2 + 0.4567^2][(s + 0.1072)^2 + 0.1574^2]$

APPROXIMATION WITH POLES AND ZEROS (ELLIPTIC FILTERS)[1]

The coefficients of the elliptic filters for one zero pair and two poles, one zero pair and three poles, and two pairs of zeros and four poles are given in Tables III-3a, b, and c, respectively [20]. The parameters A_1, A_2, and ω_s, are defined in Fig. III-1. The reader is referred to many excellent tabulations for elliptic filter data [13, 18, 21, 22].

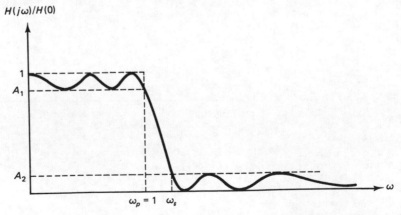

[1]Tables III-3a, b, and c are reprinted with permission from *Int. J. Comput. Elect. Eng.* [20], Copyright 1973, Pergamon Press Ltd.

TABLE III-3a   SECOND-ORDER ELLIPTIC FUNCTION PARAMETERS*

$$H(s) = \frac{H(s^2 + a_0)}{s^2 + b_1 s + b_0}$$

ω_s \ A_1	0.7	0.75	0.8	0.85	0.9	0.95	0.99
2.0	0.597566	0.672335	0.761953	0.87093	1.09079	1.28475	1.70530
	0.748566	0.807532	0.889100	1.01055	1.21614	1.67671	3.39116
	7.46410	7.46393	7.46393	7.46393	7.46410	7.46394	7.46437
	0.070208	0.081143	0.095295	0.115081	0.146639	0.213409	9.449766
1.8	0.586497	0.658788	0.744765	0.852101	0.996903	1.22172	1.49664
	0.761473	0.821030	0.903240	1.02526	1.23061	1.68388	3.25137
	5.93375	5.93377	5.93375	5.93377	5.93399	5.93377	5.93385
	0.089828	0.103773	0.121775	0.146865	0.186645	0.269588	0.542449
1.6	0.568640	0.636848	0.716947	0.814969	0.942467	1.12264	1.21673
	0.780727	0.840896	0.923621	1.04564	1.24863	1.68414	3.01139
	4.55831	4.55832	4.55842	4.55832	4.55832	4.55832	4.55836
	0.119892	0.138356	0.162086	0.194981	0.246530	0.350990	0.654022
1.4	0.535956	0.596787	0.666375	0.747996	0.845981	0.956021	0.859430
	0.811695	0.872098	0.954382	1.07401	1.26798	1.65969	2.61881
	3.33173	3.33166	3.33173	3.33167	3.33172	3.33167	3.33171
	0.170533	0.196320	0.229155	0.274010	0.342509	0.473248	0.778164
1.3	0.507505	0.562111	0.622959	0.691325	0.766598	0.829058	0.658076
	0.835122	0.894952	0.975687	1.09135	1.27415	1.62319	2.34888
	2.76980	2.76979	2.76981	2.76979	2.76982	2.76980	2.76972
	0.211054	0.242333	0.281803	0.334915	0.414008	0.556729	0.839569
1.2	0.461178	0.506162	0.553873	0.603117	0.647985	0.656811	0.450447
	0.867873	0.925546	1.00200	1.10866	1.26987	1.55118	2.02432
	2.23597	2.23595	2.23597	2.23595	2.23591	2.23595	2.23595
	0.271698	0.310453	0.358503	0.421457	0.511141	0.659054	0.896281
1.1	0.372652	0.401509	0.428498	0.450238	0.457760	0.420582	0.244714
	0.916613	0.967014	1.03128	1.11609	1.23375	1.41020	1.63605
	1.71409	1.71408	1.71409	1.71405	1.71408	1.71409	1.71394
	0.374317	0.423125	0.481308	0.553478	0.647782	0.781571	0.945011
1.05	0.285907	0.302274	0.314810	0.320161	0.310931	0.266561	0.142026
	0.951232	0.991713	1.04130	1.10322	1.18267	1.28875	1.40382
	1.43865	1.43866	1.43866	1.43866	1.43867	1.43867	1.43868
	0.462837	0.516984	0.579033	0.651806	0.739845	0.850992	0.966006

*Coefficients in each group, from top to bottom, are b_1, b_0, a_0, and $H(= A_2)$.

TABLE III-3b   THIRD-ORDER ELLIPTIC FUNCTION PARAMETERS*

$$H(s) = \frac{H(s^2 + a_1)}{(s + p)(s^2 + b_{11}s + b_{10})}$$

A_1 \ ω_s	1.05	1.10	1.15	1.20	1.30	1.40	1.60
0.99	3.00155	2.38167	2.04962	1.84049	1.59035	1.44703	1.29096
	0.085439	0.164793	0.239930	0.308389	0.423881	0.514156	0.641363
	1.17110	1.27550	1.35412	1.41484	1.50033	1.55605	1.62210
	1.20541	1.37031	1.53363	1.69962	2.04551	2.41363	3.22359
	2.91611	2.21688	1.80968	1.53210	1.16647	0.932875	0.649595
	0.835656	0.702859	0.585336	0.488077	0.346305	0.254634	0.150824
0.95	1.39312	1.16920	1.05238	0.978047	0.887854	0.834076	0.773178
	0.128759	0.208146	0.267535	0.313860	0.382084	0.429629	0.491552
	1.09401	1.12638	1.14386	1.15422	1.16538	1.17039	1.17419
	1.20541	1.37031	1.53359	1.69962	2.04548	2.41363	3.22359
	1.26436	0.961054	0.784827	0.664187	0.505756	0.404446	0.281626
	0.550613	0.393746	0.298822	0.235598	0.158103	0.113422	0.066000
0.90	0.989843	0.850207	0.776613	0.729373	0.671004	0.635840	0.595349
	0.131719	0.197936	0.243978	0.278588	0.327809	0.361387	0.404241
	1.04501	1.05130	1.05182	1.05044	1.04612	1.04183	1.03480
	1.20541	1.37031	1.53360	1.69962	2.04550	2.41363	3.22359
	0.858124	0.652271	0.532635.	0.450785	0.343166	0.274453	0.191108
	0.408593	0.279160	0.207846	0.162349	0.108001	0.077236	0.044840
0.85	0.796463	0.692067	0.636652	0.600774	0.556216	0.529099	0.497668
	0.125887	0.182370	0.220415	0.248520	0.287999	0.314599	0.348306
	1.01490	1.00922	1.00267	0.996551	0.986391	0.978507	0.967482
	1.20541	1.37031	1.53361	1.69962	2.04549	2.41362	3.22359
	0.670582	0.509696	0.416235	0.352254	0.268216	0.214500	0.149361
	0.330211	0.221527	0.163808	0.127519	0.084606	0.060434	0.035059
0.80	0.672056	0.588202	0.543408	0.514269	0.477902	0.455659	0.429765
	0.117935	0.167009	0.199450	0.223182	0.256259	0.278406	0.306340
	0.993886	0.981243	0.970727	0.962023	0.948671	0.938912	0.925798
	1.20541	1.37031	1.53361	1.69962	2.04549	2.41362	3.22359
	0.554123	0.421193	0.343958	0.291087	0.221642	0.177253	0.123426
	0.277706	0.184503	0.135945	0.105649	0.069994	0.049969	0.028976
0.75	0.580701	0.510790	0.473303	0.448768	0.418123	0.399259	0.377272
	0.109462	0.152592	0.180768	0.201218	0.229618	0.248517	0.272306
	0.978189	0.960951	0.947897	0.937548	0.922206	0.911278	0.896878
	1.20541	1.37031	1.53361	1.69962	2.04549	2.41362	3.22359
	0.471237	0.358197	0.292535	0.247550	0.188505	0.150742	0.104965
	0.238726	0.157652	0.115915	0.089987	0.059570	0.042510	0.024646
0.70	0.508377	0.448812	0.416747	0.395725	0.369369	0.353124	0.334131
	0.100986	0.139154	0.163859	0.181717	0.206412	0.222806	0.243387
	0.965965	0.945461	0.930629	0.919159	0.902442	0.890734	0.875469
	1.20541	1.37033	1.53361	1.69962	2.04549	2.41363	3.22360
	0.407390	0.309657	0.252888	0.214008	0.162958	0.130318	0.090743
	0.207883	0.136715	0.100376	0.077873	0.051520	0.036758	0.021308

*Coefficients in each group, from top to bottom, are p, b_{11}, b_{10}, a_1, H, and A_2.

TABLE III-3c FOURTH-ORDER ELLIPTIC FUNCTION PARAMETERS*

$$H(s) = \frac{H(s^2 + a_1)(s^2 + a_2)}{(s^2 + b_{11}s + b_{10})(s^2 + b_{21}s + b_{20})}$$

A_1 \ ω_s	1.05	1.075	1.1	1.15	1.20	1.25	1.3
0.99	1.24184	1.36998	1.43445	1.48461	1.49416	1.48971	1.48067
	1.76639	1.64271	1.52732	1.35343	1.23106	1.14128	1.07349
	1.15362	1.22234	1.29092	1.42978	1.57242	1.71971	1.87203
	0.073511	0.104589	0.132384	0.179552	0.218090	0.250180	0.277321
	1.09961	1.12737	1.14901	1.18196	1.20610	1.22469	1.23958
	3.31266	3.85083	4.34993	5.29789	6.22434	7.15325	8.09589
	0.503140	0.389504	0.309376	0.209067	0.150187	0.112478	0.086921
0.97	1.11694	1.15719	1.17084	1.17132	1.16143	1.14961	1.13807
	1.17034	1.05721	0.974214	0.860463	0.785950	0.733037	0.693369
	1.15363	1.22235	1.29093	1.42979	1.57244	1.71971	1.87203
	0.082382	0.109829	0.132841	0.169869	0.198774	0.222152	0.241517
	1.06041	1.07297	1.08228	1.09550	1.10463	1.11140	1.11664
	3.31252	3.85097	4.34995	5.29782	6.22442	7.15325	8.09588
	0.315149	0.233744	0.182127	0.120698	0.086036	0.064240	0.049553
0.95	1.00616	1.02456	1.02740	1.01897	1.00707	0.994981	0.983063
	0.949099	0.853237	0.785841	0.695368	0.637238	0.595833	0.563827
	1.15362	1.22235	1.29090	1.42980	1.57240	1.71971	1.87203
	0.081657	0.106240	0.126428	0.158254	0.182889	0.202523	0.218409
	1.04043	1.04722	1.05200	1.05833	1.06243	1.06526	1.06699
	3.31238	3.85097	4.34973	5.29772	6.22421	7.15328	8.09613
	0.245492	0.180326	0.139862	0.092248	0.065716	0.049015	0.037702
0.925	0.903747	0.911065	0.909212	0.897844	0.885261	0.873655	0.863732
	0.800608	0.718662	0.662419	0.587723	0.539585	0.505447	0.480059
	1.15363	1.22235	1.29093	1.42979	1.57243	1.71971	1.87204
	0.078638	0.100616	0.118456	0.146316	0.167498	0.184309	0.198090
	1.02465	1.02747	1.02921	1.03102	1.03182	1.03207	1.03216
	3.31249	3.85095	4.34992	5.29794	6.22440	7.15331	8.09589
	0.198567	0.145089	0.112302	0.073994	0.052617	0.039221	0.030239
0.9	0.825168	0.827693	0.822969	0.811096	0.799091	0.788335	0.779239
	0.709059	0.636774	0.587138	0.521773	0.479903	0.450211	0.428090
	1.15363	1.22235	1.29041	1.42981	1.57240	1.71971	1.87203
	0.075144	0.095172	0.111155	0.136283	0.155159	0.170095	0.182294
	1.01374	1.01413	1.01394	1.01299	1.01180	1.01058	1.00947
	3.31240	3.85096	4.34581	5.29794	6.22422	7.15326	8.09589
	0.169268	0.123463	0.095443	0.062763	0.044651	0.033285	0.025661
0.85	0.708408	0.706199	0.700475	0.688058	0.676966	0.667721	0.659911
	0.598882	0.537875	0.496416	0.442241	0.407509	0.383057	0.364750
	1.15363	1.22235	1.29092	1.42978	1.57243	1.71971	1.87203
	0.068333	0.085454	0.099089	0.120059	0.135776	0.148163	0.158248
	0.999163	0.996486	0.993986	0.989660	0.986076	0.983077	0.980558
	3.31250	3.85096	4.34987	5.29786	6.22434	7.15335	8.09589
	0.133093	0.096781	0.074682	0.049106	0.034894	0.026019	0.020059

(Continued)

TABLE III-3c FOURTH-ORDER ELLIPTIC FUNCTION PARAMETERS* *(continued)*

A_1 \\ ω_s	1.05	1.075	1.1	1.15	1.20	1.25	1.3
0.8	0.621079	0.617140	0.611017	0.599431	0.589557	0.581212	0.574306
	0.532447	0.478486	0.442216	0.394408	0.363967	0.342428	0.326332
	1.15363	1.22235	1.29041	1.42982	1.57240	1.71971	1.87204
	0.062131	0.077101	0.088880	0.107103	0.120623	0.131234	0.139842
	0.989514	0.984974	0.981121	0.974761	0.969765	0.965686	0.962302
	3.31250	3.85096	4.34613	5.29772	6.22423	7.15327	8.09589
	0.110293	0.080095	0.061824	0.040580	0.028851	0.021504	0.016576
0.75	0.550687	0.546069	0.540306	0.529453	0.520419	0.513005	0.506861
	0.487251	0.438112	0.405011	0.361820	0.334160	0.314640	0.300027
	1.15363	1.22235	1.29093	1.42979	1.57243	1.71971	1.87204
	0.056514	0.069762	0.080214	0.096136	0.107970	0.117237	0.124738
	0.982530	0.976711	0.971880	0.964177	0.958218	0.953427	0.949476
	3.31251	3.85096	4.34992	5.29791	6.22439	7.15327	8.09589
	0.093955	0.068176	0.052571	0.034540	0.024536	0.018289	0.014096
0.7	0.491207	0.486393	0.480919	0.470946	0.462799	0.456159	0.450655
	0.454281	0.408647	0.377979	0.337967	0.312352	0.294272	0.280747
	1.15363	1.22235	1.29092	1.42979	1.57244	1.71972	1.87203
	0.051378	0.063179	0.072459	0.086558	0.097012	0.105183	0.111795
	0.977203	0.970443	0.964919	0.956218	0.949563	0.944251	0.939898
	3.31247	3.85102	4.34989	5.29795	6.22448	7.15335	8.09607
	0.081317	0.058963	0.045465	0.029861	0.021211	0.015810	0.012188

*Coefficients in each group, from top to bottom, are b_{11}, b_{10}, a_1, b_{21}, b_{20}, a_2, and $H(= A_2)$.

Example III-1

As an illustrative example of the use of Tables III-3a to c, consider the following specifications. Let $\omega_s/\omega_p = 1.1$, the minimum stopband attenuation be ≥ 17 dB, and the attenuation in the passband $A_p = 0.45$ dB. Recall from Section 1.6 the following relations:

$$A_1 = \frac{1}{\sqrt{1 + \varepsilon^2}} = 10^{-A_p/20} \qquad A_2 = 10^{-A_s/20}$$

Hence,

$$A_1 = 10^{-0.45/20} \simeq 0.949511 \simeq 0.945$$

$$A_2 = 10^{-17/20} \simeq 0.141254 \simeq 0.141$$

From the values above corresponding to $\omega_s = 1.1$ and $A_1 = 0.95$, we see that Tables III-3a and III-3b will not satisfy the requirements, as they yield $A_2 = 0.781571$ and 0.393746. However, from Table III-3c we find (see dashed box) that $A_2 = 0.139862$, which is close to and lower than 0.141. Hence, the transfer function that meets the requirements is

$$H(s) = \frac{0.13986(s^2 + 1.29090)(s^2 + 4.34973)}{(s^2 + 1.02740s + 0.785841)(s^2 + 0.126428s + 1.05200)}$$

A plot of the magnitude response for this transfer function is shown in Fig. III-2.

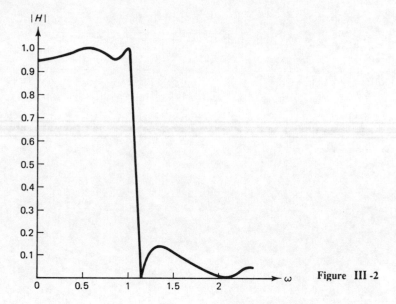

Figure III -2

TABLE III-4

n	Bessel polynomials
1	$s + 1$
2	$s^2 + 3s + 3$
3	$s^3 + 6s^2 + 15s + 15 = (s + 2.322)(s^2 + 3.678s + 6.460)$
4	$s^4 + 10s^3 + 45s^2 + 105s + 105 = (s^2 + 5.792s + 9.140)(s^2 + 4.208s + 11.488)$
5	$s^5 + 15s^4 + 105s^3 + 420s^2 + 945s + 945$ $(s + 3.6467)[(s + 3.3520)^2 + 1.7427^2][(s + 2.3247)^2 + 3.5710^2]$
6	$s^6 + 21s^5 + 210s^4 + 1260s^3 + 4725s^2 + 10395s + 10395$ $[(s + 4.2484)^2 + 0.8675^2][(s + 3.7356)^2 + 2.6263^2][(s + 2.5159)^2 + 4.4927^2]$
7	$s^7 + 28s^6 + 378s^5 + 3150s^4 + 17325s^3 + 62370s^2 + 135135s + 135135$ $(s + 4.9718)[(s + 4.7583)^2 + 1.7393^2][(s + 4.0701)^2 + 3.5172^2][(s + 2.6857)^2 + 5.4207^2]$
8	$s^8 + 36s^7 + 630s^6 + 6930s^5 + 51975s^4 + 270270s^3 + 945{,}945s^2 + 2{,}027{,}025s + 2{,}027{,}025$ $[(s + 5.5879)^2 + 0.8676^2][(s + 2.8390)^2 + 6.3539^2][(s + 4.3683)^2 + 4.1444^2]$ $[(s + 5.2048)^2 + 2.6162^2]$
9	$s^9 + 45s^8 + 990s^7 + 13{,}860s^6 + 135{,}135s^5 + 945{,}945s^4 + 4{,}729{,}752s^3 + 16{,}216{,}200s^2 + 34{,}459{,}425s + 34{,}459{,}425$ $(s + 6.2970)[(s + 6.1294)^2 + 1.7378^2][(s + 5.6044)^2 + 3.4982^2][(s + 4.6384)^2 + 5.3173^2][(s + 2.9793)^2 + 7.2915^2]$
10	$s^{10} + 55s^9 + 1{,}485s^8 + 25{,}740s^7 + 315{,}315s^6 + 2{,}837{,}835s^5 + 18{,}918{,}900s^4 + 91{,}891{,}800s^3 + 310{,}134{,}825s^2 + 654{,}729{,}075s + 645{,}729{,}075$ $[(s + 6.9220)^2 + 0.8677^2][(s + 3.1089)^2 + 8.2327^2][(s + 6.6153)^2 + 2.6116^2]$ $[(s + 5.9675)^2 + 4.3850^2][(s + 4.8862)^2 + 6.2250^2]$

Index